Target Organ Toxicology Series

NUTRITIONAL TOXICOLOGY

Second edition

Target Organ Toxicology Series

NUTRITIONAL TOXICOLOGY

Second edition

Editors

Frank N Kotsonis
Food Research Institute
University of Wisconsin-Madison
Wisconsin, USA

Maureen A Mackey
Nutrition Scientific Affairs
Monsanto Company
St. Louis, Missouri
USA

London and New York

First edition published 1994 Raven Press

This edition published in 2002 by Taylor & Francis
11 New Fetter Lane, London EC4P 4EE

Simultaneously published in the USA and Canada
by Taylor & Francis Inc,
29 West 35th Street, New York, NY 10001

Taylor & Francis is an imprint of the Taylor & Francis Group

© 2002 Frank N Kotsonis, Maureen A Mackey

Typeset in Times by
Integra Software Services Pvt. Ltd., Pondicherry, India.
Printed and bound in Great Britain by MPG Books Ltd, Bodmin

Every effort has been made to ensure that the advice and information in
this book is true and accurate at the time of going to press. However,
neither the publisher nor the authors can accept any legal
responsibility or liability for any errors or omissions that may be made.
In the case of drug administration, any medical procedure or the use of
technical equipment mentioned within this book, you are strongly advised
to consult the manufacturer's guidelines.

British Library Cataloguing in Publication Data
A catalogue record for this book is available
from the British Library

Library of Congress Cataloging-in-Publication Data
Nutritional toxicology / editors, Frank N. Kotsonis, Maureen A. Mackey—2nd ed.
p. cm.—(Target organ toxicology series)
Includes bibliographical references and index.
ISBN 0–415–24865–5 (hardcover : alk paper)
1. Food—Toxicology. 2. Drug–nutrient interactions. 3. Xenobiotics—Metabolism.
4. Nutrition. I. Kotsonis, Frank N. II. Mackey, Maureen A., 1955– III. Series.
[DNLM: 1. Food—adverse effects. 2. Food Technology. 3. Food Additives. 4. Food
Hypersensitivity. 5. Food–Drug Interactions. 6. Nutrition. WA 695 N9767 2001]
RA1258 .N876 2001
615.9′54—dc21
2001053172

CONTENTS

CONTRIBUTORS

Dr JEN Bergmans
Center of Substances and Risk
  Assessment
National Institute of Public Health
  and the Environment
3720 BA
Bilthoven, The Netherlands
Ph: 31-30-274-4195
Fax: 31-30-274-4401

David Kritchevsky, PhD
Wistar Institute
3601 Spruce Street
Philadelphia PA 19104
Ph: (215) 898-3713
Fax: (215) 898-3995
e-mail: kritchevsky@wistar.upenn.edu

Dr David Tennant
Food Chemical Risk Analysis
14 St Mary's Square
Brighton BN2 1FZ UK
Ph: (44) 0 1273 241 753
Fax: (44) 0 1273 276 358
e-mail: david-t@dircon.co.uk

M Ellin Doyle, PhD
University of Wisconsin
Food Research Institute
Dept Food Microbiology and
  Toxicology
1925 Willow Drive

Madison WI 53706
Ph: (608) 263-6934
Fax: (608) 263-1114
e-mail: medoyle@facstaff.wisc.edu

Emily K Kraczek
University of Illinois
580 Bevier Hall
905 S Goodwin Ave
Urbana, IL 61801
Ph: (217) 333-2527
Fax: (217) 333-9368
e-mail: ekraczek@staff.uiuc.edu

Frank N Kotsonis, PhD
University of Wisconsin
Food Research Institute
Dept of Food Microbiology and
  Toxicology
1925 Willow Drive
Madison, WI 53706
Ph: 847-724-7228
Fax: 847-724-0850
e-mail: frankkotsonis@msn.com

W Gary Flamm, PhD
Flamm Associates
622 Beachland Blvd
Vero Beach, FL 32963
Ph: (561) 234-0096
Fax: (561) 234-0026
e-mail: wgflamm@attglobal.net

George A Burdock, PhD
Burdock and Associates
622 Beachland Blvd
Vero Beach, FL 32963-1743
Ph: 561-234-9975
Fax: 561-234-0026
e-mail: gburdock@salvitas.com

Dr Gerrit JA Speijers
Center of Substances and Risk
  Assessment
National Institute of Public Health
  and the Environment
3720 BA
Bilthoven The Netherlands
Ph: 31 30 274 2120
Fax: 31 30 274 4401
e-mail: gja.speijers@rivm.nl

Harriett H Butchko, MD
The NutraSweet Company
699 Wheeling Road
Mt Prospect IL 60056
Ph: 847-463-1705
Fax: 847-463-1755
e-mail: harriett.h.butchko@nutrasweet.
  com

Helmut K Seitz, MD
Laboratory of Alcohol Research
Liver Disease and Nutrition
Department of Medicine
Salem Medical Center
Zeppelinstrasse 11-33
D-69121 Heidelberg, Germany
Ph: 49-0-6221 483 200
Fax: 49-0-6221 483 494
e-mail: helmut_karl.seitz@urz.
  uni-heidelberg.de

Jerry J Hjelle, PhD
Monsanto Company
800 N Lindbergh Blvd
St Louis, MO 63167

Ph: (314) 694-1442
Fax: (314) 694-4928
e-mail: jerry.j.hjelle@monsanto.com

John W Erdman, PhD
University of Illinois
580 Bevier Hall
905 S Goodwin Ave
Urbana, IL 61801
Ph: (217) 333-2527
Fax: (217) 333-9368
e-mail: j-erdman@uiuc.edu

John A Thomas, PhD
Professor Emeritus
Department of Pharmacology/
  Toxicology
The University of Texas Health Science
Center at San Antonio
San Antonio, TX 78216
Ph: (210) 494-8491
Fax: (210) 494-4881
e-mail: jat-tox@swbell.net

Maureen Mackey, PhD
Monsanto Company
800 N Lindbergh Blvd
St Louis, MO 63167
Ph: (847) 821-0403
Fax: (847) 821-0409
e-mail: maureen.a.mackey@monsanto.
  com

R Michael McClain, PhD
McClain Associates
10 Powder Horn Terrace
Randolph, NJ 07869
Ph: (973) 895-1363
Fax: (973) 895-1393
e-mail: michaelmcclain@msn.com

Michael W Pariza, PhD
University of Wisconsin
Food Research Institute

Dept Food Microbiology and
  Toxicology
1925 Willow Drive
Madison, WI 53706
Ph: (608) 263-6955
Fax: (608) 263-1114
e-mail: mwpariza@facstaff.wisc.edu

Prof Dr KJ Netter
Department of Pharmacology and
  Toxicology
Philipps University, School of Medicine
D-35033
Marburg, Germany
Ph: 49-6421-64611
Fax: 49-6421-67250
e-mail: netter@mailer.uni-marburg.de

Paolo M Suter, MD
Department of Medicine
University of Zurich Medical Policlinic,
  Hypertension Unit
Raemistrasse 100
8091 Zurich, Switzerland
Ph: 41-1-255 11 11
Fax: 41-1-311 67 07
e-mail: polpms@usz.unizh.ch

Prof Dr med Peter S Elias
Bertha Von Suttner Strasse 3A
D-76139 Karlsruhe-Waldstadt,
  Germany
Ph: 44-181-500-9584 (London)
    49-721-68 17 68 (Germany)
Fax: 49-721-68 89 60 (Germany)
e-mail: pselias@t-online.de

Richard Cotter, PhD
Assistant Vice President
Nutritional Sciences, Whitehall Robins
5 Giralda Farms
Madison, NJ 07940
Ph: (973) 660-6257
Fax: (973) 660-7390

Roy L Fuchs, PhD
Monsanto Company
700 Chesterfield Parkway
St Louis, MO 63198
Ph: 636-737-6438
Fax: 636-737-6189
e-mail: roy.l.fuchs@monsanto.com

Steve L Taylor, PhD
Food Allergy Research and Resource
  Program
University of Nebraska
143 Food Industry Bldg
Lincoln, NE 68583-0919
Ph: 402-472-2833
Fax: 402-472-1693
e-mail: staylor2@unl.edu

Susan L Hefle PhD
Food Allergy Research and Resource
  Program
University of Nebraska
351 Food Industry Bldg
Lincoln, NE 68583-0919
Ph: (402) 472-4430
Fax: (402) 472-1693
e-mail: shefle1@unl.edu

W Wayne Stargel, PharmD
The NutraSweet Company
699 Wheeling Road
Mt Prospect IL 60056
Ph: 847-463-1741
Fax: 847-463-1755
e-mail: wilford.w.stargel@nutrasweet.
  com

R William Soller, PhD
Sr VP & Director of Science &
  Technology
Consumer Healthcare Products Assn
1150 Connecticut Avenue NW
Washington, DC 20036-4193
Ph: 202-429-9260

Fax: 202-223-6835
e-mail: wsoller@CHPA-info.org

Dr M Younes
International Programme on Chemical
   Safety

World Health Organization
20 Avenue Appia
1211 Geneva 27, Switzerland
Ph: 41 22 791 3574
Fax: 41 22 791 4848
e-mail: younesm@who.ch

1

ANTIOXIDANT NUTRIENTS AND PROTECTION FROM FREE RADICALS

M. Ellin Doyle and Michael W. Pariza

Antioxidant nutrients are currently a topic of intense interest to scientists and the public, with some regarding them as the latest and best candidate for a magic bullet to stave off aging processes and prevent cancer and cardiovascular disease. Numerous epidemiological studies have demonstrated the association of diets rich in fruits and vegetables and the prevention of a variety of illnesses. Since many diseases involve oxidative stress or oxidative degradation of biological macro-molecules, it is presumed that the antioxidant micronutrients are responsible, at least in part, for the beneficial effects of plant foods.

Recent literature on antioxidants can be overwhelming with a plethora of epidemiological studies, clinical trials, and animal and in vitro tests examining the antioxidant effects of diets, specific foods, individual vitamins and micronu-trients as well as herbal medicines and other drugs. A recent Institute of Medicine Panel on Dietary Antioxidants and Related Compounds defined dietary antioxi-dants as: "substances in foods that significantly decrease the adverse effects of reactive species, such as reactive oxygen and nitrogen species, on normal physio-logical function in humans." (156). In this review, we will consider nutrients that are normally present in the diet and have been shown to exert antioxidant effects in vivo. Work published within the past 5 years will be summarized to present an overview on free radicals, markers of oxidative stress, and the antioxidants vitamin E, vitamin C, β-carotene, flavonoids, and selenium.

FREE RADICALS

Free radicals are any species, capable of independent existence, that contain one or more unpaired electrons. Radicals may be derived from single atoms such as hydrogen (H$^{\bullet}$) or chlorine (Cl$^{\bullet}$) or may involve two or more atoms for example, the hydroxyl radical (HO$^{\bullet}$) and the nitric oxide radical (NO$^{\bullet}$). Reactive oxygen species (ROS), free radicals containing oxygen, are common in living organisms and include: singlet oxygen ($^1O_2$), superoxide radical ($O_2^{\bullet-}$), and, the most reactive ROS, the hydroxyl radical (HO$^{\bullet}$). Although most living cells contend primarily with internally generated free radicals, reactive oxygen and nitrogen species are also produced by chemical reactions in the environment and by reactions of environmental contaminants with biological molecules. The major external sources of free radicals are ionizing radiation, cigarette smoke, air pollutants, and chemicals such as ethyl alcohol, ozone, and halogenated hydrocarbons.

Free radicals are normally generated in substantial amounts by internal metabolic processes in aerobic organisms, particularly during mitochondrial respiration when oxygen is reduced to water. At various steps during the electron transport chain, single electrons leak out and partially reduce oxygen to form the superoxide radical ($O_2^{\bullet-}$). Further reduction of superoxide by the addition of two more electrons produces, first, hydrogen peroxide (H_2O_2) and then the aggressive hydroxyl radical (HO$^{\bullet}$). This leakage of electrons is a normal occurrence during aerobic metabolism with an estimated 1–2% of the electrons passing down the respiratory chain leaking out to form the superoxide radical. Several other enzymatic reactions involving oxidases also produce superoxide radicals with an estimated 0.15 M $O_2^{\bullet-}$ produced in the body each day. Further details on production of free radicals are presented in some recent reviews (24,34,78,93,214).

Two free radicals can react, join their unpaired electrons, and form a non-radical product. However, more commonly free radicals will encounter non-radical compounds in the cell and will generate new radicals during addition, reduction, or oxidation reactions. For example, free radicals can abstract a hydrogen atom from polyunsaturated fatty acids (PUFA) and the resulting peroxy radical may then take a hydrogen atom from an adjacent PUFA thereby propagating a free radical chain reaction of lipid peroxidation. Free radicals may attack any of the biological macromolecules (nucleic acids, lipids, proteins, carbohydrates) altering their structure and function and causing oxidative stress. Many chronic diseases, including atherosclerosis, some types of cancer, cataracts, immune/autoimmune diseases and the degenerative conditions of old age are believed to result from an accumulation of macromolecules, damaged by free radicals, which the body has not been able to repair. Further information on the effects of free radicals can be found in several recent reviews (34,78,93,144,213,214).

In order to survive, cells have evolved a complex antioxidant defense system involving antioxidant enzymes such as catalase, superoxide dismutase, and glutathione peroxidases and small molecule antioxidants. These small molecules may be lipid soluble (α-tocopherol, β-carotene, other carotenoids) or water soluble (ascorbic acid, glutathione, urate). In healthy, young, non-smoking people, this defense system is usually adequate to counteract the deleterious effects of free radicals (38). However, as organisms age and are exposed to environmental insults, antioxidant enzyme activity and small molecule antioxidants may be insufficient to prevent mutations and injury to cells and tissues.

2

Can the consumption of antioxidant nutrients in the diet protect us from free radical induced damage? All of the dietary antioxidants considered here are readily absorbed from the diet and can therefore replenish body stores of these compounds (9,14,18,23,32,87,98,99,100,134,148,158,160,179). Data from epidemiological studies and from numerous clinical trials with different nutrients indicate that consumption of high levels of some antioxidants from the diet or from supplements does exert a protective effect against oxidative damage and some diseases but results from different studies are not always consistent (51,52,56,64,65,124, 182,198,199,216,224). Some of this inconsistency may be attributed to the practice of correlating supplement intake rather than plasma levels of the antioxidant to the outcome measure. If subjects already consume diets containing sufficient or high levels of an antioxidant nutrient, intake of supplements may not significantly increase plasma concentrations of the antioxidants and therefore there may be little difference between experimental and control groups (134).

Individual antioxidant supplements when tested alone may not appear to have protective or antioxidative effects in vivo. For example, α-tocopherol can inhibit oxidation of lipoproteins. But, in the absence of co-antioxidants, α-tocopherol radicals accumulate in lipoprotein particles and this "antioxidant" may act as a pro-oxidant (213). However, a healthy diet including many fruits and vegetables contains multiple compounds with antioxidant properties as well as other substances, such as fiber, which may interact to maintain health (68).

MARKERS OF OXIDATIVE STRESS

To determine whether oxidative stress in vivo is diminished by antioxidant nutrients, stable biomarkers of stress must be present in fluids or tissues which are easily sampled and reliable analytical methods for the quantization of these compounds must be available. Free radical induced damage to DNA, lipids, and proteins produces some characteristic compounds which have been identified as potentially useful markers. Recent reviews (144,145) provide a comprehensive discussion of useful biomarkers for assessing oxidative damage in vivo.

Lipid hydroperoxides are the initial products of free radical attack on PUFAs. Rates of these oxidation reactions can be monitored in vitro by measurements of the disappearance of PUFAs or the increase of lipid peroxides. But these molecules are less easy to determine in complex biological systems because of interfering substances. Peroxidation of PUFAs also results in production of ethane and pentane which are exhaled in the breath. But technical difficulties have also limited the use of these compounds as biomarkers. Traditionally, oxidative damage to lipids has been estimated by the increase in TBARS (thiobarbituric acid reactive substances) which reflect the production of malondialdehyde, an oxidation product of lipids. Although this assay lacks specificity, it is convenient, inexpensive, and widely used. Another marker of oxidative damage to lipids is a group of peroxidation products of arachidonic acid called F_2-isoprostanes. The most

3

frequently studied compound of this group, 8-epi $PGF_2\alpha$, has been detected in urine and plasma in both humans and animals exposed to ROS.

Oxidized LDL (low density lipoprotein) molecules appear to be important in the development of atherosclerosis because they induce production of cytokines and adhesion molecules by cells in artery walls. Oxidized lipoproteins also attract monocytes and aid in their conversion to lipid-rich foam cells (34,40,78,214). Therefore, another method to assess the possible protective effects of antioxidant nutrients is to determine their effects on the oxidizability of LDL ex vivo: LDL, isolated from subjects consuming antioxidant-rich foods or supplements, can be incubated with an oxidizing agent to determine whether the consumption of anti-oxidants has produced lipoprotein molecules resistant to oxidation.

Oxidatively damaged proteins are believed to accumulate during aging and some chronic diseases. However, they are not easy to study. Some markers of protein oxidation which have been identified are: protein carbonyl derivatives, oxidized amino acid side chains, protein fragments, and advanced glycation products (19).

Free radical damage to DNA, either in the form of strand breaks or altered bases may be the initiating event in carcinogenesis. One of the most useful biomarkers of such reactions is the presence of an altered base, 8-oxoguanine. Since hydrogen bonding of this base is altered, its presence in DNA will induce changes during DNA replication that may result in mutations and possibly initiate carcinogenesis. The DNA strand break "comet assay" which detects oxidized pyrimidine bases can also be used to assess DNA damage (66).

Research correlating the presence of biomarkers of oxidative damage with various diseases and investigating the effects of antioxidant nutrients can be used to complement clinical trials testing the effects of antioxidant supplements and epidemiological studies of disease and diet. With this knowledge, we may acquire a better understanding of the disease process and be able to answer the question of whether taking antioxidant supplements will ward off chronic diseases.

VITAMIN E

Vitamin E actually includes a group of eight structurally related lipid-soluble tocopherols and tocotrienols, the most abundant compound being α-tocopherol. The natural form of α-tocopherol is the stereoisomer designated as RRR and this isomer has the greatest biological activity (42). According to the latest report by the Panel on Dietary Antioxidants and Related Compounds of the Institute of Medicine, the RDA (Recommended Daily Allowance) for α-tocopherol is 15 mg for men and women (156). The upper limit of intake of α-tocopherol was set at 1,000 mg/day based on the adverse effect of an increased tendency for bleeding.

Dietary α-tocopherol is readily absorbed (9,32,160) and transported to cell membranes and to intracellular sites (67). Toxicity of vitamin E is very low with no mutagenic, carcinogenic, or teratogenic effects reported. Even in double-blind experiments utilizing doses of vitamin E as large as 3,200 IU/day, no adverse side

4

effects were reported (57,58,220). (1,500 IU "natural source" vitamin E = 1,100 IU "synthetic" vitamin E = 1,000 mg α-tocopherol.) Since vitamin E does affect platelet aggregation, however, it can increase the tendency for bleeding particularly in people with vitamin K deficiency (80,220). At high concentrations, α-tocopherol can also have pro-oxidant effects (214,217). There have been few studies of the safety of long term ingestion of megadoses of vitamin E.

Of all the antioxidant nutrients, vitamin E has been studied more extensively and the evidence for its beneficial effects is more convincing than that for other antioxidants. Reviews published in the past few years summarize much of the earlier work on: the function and metabolism of vitamin E (29), vitamin E and heart disease (70,122,165,201,213), vitamin E and the neurological movement disorder, tardive dyskinesia (25), and vitamin E and disorders of the central nervous system (220).

Animal studies

Antioxidative effects of α-tocopherol have been observed in numerous animal studies. Vitamin E reduced oxidative stress in laboratory rodents (207) and rhesus monkeys (192), platelet aggregation in rats (185), and lipid peroxidation caused by aflatoxin in mice (221). All of these effects may be relevant to the protective effects of vitamin E against cardiovascular disease.

Dietary vitamin E also exerts anticarcinogenic effects by preventing or reducing mutagenesis and bolstering immune function. Transgenic mice containing the oncogenes, c-*myc* and transforming growth factor alpha (TGFα), produce high levels of oxygen radicals which cause chromosomal damage in liver cells and promote hepatocarcinogenesis. Dietary vitamin E supplementation decreased the generation of oxygen radicals, stabilized chromosomes, and restricted hepatocyte proliferation in these mice (71). Oxidative DNA damage caused by the mycotoxin, fumonisin B1 (12) and mutation frequency in adipose tissue of Big Blue (TM) mice were also significantly reduced by dietary α-tocopherol (146). Immune response to challenge with influenza virus was significantly enhanced by vitamin E supplements in old, but not in young, mice (94). Because an active immune system can more efficiently detect and destroy early precancerous cells, an immune-enhancing effect of vitamin E may also be anticarcinogenic.

Cardiovascular disease

Several large epidemiological studies indicate that there is an inverse relationship between reported intake of vitamin E and incidence or death from coronary disease. Results in some studies were more definitive if the supplements had been consumed for at least 2 years. Long term followup of 34,486 post menopausal women (123), 87,245 nurses (203), 39,910 male health professionals (180), 5,133 people in Finland (118), and 11,178 people in the Established Populations for Epidemiologic Studies of the Elderly (136) indicated that higher levels of vitamin

E consumption were associated with a reduced incidence of cardiovascular disease. In another group of patients with coronary atherosclerosis, daily supplements of 400 or 800 IU α-tocopherol reduced the risk for non-fatal heart attack during an average 1.5 year followup (206). As with all epidemiological studies, there are potential problems with accurate reporting and confounding by other protective nutrients in foods as well as the caveat that an association between two events does not prove causality.

However, in some studies of patients with known heart disease, vitamin E did not always prevent further complications. Higher vitamin E intakes for 4.5 years were not beneficial in a group of subjects at high risk of heart disease (240) and vitamin E and β-carotene supplements did not decrease the incidence of further coronary events in 1,862 male smokers who had already suffered one heart attack (169). Nor did supplements of 300 mg vitamin E reduce the incidence of death, stroke or further heart attacks in patients who had already suffered one heart attack (218).

Use of vitamin E supplements has been reported to diminish oxidative changes associated with cardiovascular disease. High plasma levels of oxidized LDL (low density lipoprotein) are an indicator of coronary artery disease (40,101,213). Vitamin E supplements reduce the susceptibility of LDL to oxidation in hyper-lipidemic patients (226), men with high homocysteine levels (232), and type 2 diabetics (5), but not in type 1 diabetics (11). LDL subfractions containing higher concentrations of vitamin E were more resistant to copper-mediated oxidation ex vivo (137). However, when α-tocopherol interacts with peroxy radicals in lipoproteins, it becomes oxidized itself and must react with another antioxidant molecule. In the absence of co-antioxidants such as ascorbic acid, the α-tocopherol radical may not be reduced and may then act as a pro-oxidant (125,213).

Platelet aggregation leads to formation of blood clots and an increased risk for heart attacks and stroke. Collagen-induced platelet aggregation was inhibited by vitamin E in vitro and in vivo by the suppression of hydrogen peroxide formation (162). A dose of 75 IU α-tocopherol also decreased platelet aggregation and enhanced sensitivity to the platelet inhibitor, PGE1, in healthy subjects (139).

Persons with type 2 diabetes have much higher rates of cardiovascular disease than the general population. One reason for this is most probably the higher levels of oxidative stress reported in nearly all studies of diabetic subjects (92). Supplements of vitamin E decreased measures of oxidative stress in diabetic subjects (109,193) and increased reactivity of the brachial artery in patients with high remnant lipoprotein levels (120) and with type 2 diabetes (157). Endothelial function in response to vasodilators was not improved by vitamin E in another group of type 2 diabetics (82). In vitro studies with aortic endothelial cells and monocytic cells demonstrated that vitamin E inhibited the expression of adhesion molecules and production of inflammatory cytokines (235). These effects on endothelial function may be another mechanism by which vitamin E protects against cardiovascular disease.

6

MUTAGENESIS AND CARCINOGENESIS

Oxidative damage to DNA causes mutations, some of which may lead to uncontrolled growth or cancer. Plasma levels of antibodies to the mutant DNA base, 5-hydroxymethyl-2'-deoxyuridine (HMdU), were measured in subjects who had consumed supplements of 15, 60, or 200 mg α-tocopherol/day for 4 weeks. Only the middle dose consistently, significantly, reduced oxidative DNA damage as indicated by antibody titers (102). However, neither 1 month of 250 mg vitamin E supplements/day nor 2 months of 200–500 mg/day decreased oxidative DNA damage as measured by altered DNA in mononuclear leukocytes of male smokers and non-smokers (225) and by urinary excretion of 8-oxodG by smokers (164), respectively. DNA in lymphocytes from non-smokers consuming 800 mg vitamin E daily was resistant to in vitro damage caused by hydrogen peroxide. However, the vitamin E supplements had no effect on the endogenous level of DNA damage in lymphocytes (28).

Data from the Finnish ATBC (α-tocopherol, β-carotene) clinical trial involving over 29,000 male smokers, revealed that a daily dose of 50 mg α-tocopherol was correlated with a reduced incidence and mortality of prostate cancer but did not affect risk for colorectal cancer (95,141). In another study, supplemental vitamin E was also correlated with a reduction in prostate cancer in male smokers but not in non-smokers (35). In a population case-control study, consumption of vitamin E supplements for at least 2 years was correlated with a lower incidence of prostate cancer (119). Other published data offer evidence that consumption of vitamin E has a protective effect against esophageal, pharyngeal, gastric cancer, melanoma, and colorectal cancer (4,140).

Immune function

Efficiency of the immune system declines in many old people as manifested by a decline in lymphocyte proliferation in response to mitogens and an increased adherence of lymphocytes. Consumption of 200 mg vitamin E/day by a group of men (65–75 years old) for 3 months was found to reverse these trends and restore these measures of immune function to normal adult levels (54).

Neurological disease

Neuropathological symptoms have been observed as consequences of dietary vitamin E deficiency. Both cerebellar and peripheral nerve functions appear to be affected and patients experience difficulty in coordination of limb movements. These observations have stimulated interest in the possible role of vitamin E in the origin and progression of several neurological diseases (220).

Lipoproteins from the brain and spinal cord of patients with Alzheimer's disease (AD) were found to be much more easily oxidized than lipoproteins from controls suggesting that this increased susceptibility might be ameliorated by vitamin E supplements (189). Analysis of data on 3,385 Japanese-American men participating in the Honolulu Heart Program (HHP) indicated that regular consumption of vitamin E supplements did not have a protective effect against AD (142). However, use of vitamin E supplements was associated with a decrease in other types of vascular dementia among men in the HHP.

Other diseases

Data from the HHP (142), the ATBC Study of smokers (131) and the Health Professional Followup Study (8) indicated that consumption of vitamin E supplements did not substantially affect the incidence or mortality from stroke. Exposure of the skin to ultraviolet light may induce the formation of reactive species (photo-oxidative stress) with a resulting reddening or erythema. Supplements of α-tocopherol and β-carotene consumed for 8 weeks significantly diminished erythema in response to UV light and thus may be useful in reducing sensitivity to UV light (202). Although oxidative damage is believed to play a role in some eye diseases, vitamin E supplements had no effect on the incidence of cataracts and age-related maculopathy (209–211). However, another epidemiological study (132) did find that regular users of vitamin E supplements had a significantly reduced risk for cataracts.

VITAMIN C

Vitamin C (ascorbic acid) is an essential water-soluble micronutrient that participates in numerous enzymatic reactions as an electron donor and also functions as a chain breaking antioxidant by reacting directly with superoxide, hydroxyl radicals, and singlet oxygen. Ascorbic acid also acts as an indirect antioxidant by regenerating the lipid soluble antioxidant, vitamin E (21). Vitamin C is present in many fruits and vegetables and consumption of vegetables significantly increases human plasma concentrations of vitamin C up to a maximum of about $100\,\mu M$ (32,98). When an excess of ascorbic acid is consumed, urinary excretion of this vitamin increases to maintain an optimal physiological level (134).

According to NHANES III data, median daily dietary intake of vitamin C in the US is approximately 84 mg for men and 73 mg for women (134). Since many people also take vitamin C supplements, much higher greater amounts, up to 1,000 mg ascorbic acid/day, are consumed by some people. However, there is no evidence that very high intakes cause ill effects in healthy people. Some concern has been expressed about oxalate stone formation, uricosuria, and iron overload for people consuming excessive amounts of ascorbic acid but these conditions are generally not an issue for healthy individuals (57,84). The latest report by the Panel on Dietary Antioxidants and Related Compounds of the Institute of Medicine, set an RDA (Recommended Daily Allowance) of 75 mg vitamin C for females and 90 mg for males. At this level, neutrophil ascorbate concentrations should be near maximal with little urinary excretion. The increased oxidative stress suffered by smokers indicates that they should consume an extra 35 mg vitamin C/day (156). The upper limit of intake was set at 2 g/day based on the adverse effect of osmotic diarrhea.

Epidemiological studies have confirmed the health benefits of diets containing 200 mg or more of vitamin C (17,140,187) but inconsistent results have been obtained in trials using vitamin C supplements. Some of this inconsistency may be traced to the fact that plasma levels of ascorbate were not monitored. If subjects already consumed diets containing 100 mg ascorbic acid, intake of vitamin C supplements would not significantly increase plasma concentrations of this vitamin and therefore there may be little difference between experimental and control groups. Conversely, if subjects normally consume <100 mg ascorbate/day, the addition of supplements will make a significant difference in plasma vitamin C levels (134). Of course, another explanation for the discrepancy between epidemiological studies of dietary intake of vitamin C and trials using supplements is that the fruits and vegetables rich in vitamin C also contain other beneficial nutrients.

Vitamin C acts as an antioxidant in vitro (28,155) and can decrease oxidative stress in athletes during strenuous exercise (10,191) and urinary excretion of the F_2 prostane, 8-epi $PGF_2\alpha$, in moderate smokers (173). However, vitamin C supplements do not enhance free radical scavenging abilities under all conditions in vivo. Lipid peroxidation indices were not diminished significantly in groups of smokers given daily supplements of 500 mg vitamin C, even though plasma vitamin C concentrations increased (1,164). High doses of vitamin C also failed to suppress free radical production after primary coronary angioplasty (91). It has been suggested that vitamin C could act as a pro-oxidant in vivo by reacting with metal ions to generate hydroxyl and alkoxyl radicals. But a recent review of research on the effects of ascorbic acid on markers of oxidative DNA, protein, and lipid damage found that, in most cases, vitamin C either reduced or did not affect markers of oxidative damage (33). It appears, therefore, that vitamin C does not act as a pro-oxidant under physiological conditions.

The role that vitamin C may play in preventing cardiovascular disease was reviewed recently (77). Some trials with human volunteers demonstrated that supplements of vitamins C and E and β-carotene reduced the susceptibility of LDL to oxidation in diabetic (5) and non-diabetic (232) subjects and vitamin C and iron supplements had a similar effect in another group of healthy subjects (238). A single large dose (3 g) of vitamin C given to non-smokers counteracted the increased susceptibility of LDL to oxidation caused by exposure to cigarette smoke (219). However, in other studies vitamin C supplements had no protective effect against the ex vivo oxidation of LDL (150,186,219,227).

In other experiments related to cardiovascular disease, ascorbic acid prevented apoptosis (cell death) caused by oxidized LDL in vascular smooth muscle in vitro (195,196). This may reduce plaque instability and thrombosis in advanced cases of atherosclerosis. Increased platelet activation and stiffening of arteries are other symptoms improved by vitamin C supplements (229). Cigarette smoking reduces blood flow velocity but pretreatment of healthy volunteers with 2 g of ascorbic acid prevented this negative effect on circulation (241).

Epidemiological studies indicated that higher vitamin C intakes did not decrease risk for strokes (8) or Alzheimer's disease (142) in men. A protective effect of ascorbic acid for cognitive function was observed in elderly Japanese-American men (142) and results of another study indicated that vitamin C supplements may have a modest protective effect against Alzheimer's disease (147).

Daily vitamin C supplements did not improve pulmonary function in smokers (1) although supplements of vitamins C and E did offer some protection against the effects of ozone on lung function (90).

Although some epidemiological studies suggest a protective role for vitamin C against eye diseases (134), recently published data from studies correlating dietary or supplemental vitamin C with the incidence of cataracts requiring extraction (36,37), the incidence of nuclear cataracts (138), and the incidence of age related maculopathy (199) revealed little or no positive effect of ascorbic acid. Similarly, results of epidemiological studies indicating a protective effect of ascorbic acid against cancer are inconsistent with results of trials of the effect of supplements on carcinogenesis (69,88,190, 228).

Despite its known antioxidant capabilities, ascorbic acid supplements, by themselves, are not consistently associated with decreased oxidative damage. It may be that the water soluble nature of this antioxidant limits its ability to protect against free radicals in lipophilic compartments. Therefore, we may be more likely to observe benefits when this antioxidant is combined with lipid soluble antioxidants such as vitamin E.

CAROTENOIDS

Over 500 lipid-soluble, C-40 carotenoid compounds have been identified from plants, bacteria, insects, fish, and other plant-eating animals. β-carotene, the best known of these polyene compounds, can be enzymatically cleaved to form two molecules of retinaldehyde which can then be converted to vitamin A (retinol). Since β-carotene is not a required nutrient (if one consumes enough vitamin A), no recommended daily allowance has been set for β-carotene or any other individual carotenoids. However, the recommended daily intake of vitamin A is 800–1,000 µg retinol, which could be supplied by 4.8–6.0 mg β-carotene (3,156). Carotenoids present in vegetables are generally bioavailable and significant increases in plasma carotenoid levels can be observed as more vegetables are consumed (98,148). β-carotene is the most common carotenoid in many vegetables although lycopene, from tomatoes, is consumed in approximately the same amount as β-carotene in the US (152). Toxicity of β-carotene is low, with data indicating that daily intakes of up to 50 mg have no adverse effects except for hypercarotenemia in some subjects (57,234). For more information on the absorption, metabolism, and antioxidant potential of carotenoids, recent reviews should be consulted (55,153,167).

Besides its function as a precursor to vitamin A, β-carotene and some other carotenoids which are attached to cell membranes react directly with the free radical, singlet oxygen, quenching it and preventing its destructive oxidative effects (183,215). Consumption of tomatoes (with high lycopene concentrations) significantly decreased oxidative DNA damage in lymphocytes from healthy volunteers (172,181). A decrease in oxidative damage in lymphocytes was also noted when β-carotene supplements were given to smokers (225). However, β-carotene appears to be a weak antioxidant or even a prooxidant in some vitro test systems (13,20).

Although diets rich in fruits and vegetables have proven health benefits, some large trials of β-carotene supplements have not demonstrated a benefit in terms of reduced mortality or reduced incidence of cancer or cardiovascular disease (44,96,128). Consumption of β-carotene supplements by apparently healthy men for 12 years did not lessen the risk for subsequent development of type 2 diabetes (135). It may be that supplements of β-carotene are not beneficial in well-nourished populations.

Cancer

More than 30 epidemiological studies have demonstrated that people who eat more foods rich in carotenoids and those with higher blood β-carotene concentrations have a lower risk for developing lung cancer (3,200). In addition, persons with higher plasma β-carotene concentrations tend to have a greater lung capacity as demonstrated by measurements of vital capacity and forced expiratory volume of air (89). These encouraging epidemiological results suggested that smokers, who are known to be at increased risk for lung cancer, might benefit from taking β-carotene supplements. However, intervention studies with smokers not only failed to demonstrate an anticarcinogenic effect of carotenoids but rather indicated that β-carotene supplements enhanced development of lung cancer (151,212). It appears from recent in vitro and in vivo studies that the β-carotene molecule is unstable in the free-radical rich environment of the lungs of cigarette smokers. The oxidized products of β-carotene may actually promote carcinogenesis (222).

Results from the ATBC (α-tocopherol β-carotene) Cancer Prevention Study also demonstrated an increase in prostate cancer in smokers consuming β-carotene supplements (95). It was suggested that this was due to the interference of β-carotene with vitamin D synthesis. However, subsequent analysis of the data did not support this hypothesis (75). Another large epidemiological study of β-carotene and lung cancer did not find β-carotene to be procarcinogenic but neither did it have a protective effect against lung cancer (45). Four recent reviews discuss this research on lung cancer and point out that we should be cautious about extrapolating results from epidemiological studies directly to dietary recommendations without further research (3,51,52,166).

β-carotene supplements alone were also not effective in reducing the recurrence of skin cancer (88) or altering the risk for colorectal cancer (4,141) or pancreatic cancer (170).

11

There have been some positive results in studies of the anticarcinogenic effects of β-carotene supplements. Data from the Physicians' Health Study indicated that β-carotene may reduce the risk of prostate cancer (50) and in another study, men with high plasma lycopene levels had a reduced risk for prostate cancer (79). Results of a multi-center prospective trial demonstrated that consumption of 60 mg β-carotene/day was associated with a decrease in the severity of oral leukoplakia, a pre-malignant lesion (81). Dietary data collected from cases of colon cancer and matched controls indicated that the latter consumed higher average levels of the carotenoid, lutein (197).

Cardiovascular disease

Epidemiological studies on the relationship between carotenoids and cardiovascular disease have yielded mixed results. Data from the Rotterdam study indicate that persons with a higher intake of β-carotene from supplements and diet have a lower risk for myocardial infarction (117). But this relationship was not observed in smokers from the ATBC study who had already had one heart attack (169). Nor did these smokers have a lower incidence of mortality from stroke (131). In a subsample of the Rotterdam study, there was a modest inverse correlation between serum lycopene, but not serum β-carotene, levels and atherosclerosis (116). Data from NHANES (National Health and Nutrition Survey) III study demonstrated that persons with higher serum levels of several carotenoids had a lower risk of suffering from angina pectoris (73).

Since free radical oxidation of lipoproteins appears to be involved in atherosclerosis, effects of β-carotene and lycopene on this process were measured. Analyses of plasma lipoproteins from healthy adults revealed that lycopene and β-carotene were preferentially sequestered in LDL (161). LDL subfractions containing higher concentrations of carotenoids were more resistant to copper-mediated oxidation (137). Carotenoids in LDL also react efficiently with peroxynitrite and may be efficient scavengers of this free radical in vivo (154). In another study, only β-carotene supplements were observed to inhibit oxidative damage to LDL (61). Some carotenoids may prevent oxidation of lipoproteins in vivo and thereby retard the development of atherosclerosis.

A recent review of the relationship between β-carotene and risk of coronary artery disease concluded that this antioxidant most likely has a beneficial effect but that it probably acts in concert with other antioxidants and nutrients (208).

Other diseases

Data correlating β-carotene intake with the development of cataracts or age related maculopathy are inconsistent with some studies showing no effect (209–211) while others show a preventive effect of carotenoids (30,36,37). Two recent articles reviewed research on the effects of nutritional antioxidants on the development of cataracts and age related macular degeneration (43,108). Several human studies, which demonstrated a beneficial effect of β-carotene on immune function, have been recently reviewed (104,105).

β-Carotene is undoubtedly an important dietary nutrient and may aid in maintaining health by combining with free radicals. A number of epidemiological studies have correlated plasma β-carotene levels with the presence or development of a disease. However, Jandacek (110), citing research which reports the depletion of β-carotene in diseased tissues, regards β-carotene as the canary in the cell: Low carotenoid concentrations are the result of disease, not the cause of it.

PHENOLIC COMPOUNDS

Although phenolic compounds are not considered essential dietary constituents and no recommended daily allowance has been set for flavonoids, tannins, lignans or other phenolic compounds, they are widely distributed in plant foods and have attracted much attention in the past decade because of their antioxidant properties. Flavonoids, the best studied group, comprise a series of C-15 compounds with two phenolic rings joined by an oxygen-containing pyran ring. Thousands of individual compounds, with different substitutions on the ring carbons, have been described. Some compounds have limited bioavailability while others, including catechin and quercetin, appear to be readily absorbed from the gastrointestinal tract (16,18,27,74,87,100,129,179,188,230).

Considerable research has been concentrated on the flavonoids found in grape juice/red wine and in green tea. Berries are also excellent sources of flavonoids as are a number of other plant foods (112). Even chocolate has been found to be a significant source of flavonoids, contributing 20% to the catechin intake of a sample of the Dutch population while tea contributed 55% to catechin intake (7,179). Consumption of chocolate has been found to cause a significant increase in plasma epicatechin levels and a decrease in plasma oxidation products (174). Lignans (such as sesaminol, a major component of sesame oil), phenolic acids in apple skins, and other phenolic compounds in grapes and olive oil also exert antioxidant effects which may be beneficial to human health (85,111,113,159,194).

Antioxidant properties of these compounds are attributed to the phenolic hydroxyl groups which can donate electrons to free radicals thereby quenching their destructive activities. Comparison of superoxide scavenging effects of several flavonoids in vitro revealed that some compounds may be as much as 600 times more potent than others depending on structural features of the molecules such as the number of hydroxyl groups present (103,177,178).

According to a number of epidemiological studies, dietary flavonoids appear to have a protective effect against cardiovascular disease (83,118,239). However, these protective effects are not consistently observed in trials in which subjects consume certain amounts of flavonoids (22,31,39,64,76,231). It is likely that flavonoids, like other antioxidants, work best in vivo in the presence of other antioxidants and nutrients.

Protective effects are most likely related to antioxidant activities which may prevent oxidation of LDL and spare other antioxidants by regenerating them (53,74). Polyphenols from grape juice (114) and cocoa (175,176) inhibit platelet aggregation. Quercetin, epicatechin, olive oil and other phenolic compounds inhibited oxidation of low density lipoproteins in vitro (41,72,85,159,223). Ingestion of purple grape juice by healthy subjects reduced the susceptibility of LDL to oxidation and improved endothelial function (205). However, in studies with five healthy volunteers, a meal containing 225 g onions increased plasma quercetin levels and antioxidant activity but did not affect the susceptibility of LDL to oxidation (143). Consumption of red wine (62), black grapes (63) and green tea catechins (149) by healthy subjects also caused a significant increase in antioxidant potential in plasma and erythrocytes. But consumption of eight cups of green tea per day did not appear to enhance resistance of LDL to oxidation ex vivo (99).

Phenolic compounds appear to exert a protective effect against some types of cancer (64,126). Consumption of a diet high in flavonoids was correlated with a reduction in oxidative damage to DNA in lymphocytes from healthy subjects (130) and from type 2 diabetic subjects (127). Several polyphenol and tea extracts inhibited neoplastic transformations and formation of DNA adducts in in vitro assays (204). Protective effects of tea in carcinogenicity assays have recently been reviewed (2,60).

Some phenolic compounds also act as carcinogens in animal and in vitro assays. In a study of neoplastic changes induced in rats by heterocyclic amines, a green tea extract was inhibitory while quercetin significantly enhanced these changes (97). Therefore, it may not be wise to consume supplements of phenolic compounds before their reactions under physiological conditions are better understood.

SELENIUM

Selenium has long been recognized as an essential element but is also known to be toxic at high concentrations. The primary route of exposure to selenium for most people is the consumption of organic selenium compounds in cereals, onions, garlic, and other vegetables (14). Inorganic forms of selenium may also be consumed in drinking water but these are not as readily utilized as the seleno-amino acids present in vegetables (158). Median intakes of selenium in various countries range from a low of 30 µg/day in Turkey to a high of 130 µg/day in the US (121). However, in regions of China with selenium-rich soils, daily intake of selenium has been estimated at 1,270 µg in some individuals with overt signs of selenium toxicity (236).

Several lines of evidence indicate that selenium supplements may exert protective effects against disease by enhancing the antioxidant defenses of the body. The latest report by the Panel on Dietary Antioxidants and Related Compounds of the Institute of Medicine, proposed an RDA for selenium of 55 µg for men and women (156). The upper limit of intake was set at 400 µg/day based on the adverse effects of selenosis (pathologic nail changes, brittle hair, garlic breath). Large doses of selenium should not be consumed because the margin of safety for selenium supplements is small.

Two relatively rare diseases, Keshan disease, a type of cardiac myopathy, and Kashin–Beck disease, a degenerative osteo-articular disorder, are endemic in some selenium-deficient areas of China. These diseases may be a result of oxidative stress due to depleted stores of antioxidant enzymes. Coxsackie virus, isolated from patients with Keshan disease, appears to play a role in the development of this disease and selenium deficiency is believed to contribute to its virulence (171). Recent public health interventions providing selenium supplements to local populations have drastically decreased the incidence of Keshan and Kashin–Beck diseases (14,236).

According to recent reviews (158,171), the principal effects of selenium in the body can be attributed to its antioxidant effects and to its roles in maintaining and regulating immune function, thyroid metabolism, fertility, cell growth, and eicosanoid synthesis. Selenium is a component of iodothyronine deiodinase, an enzyme which controls the synthesis and degradation of the active form of the thyroid hormone (T-3) (6). Selenium supplements also stimulate the immune system and selenium deficiency has been correlated with an increase in incidence and severity of some viral infections (15,133,171). However, elevated levels of dietary selenium can have adverse effects and have been reported to induce an increase in superoxide radicals (115).

Selenium by itself does not act as an antioxidant but rather is an essential component of some antioxidant enzymes including glutathione peroxidases and thioredoxin reductase. Selenocysteine, present in the active sites of these enzymes, is responsible for their antioxidative effects. Glutathione peroxidases protect lipids and proteins from the destructive effects of hydrogen peroxide and organic peroxides which cause oxidative degradation (14,163). Activity of glutathione peroxidase in plasma of healthy volunteers increases with indicators of oxidative stress such as malondialdehyde concentrations, indicating that activity of selenium-dependent antioxidative enzymes can be adjusted according to the extent of oxidative stress in the organism (184). Oxidative stress is elevated in diabetic patients and contributes to many disorders, such as kidney disease, that are associated with progression of this disease. Selenium supplements added to the diet of diabetic rats were found to suppress the development of renal lesions (59).

Oxidation of low density lipoproteins appears to be an important factor in the development of cardiovascular disease. Mean selenium levels in erythrocytes of a group of patients who suffered myocardial infarctions were found to be significantly lower than those in a group of healthy controls (26). Since erythrocyte selenium levels reflect selenium status over a period of several weeks, it appears that the patients had a decreased antioxidant defense system for some time before their heart attacks occurred.

Several studies indicate that selenium supplements exert a protective effect against some types of cancer in humans, including prostate and colon cancer (46,47,48,

168). Dietary selenium, administered as Brazil nuts or as selenium-rich garlic and onion, is anticarcinogenic, effectively reducing mammary cancer in rats (106,107).

The mechanism for the anticarcinogenic effects of selenium is unknown but may involve both protection of DNA from oxidative degradation and alterations in carcinogen metabolism, immune function, cell cycle regulation, and apoptosis. Some data demonstrate that antitumorigenic effects of selenium are exerted at intake levels above those required to prevent expression of selenium deficiency symptoms (49). Other experiments on colon cancer in rats indicated that the chemopreventive effects of selenium were associated with a down-regulation of protein kinase C and tyrosine protein kinase activities and an up-regulation of diacylglycerol activity in colon tissues (168). A study of pancreatic cancer in rats indicated that selenium supplements were more effective as anticarcinogens when given during the promotion phase of carcinogenesis (233). In addition, selenium may inhibit the promotion of cancer by helping to maintain adequate immune function (86,104,237). Two large scale trials of the anticarcinogenic effects of selenium, PRECISE in Europe and SELECT in the US, are each enrolling more than 30,000 participants and should provide more useful data in the future (171).

Epidemiological studies in many countries with various groups of people have repeatedly demonstrated the health benefits of consuming diets rich in fruits and vegetables. Analyses of these foods to identify the active constituents have focused on antioxidants since damage caused by free radicals is apparently an integral part of many chronic diseases and the aging process in animals.

For the bioavailable antioxidant nutrients normally found in foods, epidemiological evidence indicates that diets rich in vitamins A and C, β-carotene, flavonoids, and selenium reduce the risk of developing cardiovascular disease, cancer, and some other conditions such as cataracts, immune dysfunction and possibly neurological diseases. However, trials testing the effects of individual antioxidants often yield inconsistent results. This may be a result of problems in methodology but is likely also due to the fact that optimal functioning of these antioxidants requires the presence of other antioxidants (for regeneration) and other micronutrients. Investigations of the redox cycles of ascorbate, α-tocopherol, and caffeic acid indicate that they act synergistically to protect biological molecules from oxidation (125). Evidence for the positive effects of antioxidant supplement use is most convincing for α-tocopherol. But, generally, it's probably best to consume antioxidants in fruits and vegetables where other micronutrients and cofactors are also present.

REFERENCES

1. Aghdassi E, Royall D, Allard JP. Oxidative stress in smokers supplemented with vitamin C. *Int J Vit Nutr Res* 1999; 69(1):45–51.
2. Ahmad N, Mukhtar H. Green tea polyphenols and cancer: biologic mechanisms and practical implications. *Nutr Rev* 1999; 57(3):78–83.

16

3. Albanes D. Beta-carotene and lung cancer: a case study. *Am J Clin Nutr* 1999; 69(6):1345S–50S.
4. Albanes D, Malila N, Taylor PR, Huttunen JK, Virtamo J, Edwards BK, Rautalahti M, Hartman AM, Barrett MJ, Pietinen P, Hartman TJ, Sipponen P, Lewin K, Teerenhovi L, Hietanen P, Tangrea JA, Virtanen M, Heinonen OP. Effects of supplemental alpha-tocopherol and beta-carotene on colorectal cancer: results from a controlled trial (Finland). *Cancer Causes Control* 2000; 11(3):197–205.
5. Anderson JW, Gowri MS, Turner J, Nichols L, Diwadkar VA, Chow CK, Oeltgen PR. Antioxidant supplementation effects on low-density lipoprotein oxidation for individuals with type 2 diabetes mellitus. *J Am Coll Nutr* 1999; 18(5):451–61.
6. Arthur JR, Beckett GJ. Thyroid function. *Brit Med Bull* 1999; 55(3):658–68.
7. Arts ICW, Hollman PCH, Kromhout D. Chocolate as a source of tea flavonoids. *Lancet* 1999; 354(9177):488.
8. Ascherio A, Rimm EB, Hernan MA, Giovannucci E, Kawachi I, Stampfer MJ, Willett WC. Relation of consumption of vitamin E, vitamin C, and carotenoids to risk for stroke among men in the United States. *Ann Intern Med* 1999; 130(12): 963–70.
9. Ascherio A, Stampfer MJ, Colditz GA, Rimm EB, Litin L, Willett WC. Correlations of vitamin A and E intakes with the plasma concentrations of carotenoids and tocopherols among American men and women. *J Nutr* 1992; 122(9):1792–801.
10. Ashton T, Young IS, Peters JR, Jones E, Jackson SK, Davies B, Rowlands CC. Electron spin resonance spectroscopy, exercise, and oxidative stress: an ascorbic acid intervention study. *J Appl Physiol* 1999; 87(6):2032–36.
11. Astley S, Langrish-Smith A, Southon S, Sampson M. Vitamin E supplementation and oxidative damage to DNA and plasma LDL in type 1 diabetes. *Diabetes Care* 1999; 22(10):1626–31.
12. Atroshi F, Rizzo A, Biese I, Veijalainen P, Saloniemi H, Sankari S, Andersson K. Fumonisin B-1-induced DNA damage in rat liver and spleen: Effects of pretreatment with coenzyme Q(10), L-carnitine, alpha-tocopherol and selenium. *Pharmacol Res* 1999; 40(6):459–67.
13. Baker DL, Krol ES, Jacobsen N, Liebler DC. Reactions of beta-carotene with cigarette smoke oxidants. Identification of carotenoid oxidation products and evaluation of the prooxidant antioxidant effect. *Chem Res Toxicol* 1999; 12(6):535–43.
14. Barceloux DG. Selenium. *J Toxicol – Clin Toxicol* 1999; 37(2):145–72.
15. Beck MA. Selenium and host defence towards viruses. *Proc Nutr Soc* 1999; 58(3):707–11.
16. Bell JRC, Donovan JL, Wong R, Waterhouse AL, German JB, Walzem RL, Kasim-Karakas SE. (+)-Catechin in human plasma after ingestion of a single serving of reconstituted red wine. *Am J Clin Nutr* 2000; 71(1):103–8.
17. Bendich A, Langseth L. The health effects of vitamin C supplementation: a review. *J Am Coll Nutr* 1995; 14:124–36.
18. Benzie IFF, Szeto YT, Strain JJ, Tomlinson B. Consumption of green tea causes rapid increase in plasma antioxidant power in humans. *Nutr Cancer* 1999; 34(1):83–7.
19. Berlett BS, Stadtman ER. Protein oxidation in aging, disease, and oxidative stress. *J Biol Chem* 1997; 272(33):20313–16.
20. Bestwick CS, Milne L. Effects of beta-carotene on antioxidant enzyme activity, intracellular reactive oxygen and membrane integrity within post confluent Caco-2 intestinal cells. *Biochem Biophys Acta* 2000; 1474(1):47–55.

21. Beyer RE. The role of ascorbate in antioxidant protection of biomembranes: interaction with vitamin E and coenzyme Q. *J Bioenergetics Biomembranes* 1994; 26:349–58.

22. Blostein-Fujii A, DiSilvestro RA, Frid D, Katz C. Short term citrus flavonoid supplementation of type II diabetic women: no effect on lipoprotein oxidation tendencies. *Free Rad Res* 1999; 30(4):315–20.

23. Böhm V, Bitsch R. Intestinal absorption of lycopene from different matrices and interactions to other carotenoids, the lipid status, and the antioxidant capacity of human plasma. *European J Nutr* 1999; 38(3):118–25.

24. Bonorden WR, Pariza MW. Antioxidant nutrients and protection from free radicals. In: Kotsonis FN, Mackey M, Hjelle J, eds. *Nutritional Toxicology*. New York: Raven Press; 1994; 19–48.

25. Boomershine KH, Shelton PS, Boomershine JE. Vitamin E in the treatment of tardive dyskinesia. *Ann Pharmacother* 1999; 33(11):1195–202.

26. Bor MV, Cevik C, Uslu I, Guneral F, Duzgun E. Selenium levels and glutathione peroxidase activities in patients with acute myocardial infarction. *Acta Cardiol* 1999; 54(5):271–6.

27. Bourne LC, Rice-Evans CA. Detecting and measuring bioavailability of phenolics and flavonoids in humans: pharmacokinetics of urinary excretion of dietary ferulic acid. *Oxidants and Antioxidants*, Pt A. 1999; 299:91–106.

28. Brennan LA, Morris GM, Wasson GR, Hannigan BM, Barnett YA. The effect of vitamin C or vitamin E supplementation on basal and H_2O_2-induced DNA damage in human lymphocytes. *Brit J Nutr* 2000; 84(2):195–202.

29. Brigelius-Flohe R, Traber MG. Vitamin E: function and metabolism. *FASEB J* 1999; 13(10):1145–55.

30. Brown L, Rimm EB, Seddon JM, Giovannucci EL, Chasan-Taber L, Spiegelman D, Willett WC, Hankinson SE. A prospective study of carotenoid intake and risk of cataract extraction in US men. *Am J Clin Nutr* 1999; 70(4):517–24.

31. Caccetta RAA, Croft KD, Beilin LJ, Puddey IB. Ingestion of red wine significantly increases plasma phenolic acid concentrations but does not acutely affect ex vivo lipoprotein oxidizability. *Am J Clin Nutr* 2000; 71(1):67–74.

32. Calzada C, Bizzotto M, Paganga G, Miller NJ, Bruckdorfer KR, Diplock AT, Rice-Evans CA. Levels of antioxidant nutrients in plasma and low density lipoproteins: a human volunteer supplementation study. *Free Rad Res* 1995; 23(5): 489–503.

33. Carr A, Frei B. Does vitamin C act as a pro-oxidant under physiological conditions? *FASEB J* 1999; 13(9):1007–24.

34. Carr AC, McCall MR, Frei B. Oxidation of LDL by myeloperoxidase and reactive nitrogen species – Reaction pathways and antioxidant protection. *Arteriosclerosis Thrombosis & Vascular Biol* 2000; 20(7):1716–23.

35. Chan JM, Stampfer MJ, Ma J, Rimm EB, Willett WC, Giovannucci EL. Supplemental vitamin E intake and prostate cancer risk in a large cohort of men in the United States. *Cancer Epidemiol Biomarkers Prevent* 1999; 8(10):893–9.

36. Chasan-Taber L, Willett WC, Seddon JM, Stampfer MJ, Rosner B, Colditz GA, Hankinson SE. A prospective study of vitamin supplement intake and cataract extraction among US women. *Epidemiology* 1999; 10(6):679–84.

37. Chasan-Taber L, Willett WC, Seddon JM, Stampfer MJ, Rosner B, Colditz GA, Speizer FE, Hankinson SE. A prospective study of carotenoid and vitamin A intakes and risk of cataract extraction in US women. *Am J Clin Nutr* 1999; 70(4):509–16.

38. Chaudiere J, Ferrari-Iliou R. Intracellular antioxidants: from chemical to biochemical mechanisms. *Food Chem Toxicol* 1999; 37(9–10):949–62.
39. Cherubini A, Beal MF, Frei B. Black tea increases the resistance of human plasma to lipid peroxidation in vitro, but not ex vivo. *Free Rad Biol Med* 1999; 27(3–4):381–7.
40. Chisolm GM, Steinberg D. The oxidative modification hypothesis of atherogenesis: An overview. *Free Rad Biol Med* 2000; 28(12):1815–26.
41. Chopra M, Fitzsimons PEE, Strain JJT, Thurnham DI, Howard AN. Nonalcoholic red wine extract and quercetin inhibit LDL oxidation without affecting plasma antioxidant vitamin and carotenoid concentrations. *Clin Chem* 2000; 46(8 Part 1):1162–70.
42. Chopra RK, Bhagavan HN. Relative bioavailabilities of natural and synthetic vitamin E formulations containing mixed tocopherols in human subjects. *Int J Vit Nutr Res* 1999; 69(2):92–5.
43. Christen WG. Antioxidant vitamins and age-related eye disease. *Proc Assoc Am Physicians* 1999; 111(1):16–21.
44. Christen WG, Buring JE, Manson JE, Hennekens CH. Beta-carotene supplementation: a good thing, a bad thing, or nothing? *Curr Opin Lipidol* 1999; 10(1):29–33.
45. Christen WG, Gaziano JM, Hennekens CH. Design of physicians' health study II – a randomized trial of beta-carotene, vitamins E and C, and multivitamins, in prevention of cancer, cardiovascular disease, and eye disease, and review of results of completed trials. *Ann Epidemiol* 2000; 10(2):125–34.
46. Clark LC, Alberts DS. Selenium and cancer: risk or protection? *J Natl Cancer Inst* 1995; 87:473–5.
47. Clark LC, Combs GF, Turnbull BW, Slate EH, Chalker DK, Chow J, Davis LS, Glover RA, Graham GF, Gross EG, Krongrad A, Lesher JL, Park HK, Sanders BB, Smith CL, Taylor JR. Effects of selenium supplementation for cancer prevention in patients with carcinoma of the skin – a randomized controlled trial. *J Am Med Assoc* 1996; 276(24):1957–63.
48. Clark LC, Dalkin B, Krongrad A, Combs GF, Turnbull BW, Slate EH, Witherington R, Herlong JH, Janosko E, Carpenter D, Borosso C, Falk S, Rounder J. Decreased incidence of prostate cancer with selenium supplementation – results of a double-blind cancer prevention trial. *Brit J Urol* 1998; 81(5):730–4.
49. Combs GF. Chemopreventive mechanisms of selenium. *Medizinische Klinik* 1999; 94(Suppl 3):18–24.
50. Cook NR, Stampfer MJ, Ma J, Manson JE, Sacks FM, Buring JE, Hennekens CH. Beta-carotene supplementation for patients with low baseline levels and decreased risks of total and prostate carcinoma. *Cancer* 1999; 86(9):1783–92.
51. Cooper DA, Eldridge AL, Peters JC. Dietary carotenoids and certain cancers, heart disease, and age-related macular degeneration: a review of recent research. *Nutr Rev* 1999; 57(7):201–14.
52. Cooper DA, Eldridge AL, Peters JC. Dietary carotenoids and lung cancer: a review of recent research. *Nutr Rev* 1999; 57(5 Part 1):133–45.
53. da Silva EL, Abdalla DSP, Terao J. Inhibitory effect of flavonoids on low-density lipoprotein peroxidation catalyzed by mammalian 15-lipoxygenase. *Iubmb Life* 2000; 49(4):289–95.
54. De la Fuente M, Victor VM. Anti-oxidants as modulators of immune function. *Immunol Cell Biol* 2000; 78(1):49–54.
55. Deming DM, Erdman JW. Mammalian carotenoid absorption and metabolism. *Pure Appl Chem* 1999; 71(12):2213–23.

56. De Stefani E, Boffetta P, Deneo-Pellegrini H, Mendilaharsu M, Carzoglio JC, Ronco A, Olivera L. Dietary antioxidants and lung cancer risk: a case-control study in Uruguay. *Nutr Cancer* 1999; 34(1):100–10.

57. Diplock AT. Safety of antioxidant vitamins and beta-carotene. *Am J Clin Nutr* 1995; 62(6 Suppl):1510S–16S.

58. Diplock AT, Charleux JL, Crozier-Willi G, Kok FJ, Rice-Evans C, Roberfroid M, Stahl W, Vina-Ribes J. Functional food science and defence against reactive oxidative species. *Brit J Nutr* 1998; 80(Suppl 1):S77–112.

59. Douillet C, Tabib A, Bost M, Accominotti M, Borson-Chazot F, Ciavatti M. Selenium in diabetes: effects of selenium on nephropathy in type 1 streptozotocin-induced diabetic rats. *J Trace Elem Exp Med* 1999; 12(4):379–92.

60. Dreosti IE, Wargovich MJ, Yang CS. Inhibition of carcinogenesis by tea – the evidence from experimental studies. *Crit Rev Food Sci Nutr* 1997; 37(8):761–70.

61. Dugas TR, Morel DW, Harrison EH. Dietary supplementation with beta-carotene, but not with lycopene, inhibits endothelial cell-mediated oxidation of low-density lipoprotein. *Free Rad Biol Med* 1999; 26(9–10):1238–44.

62. Durak I, Cimen MYB, Buyukkocat S, Kacmaz M, Ozturk HS. The effect of red wine on blood antioxidant potential. *Curr Med Res Opinion* 1999; 15(3):208–13.

63. Durak I, Koseoglu MH, Kacmaz M, Buyukkocak S, Cimen MYB, Ozturk HS. Black grape enhances plasma antioxidant potential. *Nutr Res* 1999; 19(7):973–7.

64. Duthie GG, Duthie SJ, Kyle JAM. Plant polyphenols in cancer and heart disease: implications as nutritional antioxidants. *Nutr Res Rev* 2000; 13(1):79–106.

65. Duthie GG, Bellizzi MC. Effects of antioxidants on vascular health. *Brit Med Bull* 1999; 55(3):568–77.

66. Duthie SJ, Ma AG, Ross MA, Collins AR. Antioxidant supplementation decreases oxidative DNA damage in human lymphocytes. *Cancer Res* 1996; 56(6): 1291–95.

67. Dutta-Roy AK. Molecular mechanism of cellular uptake and intracellular translocation of alpha-tocopherol: Role of tocopherol-binding proteins. *Food Chem Toxicol* 1999; 37(9–10):967–71.

68. Eastwood MA. Interaction of dietary antioxidants in vivo: how fruit and vegetables prevent disease? *QJM* 1999; 92(9):527–30.

69. Eichholzer M, Stahelin HB, Ludin E, Bernasconi F. Smoking, plasma vitamins C, E, retinol, and carotene, and fatal prostate cancer: seventeen-year follow-up of the prospective Basel study. *Prostate* 1999; 38(3):189–98.

70. Emmert DH, Kirchner JT. The role of vitamin E in the prevention of heart disease. *Arch Fam Med* 1999; 8(6):537–42.

71. Factor VM, Laskowska D, Jensen MR, Woitach JT, Popescu NC, Thorgeirsson SS. Vitamin E reduces chromosomal damage and inhibits hepatic tumor formation in a transgenic mouse model. *Proc Natl Acad Sci USA* 2000; 97(5):2196–201.

72. Fito M, Covas MI, Lamuela-Raventos RM, Vila J, Torrents J, de la Torre C, Marrugat J. Protective effect of olive oil and its phenolic compounds against low density lipoprotein oxidation. *Lipids* 2000; 35(6):633–8.

73. Ford ES, Giles WH. Serum vitamins, carotenoids, and angina pectoris: findings from the National Health and Nutrition Examination Survey III. *Ann Epidemiol* 2000; 10(2):106–16.

74. Frankel EN. Food antioxidants and phytochemicals: present and future perspectives. *Fett-Lipid* 1999; 101(12):450–5.

75. Freedman DM, Tangrea JA, Virtamo J, Albanes D. The effect of beta-carotene supplementation on serum vitamin D metabolite concentrations. *Cancer Epidemiol Biomarkers Prevent* 1999; 8(12):1115–16.

76. Freese R, Basu S, Hietanen E, Nair J, Nakachi K, Bartsch H, Mutanen M. Green tea extract decreases plasma malondialdehyde concentration but does not affect other indicators of oxidative stress, nitric oxide production, or hemostatic factors during a high-linoleic acid diet in healthy females. *Eur J Nutr* 1999; 38(3): 149–57.

77. Frei B. On the role of vitamin C and other antioxidants in atherogenesis and vascular dysfunction. *Proc Soc Exper Biol Med* 1999; 222(3):196–204.

78. Frei B. Reactive oxygen species and antioxidant vitamins: mechanisms of action. *Am J Med* 1994; 97(3A):5S–13S.

79. Gann PH, Ma J, Giovannucci E, Willett W, Sacks FM, Hennekens CH, Stampfer MJ. Lower prostate cancer risk in men with elevated plasma lycopene levels: results of a prospective analysis. *Cancer Res* 1999; 59(6):1225–30.

80. Garewal HS, Diplock AT. How "safe" are antioxidant vitamins? *Drug Safety* 1995; 13(1):8–14.

81. Garewal HS, Katz RV, Meyskens F, Pitcock J, Morse D, Friedman S, Peng Y, Pendrys DG, Mayne S, Alberts D, Kiersch T, Graver E. Beta-carotene produces sustained remissions in patients with oral leukoplakia – results of a multicenter prospective trial. *Arch Otolaryngol – Head & Neck Surg* 1999; 125(12):1305–10.

82. Gazis A, White DJ, Page SR, Cockcroft JR. Effect of oral vitamin E (alpha-tocopherol) supplementation on vascular endothelial function in type 2 diabetes mellitus. *Diabetic Med* 1999; 16(4):304–11.

83. Geleijnse JM, Launer LJ, Hofman A, Pols HAP, Witteman JCM. Tea flavonoids may protect against atherosclerosis – the Rotterdam study. *Arch. Int Med* 1999; 159(18): 2170–74.

84. Gerster H. High-dose vitamin C: a risk for persons with high iron stores? [Review]. *Int J Vit Nutr Res* 1999; 69(2):67–82.

85. Giovannini C, Straface E, Modesti D, Coni E, Cantafora A, De Vincenzi M, Malorni W, Masella R. Tyrosol, the major olive oil biophenol, protects against oxidized-LDL-induced injury in Caco-2 cells. *J Nutr* 1999; 129(7):1269–77.

86. Girodon F, Galan P, Monget AL, Boutron-Ruault MC, Brunet-Lecomte P, Preziosi P, Arnaud J, Manuguerra JC, Hercberg S. Impact of trace elements and vitamin supplementation on immunity and infections in institutionalized elderly patients – a randomized controlled trial. *Arch Intern Med* 1999; 159(7):748–54.

87. Graefe EU, Derendorf H, Veit M. Pharmacokinetics and bioavailability of the flavonol quercetin in humans. *Int J Clin Pharmacol Ther* 1999; 37(5):219–33.

88. Greenberg ER, Baron JA, Stukel TA, Stevens MM, Mandel JS, Spencer SK, Elias PM, Lowe N, Nierenberg DW, Bayrd G, *et al.* A clinical trial of beta carotene to prevent basal-cell and squamous-cell cancers of the skin. The skin cancer prevention study group. *New Engl J Med* 1990; 323(12):789–95.

89. Grievink L, Smit HA, van't Veer P, Brunekreef B, Kromhout D. Plasma concentrations of the antioxidants beta-carotene and alpha-tocopherol in relation to lung function. *European J Clin Nutr* 1999; 53(10):813–17.

90. Grievink L, Zijlstra AG, Ke XD, Brunekreef B. Double-blind intervention trial on modulation of ozone effects on pulmonary function by antioxidant supplements. *Am J Epidemiol* 1999; 149(4):306–14.

91. Guan WP, Osanai T, Kamada T, Ishizaka H, Hanada H, Okumura K. Time course of free radical production after primary coronary angioplasty for acute myocardial infarction and the effect of vitamin C. *Japan Circulation J* 1999; 63(12):924–8.

92. Haffner SM. Clinical relevance of the oxidative stress concept. *Metabolism: Clin Exp* 2000; 49(2 Suppl 1):30–4.

93. Halliwell B, Murcia MA, Chirico S, Aruoma OI. Free radicals and antioxidants in food and in vivo: what they do and how they work. *Crit Rev Food Sci Nutr* 1995; 35(1–2):7–20.

94. Han SN, Meydani SN. Vitamin E and infectious diseases in the aged. *Proc Nutr Soc* 1999; 58(3):697–705.

95. Heinonen OP, Albanes D, Virtamo J, Taylor PR, Huttunen JK, Hartman AM, Haapa-koski J, Malila N, Rautalahti M, Ripatti S, Maenpaa H, Teerenhovi L, Koss L, Viro-lainen M, Edwards BK. Prostate cancer and supplementation with alpha-tocopherol and beta-carotene: incidence and mortality in a controlled trial. *J Natl Cancer Inst* 1998; 90(6):440–6.

96. Hennekens CH, Buring JE, Manson JE, Stampfer M, Rosner B, Cook NR, Belanger C, LaMotte F, Gaziano JM, Ridker PM, Willett W, Peto R. Lack of effect of long-term supplementation with beta carotene on the incidence of malignant neoplasms and cardiovascular disease. *New Engl J Med* 1996; 334(18):1145–9.

97. Hirose M, Takahashi S, Ogawa K, Futakuchi M, Shirai T. Phenolics: Blocking agents for heterocyclic amine-induced carcinogenesis. *Food Chem Toxicol* 1999; 37(9–10):985–92.

98. Hof KHV, Tijburg LBM, Pietrzik K, Weststrate JA. Influence of feeding different vegetables on plasma levels of carotenoids, folate and vitamin C. Effect of disruption of the vegetable matrix. *Brit J Nutr* 1999; 82(3):203–12.

99. van het Hof, KH, Wiseman SA, Yang CS, Tuburg LBM. Plasma and lipoprotein levels of tea catechins following repeated tea consumption. *Proc Soc Exp Biol Med* 1999; 220(4):203–9.

100. Hollman PCH, Katan MB. Dietary flavonoids: Intake, health effects and bioavail-ability. *Food Chem Toxicol* 1999; 37(9–10):937–42.

101. Holvoet P, Vanhaecke J, Janssens S, Van de Werf F, Collen D. Oxidized LDL and malondialdehyde-modified LDL in patients with acute coronary syndromes and stable coronary artery disease. *Circulation* 1998; 98(15):1487–94.

102. Hu JJ, Chi CX, Frenkel K, Smith BN, Henfelt JJ, Berwick M, Mahabir S, D'Agostino RB. Alpha-tocopherol dietary supplement decreases titers of antibody against 5-hydroxymethyl-2′-deoxyuridine (HMdU). *Cancer Epidemiol Biomarkers Prevent* 1999; 8(8):693–8.

103. Hu JP, Calome M, Lasure A, De Bruyne T, Pieters T, Vlietinck A, Vanden Berghe DA. Structure–activity relationship of flavonoids with superoxide scavenging activity. *Biol Trace Element Res* 1995; 47:327–31.

104. Hughes DA. Effects of dietary antioxidants on the immune function of middle-aged adults. *Proc Nutr Soc* 1999; 58(1):79–84.

105. Hughes DA. Effects of carotenoids on human immune function. *Proc Nutr Soc* 1999; 58(3):713–18.

106. Ip C, Lisk DJ. Enrichment of selenium in allium vegetables for cancer prevention. *Carcinogenesis* 1994; 15:1881–5.

107. Ip C, Lisk DJ. Bioactivity of selenium from Brazil nut for cancer prevention and selenoenzyme maintenance. *Nutr Cancer* 1994; 21:203–12.

108. Jacques PF. The potential preventive effects of vitamins for cataract and age-related macular degeneration. *Int J Vit Nutr Res* 1999; 69(3):198–205.

109. Jain SK, McVie R, Smith T. Vitamin E supplementation restores glutathione and malondialdehyde to normal concentrations in erythrocytes of type 1 diabetic children. *Diabetes Care* 2000; 23(9):1389–94.

110. Jandacek RJ. The canary in the cell: a sentinel role for beta-carotene. *J Nutr* 2000; 130(3):648–65.

111. Ju ZG, Bramlage WJ. Phenolics and lipid-soluble antioxidants in fruit cuticle of apples and their antioxidant activities in model systems. *Postharvest Biol Technol* 1999; 16(2):107–18.

112. Kahkonen MP, Hopia AI, Vuorela HJ, Rauha JP, Pihlaja K, Kujala TS, Heinonen M. Antioxidant activity of plant extracts containing phenolic compounds. *J Agr Food Chem* 1999; 47(10):3954–62.

113. Kang MH, Naito M, Sakai K, Uchida K, Osawa T. Mode of action of sesame lignans in protecting low-density lipoprotein against oxidative damage in vitro. *Life Sci* 2000; 66(2):161–71.

114. Keevil JG, Osman HE, Reed JD, Folts JD. Grape juice, but not orange juice or grape-fruit juice, inhibits human platelet aggregation. *J Nutr* 2000; 130(1):53–6.

115. Kitahara J, Seka Y, Imura N. Possible involvement of active oxygen species in selenite toxicity in isolated rat hepatocytes. *Arch Toxicol* 1993; 67:497–501.

116. Klipstein-Grobusch K, Launer LJ, Geleijnse JM, Boeing H, Hofman A, Witteman JCM. Serum carotenoids and atherosclerosis – the Rotterdam study. *Atherosclerosis* 2000; 148(1):49–56.

117. Klipstein-Grobusch K, Geleijnse JM, den Breeijen JH, Boeing H, Hofman A, Grobbee DE, Witteman JCM. Dietary antioxidants and risk of myocardial infarction in the elderly: the Rotterdam Study. *Am J Clin Nutr* 1999; 69:261–6.

118. Knekt P, Reunanen A, Jarvinen R, Seppanen R, Heliovaara M, Aromaa A. Anti-oxidant vitamin intake and coronary mortality in a longitudinal population study. *Am J Epidemiol* 1994; 139(12):1180–9.

119. Kristal AR, Stanford JL, Cohen JH, Wicklund K, Patterson RE. Vitamin and mineral supplement use is associated with reduced risk of prostate cancer. *Cancer Epidemiol Biomarkers Prevent* 1999; 8(10):887–92.

120. Kugiyama K, Motoyama T, Doi H, Kawano H, Hirai N, Soejima H, Miyao Y, Takazoe K, Moriyama Y, Mizuno Y, Tsunoda R, Ogawa H, Sakamoto T, Sugiyama S, Yasue H. Improvement of endothelial vasomotor dysfunction by treatment with alpha-tocopherol in patients with high remnant lipoproteins levels. *J Am Coll Cardiol* 1999; 33(6):1512–18.

121. Kumpulainen JT. Selenium in foods and diets of selected countries. *J. Trace Elem Electrolytes Health Dis* 1993; 7:107–8.

122. Kushi LH. Vitamin E and heart disease: a case study. *Am J Clin Nutr* 1999; 69(6):1322S–9S.

123. Kushi LH, Folsom AR, Prineas RJ, Mink PJ, Wu Y, Bostick RM. Dietary antioxidant vitamins and death from coronary heart disease in postmenopausal women. *New Engl J Med* 1996; 334(18):1156–62.

124. Lampe JW. Health effects of vegetables and fruit: assessing mechanisms of action in human experimental studies. *Am J Clin Nutr* 1999; 70(3 Suppl S):475S–90S.

125. Laranjinha J, Cadenas E. Redox cycles of caffeic acid, alpha-tocopherol, and ascor-bate: implications for protection of low-density lipoproteins against oxidation. *Iubmb Life* 1999; 48(1):57–65.

126. Le Marchand L, Murphy SP, Hankin JH, Wilkens LR, Kolonel LN. Intake of flavonoids and lung cancer. *J Nat Cancer Inst* 2000; 92(2):154–60.

127. Lean MEJ, Noroozi M, Kelly I, Burns J, Talwar D, Sattar N, Crozier A. Dietary flavonols protect diabetic human lymphocytes against oxidative damage to DNA. *Diabetes* 1999; 48(1):176–81.

128. Lee IM, Cook NR, Manson JE, Buring JE, Hennekens CH. Beta-carotene supplementation and incidence of cancer and cardiovascular disease: the Women's Health Study. *J Natl Cancer Inst* 1999; 91(24):2102–6.

129. Leenen R, Roodenburg AJC, Tijburg LBM, Wiseman SA. A single dose of tea with or without milk increases plasma antioxidant activity in humans. *Eur J Clin Nutr* 2000; 54(1):87–92.

130. Leighton F, Cuevas A, Guasch V, Perez DD, Strobel P, San Martin A, Urzua U, Diez MS, Foncea R, Castillo O, Mizon C, Espinoza MA, Urquiaga I, Rozowski J, Maiz A, Germain A. Plasma polyphenols and antioxidants, oxidative DNA damage and endothelial function in a diet and wine intervention study in humans. *Drugs Exp Clin Res* 1999; 25(2–3):133–41.

131. Leppala JM, Virtamo J, Fogelholm R, Huttunen JK, Albanes D, Taylor PR, Heinonen OP. Controlled trial of alpha-tocopherol and beta-carotene supplements on stroke incidence and mortality in male smokers. *Arterioscler Thromb Vasc Biol* 2000; 20(1):230–5.

132. Leske MC, Chylack LT, He QM, Wu SY, Schoenfeld E, Friend J, Wolfe J. Antioxidant vitamins and nuclear opacities – the longitudinal study of cataract. *Ophthalmology* 1998; 105(5):831–6.

133. Levander OA, Beck MA. Selenium and viral virulence. *Brit Med Bull* 1999; 55(3):528–33.

134. Levine M, Rumsey SC, Daruwala R, Park JB, Wang YH. Criteria and recommendations for vitamin C intake. *J Am Med Assoc* 1999; 281(15):1415–23.

135. Liu SM, Ajani U, Chae C, Hennekens C, Buring JE, Manson JE. Long-term beta-carotene supplementation and risk of type 2 diabetes mellitus – a randomized controlled trial. *J Am Med Assoc* 1999; 282(11):1073–5.

136. Losonczy KG, Harris TB, Havlik RJ. Vitamin E and vitamin C supplement use and risk of all-cause and coronary heart disease mortality in older persons: the established populations for epidemiologic studies of the elderly. *Am J Clin Nutr* 1996; 64(2):190–6.

137. Lowe GM, Bilton RF, Davies IG, Ford TC, Billington D, Young AJ. Carotenoid composition and antioxidant potential in subfractions of human low-density lipoprotein. *Ann Clin Biochem* 1999; 36(Part 3):323–32.

138. Lyle BJ, Mares-Perlman JA, Klein BEK, Klein R, Greger JL. Antioxidant intake and risk of incident age-related nuclear cataracts in the Beaver Dam eye study. *Am J Epidemiol* 1999; 149(9):801–9.

139. Mabile L, Bruckdorfer KR, Rice-Evans C. Moderate supplementation with natural alpha-tocopherol decreases platelet aggregation and low-density lipoprotein oxidation. *Atherosclerosis* 1999; 147(1):177–85.

140. Machlin LJ. Critical assessment of the epidemiological data concerning the impact of antioxidant nutrients on cancer and cardiovascular disease. *Crit Rev Food Sci Nutr* 1995; 35:41–50.

141. Malila N, Virtamo J, Virtanen M, Albanes D, Tangrea JA, Huttunen JK. The effect of alpha-tocopherol and beta-carotene supplementation on colorectal adenomas in middle-aged male smokers. *Cancer Epidemiol Biomarkers Prevent* 1999; 8(6):489–93.

142. Masaki KH, Losonczy KG, Izmirlian G, Foley DF, Ross GW, Petrovitch H, Havlik R, White LR. Association of vitamin E and C supplement use with cognitive function and dementia in elderly men. *Neurology* 2000; 54:1265–72.

143. McAnlis GT, McEneny J, Pearce J, Young IS. Absorption and antioxidant effects of quercetin from onions in man. *Eur J Clin Nutr* 1999; 53(2):92–6.

144. McCall MR, Frei B. Can antioxidant vitamins materially reduce oxidative damage in humans? *Free Rad Biol Med* 1999; 26(7–8):1034–53.

145. Meagher EA, Fitzgerald GA. Indices of lipid peroxidation in vivo: Strengths and limitations. *Free Radical Biol Med* 2000; 28(12):1745–50.

146. Moore SR, Hill KA, Heinmoller PW, Halangoda A, Kunishige M, Buettner VL, Graham KS, Sommer SS. Spontaneous mutation frequency and pattern in Big Blue (R) mice fed a vitamin E-supplemented diet. *Environ Mol Mutagen* 1999; 34(2–3):195–200.

147. Morris MC, Beckett LA, Scherr PA, Hebert LE, Bennett DA, Field TS, Evans DA. Vitamin E and vitamin C supplement use and risk of incident Alzheimer Disease. *Alzheimer Dis Assoc Disorders* 1998; 12(3):121–6.

148. Muller H, Bub A, Watzl B, Rechkemmer G. Plasma concentrations of carotenoids in healthy volunteers after intervention with carotenoid-rich foods. *European J Nutr* 1999; 38(1):35–44.

149. Nakagawa K, Ninomiya M, Okubo T, Aoi N, Juneja LR, Kim M, Yamanaka K, Miyazawa T. Tea catechin supplementation increases antioxidant capacity and prevents phospholipid hydroperoxidation in plasma of humans. *J Agr Food Chem* 1999; 47(10):3967–73.

150. Nyyssonen K, Poulsen HE, Hayn M, Agerbo P, Porkkalasarataho E, Kaikkonen J, Salonen R, Salonen, JT. Effect of supplementation of smoking men with plain or slow release ascorbic acid on lipoprotein oxidation. *Eur J Clin Nutr* 1997; 51(3):154–63.

151. Omenn GS, Goodman GE, Thornquist MD, Balmes J, Cullen MR, Glass A, Keogh JP, Meyskens FL, Valanis B, Williams JH, Barnhart S, Cherniack MG, Brodkin CA, Hammar S. Risk factors for lung cancer and for intervention effects in CARET, the beta-carotene and retinol efficacy trial. *J Natl Cancer Inst* 1996; 88(21): 1550–9.

152. Paiva SAR, Russell RM. Beta-carotene and other carotenoids as antioxidants. *J Am Coll Nutr* 1999; 18(5):426–33.

153. Palace VP, Khaper N, Qin QI, Singal PK. Antioxidant potentials of vitamin A and carotenoids and their relevance to heart disease. *Free Rad Biol Med* 1999; 26(5–6): 746–61.

154. Panasenko OM, Sharov VS, Briviba K, Sies H. Interaction of peroxynitrite with carotenoids in human low density lipoproteins. *Arch Biochem Biophys* 2000; 373(1): 302–5.

155. Panda K, Chattopadhyay R, Ghosh MK, Chattopadhyay DJ, Chatterjee IB. Vitamin C prevents cigarette smoke induced oxidative damage of proteins and increased proteolysis. *Free Rad Biol Med* 1999; 27(9–10):1064–79.

156. Panel on Dietary Antioxidants and Related Compounds. Dietary reference intakes for vitamin C, vitamin E, selenium, and carotenoids. Washington, DC: National Academy Press; 2000.

157. Paolisso G, Tagliamonte MR, Barbieri M, Zito GA, Gambardella A, Varricchio G, Ragno E, Varricchio M. Chronic vitamin E administration improves brachial reactivity and increases intracellular magnesium concentration in type 2 diabetic patients. *J Clin Endocrinol Metab* 2000; 85(1):109–15.

158. Patching SG, Gardiner PHE. Recent developments in selenium metabolism and chemical speciation: a review. *J Trace Elem Med Biol* 1999; 13(4):193–214.

159. Pearson DA, Tan CH, German JB, Davis PA, Gershwin ME. Apple juice inhibits human low density lipoprotein oxidation. *Life Sci* 1999; 64(21):1913–20.

160. Perugini C, Bagnati M, Cau C, Bordone R, Paffoni P, Re R, Zoppis E, Albano E, Bellomo G. Distribution of lipid-soluble antioxidants in lipoproteins from healthy subjects. II. Effects of in vivo supplementation with alpha-tocopherol. *Pharmacol Res* 2000; 41(1):67–74.

161. Perugini C, Bagnati M, Cau C, Bordone R, Zoppis E, Paffoni P, Re R, Albano E, Bellomo G. Distribution of lipid-soluble antioxidants in lipoproteins from healthy subjects. I. Correlation with plasma antioxidant levels and composition of lipoproteins. *Pharmacol Res* 2000; 41(1):55–65.

162. Pignatelli P, Pulcineli FM, Lenti L, Gazzaniga PP, Violi F. Vitamin E inhibits collagen-induced platelet activation by blunting hydrogen peroxide. *Arteriosclerosis Thrombosis & Vascular Biol* 1999; 19(10):2542–7.

163. Plecko T, Rukgauer M, Kruse-Jarres JD. Distribution of human plasma selenium and its role in the antioxidant system. *Metal Ions in Biol Med* 1998; 5:385–9.

164. Prieme H, Loft S, Nyyssonen K, Salonen JT, Poulsen HE. No effect of supplementation with vitamin E, ascorbic acid, or coenzyme Q10 on oxidative DNA damage estimated by 8-oxo-7,8-dihydro-2'-deoxyguanosine excretion in smokers. *Am J Clin Nutr* 1997; 65(2):503–7.

165. Pryor WA. Vitamin E and heart disease: basic science to clinical intervention trials. *Free Rad Biol Med* 2000; 28(1):141–64.

166. Pryor WA, Stahl W, Rock CL. Beta carotene: from biochemistry to clinical trials. *Nutr Rev* 2000; 58(2 Part 1):39–53.

167. Rao AV, Agarwal S. Role of lycopene as antioxidant carotenoid in the prevention of chronic diseases: a review [review]. *Nutr Res* 1999; 19(2):305–23.

168. Rao CV, Simi B, Hirose Y, Upadhyaya P, El-Bayoumy K, Reddy BS. Mechanisms in the chemoprevention of colon cancer: modulation of protein kinase c, tyrosine protein kinase and diacylglycerol kinase activities by 1,4-phenylenebis (methylene)selenocyanate and impact of low-fat diet. *Int J Oncol* 2000; 16(3): 519–27.

169. Rapola JM, Virtamo J, Ripatti S, Huttunen JK, Albanes D, Taylor PR, Heinonen OP. Randomised trial of alpha-tocopherol and beta-carotene supplements on incidence of major coronary events in men with previous myocardial infarction. *Lancet* 1997; 349(9067):1715–20.

170. Rautalahti MT, Virtamo JRK, Taylor PR, Heinonen OP, Albanes D, Haukka JK, Edwards BK, Karkkainen PA, Stolzenberg-Solomon RZ, Huttunen J. The effects of supplementation with alpha-tocopherol and beta-carotene on the incidence and mortality of carcinoma of the pancreas in a randomized, controlled trial. *Cancer* 1999; 86(1):37–42.

171. Rayman MP. The importance of selenium to human health. *Lancet* 2000; 356(9225):233–41.

172. Rehman A, Bourne LC, Halliwell B, Rice-Evans CA. Tomato consumption modulates oxidative DNA damage in humans. *Biochem Biophys Res Commun* 1999; 262(3):828–31.

173. Reilly M, Delanty N, Lawson JA, FitzGerald GA. Modulation of oxidant stress in vivo in chronic cigarette smokers. *Circulation* 1996; 94(1):19–25.

174. Rein D, Lotito S, Holt RR, Keen CL, Schmitz HH, Fraga CG. Epicatechin in human plasma: In vivo determination and effect of chocolate consumption on plasma oxidation status. *J Nutr* 2000; 130(8 Suppl S):2109S–14S.

175. Rein D, Paglieroni TG, Pearson DA, Wun T, Schmitz HH, Gosselin R, Keen CL. Cocoa and wine polyphenols modulate platelet activation and function. *J Nutr* 2000; 130(8 Suppl S):2120S–6S.

176. Rein D, Paglieroni TG, Wun T, Pearson DA, Schmitz HH, Gosselin R, Keen CL. Cocoa inhibits platelet activation and function. *Am J Clin Nutr* 2000; 72(1):30–5.

177. Rice-Evans C. Implications of the mechanisms of action of tea polyphenols as antioxidants in vitro for chemoprevention in humans. *Proc Soc Exp Biol Med* 1999; 220(4):262–6.

178. Rice-Evans CA, Miller NJ, Bolwell PG, Bramley PM, Pridham JB. The relative antioxidant activities of plant-derived polyphenolic flavonoids. *Free Rad Res* 1995; 22(4):375–83.

179. Richelle M, Tavazzi I, Enslen M, Offord EA. Plasma kinetics in man of epicatechin from black chocolate. *European J Clin Nutr* 1999; 53(1):22–6.

180. Rimm EB, Stampfer MJ, Ascherio A, Giovannucci E, Colditz GA, Willett WC. Vitamin E consumption and the risk of coronary heart disease in men. *New Engl J Med* 1993; 328(20):1450–6.

181. Riso P, Pinder A, Santangelo A, Porrini M. Does tomato consumption effectively increase the resistance of lymphocyte DNA to oxidative damage? *Am J Clin Nutr* 1999; 69(4):712–18.

182. Ronco A, De Stefani E, Boffetta P, Deneo-Pellegrini H, Mendilaharsu M, Leborgne P. Vegetables, fruits, and related nutrients and risk of breast cancer: a case-control study in Uruguay. *Nutr Cancer* 1999; 35(2):111–19.

183. Rouseff RL, Nagy S. Health and nutritional benefits of citrus fruit components. *Food Technol* 1994; 48(11):125–32.

184. Rukgauer M, Neugebauer R, Plecko T, Kruse-Jarres JD. The selenium-dependent antioxidative system. *Metal Ions Biol Med* 1998; 5:400–4.

185. Saldeen T, Li DY, Mehta JL. Differential effects of alpha- and gamma-tocopherol on low-density lipoprotein oxidation, superoxide activity, platelet aggregation and arterial thrombogenesis. *J Am Coll Cardiol* 1999; 34(4):1208–15.

186. Samman S, Brown AJ, Beltran C, Singh S. The effect of ascorbic acid on plasma lipids and oxidisability of LDL in male smokers. *Eur J Clin Nutr* 1997; 51(7):472–7.

187. Sauberlich HE. Pharmacology of vitamin C. *Annu Rev Nutr* 1994; 14:371–91.

188. Scalbert A, Williamson G. Dietary intake and bioavailability of polyphenols. *J Nutr* 2000; 130(8 Suppl S):2073S–85S.

189. Schippling S, Kontush A, Arlt S, Buhmann C, Sturenburg HJ, Mann U, Muller-Thomsen T, Beisiegel U. Increased lipoprotein oxidation in Alzheimer's disease. *Free Rad Biol Med* 2000; 28(3):351–60.

190. Schorah CJ. Micronutrients, vitamins, and cancer risk. *Vitamins and Hormones – Advances in Research and Applications*, 1999; 571–623.

191. Schroder H, Navarro E, Tramullas A, Mora J, Galiano D. Nutrition antioxidant status and oxidative stress in professional basketball players: effects of a three compound antioxidative supplement. *Int J Sports Med* 2000; 21(2):146–50.

192. Sharma N, Desigan B, Ghosh S, Sanyal SN, Ganguly NK, Majumdar S. Effect of antioxidant vitamin E as a protective factor in experimental atherosclerosis in rhesus monkeys. *Ann Nutr Metab* 1999; 43(3):181–90.

193. Sharma A, Kharb S, Chugh SN, Kakkar R, Singh GP. Evaluation of oxidative stress before and after control of glycemia and after vitamin E supplementation in diabetic patients. *Metabolism: Clin Exp* 2000; 49(2):160–2.

194. Shrikhande AJ. Wine by-products with health benefits. *Food Res Int* 2000; 33(6): 469–74.

195. Siow RCM, Richards JP, Pedley KC, Leake DS, Mann GE. Vitamin C protects human vascular smooth muscle cells against apoptosis induced by moderately oxi-dized LDL containing high levels of lipid hydroperoxides. *Arterioscler Thromb Vasc Biol* 1999; 19(10):2387–94.

196. Siow RCM, Sato H, Leake DS, Ishii T, Bannai S, Mann GE. Induction of antioxidant stress proteins in vascular endothelial and smooth muscle cells: Protective action of vitamin C against atherogenic lipoproteins. *Free Rad Res* 1999; 31(4):309–18.

197. Slattery ML, Benson J, Curtin K, Ma KN, Schaeffer D, Potter JD. Carotenoids and colon cancer. *Am J Clin Nutr* 2000; 71(2):575–82.

198. Smith MJ, Inserra PF, Watson RR, Wise JA, O'Neill KL. Supplementation with fruit and vegetable extracts may decrease DNA damage in the peripheral lymphocytes of an elderly population. *Nutr Res* 1999; 19(10):1507–18.

199. Smith W, Mitchell P, Webb K, Leeder SR. Dietary antioxidants and age-related maculopathy – the blue mountains eye study. *Ophthalmology* 1999; 106(4):761–7.

200. Speizer FE, Colditz GA, Hunter DJ, Rosner B, Hennekens C. Prospective study of smoking, antioxidant intake, and lung cancer in middle-aged women (USA). *Cancer Causes & Control* 1999; 10(5):475–82.

201. Spencer AP, Carson DS, Crouch MA. Vitamin E and coronary artery disease. *Arch Intern Med* 1999; 159(12):1313–20.

202. Stahl W, Heinrich U, Jungmann H, Sies H, Tronnier H. Carotenoids and carotenoids plus vitamin E protect against ultraviolet light-induced erythema in humans. *Am J Clin Nutr* 2000; 71(3):795–8.

203. Stampfer MJ, Hennekens CH, Manson JE, Colditz GA, Rosner B, Willett WC. Vitamin E consumption and the risk of coronary disease in women. *New Engl J Med* 1993; 328(20):1444–9.

204. Steele VE, Kelloff GJ, Balentine D, Boone CW, Mehta R, Bagheri D, Sigman CC, Zhu SY, Sharma S. Comparative chemopreventive mechanisms of green tea, black tea and selected polyphenol extracts measured by in vitro bioassays. *Carcinogenesis* 2000; 21(1):63–7.

205. Stein JH, Keevil JG, Wiebe DA, Aeschlimann S, Folts JD. Purple grape juice improves endothelial function and reduces the susceptibility of LDL cholesterol to oxidation in patients with coronary artery disease. *Circulation* 1999; 100(10):1050–55.

206. Stephens NG, Parsons A, Schofield PM, Kelly F, Cheeseman K, Mitchinson MJ. Randomised controlled trial of vitamin E in patients with coronary disease: Cambridge Heart Antioxidant Study (CHAOS). *Lancet* 1996; 347(9004):781–6.

207. Suarez A, Ramirez-Tortosa M, Gil A, Faus MJ. Addition of vitamin E to long-chain polyunsaturated fatty acid-enriched diets protects neonatal tissue lipids against peroxidation in rats. *European J Nutr* 1999; 38(4):169–76.

208. Tavani A, La Vecchia C. Beta-carotene and risk of coronary heart disease. A review of observational and intervention studies. *Biomed Pharmacother* 1999; 53(9): 409–16.

209. Teikari JM, Laatikainen L, Virtamo J, Haukka J, Rautalahti M, Liesto K, Albanes D, Taylor P, Heinonen OP. Six-year supplementation with alpha-tocopherol and beta-carotene and age-related maculopathy. *Acta Ophthal Scand* 1998; 76(2):224–9.

210. Teikari JM, Rautalahti M, Haukka J, Jarvinen P, Hartman AM, Virtamo J, Albanes D, Heinonen O. Incidence of cataract operations in Finnish male smokers unaffected by α tocopherol or beta carotene supplements. *J Epidemiol Commun Health* 1998; 52(7):468–72.

211. Teikari JM, Virtamo J, Rautalahti M, Palmgren J, Liesto K, Heinonen OP. Long-term supplementation with alpha-tocopherol and beta-carotene and age-related cataract. *Acta Ophthal Scand* 1997; 75(6):634–40.

212. Teikari JM, Virtamo J, Rautalahti M, Palmgren J, Liesto K, Heinonen OP. The alpha-tocopherol, beta-carotene cancer prevention study group. The effect of vitamin E and beta carotene on the incidence of lung cancer and other cancers in male smokers. *New Engl J Med* 1994; 330(15):1029–35.

213. Thomas SR, Stocker R. Molecular action of vitamin E in lipoprotein oxidation: Implications for atherosclerosis. *Free Radical Biol Med* 2000; 28(12):1795–805.

214. Thomas MJ. The role of free radicals and antioxidants: how do we know that they are working? *Crit Rev Food Sci Nutr* 1995; 35(1–2):21–39.

215. Tinkler JH, Böhm F, Schalch W, Truscott TG. Dietary carotenoids protect human cells from damage. *J Photochem Photobiol B: Biology* 1994; 26:283–5.

216. Tribble DL. AHA science advisory. Antioxidant consumption and risk of coronary heart disease: emphasis on vitamin C, vitamin E, and beta-carotene: a statement for healthcare professionals from the American Heart Association. *Circulation* 1999; 99(4):591–5.

217. Upston JM, Terentis AC, Stocker R. Tocopherol-mediated peroxidation of lipoproteins: implications for vitamin E as a potential antiatherogenic supplement. *FASEB J* 1999; 13(9):977–94.

218. Valagussa F, Franzosi MG, Geraci E, Mininni N, Nicolosi GL, Santini M, Tavazzi L, Vecchio C, Marchioli R, Bomba E, Chieffo C, Maggioni AP, Schweiger C, Tognoni G, Barzi F, Flamminio AV, Marfisi RM, Olivieri M, Pera C, Polidoro A, Santoro E, Zama R, Pagliaro L, Correale E, Del Favero A, *et al.* Dietary supplementation with n-3 polyunsaturated fatty acids and vitamin E after myocardial infarction: results of the GISSI-prevenzione trial. *Lancet* 1999; 354(9177):447–55.

219. Valkonen MM, Kuusi T. Vitamin C prevents the acute atherogenic effects of passive smoking. *Free Rad Biol Med* 2000; 28(3):428–36.

220. Vatassery GT, Bauer T, Dysken M. High doses of vitamin E in the treatment of disorders of the central nervous system in the aged. *Am J Clin Nutr* 1999; 70(5): 793–801.

221. Verma RJ, Nair A. Vitamin E prevents aflatoxin-induced lipid peroxidation in the liver and kidney. *Med Sci Res* 1999; 27(4):223–6.

222. Wang XD, Russell RM. Procarcinogenic and anticarcinogenic effects of beta-carotene. *Nutr Rev* 1999; 57(9 Part 1):263–72.

223. Wang WQ, Goodman MT. Antioxidant property of dietary phenolic agents in a human LDL-oxidation ex vivo model: interaction of protein binding activity. *Nutr Res* 1999; 19(2):191–202.

224. Weisburger JH. Mechanisms of action of antioxidants as exemplified in vegetables, tomatoes and tea. *Food Chem Toxicol* 1999; 37(9–10):943–8.

225. Welch RW, Turley E, Sweetman SF, Kennedy G, Collins AR, Dunne A, Livingstone MBE, McKenna PG, McKelvey-Martin VJ, Strain JJ. Dietary antioxidant supplementation and DNA damage in smokers and nonsmokers. *Nutr Cancer* 1999; 34(2):167–72.

226. Wen Y, Killalea S, Norris LA, Cooke T, Feely J. Vitamin E supplementation in hyperlipidaemic patients: effect of increasing doses on in vitro and in vivo low-density lipoprotein oxidation. *European J Clin Invest* 1999; 29(12):1027–34.

227. Wen Y, Cooke T, Feely J. The effect of pharmacological supplementation with vitamin C on low-density lipoprotein oxidation. *Br Clin Pharmacol* 1997; 44(1):94–7.

228. Whelan RL, Horvath KD, Gleason NR, Forde KA, Treat MD, Teitelbaum SL, Bertram A, Neugut AI. Vitamin and calcium supplement use is associated with decreased adenoma recurrence in patients with a previous history of neoplasia. *Dis Colon Rectum* 1999; 42(2):212–17.

229. Wilkinson IB, Megson IL, MacCallum H, Sogo N, Cockcroft JR, Webb DJ. Oral vitamin C reduces arterial stiffness and platelet aggregation in humans. *J Cardio Pharmacol* 1999; 34(5):690–3.

230. Wiseman H. The bioavailability of non-nutrient plant factors: dietary flavonoids and phyto-oestrogens. *Proc Nutr Soc* 1999; 58(1):139–46.

231. Wollgast J, Anklam E. Polyphenols in chocolate: is there a contribution to human health? *Food Res Int* 2000; 33(6):449–59.

232. Woodside JV, Young IS, Yarnell JWG, Roxborough HE, McMaster D, McCrum EE, Gey KF, Evans A. Antioxidants, but not B-group vitamins increase the resistance of low-density lipoprotein to oxidation: a randomized, factorial design, placebo-controlled trial. *Atherosclerosis* 1999; 144(2):419–27.

233. Woutersen RA, Appel MJ, Van Garderen-Hoetmer A. Modulation of pancreatic carcinogenesis by antioxidants. *Food Chem Toxicol* 1999; 37(9–10):981–4.

234. Woutersen RA, Wolterbeek APM, Appel MJ, van den Berg H, Goldbohm RA, Feron VJ. Safety evaluation of synthetic beta-carotene. *Crit Rev Toxicol* 1999; 29(6):515–42.

235. Wu DY, Koga T, Martin KR, Meydani M. Effect of vitamin E on human aortic endothelial cell production of chemokines and adhesion to monocytes. *Atherosclerosis* 1999;147(2):297–307.

236. Yang C, Wolf E, Röser, *et al*. Selenium deficiency and fulvic acid supplementation induces fibrosis of cartilage and disturbs subchondral ossification in knee joints of mice: an animal model study of Kashin–Beck disease. *Virch Archiv A* 1993; 423: 483–91.

237. Yang CS. Vitamin nutrition and gastroesophageal cancer. *J Nutr* 2000; 130(2 Suppl S): 338S–9S.

238. Yang M, Collis CS, Kelly M, Diplock AT, Rice-Evans C. Do iron and vitamin C co-supplementation influence platelet function or LDL oxidizability in healthy volunteers? *European J Clin Nutr* 1999; 53(5):367–74.

239. Yochum L, Kushi LH, Meyer K, Folsom AR. Dietary flavonoid intake and risk of cardiovascular disease in postmenopausal women. *Am J Epidemiol* 1999; 149(10): 943–9.

240. Yusuf S, Phil D, Dagenais G, Pogue J, Bosch J, Sleight P. Vitamin E supplementation and cardiovascular events in high-risk patients. *N E J Med* 2000; 342(3):154–60.

241. Zhang J, Ying X, Lu Q, Kallner A, Xiu RJ, Henriksson P, Bjorkhem I. A single high dose of vitamin C counteracts the acute negative effect on microcirculation induced by smoking a cigarette. *Microvascular Res* 1999; 58(3):305–11.

2

CAROTENOIDS AND HEALTH RISK

Emily K. Kraczek and John W. Erdman, Jr.

INTRODUCTION

In recent years, carotenoids have both been hailed as miracle nutrients and viewed with skepticism regarding their actual impact on health. Although observational epidemiologic studies show a consistent inverse association between fruit and vegetable intake and the risk of several cancers, cardiovascular disease, and age-related macular degeneration, β-carotene (BC) intervention trials in smokers did not result in reduced risk of lung cancer or cardiovascular disease. This mixture of optimism and uncertainty stems from the limited knowledge about the absorption, transport, utilization, and effect of carotenoids in human tissues. The study of carotenoids has proceeded steadily, but slowly, as carotenoids are delicate compounds making research on their chemical structure and behavior arduous. Also, observational epidemiologic studies evaluating the consumption of whole foods do not allow us to know with certainty whether the results are due to the carotenoid content of certain foods, other components of the whole food, or a combination of interacting factors. In addition, the human absorption of intact carotenoids is unique and not easily replicated in animal models.

More than 600 different carotenoids occur in nature of which only 50 have provitamin A activity (1). Thirty-four carotenoids can be found in human serum and breast milk (2). Information about carotenoid absorption and transport is mainly limited to BC, which is an important carotenoid but not the only one thought to affect health status. In general we know very little about the actual vitamin A (VA) value of carotenoid containing foods and Thomas Moore's conclusion in 1957 still holds true today: "There are many complicating factors, both chemical and physiological, which will make it difficult to give an account of the absorption of VA and its provitamins, which is both clear and reasonably comprehensive" (3). More work is needed to understand the effects of food matrix and interactions between carotenoids and other nutrients on carotenoid absorption and transport.

This chapter will provide only a brief overview of carotenoid chemistry, absorption, and transport as these topics have recently been reviewed by Boileau *et al.* (4). The principle focus of this chapter will be a review of the effects of

31

carotenoids on lung and prostate cancers, cardiovascular disease, and the potential toxicity of carotenoids alone and in combination with smoking and ethanol. Brief coverage will be given to other cancers and diseases. The reader is referred to reviews by Cooper *et al.* (5), Tavani *et al.* (6), and Clinton (7) for further information about the influence of carotenoids on other health outcomes.

Carotenoid chemistry

Carotenoids are a class of hydrocarbons (carotenes) and their oxygenated derivatives (xanthophylls) generally comprising eight isoprenoid units (Figure 1). These units are joined in such a manner that the arrangement is reversed at the center of the molecule, placing the two central methyl groups in a 1,6 relationship. Caro-

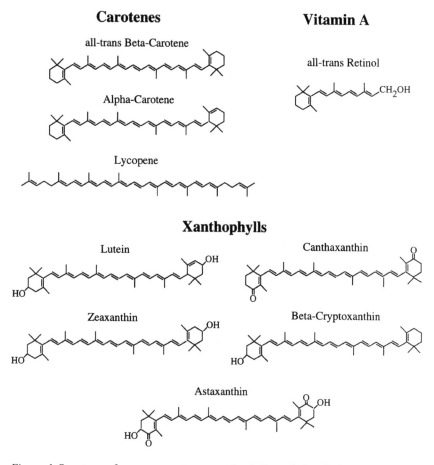

Figure 1 Structures of common carotenes, xanthophylls, and vitamin A.

tenoids can be derived from the acyclic $C_{40}H_{56}$ structure by hydrogenation, dehydrogenation, oxidation, cyclization, or combinations of these. Modifications of the basic carbon skeleton are reflected in the nomenclature, and the configuration about all double bonds are assumed to be *trans* unless *cis* and location numbers appear.

Carotenes are the class of carotenoids most noted for their provitamin A activity, of which BC has the highest provitamin A value. Central cleavage of BC will theoretically yield two molecules of retinol. α-Carotene and γ-carotene each yield just one molecule of VA, as do β-cryptoxanthin, β-apo-8$'$-carotenal, and β-zeacarotene. Common food carotenoids that do not exhibit provitamin A activity include lycopene (the Ψ,Ψ-carotene precursor to BC found in tomatoes and pink grapefruit), astaxanthin (an oxycarotenoid found in lobster, shrimp, and salmon), canthaxanthin (an oxycarotenoid found in mushrooms), and lutein and zeaxanthin (xanthophylls found in egg yolks, potatoes, spinach, broccoli, and wheat).

Isolating and identifying carotenoids requires great care as these compounds are sensitive to light, heat, oxygen, acid, and alkaline substances. In fact, concern that some potentially toxic and possibly carcinogenic compounds could be formed when carotenoids are heated led to extensive research on carotenoid decomposition. To date, studies have shown that the compounds formed during the heating of carotenoids at the temperatures characteristic of food processing have quite different structures than those suspected of being carcinogens (8).

Carotenoid absorption and transport

The absorption and transport of carotenoids are quite complex processes that are not well understood. Humans are unique in their ability to absorb a wide range of carotenoids intact, an ability not known to be completely shared by any animal. Thus animal studies may not precisely represent human absorption and metabolism (9). Both dietary and nondietary factors affect carotenoid absorption, defined as the uptake of carotenoids by the intestinal mucosal cells and then the movement of carotenoids to the lymphatic or portal circulation.

Food preparation plays a significant role in carotenoid absorption. Carotenoids are more bioavailable following mild heating and after chopping or pureeing, which allows enzymes to penetrate the food matrix and increase digestibility (10,11). Carotenoid absorption from uncooked foods and green leafy vegetables can be quite poor (12–14). Adequate amounts of fat, and perhaps specific types of fatty acids, are needed to ensure absorption of carotenoids (10). Dietary fat stimulates bile flow, which facilitates the emulsification of fat and fat-soluble vitamins into micelles in the small intestine.

Studies have shown that crystalline BC in oil, commercial beadlet form, or dietary supplements are much more bioavailable than carotenoids from foods (11,12,14). These highly absorbed forms of BC can result in a 2:1 weight-for-weight conversion efficiency whereas the Recommended Dietary Allowances use conversion factors of 6:1 and 12:1 for BC and other provitamin A carotenoids, respectively, to VA (15).

Carotenoid absorption can be affected by interactions among various carotenoids and interactions with other nutrients. Some studies suggest that consumption of VA with BC may promote better absorption and utilization (16). Lycopene has little effect on BC uptake (17), while high levels of canthaxanthin and lutein reduce BC uptake (18,19). Protein and zinc deficiencies are detrimental to VA and carotenoid metabolism (11). Optimal vitamin E intake may help protect BC from oxidation, but in megadoses may interfere with the conversion of BC to VA in the mucosal cells (11).

It is not surprising that malabsorbed food components, intestinal parasites, and certain drugs (such as the bile acid sequesterant, cholestyramine) can interfere with carotenoid absorption (20). High amounts of sucrose polyester (Olestra^{TM}), consumed in the same meal as carotenoids will lower absorption (21,22). Ethanol consumption does not reduce absorption of VA or BC, however both acute and chronic alcohol consumption reduce liver storage of VA and increase its metabolism, resulting in lower VA and BC status (23,24).

Uptake of available BC by mucosal cells ranges from 2% to 50% and depends on digestion of the food matrix, the formation of lipid micelles, and the health of the gastrointestinal tract (4,11). Once inside the mucosal cell, intact BC molecules can be cleaved centrally to form two retinol molecules which are esterified prior to incorporation into chylomicrons or can be cleaved asymmetrically, resulting in the formation of β-apo-carotenal (released into portal blood) and retinol. From the intestinal mucosal cells, carotenoids and their metabolic products are incorporated into chylomicrons and enter the lymphatic system and portal blood where they are transported to tissues in the body. The metabolism of absorbed carotenoids has not been extensively studied, and most assumptions are derived from studies of plasma appearance and disappearance of carotenoids and analysis of a limited number of surgical and autopsy tissues. Generally, the appearance of BC in lipoprotein fractions follows the same time course as newly absorbed cholesterol from the same meal (25). Other carotenoids, although less extensively studied, are generally assumed to follow the same absorption pathway as BC.

Carotenoid concentrations vary widely among tissue types (Table 1), with adrenal and reproductive tissues having the highest levels of BC, possibly due to their large number of low-density lipoprotein receptors. Some BC accumulates in the liver, but most is mobilized into the blood stream as part of the very low-, low-, and high-density lipoprotein cholesterol fractions. In the liver, BC may be converted to VA or to other metabolites. BC may accumulate in extra hepatic tissues but there is a question whether carotenes in these tissues can be mobilized and converted to VA (26). The macular pigment of the eye contains primarily lutein and zeaxanthin, which absorb blue light (~240 nm), the light range most damaging to the eye.

In one study, correlations between diet and plasma/tissue (buccal mucosal cells and skin) concentrations of seven carotenoids were generally not as strong as the plasma–tissue relationships; the diet–plasma and diet–tissue relationships of the carotenoids were particularly poor in smokers (17). In fact, plasma and tissue

Table 1 Estimated concentration of β-carotene and total carotenoids in selected tissues of human adults (mg/g)

Organ	β-carotene concentration	Total carotenoid concentration
Corpus luteum	60.0	60.0
Adrenal	5.0–23.0	18.3–?
Testes	2.3	14.3
Liver	1.0–10.0	3.0–27.0
Pancreas	0.7	2.0
Ovary	0.5	1.4
Prostate	0.5	1.4
Adipose	0.2–?	1.8–?
Kidney	0.2–0.3	0.5–1.6
Buccal mucosa	0.1–8.7	0.3–20.2
Spleen	0.2	0.5
Lung	0.1	1.0
Heart	0.1	0.4
Thyroid	0.1	0.4
Skin	0.02–0.09	0.05–0.5

Source: (97–101).

concentrations of most micronutrients studied were lower in smokers than in non-smokers and higher in vitamin supplement users than in nonsupplement users. These differences remained significant after adjustment for age, sex, and diet intake estimates. Among the seven carotenoids examined, lycopene was unique because its serum concentration was not lower in smokers or higher in supplement users but was inversely associated with age (17).

ROLE OF CAROTENOIDS IN HEALTH

Carotenoids are primarily regarded as contributors to the VA requirement and thus are promoted worldwide for preventing blindness and reducing measles risk and mortality in children. Beyond the formation of retinol, both provitamin A and non-vitamin A carotenoids may offer additional health benefits. Research on the ability of carotenoids to prevent cancer, heart disease, and age-related macular degeneration is well underway, as is work on the effect of carotenoids on immune response. Many epidemiological studies show that consuming a diet high in fruits and vegetables, particularly those containing carotenoids, may reduce the risk for certain cancers, cardiovascular disease, and degenerative visual diseases such as age-related macular degeneration and cataracts. Low levels of serum and tissue carotenoids have been associated with disease states, however Jandacek (27) believes that low levels of BC may reflect the effect of a disease rather than be the cause of the disease. Possible mechanisms by which carotenoids may reduce disease risk include acting as antioxidants, up-regulating gap-junction cell–cell

Table 2 Major intervention trials of antioxidant supplementation for the prevention of disease

Study/country/ reference	Study group	Principal endpoints	Treatment	Duration	Outcomes
Linxian/China/ (69,70)	Almost 30,000 men and women from an area with unusually high rates of esophageal and stomach cancer, 30% smoked for 6+ months	Cancer mortality	BC (15 mg), vitamin E (30 mg), and selenium (50 µg), administered together daily	5 years	Subjects taking the supplement had 9% lower total mortality, 13% lower cancer mortality, and 21% lower stomach cancer mortality. Lung cancer and heart disease were not specifically examined.
ATBC trial/ Finland/ (50–52,102)	29,133 male smokers, aged 50–69 years at enrollment	Lung cancer incidence	BC (20 mg) and vitamin E (50 mg), separately and together	5–8 years, 6.1 years median	Subjects taking BC showed an 18% increase in lung cancer, an 8% increase in total mortality, and a trend (not statistically significant) for a greater incidence of ischemic heart disease.
Physicians' Health Study/ USA/ (56)	22,071 male physicians, 50% never smoked, 39% former smokers, 11% current smokers	All cancers and cardiovascular diseases	BC (50 mg) every other day, with and without aspirin	12 years	BC had no significant effect, positive or negative, on cancer incidence, heart disease, or mortality

Study/country (ref)	Subjects	Endpoint	Intervention	Duration	Results
CARET/USA/ (55)	18,314 current and former smokers and asbestos-exposed workers	Lung cancer	BC (30 mg) and vitamin A (25,000 IU) daily	Planned for 6 years, discontinued after 4 years	Trial discontinued when preliminary results showed subjects taking BC had statistically non-significant increases in lung cancer (28%), cardiovascular disease (26%), and mortality (17%)
Women's Health Study/ USA/ (57)	39,876 women aged 45 and above	All cancers and cardiovascular diseases	BC (50 mg) every other day, vitamin E (600 mg), aspirin	BC component was stopped early. Median treatment of 2.1 yrs.	No significant benefit or harm in the incidence of cancer or cardiovascular disease in subjects taking BC
SU.VI.M.AX Study/France/ (103–105)	12,735 men and women representative of the general population of France (women aged 35–60 and men aged 45–60)	All cancers and cardiovascular diseases	BC (6 mg), vitamin C (120 mg), vitamin E (30 mg), selenium (100 µg), zinc (20 mg)	8 years	Study in progress
Physicians' Health Study II/USA/ (106)	15,000 physicians aged 55 years and above	All cancers, cardiovascular disease, and eye disease	BC and vitamin E every other day and daily vitamin C and multivitamin		Study planned

communication, modulating the immune system, and acting as photoprotectants in the skin and the eye.

Carotenoids act as antioxidants, are excellent singlet oxygen quenchers, and can also quench other reactive species, which may be one mechanism for their purported preventive benefits (28–33). BC and lutein have been shown to protect HepG2 cells from oxidative damage induced by *tert*-butylhydroperoxide (34). Of all the carotenoids, lycopene has been reported as the most potent biological quencher of singlet oxygen because of it long polyene chain length, alternating double bonds, open chain, and lack of oxygen substitutents (29).

BC, canthaxanthin, and lycopene all have been shown to increase cell–cell communication in culture via up-regulation of gap-junction proteins. Specifically, the expression of connexin-43, a six sub-unit membrane-bound protein that forms pores in the cell membrane and allows for the transfer of small molecules between adjacent cells, is increased with carotenoid treatment (30,35–38). This may be a mechanism by which carotenoids inhibit neoplastic transformation in vitro and help maintain controlled, uniform growth of cells (39).

Adequate nutrition has long been known to influence the activity of the immune system. Recently, studies have began to uncover the mechanisms by which nutrients, including carotenoids, modulate the immune response (40–43). Supplementation with 15 mg of BC has been shown to increase monocyte expression of adhesion molecules, increase secretion of tumor necrosis factor-α, and increase monocyte expression of MHC II in subjects receiving BC as compared to those receiving a placebo (42). Also, astaxanthin, a carotenoid without VA activity, has been shown to enhance T-cell-dependent immune responses (44,45).

Although it is apparent that carotenoids have numerous beneficial effects, one must use caution when interpreting the results of food frequency surveys or recommending carotenoid supplementation to decrease disease risk. As will be detailed in the next section of this chapter, studies have shown that smokers taking BC supplements had a higher incidence of lung cancer and death than smokers taking a placebo. The metabolism of carotenoids can also be affected by ethanol resulting in hepatotoxic alcohol-BC interactions. Thus while research consistently shows that a diet high in carotenoid-containing fruits and vegetables reduces disease risk, further research is needed to determine the efficacy and safety of carotenoid supplementation.

CAROTENOIDS AND CANCER

Several nutrition surveys have noted that increased consumption of carotenoid-rich fruits and vegetables is associated with decreased risk of certain types of cancer. Out of 170 studies, 132 showed a significant protective effect for fruits and vegetables – many rich in carotenoids, and only a few showed any adverse results (46). However, observational studies cannot resolve whether the beneficial associations are due to a specific carotenoid, to other components in fruits and vegetables, or a combination of factors. Although some cancer studies have been carried out in rats and mice, interpretation of these feeding studies is hampered because the metabolism of carotenoids in most animals differs notably from that in humans (9).

Large scale human intervention trials of BC in the prevention of lung cancer, other cancers, and heart disease (Table 2) were carried out based on the consistency of the literature from observational epidemiologic studies regarding carotenoid intake and cancer prevention. BC has been the carotenoid used in large-scale supplement trials as until recently it has been the only carotenoid readily available in pill form.

Lung cancer

From epidemiologic studies, the association between consuming a diet high in carotenoid-rich fruits and vegetables and a lower risk of lung cancer is particularly strong. For example, one review reported a decreased risk for lung cancer with a high fruit and vegetable intake in 8 of 8 prospective studies and in 18 of 20 retrospective studies (47). In a serologic study in the United States, previously drawn and frozen blood samples were analyzed for 99 individuals who later developed lung cancer and for 196 matched case controls (48). The relative odds for developing lung cancer (squamous cell carcinoma) was 4.3 times as great among those with the lowest serum BC level compared with those with the highest level. Out of four prospective case-control studies reviewed by Ziegler (49), analyzing serum or plasma carotenoid levels and lung cancer risk, three showed a statistically significant decreased risk for developing lung cancer at higher levels of blood BC.

The widely reported results of two intervention studies, in which BC supplements were given to high-risk smokers, have dimmed the excitement regarding the protective effects of supplements and raises the issue of the safety of high-dose BC supplementation for heavy smokers. The Alpha-Tocopherol and Beta-Carotene Intervention Study (ATBC) of 29,133 long-term male smokers in Finland, aged 50–69, average smoking time of 36 years, showed that men receiving 20 mg/day of BC for an average of 6 years were found to have an 8% increase in total mortality, an 18% increase in lung cancer incidence, and an increase, although not statistically significant, of ischemic heart disease compared to subjects taking vitamin E, BC + vitamin E, or a placebo (50,51). The researchers are continuing to examine these unexpected results and have reported that the increase in lung cancer in the BC-supplemented group may be related to the level of alcohol intake, as in the BC group, subjects with higher alcohol intake showed higher risk of lung cancer (52). Ethanol induces activation of a specific cytochrome P-450, the enzyme that starts VA on its degradation/oxidation pathway, so an adverse interaction among ethanol, smoking, and BC may have occurred (53). It has also been hypothesized that the free-radical rich atmosphere of a smoker's lungs may cause BC to act a pro-oxidant (54).

A second intervention study, The Beta-Carotene and Retinol Efficacy Trial (CARET), was terminated 21 months early due to increased incidence of lung cancer among the BC supplemented participants. In that trial, 30 mg of BC and 20,000 IU of

VA or a placebo were given to 18,314 randomly divided smokers, former smokers, and men with occupational asbestos exposure. The group supplemented with BC and VA showed non-significant increases of 28% greater incidence of lung cancer, 26% higher incidence of cardiovascular disease, and 17% more deaths (55). Perhaps with both the ATBC trial and the CARET trial, the very late intervention of BC supplementation in long-time heavy smokers was not sufficient to stop the long progression of events that lead to the development of lung cancer (53).

The Physicians' Health Study (PHS) showed that supplementation with 50 mg of BC every other day for 12 years had no significant effects, positive or negative, on cancer incidence, heart disease, or mortality as compared to control groups (56). It must be noted however, that of the 22,071 male physicians in the study, 50% had never smoked, 39% were former smokers, and only 11% were current smokers. The results of the Women's Health Study (WHS) agree with the PHS. In the WHS, examining a cohort of 39,876 women aged 45 and above, one group was supplemented with 50 mg of BC every other day for a median of 2.1 years. The BC component of the study was terminated early but no significant benefit or harm on the incidence of cancer or cardiovascular disease was observed in subjects taking BC (57).

A recent study by Wang *et al.* (54) using ferrets exposed to tobacco smoke and/or supplemented with BC helps to identify a possible mechanism for the increase in lung cancer seen with BC supplementation in the ATBC trial and the CARET trial. In this study, ferrets (an animal model that absorbs BC intact similar to humans (58)) were given a BC supplement, exposed to cigarette smoke, or both for 6 months. All BC supplemented animals showed a strong proliferative response in lung tissue and squamous metaplasia, a response that was enhanced by tobacco smoke exposure. As compared to control animals, all treatment groups had statistically significantly lower concentrations of retinoic acid in lung tissue and reductions of retinoic acid receptor-β (RARβ) gene expression, a gene that plays an important role in normal lung development and is often not expressed in lung cancer cell lines or in lung tumors. Wang *et al.* also showed that in vitro incubation of all-*trans*-BC with postnuclear fractions of lung tissue from ferrets exposed to cigarette smoke increased the formation of excentric cleavage oxidative metabolites of BC. These cleavage products include weak ligands (and/or agonists) for the RARβ and β-apo-8'-carotenal, a strong inducer of cytochrome P-450 enzymes. It is hypothesized that the free-radical rich atmosphere in the lungs of smokers may cause diminished retinoid signaling, enhanced lung cell proliferation, increased excentric cleavage of BC, and potential tumorigenesis. Although epidemiologic studies show that a diet high in fruits and vegetables and/or high serum BC levels decreases cancer risk (47), BC supplementation in smokers is contraindicated by the results of both this recent animal study and human studies.

Prostate cancer

Prostate cancer is the most common cancer among males in the United States. A diet high in vegetables and fruits has been associated with a decreased risk of prostate cancer. Giovannucci (59) reviewed 72 studies regarding cancer risk and the intake of tomatoes and tomato-based products or blood lycopene level (the predominate carotenoid in tomatoes). Of the 72 studies, 57 reported inverse associations between tomato intake or blood lycopene level and cancer risk at defined sites; 35 of these inverse associations were statistically significant. No study found consumption of tomatoes or high blood lycopene levels to be associated with an increased risk of cancer. Lycopene may account for or contribute to the decreased cancer risk observed; however, since the data are from observational studies, a cause–effect relationship cannot be definitively established. The biological role for lycopene in reducing cancer risk may be as an antioxidant, particularly as a quencher of singlet oxygen. Lycopene in vitro is the most powerful antioxidant of all carotenoids tested (29).

In a cohort study of diet and prostate cancer incidence in 14,000 Seventh-day Adventist men, only intakes of tomatoes, beans, lentils, and peas were statistically related to lower prostate cancer risk (60). In a subsequent cohort study, Giovannucci *et al.* (61) reported an inverse relationship between the consumption of tomatoes and tomato based products and prostate cancer risk. In this report of the Health Professionals Follow-up Study (HPFS), dietary intake was assessed for 1 year by a semi-quantitative food frequency questionnaire for 47,894 males. Follow-up questionnaires to assess disease incidence were collected from these subjects every 2 years for 6 years. No consistent association was found between dietary retinol, BC, α-carotene, lutein, or β-cryptoxanthin and prostate cancer risk. However, lycopene intake was related to lower prostate cancer risk (age- and energy-adjusted RR = 0.79; 95% CI = 0.64–0.99 for high vs. low quintile of intake). Of 46 vegetables and fruits studied, only tomato sauce (P for trend = 0.001), tomatoes (P for trend = 0.03), pizza (P for trend = 0.05), and strawberries were significantly associated with a lower risk of prostate cancer (61).

Of four case-control studies examining the relationship between diet and prostate cancer, two found an inverse association between tomato intake and risk of prostate cancer (62,63), however the results were not statistically significant. Case-control studies by Le Marchand (64) in Hawaii and Key (65) in the UK found no association between the consumption of tomatoes and the risk of prostate cancer.

Serum lycopene levels have been found to be lower in patients with prostate cancer than in matched controls. Gann *et al.* (66) reported that lycopene was the only antioxidant found at significantly lower mean levels in prostate cancer cases than in matched controls. They examined plasma from 581 case subjects from the PHS for antioxidant concentration. Rao *et al.* (67) also investigated levels of major carotenoids in serum and prostate tissue in cancer patients and in matched controls. They found that only serum and tissue lycopene levels were significantly lower in the cancer patients than in the controls, whereas other carotenoids did not significantly differ between the two groups.

In a case-control study by Kristal *et al.* (68), supplement use over 2 years prior to diagnosis was examined in 697 men with prostate cancer and in 666 matched controls. Results suggested that multivitamin use was not associated with prostate cancer, but individual supplements of zinc, vitamin C, and vitamin E may be protective. Vitamin A was not associated with prostate cancer risk and carotenoids were not studied. In addition, they found that tomatoes had no statistical effect on prostate cancer incidence but broccoli appeared to have preventative benefits.

Other cancers

In China, which has some of the highest rates of stomach cancer in the world, serum levels of BC and other micronutrients were measured along with examinations for chronic atrophic gastritis, intestinal metaplasia, and intestinal dysplasia, all of which generally precede the onset of gastric cancer. Concentrations of BC were significantly lower among individuals with intestinal metaplasia. During a nutrition intervention study conducted in Linxian County, China, use of a combined supplement of BC, vitamin E, and selenium in populations with a variety of marginal nutritional intake problems was associated with significant reductions in total mortality (relative risk = 0.91), cancer incidence (relative risk = 0.87), and stomach cancer mortality (relative risk = 0.79), with the reduced risk first noted about 1–2 years after the start of supplementation (69,70). No significant effects on total mortality rates were found for supplementation with retinol and zinc, riboflavin and niacin, or vitamin C and molybdenum.

The incidence of oral and pharyngeal cancer may also be inversely related to serum carotenoid levels. In a case-control serologic study using previously drawn and frozen blood samples, serum levels of all carotenoids, particularly BC, were lower among subjects who developed oral and pharyngeal cancer 1–15 years after the blood was drawn (71). The risks of this malignancy decreased substantially with increasing serum level of each individual carotenoid. Persons in the highest tertile of total carotenoids had about one-third the cancer risk as those in the lowest tertile.

In a comprehensive review of epidemiologic literature concerning diet and breast cancer by Clavel-Chapelon *et al.* (72), higher BC consumption was associated with a lower risk of breast cancer in 11 of 16 case-control studies but the results were significant in only four of these studies. Of three cohort studies, one showed a non-significant decreased risk of breast cancer with high BC consumption, while the results of the other two studies were inconclusive. Since the Clavel-Chapelon review, a review by Cooper *et al.* (5) examined eight additional case-control studies and three additional cohort studies. Three of the eight case-control studies found no association between dietary carotenoid intake and breast cancer risk. One study found that α-carotene had a significant protective effect against breast cancer. The other four studies reviewed found a non-significant decreased risk of breast cancer with high intakes of lycopene and β-cryptoxanthin, BC and lycopene, total carotenoids, or BC, lutein, and zeaxanthin. Of the three cohort studies reported, there was no association found between dietary carotenoid intake and decreased risk for breast cancer. As concluded by Clavel-Chapelon *et al.*, it is probable that diet plays an important role in the etiology of breast cancer, however there is currently insufficient evidence to support specific dietary recommendations regarding carotenoid intake and breast cancer.

CAROTENOIDS AND CARDIOVASCULAR DISEASE

One of the best understood processes contributing to the development of cardiovascular disease (CVD) is the oxidation of low-density lipoprotein (LDL). Several studies indicate that antioxidants such as vitamin E, BC, and vitamin C may delay or reduce the oxidation of LDL (73). LDL particles that either have their lipid or protein portions chemically altered through oxidation, acetylation, or glycosylation, are particularly atherogenic. These altered particles are taken up by activated monocytes, which, when laden with cholesterol, become less mobile and remain in the arterial wall as foam cells. The antioxidant action of BC may also play a role in reducing oxidative damage (e.g.: lipid peroxidation, DNA damage, and other effects of free radical production) associated with reperfusion injury which occurs when blood flow is restored following an ischemic incident. Data from one intervention study (74) suggest that consuming an antioxidant-rich diet following acute myocardial infarction may reduce the plasma levels of lipid peroxides and cardiac enzyme levels, suggesting a reduction in myocardial necrosis and reperfusion injury induced by oxygen free radicals.

Several studies have shown that a high dietary intake of fruits and vegetables is inversely related to the risk of CVD and stroke (75,76). The Nurses Health Study found that women in the highest quintile of BC consumption had a relative risk of 0.78 for coronary heart disease (77). Some studies also support an inverse association between serum carotenoid levels and CVD morbidity and mortality. Analysis of a cohort from the Lipid Research Clinics Coronary Primary Prevention Trial and Follow-up Study examined the relationship between total serum carotenoid levels and the risk of subsequent coronary heart disease events (78). After adjustment for known risk factors, serum carotenoid levels were inversely related to CVD events. Men in the highest quartile of serum carotenoids had an adjusted relative risk of 0.64 compared with the lowest quartile; for men who never smoked, this relative risk was 0.28.

Case-control data using adipose tissue BC levels from The European Community Multicenter Study on Antioxidants, Myocardial Infarction, and Breast Cancer (EURAMIC) show that the age-adjusted and center-adjusted odds ratio for risk of myocardial infarction in the lowest quintile of BC as compared with the highest was 2.62. Additional control for body mass index and smoking reduced the odds ratio to 1.78, while other established risk factors did not substantially alter this ratio (79). The overall multivariate odds ratio for low (10th percentile) vs. high (90th percentile) BC was 1.98, with the strength of this inverse association with myocardial infarction dependent on polyunsaturated fatty acid (PUFA) levels.

For low PUFA the odds ratio of low vs. high BC was 1.79, for medium PUFA the odds ratio was 1.76, and for high PUFA 3.47 (79). Later analysis of α-carotene, BC, and lycopene in adipose tissue samples from the same population found that the adjusted odds ratio for risk of myocardial infarction was only significantly reduced with high lycopene concentrations but not α-carotene or BC (80).

Although numerous epidemiological studies indicate that high dietary intake of carotenoid-containing fruits and vegetables and also high serum carotenoid levels have a protective effect against CVD, some intervention trials do not support this association. The ATBC trial showed a non-significant trend for a greater incidence of ischemic heart disease in subjects receiving BC supplementation (52). In the CARET trial, subjects supplemented with BC had a 26% increase (non-significant) in CVD (55). Analysis of serum cholesterol and triglyceride levels among the participants of the CARET trial showed that there was a small, non-significant increase in serum triglyceride levels during intervention in the BC supplemented group. After the intervention was discontinued, a small decrease was observed in the BC group (81). Two intervention trials using BC supplementation, the PHS and the WHS, showed no significant benefit or harm in subjects taking 50 mg of BC every other day (56,57). In both the ATBC and CARET trials the subjects were current or former smokers which would indicate that BC supplementation to reduce the risk of CVD is contraindicated in a smoking population.

CAROTENIODS AND VISUAL PROTECTION

Age-related macular degeneration

In the United States, age-related macular degeneration is the leading cause of visual diminution in adults over age 65. The point at which vision is most focused, the macula lutea or yellow spot, has a central depression, the fovea centralis, which is exposed to the highest intensity of light and is very susceptible to oxidative damage due to the high blood flow. Lutein and zeaxanthin are concentrated there and it is hypothesized that these carotenoids may prevent retinal damage by acting as antioxidants or by absorbing high-energy blue light. It is not known whether increased dietary intake of these carotenoids can prevent macular pigment degeneration, however in a recent review by Pratt (82) it is emphasized that high dietary intake of carotenoids, particularly lutein and zeaxanthin, may reduce the risk of age-related macular degeneration.

The multicenter Eye Disease Case-Control Study examined 356 subjects with age-related macular degeneration and 520 control subjects aged 55–80 years. The relative risk for age-related macular degeneration was estimated according to dietary indicators of antioxidant status, controlling for smoking and other risk factors by using multiple logistic-regression analyses. After adjusting for risk factors, the researchers found that subjects in the highest quintile of carotenoid intake had a 43% lower risk for age-related macular degeneration compared with those in the lowest quintile (83). Among the specific carotenoids, lutein and zeaxanthin were most strongly associated with a reduced risk for age-related macular degeneration. This same group found similar results in an earlier study of 421 patients with neovascular age-related macular degeneration and 615 controls where subjects were classified by blood level of micronutrient (low, medium, and high). Persons with carotenoid levels in the medium and high groups had risk of neovascular age-related macular degeneration reduced to one half and one third, respectively, compared with those in the low group (84). However, the Beaver Dam Eye Study found no relationship between the risk of age-related macular degeneration and serum concentrations of α-carotene, BC, lutein, zeaxanthin, or β-cryptoxanthin. A 2.2-fold higher risk for age-related macular degeneration was found for the lowest quintile of serum lycopene (85). Analysis of a subgroup of subjects participating in the ATBC study found that supplementation with BC was not associated with reduced risk for age-related macular degeneration (86).

Cataracts

The lens of the eye is also highly susceptible to oxidative damage, which may play a central role in the development of age-related cataracts. Out of 10 studies looking at plasma concentrations of carotenoids or dietary intakes of carotenoids and cataract risk, seven studies found a decreased risk of cataract with elevated plasma concentrations or dietary intake of carotenoids (87). A cohort of 77,466 female nurses aged 45–71 participated in the Nurses' Health Study (NHS) and 12 year follow-up in which nutrient intake was assessed by repeated administration of a food-frequency questionnaire. After controlling for age and smoking, those with the highest intake of lutein and zeaxanthin had a 22% decreased risk of cataracts needing surgical extraction as compared to those in the lowest quintile (87). In this study, α-carotene, BC, lycopene, β-cryptoxanthin, and VA were not associated with incidence of cataract extraction.

CAROTENOID SAFETY AND TOXICITY

Very high doses of VA are associated with numerous toxic effects including reduced food intake, weight loss, weakness, reduced motor activity, bone and skin lesions, alopecia, hepatic failure, decreased fertility, teratogenic effects to fetal development, and adverse effects to the musculoskeletal system, eye, and central nervous system (88). However, very high doses of BC and other carotenoids do not appear to produce these same toxic symptoms. A recent review of current literature concerning the safety of synthetic BC (89) did not find BC to be geno-toxic, reprotoxic, or teratogenic. Also, BC showed no signs of producing organ toxicity in subacute, subchronic, or chronic oral toxicity studies in experimental animals receiving doses of up to 1,000 mg/day per kg body weight. Synthetic BC was not found to be carcinogenic in Sprague-Dawley rats or CD1 mice, however in human supplementation trials BC was associated with an increased risk of lung cancer and CVD in smokers.

Carotodermia, a yellow discoloration of the skin due to very high BC intake has for decades been reported in medical literature. This discoloration, however, appears to be benign (90). No serious medical complaints, increased genotoxicity, or birth defects have been associated with carotodermia (91–93). Doses of BC as high as 300 mg/day or more have been used in humans for over 25 years to treat erythropoetic propoporphyria (EPP), a genetic disease characterized by a defect in the heme biosynthetic pathway and extreme photosensitivity (90,92). There appears to be no acute or long-term toxicity associated with BC treatment of EPP. However, from the ATBC and CARET trials, we can see that BC supplementation is not without risk for high risk groups of smokers and asbestos exposed workers. More research is needed to determine potential risks and benefits of carotenoid supplementation.

Interactions between carotenoids and ethanol

Numerous studies have shown that chronic alcohol consumption leads to hepatic VA depletion and can cause VA deficiency. Correction of this deficiency through VA supplementation is complicated by the fact that the intrinsic hepatotoxicity of VA is potentiated by concomitant alcohol consumption (24). BC, a VA precursor that until recently was presumed to be innocuous, has also been tried in supplement form to replete VA in alcoholics. However, ethanol interferes with the conversion of BC to VA and the combination of BC and ethanol also results in hepatotoxicity and possibly other toxic effects (24).

Baboons fed daily with a standard amount of BC (200 g carrot/day) and ethanol (50% of total energy) for 2–5 years had significantly higher levels of BC in both plasma and liver than controls fed an isocaloric carbohydrate (94). On cessation of BC treatment, plasma clearance of BC was delayed in alcohol-fed baboons as compared to control animals. Further studies in both baboons and rats showed that BC supplementation using BC beadlets further raised hepatic BC status but did not effectively restore normal VA concentrations (95). BC beadlets and high BC plasma levels in alcohol-fed animals were also associated with increased hepatotoxicity as shown by leakage of mitochondrial enzymes, microscopic lesions, and proliferation of the smooth endoplasmic reticulum in the liver (95).

Concomitant BC supplementation and high ethanol consumption has been associated with an increased risk of lung cancer in smokers. In both the ATBC and CARET studies, the increased risk of lung cancer in the BC supplemented individuals was also associated with higher alcohol consumption (52,96). These trials also showed that BC supplementation in smokers may be associated with an increased risk for cardiovascular disease, particularly in heavy drinkers. The mechanisms for these increased risks are not yet known.

CONCLUSIONS

Carotenoids are traditionally considered as contributors to the VA requirement, and rightly so. However, the more we learn about these abundant and versatile compounds, the more we realize that carotenoids may have other functions as well. So far, carotenoids have been identified as having possible protective roles in lowering the incidence of cancer, CVD, age-related macular degeneration, and cataracts. As we study their absorption and metabolism in greater detail, we hope to discover the mechanisms for the health benefits and risks associated with carotenoids. More importantly, we will learn more about how to "fine-tune" our diet to optimize the intake and use of carotenoids. Observational epidemiological studies consistently show that consuming a diet high in carotenoid containing fruits and vegetables may lower the risk for chronic diseases.

Evidence for the benefit of carotenoid supplementation as a prophylactic measure against disease is inconsistent. However, both animal and human intervention studies show that BC supplementation in heavy smokers and heavy drinkers can increase the risk for lung cancer, CVD, and hepatotoxicity. However, this observation is not consistent in BC supplementation trials conducted in populations of non-smokers and moderate drinkers. There is strong evidence to support the dietary guidelines published by the American Heart Association, the American Cancer Society, and the United States Recommended Dietary Allowances which all recommend a balanced diet high in fruits, vegetables, and whole grains to obtain optimal health and reduce disease risk.

We conclude that there is evidence to support the consumption of carotenoid-rich foods as a means to lower the risk of certain cancers, cardiovascular disease, and age-related macular degeneration. Use of carotenoid supplements, other than to treat VA deficiency, is not warranted at this time. As is often concluded when discussing nutrition and health, additional research is needed on the absorption, metabolism, and health impact of dietary carotenoids.

REFERENCES

1. Olson JA, Krinsky NI. Introduction: the colorful, fascinating world of the carotenoids: important physiological modulators. *FASEB Journal* 1995; 9:1547–50.
2. Khachik F, Spangler CJ, Cecil Smith JJ. Identification, quantification, and relative concentrations of carotenoids and their metabolites in human milk and serum. *Analytical Chemistry* 1997; 69:1873–81.
3. Moore T. *Vitamin A*. New York: Elsevier Publishing Company; 1957; 643.
4. Boileau TWM, Moore AC, John WE Jr. Carotenoids and Vitamin A. In: Papas AM, ed. *Antioxidant Status, Diet, Nutrition, and Health*, Boca Raton, FL: CRC Press; 1999; 133–58.
5. Cooper DA, Eldridge AL, Peters JC. Dietary carotenoids and certain cancers, heart disease, and age-related macular degeneration: A review of recent research. *Nutrition Reviews* 1999; 57:201–14.

6. Tavani A, LaVecchia C. Beta-carotene and risk of coronary heart disease. A review of observational and intervention studies. *Biomed Pharmacother* 1999; 53:409–16.

7. Clinton SK. Lycopene: chemistry, biology, and implications for health and disease. *Nutrition Reviews* 1998; 56:35–51.

8. Schwartz JL, Shklar G. Retinoid and carotenoid angiogenesis: a possible explanation for enhancement of oral carcinogenesis. *Nutrition and cancer* 1997; 27:192–9.

9. Lee CM, Boileau AC, Boileau TWM, Williams AW, Swanson KS, *et al.* Review of animal models in carotenoid research. *J Nutr*, 1999; 129.

10. Van Het Hof KH, West CE, Westrate JA, Hautvast JGAJ. Dietary factors that affect the bioavailability of carotenoids. *Journal of Nutrition* 2000; 130:503–6.

11. Erdman JW Jr, Bierer TL, Gugger ET. Absorption and transport of carotenoids. *Annals of the New York Academy of Sciences* 1993; 691:76–85.

12. de Pee S, West CE, Karyadi MD, Hautvast GAJ. Lack of improvement in vitamin A status with increased consumption of dark-green leafy vegetables. *Lancet* 1995; 346:75–81.

13. de Pee S, West CE, Permaesih D, Martuti SM, *et al.* Orange fruit is more effective than are dark-green, leafy vegetables in increasing serum concentrations of retinol and beta-carotene in schoolchildren in Indonesia. *Am J Clin Nutr* 1998; 68:1058–67.

14. Van Het Hof KH, Tijburg LB, Pietrzik K, Westrate JA. Infulence of feeding different vegetables on plamsa levels of carotenoids, folate and vitamin C. Effect of disruption of the vegetable matrix. *Br J Nutr* 1999; 82:203–12.

15. *Recommended Dietary Allowances*. 10th edn. Washington, DC: National Research Council, National Academy Press; 1989; 283.

16. Solomons NW. Plant sources of Vitamin A and human nutrition: renewed strategies. *Nutrition Reviews* 1996; 54:89–91.

17. Johnson EJ, Qin J, Krinsky NI, Russell RM. Ingestion by men of a combined dose of β-carotene and lycopene does not affect the absorption of β-carotene but improves that of lycopene. *Journal of Nutrition* 1997; 127:1833–7.

18. White WS, Peck KM, Bierer TL, Erdman JW Jr. Interactions of oral β-carotene and canthaxanthin in ferrets. *Journal of Nutrition* 1993; 123:1405–13.

19. White WS, Stacewicz-Sapuntzakis M, Erdman JJW, Bowen PE. Pharmacokinetics of β-carotene and canthaxanthin after ingestion of individual and combined doses by human subjects. *Journal of the American College of Nutrition* 1994; 13:665–71.

20. Elinder LS, Hadell K, Johansson J, Holme JM, Olsson AG, *et al.* Probucol treatment decreases serum concentrations of diet-derived antioxidants. *Arteriosclerosis, Thrombosis and Vascular Biology* 1995; 15:1057–63.

21. Weststrate JA, Hof KH. Sucrose polyester and plasma carotenoid concentrations in healthy subjects. *American Journal of Clinical Nutrition* 1995; 62:591–7.

22. Peters JC, Lawson KD, Middleton SJ, Triebwasser KC. Assesment of the nutritional effects of Olestra, a nonabsorbable fat replacement: summary. *The Journal of Nutrition* 1997; 8S:1719s–28s.

23. Grummer MA, Erdman JW Jr. Effect of chronic alcohol consumption and moderate fat diet in vitamin A status in rats fed either vitamin A or beta-carotene. *J Nutr* 1983; 113:350–64.

24. Leo MA, Lieber CS. Alcohol, vitamin A, and beta-carotene: adverse interactions, including hepatotoxicity and carcinogenicity. *American Journal of Clinical Nutrition* 1999; 69:1071–85.

25. Bierer TL, Merchen NR, Erdman JW Jr. Comparative absorption and transport of five common carotenoids in preruminant calves. *Journal of Nutrition* 1995; 125: 1569–77.

26. Thatcher AJ, Lee CM, Erdman JW Jr. Tissue stores of beta-carotene are not conserved for later use as a source of vitamin A during compromised vitamin A status in Mongolian gerbils (*Meriones unguiculatus*). *J Nutr* 1998; 128:1179–85.

27. Jandacek RJ. The canary in the cell: a sentinel role for beta-carotene. *Journal of Nutrition* 2000; 130:648–51.

28. Hirayama O, Nakamura K, Hamada S, Kobayasi Y. Singlet oxygen quenching ability of naturally occuring carotenoids. *Lipids* 1994; 29:149–50.

29. Di Mascio P, Kaiser S, Sies H. Lycopene as the most efficient biological carotenoid singlet oxygen quencher. *Archives of Biochemistry and Biophysics* 1989; 274:532–8.

30. Stahl W, Nicolai S, Briviba K, Hanusch M, *et al*. Biological activities of natural and synthetic carotenoids: induction of gap junctional communication and singlet oxygen quenching. *Carcinogenesis* 1997; 18:89–92.

31. Levin G, Yeshurun M, Mokady S. In vivo antiperoxidative effect of 9-cis β-carotene compared with that of the all-trans isomer. *Nutrition and Cancer* 1997; 27:293–7.

32. Liebler DC. Antioxidant reactions of carotenoids. *Annals of the New York Academy of Sciences* 1993; 691:20–31.

33. Packer L. Antioxidant action of carotenoids in vitro and in vivo and protection against oxidation of human low-density lipoproteins. *Annals of The New York Academy of Sciences* 1993; 691:48–60.

34. Martin KR, Failla ML, Smith JC Jr. β-carotene and lutein protect HepG2 human liver cells against oxidant-induced damage. *Journal of Nutrition* 1996; 126:2098–106.

35. Bertram J. Cancer prevention by carotenoids: mechanistic studies in cultured cells. *Annals of The New York Academy of Sciences* 1993; 691:177–91.

36. Bertram JS, Bortkiewicz H. Dietary carotenoids inhibit neoplastic transformation and modulate gene expression in mouse and human cells. *American Journal of Clinical Nutrition* 1995; 62:1327s–36s.

37. Hanusch M, Stahl W, Schulz WA, Sies H. Induction of gap junctional communication by 4-oxoretinoic acid generated from its precursor canthaxanthin. *Archives of Biochemistry and Biophysics* 1995; 317:423–8.

38. Zhang L-X, Cooney RV, Bertram JS. Carotenoids enhance gap junctional communication and inhibit lipid peroxidation in C3H/10T1/2 cells: relationship to their cancer chemopreventive action. *Carcinogenesis* 1991; 12:2109–14.

39. Bertram JS, Pung A, Churley M, Kappock TJ, Wilkins LR, *et al*. Diverse carotenoids protect against chemically induced neoplastic transformation. *Carcinogenesis* 1991; 12.

40. Bendich A. Carotenoids and the immune response. *Journal of Nutrition* 1989; 119: 112–15.

41. Bendich A. Biological functions of dietary carotenoids. *Annals of the New York Academy of the Sciences* 1993; 691:61–7.

42. Hughes DA, Wright AJA, Finglas PM, Peerless ACJ, Bailey AL, *et al*. The effect of β-carotene supplementation on the immune function of blood monocytes from healthy male nonsmokers. *Journal of Laboratory and Clinical Medicine* 1997; 129:309–17.

43. Baker KR, Meydani M. β-carotene as an antioxidant in immunity and cancer. *Journal of Optimal Nutrition* 1994; 3:39–50.

44. Jyonouchi H, Sun S, Gross M. Effect of carotenoids on in vitro immunoglobin production by human peripheral blood mononuclear cells: astaxanthin, a carotenoid without

vitamin A activity, enhances in vitro response to a T-dependent stimulant antigen. *Nutrition and Cancer* 1995; 23:171–83.

45. Jyonouchi H, Sun S, Tomita Y, Gross MD. Astaxanthin, a carotenoid without vitamin A activity, augments antibody responses in cultures including T-helper cell clones and suboptimal doses of antigen. *Journal of Nutrition* 1995; 125:2483–92.

46. Block G, Patterson B, Subar A. Fruit, vegetables, and cancer prevention: a review of the epidemiological evidence. *Nutrition and Cancer* 1992; 18:1–29.

47. Ziegler RG, Mayne ST, Swanson CA. Nutrition and lung cancer. *Cancer Causes and Control* 1996; 7:157–77.

48. Menkes MS, Comstock GW, Vuilleumier JP, Helsing KJ, Rider AA, *et al*. Serum beta-carotene, vitamins A and E, selenium, and the risk of lung cancer. *N Engl J Med* 1986; 315:1250–4.

49. Ziegler RG. A review of epidemiologic evidence that carotenoids reduce the risk of cancer. *J Nutr* 1989; 119:116–22.

50. The alpha tocopherol β-carotene cancer prevention study group. The effect of vitamin E and beta-carotene on the incidence of lung cancer and other cancers in male smokers. *The New England Journal of Medicine* 1994; 330:1029–35.

51. The alpha tocopherol β-carotene cancer prevention study group. Cancer prevention study: Design, methods, participants characteristics, and compliance. *Ann Epidemiol* 1994; 4:1–10.

52. Albanes D, Heinonen O, Taylor P, Virtamo J, Edwards B, *et al*. Alpha-tocopherol and β-carotene supplements and lung cancer incidence in the alpha-tocopherol, β-carotene cancer prevention study: effects of baseline characteristics. *Journal of the National Cancer Institute* 1996; 88:1560–70.

53. Erdman JW Jr, Russell RM, Rock CL, Barua AB, Bowen PE, *et al*. Beta-carotene and the caroteniods: beyond the intervention trials. *Nutr Rev* 1996; 54:185–8.

54. Wang X-D, Liu C, Bronson RT, Smith DE, Krinsky NI, *et al*. Retinoid signaling and activator protein-1 expression in ferrets given beta-carotene supplements and exposed to tobacco smoke. *Journal of the National Cancer Institute* 1999; 91:60–6.

55. Omenn G, Goodman G, Thornquist M, Balmes J, Cullen M, *et al*. Efftects of a combination of beta carotene and vitamin A on lung cancer and cardiovascular disease. *The New England Journal of Medicine* 1996; 334:1150–5.

56. Hennekens CH, Buring JE, Manson JE, Stampfer M, Rosner B, *et al*. Lack of effect of long-term supplementation with beta carotene on the incidence of malignant neoplasms and cardiovascular disease. *The New England Journal of Medicine* 1996; 334:1145–9.

57. Lee IM, Cook NR, Manson JE, Buring JE, Hennekens CH. Beta-carotene supplementation and incidence of cancer and cardiovascular disease: the Women's Health Study. *J Natl Cancer Inst* 1999; 91:2102–6.

58. Gugger ET, Bierer TL, Henze TM, White WS, Erdman JJW. β-carotene uptake and tissue distribution in ferrets (*Mustela putorius furo*). *Journal of Nutrition* 1992; 122:115–19.

59. Giovannucci E. Tomatoes, tomato-based products, lycopene, and cancer: review of the epidemiologic literature. *Journal of the National Cancer Institute* 1999; 91:317–31.

60. Mills PK, Beeson L, Phillips RL, Fraser GE. Cohort study of diet, lifestyle, and prostate cancer in adventist men. *Cancer* 1989; 64:598–604.

61. Giovannucci E, Ascherio A, Rimm EB, Stampfer MJ, Colditz GA, *et al*. Intake of carotenoids and retinol in relation to prostate cancer risk. *Journal of the National Cancer Institute* 1995; 87:1767–76.

62. Schuman LM, Mandel JS, Radke A, Seal U, Halberg F. Some selected features of the epidemiology of prostatic cancer: Minneapolis-St. Paul, Minnesota case-control study, 1976–1979. In: Magnus K, ed. *Trends in cancer incidence: causes and pratical implications*; Washington DC: Hemisphere Publishing Corp; 1982; 345–54.

63. Cohen JH, Kristal AR, Stanford JL. Fruit and vegetable intakes and prostate cancer risk. *Journal of the National Cancer Institute* 2000; 92:61–8.

64. LeMarchand L, Hankin JH, Kolonel LN, Wilkens LR. Vegetable and fruit consumption in relation to prostate cancer risk in Hawaii: a reevaluation of the effect of dietary beta-carotene. *Am J Epidemiol* 1991; 133:215–19.

65. Key TJA, Silcocks PB, Davey GK. A case-control study of diet and prostate cancer. *Br J Cancer* 1997; 76:678–87.

66. Gann PH, Ma J, Giovannucci E, Willet W, Sacks FM, *et al.* Lower prostate cancer risk in men with elevated plasma lycopene levels: results of a prospective analysis. *Cancer Research* 1999; 59:1225–30.

67. Rao AV, Fleshner N, Agarwal S. Serum and tissue lycopene and biomarkers of oxidation in prostate cancer patients: a case-control study. *Nutr Cancer* 1999; 33:159–64.

68. Kristal AR, Stanford JL, Cohen JH, Wicklund K, Patterson RE. Vitamin and mineral supplement use is associated with reduced risk of prostate cancer. *Cancer Epidemiol Biomarkers Prev* 1999; 8:887–92.

69. Blot WJ, Li J-Y, Taylor PR, Guo W, *et al.* Nutrition intervention trials in Linxian, China: Supplementaion with specific vitamin/mineral combinations, cancer incidence, and disease-specific mortality in the general population. *Journal of National Cancer Institute* 1993; 85:1483–92.

70. Blot, W, Li J, Taylor P, Guo W, Dawsey S, *et al.* The Linxian trials: mortality rates by vitamin-mineral intervention group. *Amer J Clin Nutr* 1995; 62:1424S–6S.

71. Zheng W, Blot WJ, Diamond EL, Norkus EP, Spate V, *et al.* Serum micronutrients and the subsequent risk of oral pharyngeal cancer. *Cancer Res* 1993; 53:795–8.

72. Clavel-Chapelon F, Niravong M, Joseph RR. Diet and breast cancer: review of the epidemiologic literature. *Cancer Detect Prev* 1997; 21:426–40.

73. Holvoet P, Collen D. Oxidized lipoproteins in atherosclerosis and thrombosis. *FASEB J* 1994; 8:1279–84.

74. Singh RB, Niaz MA, Agarwal P, Begom R, Rastogi SS. Effect of antioxidant-rich foods on plasma ascorbic acid, cardiac enzyme, and lipid peroxide levels in patients hospitalized with acute myocardial infarction. *Journal of the American Dietetic Association* 1995; 95:775–80.

75. Steinmetz KA, Potter JD. Vegetables, fruit, and cancer prevention: A review. *Journal of the American Dietetic Association* 1996; 96:1027–39.

76. van Poppel G. Epidemiological evidence for beta-carotene in prevention of cancer and cardiovascular disease. *Eur J Clin Nutr* 1996; 50:S57–S61.

77. Stampfer MJ, Hennekens CH, Manson JE, Colditz GA, Rosner B, *et al.* Vitamin E consumption and the risk of conorary heart disease in women. *N Engl J Med* 1993; 328:1444–9.

78. Morris DL, Kritchevsky SB, Davis CE. Serum carotenoids and coronary heart disease: the lipid research clinics coronary primary prevention trial and follow-up study. *JAMA* 1994; 272:1439–41.

79. Kardinaal AFM, Aro A, Kark JD, Riemersma RA, van'tVeer P, *et al.* Association between beta-carotene and acute myocardial infarction depends on polyunsaturated fatty acid status. *Arterioscler Thromb Vasc Biol* 1995; 15:726–32.

80. Kohlmeier L, Kark JD, Gomez-Garcia E, Martin BC, *et al.* Lycopene and myocardial infarction risk in the EURAMIC Study. *American Journal of Epidemiology* 1997; 146:618–26.

81. Redlich CA, Chung JS, Cullen MR, Blaner WS, VanBennekum AM, *et al.* Effect of long-term beta-carotene and vitamin A on serum cholesterol and triglyceride levels among participants in the Carotene and Retinol Efficacy Trial. *Atherosclerosis* 1999; 145:425–32.

82. Pratt S. Dietary prevention of age-related macular degeneration. *Journal of the American Optometric Association* 1999; 70:39–47.

83. Seddon JM, Ajani UA, Sperduto RD, Hiller R, *et al.* Dietary carotenoids, vitamins A, C, and E, and advanced age-related macular degeneration. *JAMA* 1994; 9272:1413–20.

84. Eye Disease Case-Control Study Group. Antioxidant status and neovascular age-related macular degeneration. *Arch Ophthalmol* 1993; 111:104–9.

85. Mares-Perlman JA, Brady WE, Klein R, *et al.* Serum antioxidants and age-related macular degeneration is a population-based case-control study. *Arch Ophthalmol* 1995; 113:1518–23.

86. Teikari JM, Laatikainen L, Virtamo J, *et al.* Six-year supplementation with alpha-tocopherol and beta-carotene and age-related maculopathy. *Acta Ophthamol Scand* 1998; 76:224–9.

87. Chasan-Taber L, Willett WC, Seddon JM, Stampfer MJ, Rosner B, *et al.* A prospective study of carotenoid and vitamin A intakes and risk of cataract extraction in US women. *American Journal of Clinical Nutrition* 1999; 70:509–516.

88. Armstrong RB, Ashenfelter KO, Eckhoff C, Levin AA, Shapiro SS. General and Reproductive Toxicology of Retinoids. In: Sporn MB, Roberts AB, Goodman DS, eds. *The Retinoids: Biology, Chemistry, and Medicine*, 2nd edn, New York: Raven Press, Ltd; 1994; 545–72.

89. Woutersen RA, Wolterbeek AP, Appel MJ, vandenBerg H, Goldbohm RA, *et al.* Safety evaluation of synthetic beta-carotene. *Crit Rev Toxicol* 1999; 29:515–42.

90. Burri BJ. Beta-carotene and human health: a review of current research. *Nutrition Research* 1997; 17:547–80.

91. Diplock A. Safety of antioxidant vitamins and beta-carotene. *Am J Clin Nutr* 1995; 62: 1510S–16S.

92. Mathews-Roth M. Beta-carotene therapy for erythropoetic protoporphhyria and other photosensitivity diseases. *Biochimie* 1986; 68:875–84.

93. Mathews-Roth M. Lack of genotoxicity with beta-carotene. *Toxicol Lett* 1988; 41:185–91.

94. Leo MA, Kim C, Lowe N, Lieber CS. Interaction of ethanol with beta-carotene: delayed blood clearance and enhanced hepatotoxicity. *Hepatology* 1992; 15:883–91.

95. Leo MA, Aleynik SI, Aleynik MK, Lieber CS. Beta-carotene beadlets potentiate hepatotoxicity of alcohol. *American Journal of Clinical Nutrition* 1997; 66:1461–9.

96. Omenn GS, Goodman GE, Thornquist MD, Balmes J, Cullen MR, *et al.* Risk factors for lung cancer and for intervention effects in CARET, the Beta-Carotene and Retinol Efficacy Trial. *J Natl Cancer Inst* 1996; 88:1550–9.

97. Clinton SK, Emenhiser C, Schwartz SJ, Bostwick DG, Williams AW, *et al. cis–trans* lycopene isomers, carotenoids, and retinol in the human prostate. *Cancer Epidemiology, Biomarkers & Prevention* 1996; 5:823–33.

98. Kaplan LA, Lau JM, Stein EA. Carotenoid composition, concentrations and relationships in various human organs. *Clinical Physiology and Biochemistry* 1990; 8:1–10.

99. Schmitz HH, Poor CL, Wellman RB, Erdman JW Jr. Concentrations of selected carotenoids and vitamin A in human liver, kidney and lung tissue. *Journal of Nutrition* 1991; 121:1613–21.

100. Peng YM, Peng YS, Lin Y. A nonsaponification method for the determination of carotenoids, retinoids, and tocopherols in solid human tissues. *Cancer Epidemiol Biomarkers Prev* 1993; 2:139–44.

101. Peng YS, Peng YM, McGee DL, Alberts DS. Carotenoids, tocopherols, and retinoids in human buccal mucosal cells: intra- and interindividual variability and storage stability. *Am J Clin Nutr* 1994; 59:636–43.

102. Albanes D, Heinonen O, Huttunen J, Taylor P, Virtamo J, *et al.* Effects of alpha-tocopherol and beta-carotene supplements on cancer incidence in the Alpha-Tocopherol Beta-Carotene Cancer Prevention Study. *Am J Clin Nutr* 1995; 62:1427S–30S.

103. Hercberg S, Preziosi P, Galan P, Faure H, Arnaud J, *et al.* "The SU.VI.MAX Study": a primary prevention trial using nutritional doses of antioxidant vitamins and minerals in cardiovascular diseases and cancers. SUpplementation on VItamines et Mineraux AntioXydants. *Food Chem Toxicol* 1999; 37:925–30.

104. Hercberg S, Galan P, Preziosi P, Roussel AM, Arnaud J, *et al.* Background and rational behind the SU.VI.MAX Study, a prevention trial using nutritional doses of a combination of antioxidant vitamins and minerals to reduce cardiovascular diseases and cancers. *Int J Vitam Nutr Res* 1998; 68:3–20.

105. Hercberg S, Preziosi P, Briancon S, Galan P, Triol I, *et al.* A primary prevention trial using nutritional doses of antioxidant vitamins and minerals in cardiovascular diseases and cancers in a general population: the SU.VI.MAX study – design, methods, and participant characteristics. *Control Clin Trials* 1998; 19:336–51.

106. Christen WG, Gaziano JM, Hennekens CH. Design of Physicians' Health Study II – a randomized trial of beta-carotene, vitamins E and C, and multivitamins, in prevention of cancer, cardiovascular disease, and eye disease, and review of results of completed trials. *Ann Epidemiol* 2000; 10:125–34.

3

THE ROLE OF NUTRIENTS IN DETOXIFICATION MECHANISMS

K.J. Netter

INTRODUCTION

This chapter addresses the influence of nutrition on the enzymatic reactions that lead to inactivation, and possibly also metabolic activation, of foreign compounds. As most toxicants are substrates of inactivating enzymes, a consideration of nutritional influences on drug metabolic reactions is at the same time a consideration of nutritional influences on detoxification.

GENERAL CONSIDERATIONS

The basic enzymology of chemical detoxification

The enzymatic mechanisms of detoxification consist mainly of two types, namely oxidative (phase I) and conjugative (phase II) detoxification reactions. The oxidative detoxification mechanisms depend on the catalytic activity of the microsomal mixed-function oxidase system, in which one atom of molecular oxygen is incorporated into the substrate and the necessary energy derived from the NADPH-facilitated reduction of oxygen to water. The hemoprotein catalyst, cytochrome P-450, undergoes a reaction cycle beginning with the binding of the substrate to a mainly lipophilic binding site, followed by successive additions of oxygen and electrons, and leading to the liberation of the hydroxylated product and water. A first-approach description of this process can be found in the work of Benet *et al.* (1). The last decades have witnessed an explosive growth of knowledge about cytochrome P-450 and, in particular, about its numerous isoenzymes. Concerning the techniques alone there is a whole volume of *Methods in Enzymology* devoted to this enzyme (2). There are about 150 different genes for this protein, which can be classified into various families and subfamilies. Also there is much information on its evolution; it is a well-conserved protein. The multitude of cytochrome P-450 isoenzymes has been classified, and the new nomenclature is now widely accepted (3). The regulation of gene expression is subject to many influences, environmental and also nutritional (4), which leads to the important phenomenon that particular treatments result in an alteration of the constitutive pattern and hence in an alteration of the selection of substrates that are preferentially metabolized. Further information on cytochromes P-450 can be obtained from Coon *et al.* (5). This field is also covered by other reviews, such as those by Gonzalez (6) and Nebert and Gonzalez (7).

Not all cytochrome P-450 families are involved in xenobiotic metabolism (8); for a number of subfamilies the hydroxylation of steroids is their prime biologic purpose. On the other hand, there are about 67 different forms in the cytochrome P-450 families 1, 2, and 3 (CYP 1, 2, and 3) that are involved in the metabolism of toxicants. In particular, the CYP 2 family contributes the bulk of the iso-enzymes and also shows great inter- and intraspecies variability in its catalytic activities. While originally cytochromes P-450 were involved in steroid and bile acid metabolism, the xenobiotic-metabolizing cytochromes P-450 have evolved from these during phylogeny, mainly with the teleologic purpose of oxidatively degrading dietary compounds that could not be eliminated otherwise because of their lipophilicity.

In humans, about 20 different xenobiotic-metabolizing cytochromes P-450 have been described (8), of which the CYP 1 family, as the evolutionarily best-conserved family, plays an important role in the detoxification of xenobiotics. But in a similar way members of the CYP 2 and CYP 3 families also contribute to drug and toxin inactivation. As the pattern of cytochromes P-450 varies between species, there is the particular dilemma that one cannot easily apply the analogy to humans. Furthermore, even within the human species racial distribution differences lead to genetically determined disparities in xenobiotic handling, a phenomenon that has long been described by pharmacogeneticists.

A comprehensive overview on the modulation of cytochromes P-450 by Ioannides (9) appeared in 1998. Providing a large bibliography, it deals with the distribution of cytochrome P-450 isoforms and methods of assessing them in various species. It discusses the influences of diet, caloric restriction and nutrients on drug metabolism. In concluding, it arrives at the opinion that the mechanisms of nutritional alteration of cytochrome P-450 patterns have not in all cases been extensively studied and elucidated, although certain *influences* such as enzyme induction by grilled food have been standard knowledge for more than two decades.

Phase II reactions are catalyzed by various transferases that transfer an activated conjugating group to the substrate, the recipient being mostly a hydroxyl or amino group. The *glucuronidation* of substrates has been reviewed by Burchell and Coughtrie (10). The family of uridine 5'-diphosphate [UDP]-glucuronyl transferases has been reviewed by Tephly and Burchell (11). Here we again deal with a family of enzymes, the members of which have developed discrete specificities toward groups of substrates during their evolution. The various members of the UDP-glucuronyl transferase family have been characterized mainly by their isolation. These proteins are attached to endoplasmic membranes from which they can be released by detergents with the consequence of great variations in activity. It is to be noted that glucuronidation does not always lead to inactivation, but rather can produce active products.

Another conjugative coupling mechanism is that of *sulfation* by activated sulfate groups; this has been recently reviewed by Mulder and Jakoby (12). Currently there are only three cytosolic sulfotransferases in human liver, a figure that contrasts with the multitude of other drug-metabolizing enzymes. Sulfotransferases can be categorized into two enzymes, sulfating phenols and monoamines, and one steroid sulfotransferase. Interestingly, phenol sulfotransferase activates the potassium channel-blocker antihypertensive minoxidil by sulfation of an N-O group; sulfation in the hair follicles leads to the sometimes desirable side effect of hypertrichosis. For a review of sulfotransferases, see the work of Falany (13).

Acetylation is a process for detoxification of certain nitrogen-containing xenobiotics, such as sulfonamides and hydrazines. For a review of acetylation, see the work of Netter (14).

Glutathione is present in cells in markedly high concentrations (about 5 mmol/L) and serves as an intracellular pool of a detoxifying ligand. The enzymatic basis of this conjugative inactivation has been reviewed by Jakoby and Habig (15). The molecular biology and biologic function of glutathione-S-transferases also have been described (16). It is also to be noted that in this case conjugation can cause "erroneous" metabolic activation, leading to renal damage by the resulting sulfonium ion of dichlorovinylcysteine.

This discussion cannot be concluded without mentioning two basic phenomena concerning xenobiotic metabolizing enzymes, namely *induction* and *inhibition*. As pointed out earlier (17), there are a number of inducing substances that increase the activity of hepatic microsomal mixed-function oxidases and that have attained the character of such classic inducers as phenobarbital and 3-methylcholanthrene. Besides these there are selective inducers, such as nitrogen-containing heterocyclic aromatic compounds (17) and metals, that cause increased levels of coumarin hydroxylation (18). The opposite phenomenon, namely the inhibition of oxidative drug metabolism by a wide variety of substances, has also been widely investigated (19,20). Interestingly grapefruit juice contains a strongly inhibiting substance (naringin) that influences the inactivation of the calcium channel-blocker nitrendipin to such an extent, that its area under the plasma concentration-time curve can be doubled (21).

Interaction of nutrition with xenobiotic metabolism

The interrelationship between nutrition and xenobiotic metabolism has been the subject of intensive consideration. For earlier reviews see Campbell and Hayes (22), and Parke and Ioannides (23); for monographs or symposia see Hathcock and Coon (24), and Hathcock (25). Toxicodietetics, as manifested by dietary alterations in toxicity, was treated in 1979, 1986, and 1992 (26,27). The role of nutrition in the generation of liver tumors in mice has been investigated in 1987 (29), while the bioactivation of chemicals has been evaluated in 1984 (30) and in 1986 (31). More recent reviews investigating dietary effects on the various members of the cytochrome P-450 families have incorporated the progress in understanding inductive and inhibitory influences on this and other detoxifying systems (32–35). Nutritional influences, not only on detoxification but also on causation and prevention of chronic diseases, have been the subject of an authoritative survey (36). This chapter does not address the effects of micronutrients such as vitamins and essential elements. A table summarizing their effects on xenobiotic metabolism can be found in the review of Yang *et al.* (35).

The interaction of nutrition with enzymatic mechanisms of detoxification (and toxification) can occur at various levels. Interactions with absorption and distribution have been treated in other parts of this book and also in the work of Netter (28).

The principal changes that can be affected by nutrition are outlined in the figures. Figure 1 shows the reaction cycle of cytochrome P-450, leading from substrate to oxidized product with the consumption of reducing equivalents and oxygen. The points of possible intervention are indicated in the figure by the following numbers:

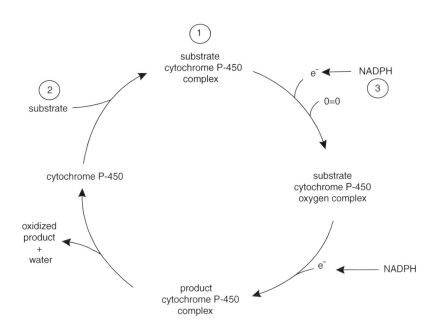

Figure 1 The cytochrome P-450 reaction cycle. The numbers give the points of interference by nutritional factors as they are discussed in the text.

1   The amount of hemoprotein can clearly be affected by nutritional influences, whereas specific isoenzymes of the P-450 superfamily can be differentially affected. Binding of substrate to the cytochrome is not likely to be influenced because this would occur mainly by competition for the binding sites only, which would require alternative substrates with chemical structures that are normally not found in nutrients, and if they were (short-chain fatty acids) would not be so abundant as to displace xenobiotic substrates.

2   A very likely point of interference develops by changing the accessibility of a substrate at different points, mainly by interference with toxicokinetics and possibly perturbation of the membrane composition (see also Figure 3).

3   Depletion of reducing equivalents seems to be rare, occurring not even during starvation, but should be mentioned here when discussing principles of interference. Experiments in our own laboratory (37) have shown that, conversely, intoxication with a redox ion, namely vanadate, shuts down xenobiotic metabolism immediately and drastically, but does not equally as drastically lower the intracellular level of NADPH; rather, vanadate seems to interact with the oxidoreduction of folate (38).

Figure 2 describes the main steps leading to a detoxified conjugate and the possible points of interference by nutrition. The activation of the conjugating group has not been considered to be under dietary control because the levels of cellular ATP are kept constant under all circumstances. The following two points, however, may be subject to nutritional influence:

4   The availability of the conjugating groups could be limited by respective deprivation. As glucose-l-phosphate and also acetyl-CoA are omnipresent compounds in cellular metabolism, only sulfate could become a rate-limiting factor in conjugative detoxification. Similarly, glutathione or cysteine could

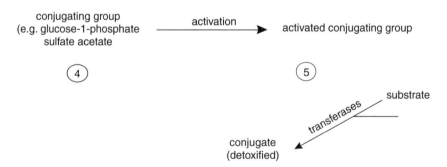

Figure 2 A general scheme of the two basic reactions in conjugation. The respective conjugated moieties must be metabolically activated before the transferases catalyse the actual group transfer. The numbers refer to the discussion in the text.

conceivably be critically depleted by nutritional deficiencies, but solid data on this are lacking.

5 Transferases can be induced by a number of hydrophobic compounds.

Figure 3 depicts possible nutritional interactions in the formation and activity of cytochromes P-450, including the following:

6 The enhancement of transcription leading to increased P-450 levels is an early observed phenomenon, particularly for the induction of intestinal cytochromes P-450 by respective food components.

7 With lesser probability one can expect influences on the degradation of messenger RNA (mRNA), but theoretically this also could be an example of nutritional control.

8 The supply of amino acids and their balance can become important in the biosynthesis of the enzyme, yet starvation experiments have led to varied experiences regarding the enzymatic activities toward various substrates. Although toxic effects of amino acids have been described (39), there is so far no evidence of direct damage to cytochrome P-450 by excessive amino acid feeding.

9 The lipid composition of the endoplasmic membranes greatly influences the activity and function of cytochromes P-450 but, on the other hand, is subject to change by the lipid level in the diet.

10 In vegetable diets there are a number of natural ingredients that act at this point as inhibitors of xenobiotic metabolism.

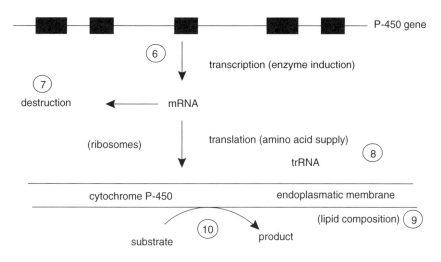

Figure 3 A survey of the main factors determining cytochrome P-450 biosynthesis, induction, membrane-facilitated function, and inhibition. The numbers refer to the discussion of these points in the text.

The above points have been made in order to illustrate the principles of nutritional interaction, and it is obvious that this list might not be complete nor may it be stringently correct given the complexity of nutritional influences.

In the following sections the currently known effects of diet on the enzymatic mechanisms of xenobiotic activation or inactivation will be reviewed without any distinction between macronutrients and micronutrients.

LEVELS AND ACTIVITIES OF CYTOCHROMES P-450

Starvation is the most drastic form of dietary modification, and it has long been known that fasting increases the susceptibility to hepatotoxic injury. Thus, it has been described (40) that in starved mice liver glycogen decreases to less than one-tenth and reduced glutathione (GSH) to about half of their original values; in contrast, liver triglycerides triple to about 30 mg/g. When, under these conditions, animals are exposed to toxic agents such as paracetamol (acetaminophen) or carbon tetrachloride, they develop liver damage that is much more severe. In this case the serum transaminases increase after a dose of paracetamol or carbon tetrachloride. The same can be seen for large doses of bromobenzene (41).

More recently, with the recognition of many isoenzymes of P-450, the situation has become more diversified: Fasting not only decreases the glycogen stores in the liver and thus decreases resistance against cytotoxic agents, but it also increases the amount of certain P-450 isoenzymes. By measuring liver microsomal N-nitrosodimethylamine demethylation in rats, the conclusion has been reached that cytochrome P-450 increases by 60% and 116% after 24 and 48 hours of fasting, respectively. The important point is that the enzyme protein *per se*, not just its activity, is increased, possibly by better access of substrate to the enzyme. Immunoblotting and complementary DNA (cDNA) probing have shown an increase in the respective mRNA resulting from increased gene expression (42). Other authors (43) have described not only a 2.5-fold increase in cytochrome P-450 but also a decrease of typical male-specific forms that preferentially hydroxylate testosterone in the 2 and 16 positions; on the other hand, there is a 12-fold induction of phenobarbital-inducible P-450. The changes in activities are caused by changes in hemoprotein, as evidenced by carbon monoxide-binding spectra. In another experiment (44) protein-energy malnutrition reduced the microsomal testosterone hydroxylation rate to about half of the control value; however, there was great variation in the response of different testosterone hydroxylases to starvation. Streptozotocin-induced diabetes mellitus had an effect very similar to that of fasting on hepatic cytochrome levels, and the acetone/ethanol-inducible form increased four- to five-fold in diabetic and in fasting rats. In contrast, other isoenzymes decreased very drastically in diabetic rats and increased in fasting rats. These immunoquantitative results are paralleled by the activities toward the respective substrates (45). These results show that starvation (i.e. glycogen depletion and acetonemia) differentially affects the biosynthesis of cytochromes P-450. However, it is not known why the same basic phenomenon acts so differently on the biosynthesis of specific proteins.

The opposite treatment, namely hyperalimentation by parenteral infusion of glucose and crystalline amino acids, also affects P-450 levels (46): A decrease in cytochrome P-450 PCN-e (α-cyanopregnenolone-inducible) is observed under these conditions. It is interesting that oral administration of the same nutrient solution does not yield these effects; it is hypothesized that this opposite influence on hemoprotein synthesis may be due to differences in the blood composition in the portal vein, suggesting something like a "first-pass effect." In this connection it is interesting to speculate that tetrachlorodibenzo-*p*-dioxin (TCDD) intoxication, that seems to cause a disturbance of the satiety signaling mechanism which leads to diminished food uptake (wasting syndrome) (47), may also cause changes in the cytochromes P-450.

The composition of a calorically sufficient diet also influences the metabolism of xenobiotics. Thus the administration of a high-fat/low-carbohydrate diet containing corn oil for 4 days increases hepatic microsomal P-450 2E1 over those rats with lower fat/carbohydrate ratios (48). Unsaturated fats have a greater effect than lard or olive oil diets. The increase in hemoprotein and enzymatic activity is about two-fold. Dietary lipids have been studied by comparing the effects of a fat free and a 20% corn oil diet with respect to induction of hemoproteins (49). The corn oil admixture did not affect the constitutive level of P-450 2B, but it doubled the inductive effect of phenobarbital as measured by activity, apoprotein, and mRNA. In contrast, corn oil increased the constitutive level of P-450 2E (ethanol-inducible). However, with this isoenzyme there is no influence of dietary fat on its induction by acetone. Dietary corn oil also increased the testosterone 6-β-hydroxylase activity, suggesting an increase in other isoenzymes. Similarly, there is an increase in arylhydrocarbon hydroxylase and 7-ethoxy-coumarin *O*-deethylase activities in the lung of food-restricted rats and a decrease after high-fat diet, but this does not occur in liver (50).

Inducibility of cytochrome P-450 is dependent on the supply of dietary lipids, as has been shown by Marshall and McLean (51) for the induction of rat liver microsomal activities by phenobarbital. The necessity of polyunsaturated fatty acids for oxidative demethylation as well as inducibility by phenobarbital has been emphasized (52), whereas dietary lipid peroxides have been shown to increase these two parameters. The same was shown (53) by measuring carbon monoxide binding spectra after giving either a fat-free diet or one consisting of 20% corn oil; the fat-free rats had only an increase of about 20% after phenobarbital, while in the corn oil group cytochrome P-450 was doubled. The lack of biosynthesis of new cytochrome P-450 hemoproteins in fat-free nutrition was attributed to two possible but as yet unexplored phenomena: insensitivity toward the inducer and the paucity of utilizable fatty acids needed to synthesize the phospholipid matrix for the support and proper juxtapositioning of the active proteins. These results show that for unknown reasons dietary fat, and in particular unsaturated fatty acids, facilitate the response to inducing agents and their specific stimulatory action on protein synthesis.

Male Wistar rats were pair-fed on isocaloric diets containing 5%, 15%, or 40% casein, and tumorigenesis was initiated with a single dose of aflatoxin B1. Tumor initiation was achieved by 2-acetylaminofluorene (54). Under these conditions no significant differences in the total microsomal P-450 could be found but a higher rate of steroid 16-α-hydroxylation was noted. No change in the expression of oncogenes (c-rasHa, c-myc or c-fos) was observed; however, γ-glutamyl-trans-ferase and glutathione-S-transferase in precancerous foci were increased in proportion to the dietary protein. This seems to show that dietary protein has lesser influence on the drug metabolic enzymes than dietary fats.

Yet another component of nutrition, namely sodium, may have influences on cytochrome P-450 content and glutathione peroxidase activity (55). After a year's feeding of an 8% sodium chloride diet, cytochrome P-450 content decreased and glutathione peroxidase activity increased, resulting in a lowered capacity to activ-ate benzo[a]pyrene to Ames-positive mutagens. These experiments suggest that long-term excessive sodium intake could reduce the ability to detoxify xenobiotic compounds.

Finally, imidazole derivatives such as 2-amino-3,8-dimethylimidazo(4,5-f) quinoxaline (MeIQx), have been found to be promutagenic. As a result, they have been tested for their ability to induce cytochrome P-450 and were found to double the ethoxyresorufin O-deethylase activity. Furthermore, the levels of cyto-chrome P-450 1A proteins were increased, however, the corresponding mRNA level was not found to be increased. Interestingly, it also has been established that the induction mechanism does not operate through the TCDD receptor.

SUBSTRATE ACCESS TO THE ENZYME BINDING SITE

As pointed out in Figure 1, nutritional influences could be suspected in the accessibility of the binding site of cytochrome P-450 (or conjugates) for the respective substrate. Short-chain fatty acids are hydroxylated by cytochromes P-450. Therefore, it is conceivable that a nutritional increase in intracellular fatty acids might cause competition between fatty acids and substrates for the lipophilic binding site. However, this will be difficult to prove in vivo, and also one might suspect that the above-described changes in cytochrome P-450 biosynthesis might obscure the picture. There is, however, one study in isolated hepatocytes measur-ing the rates of benzo[a]pyrene metabolism in relation to the fat content of the preceding diet (57). Rats were maintained on control, high-fat or food-restricted diets; restriction meant 65% of food consumed by the control group fed ad libitum. The high-fat group had free access to a diet in which the amount of corn oil was increased four-fold at the expense of carbohydrates. The triacylglycerol content in the isolated hepatocytes varied proportionally to dietary fat and calories and was 66, 105, and 192 nmol/mg dry wt after restricted, control, and high-fat diets, respectively. The rate of benzo[a]pyrene metabolism was inversely proportional to the triacylglycerol content, showing rates of 1,324, 1,150, and 829 pmol/mg dry wt/hour at a substrate concentration of 4,040 mmol/L. At smaller substrate concentrations the difference between the extreme rates increased and became more than two-fold. The necessary control experiment showed that the specific activities of arylhydrocarbon hydroxylase or UDP-glucuronyltransferase were not affected by the diets, and consequently a more hydrophilic test compound did not show the above differences. From these studies the conclusion can be drawn that an increase in triacylglycerol in the diet can modify the access of the highly lipophilic substrate benzo[a]pyrene to the binding site of cytochrome P-450.

PROVISION OF REDUCING EQUIVALENTS FOR THE MIXED-FUNCTION OXIDASE

The operation of the cytochrome P-450 reaction cycle requires the supply of two electrons from NADPH, which is facilitated by cytochrome c (P-450) reductase. There is a vast literature on dietary influences on this enzyme, and it seems that it largely behaves similarly to cytochromes P-450 themselves. But the response of the cytochrome c (P-450) reductase to various dietary challenges shall not be further discussed here because this enzyme obviously is not a rate-limiting step in the detoxification reaction; instead it seems more interesting to assess whether the actual intracellular level of NADPH can undergo nutritional changes. There is very little information to answer this question. NADPH-producing enzymes have been investigated in an aquatic animal, the American eel, under the influence of fasting and a diet of worms or beef liver (58). The major source of reducing equivalents in these fish is isocitrate dehydrogenase (IDH). There were no differences in several NADPH-producing enzymes between eels fed the beef liver diet vs. fasting; however, the worm diet led to significantly greater glucose-6-phosphate dehydrogenase (G6PDH) activity. In rats, feed restriction caused a significant increase in the activities of NADPH-generating enzymes in the liver; the enzymes of drug metabolism also were increased by feed restriction (59). A detailed biochemical investigation into pentose-cycle intermediates and the activity of the three major cytoplasmic NADPH-producing enzymes in starved, fed ad libitum, and meal-fed rats has been carried out (60). Actual measurements of NAD (which equilibrates with NADP) under the influence of dietary manipulation are known only for a high-tryptophan diet in mice, which leads to a slight increase of NAD in the spleen (61). NADPH itself has not been measured. As pointed out before (37), there is a discrepancy between a sharply depressed mixed-function oxidase activity and an only very slight reduction in NADPH, so that a possible correlation between NADPH concentration and detoxifying activity may not be stringent. In conclusion, it does not seem that reducing equivalents are major factors in the manipulation of xenobiotic metabolism by nutrition.

AVAILABILITY OF CONJUGATING GROUPS

There is very little information about dietary sulfate deficiency and the consequent lack of sulfate conjugation; as long as sulfate conjugation is not overburdened by a respective substrate (in which case glucuronidation will substitute), there will be sufficient sulfate available from the breakdown of cysteine and methionine. For similar reasons there will be no lack of dietary glucose-1-phosphate, the principal source of UDP-glucuronic acid, as the activated conjugate for glucuronidation. The same seems to be true for acetyl-CoA as the activated conjugate for acetylation. In contrast, liver glutathione content is variable (62,63). First, it undergoes a diurnal variation with a maximum of 62 nmol/mg protein at 6–10 a.m. in the morning and a minimum of 42 nmol/mg protein 12 hours later, as measured in mice. Second, this level can be decreased to about 20 nmol/mg protein by starvation for 24 hours or more. Consequently animals show a nutrition-dependent susceptibility to compounds that are preferentially conjugated to glutathione (e.g. paracetamol). On the other hand, administration of rather large doses of the antioxidants butylated hydroxyanisole (BHA) and hydroxytoluene (BHT) for 5 days increases the content of reduced glutathione by about 50% to 100% in liver. Thus, we have an example of a manipulatable intracellular protective principle. One interesting point must be raised here, namely nutritional danger from overeating on a diet that induces obesity (64). Reduced and oxidized glutathione decrease to half of the original value after feeding rats and mice a diet that induces obesity. In conclusion, it can be said that activated conjugating ligands are normally sufficiently available for phase II detoxication, except for glutathione.

DIETARY MANIPULATION OF TRANSFERASES

There is extensive literature on the induction of glucuronyltransferase and glutathione-S-transferase by a great number of chemicals. However, there is very sparse literature on nutritional effects on these enzymes. Feeding, starvation, and refeeding influence the ability to glucuronidate xenobiotics as measured in the isolated perfused rat liver (65). p-Nitrophenol glucuronide in the efflux of perfused liver of 24-hour-fasted rats indicates a glucuronidation capacity of less than 2 μmol/g/hour; this value is 5 μmol/g/hour in fed rats and does not correlate directly with the 10-fold change in glycogen content between fed and starving rats. Animals re-fed after a starving period show something like an "overshoot phenomenon" and have an even higher glucuronidation capacity. This finding conflicts somewhat with other observations that food restriction enhances enzyme activity, but here the lability of the enzyme in the membrane and its influence on its activity become apparent.

With respect to nutritional influences a 20% corn oil diet increases one form of glutathione-S-transferase; namely GST-B, but not that of GST-A (49,50). With a low-protein diet there is an increase in both GST-C and in glucuronyl transferase, mainly in the intestine (66). A curious result was obtained when feeding rats a magnesium-deficient diet for 10 days and measuring glucuronide formation in isolated hepatocytes (67). There was a decrease by about 50%. Also, in humans a similar phenomenon was observed (68) by measuring a 30% increase in urinary recovery of oxazepam glucuronide when changing from a high-protein/low-carbohydrate diet to a low-protein/high-carbohydrate diet. In conclusion, it appears transferases are influenced in a way similar to P-450 in terms of activation by either food restriction or fat-containing diets.

TRANSCRIPTIONAL INFLUENCES

Ultimately, all data and observations about induction of cytochromes P-450 (or transferases) must be classified under a heading concerned with gene expression in the widest sense. In Figure 3, for example, interference with the level of mRNA by food components is discussed. As a *pars pro toto* example, the induction by vegetable components such as indole-3-carbinol, that occur in Brussels sprouts, cabbage, and other vegetables will be briefly discussed. After administration of an oral dose of this compound to rats the mRNA for cytochrome P-450 1A1 was elevated both in the colon and liver (69). This process requires several hours and involves the binding of the inducer to the Ah receptor. The possible biologic consequences of this interference are protection from carcinogens by more rapid inactivation or, equally possible, increased activation of suitable substrates. Currently information on the food-induced variation of mRNA levels is rapidly increasing. In conclusion, elevation of cytochromes P-450 mRNA levels is an important mechanism by which food ingredients stimulate xenobiotic metabolism.

POSTTRANSCRIPTIONAL INFLUENCES

Decreased elimination of mRNA may also be responsible for induction of cytochromes P-450. In one case, such posttranscriptional events are hypothesized for induction by pyrazole and others (70). Posttranscriptional stabilization of mRNA may be a consequence of interaction with food components.

AMINO ACID SUPPLY

In principle, undernourishment with proteins could lead to an insufficient supply of amino acids, in other words, starvation. So far it seems that neither starvation nor food restriction impairs the biosynthesis of cytochromes P-450. Hence this discussion will be fairly brief.

A low-protein diet of 6% casein has been described (71) as inducing very marked changes in the amino acid composition of purified forms of rat liver cytochrome P-450. Valine, isoleucine, and phenylalanine were significantly increased, while glutamine and tyrosine were decreased. Cytochrome P-450 UT50 from the control group contained 40% hydrophobic, 31% polar, 14% negatively charged, and 15% positively charged amino acids. In the same cytochrome from the low-protein group the level of hydrophobic amino acids increased from 40 to 48% and that of the polar amino acids decreased from 31 to 26% while the total number of amino acids increased from 435 to 439. These results show that manipulation of the protein intake causes changes in the amino acid composition of a biosynthetic apoprotein of the cytochrome P-450 family. This may lead to changes in, for example, substrate binding or catalytic activity, as one may assume that the hydrophobic amino acid region in particular is involved in the interaction of the enzyme with the lipid environment of the membrane as well as with substrate binding. In conclusion, we have to recognize that protein restriction leads to imbalances in the amino acid composition of a cytochrome P-450; to achieve this, it is not even necessary to feed the animals an artificial diet having an imbalanced amino acid composition.

NUTRITIONAL INFLUENCE ON ENDOPLASMATIC MEMBRANES

It is known that attachment to endoplasmic lipoid membranes to a large extent determines the catalytic activity of phase I and phase II xenobiotic enzymes. This again implicitly emphasizes the role of dietary lipids in determining the physical characteristics of membranes. The effects of dietary fat on drug metabolism have been reviewed (72). Generally speaking, there seems to be a negative correlation between supply of fats, in particular of unsaturated oils, and hepatic xenobiotic metabolism rates. Hyper-alimentation with a pure amino acid-glucose solution leads to an increase in fluorescence anisotropy and significant decreases in detoxification rates (73). Other authors have investigated the relationship among lipid composition, physical properties of microsomal phospholipids, and the kinetics of liver UDP-glucuronyl transferase. Fluorescent probes show changes with fat-free diets, particularly with regard to the gel–liquid–crystalline modifications as measured by the temperature dependency of phase separation and Arrhenius plots of the glucuronyl transferase (74).

Unsaturated essential fatty acids are necessary to maintain membrane function, and it appears that high intake of polyunsaturated fatty acids stimulates the liver microsomal mixed-function oxidase (75,76). Menhaden oil from a herring type fish increases hepatic cytochrome P-450 but also increases catalytic capacity and the affinities for certain substrates; however, this does not apply to every form of cytochrome P-450 (77,78). In the meantime, the effects of fat in the diet have been further investigated in more detail (79–81). Briefly, lipids affect the inducibility of P-450 2B while increasing the constitutive level of the ethanol inducible P-450 2E. Also, there are distinctive effects on glutathione-S-transferases. In particular, Menhaden oil produces higher P-450 2El activity, which then leads to clearly enhanced metabolic rates. The level of constitutive hepatic cytochrome P-450 is differently influenced by dietary lipid. For example, P-450 1A2, 2B2, 2E1, and 3A are increased but pulmonary P-450 2B1 is decreased in rats given 20% corn oil in the diet for 4 days.

In conclusion, dietary lipids have a profound effect on the enzymatic instruments of xenobiotic detoxication, but unfortunately, we do not yet understand the exact mechanisms and therefore cannot predict which particular apoenzyme will be affected.

PLANT CONSTITUENTS – DIETARY INHIBITORS

After it became known that cruciferous vegetables contain inducing substances, a body of information began to accumulate on the biologic effects of these often very unusual structures that occur in low concentrations in food. Three examples will be given here to illustrate this currently expanding field.

Garlic oil has long been considered both a curse and boon to mankind, the latter more or less superstitiously derived from undocumented experiences. This has now changed, because the main constituent, diallyl sulfide (DAS), is found to inhibit cytochrome P-450-catalyzed oxidations. Earlier, a hypolipidemic effect of garlic oil had been ascribed to an activation of lipase and a deprivation of NADPH (82). DAS, which is itself oxidized, competitively inhibits cytochrome P-450 2El and also inactivates this cytochrome by a suicide mechanism of DAS-O_2(83). This inhibitory action minimizes the metabolic activation of the carcinogen 4-(methylnitrosamino)-1-(3-pyridyl)-l-butanone (NNK) in mice (84). Similarly, extracts from tea reduce tumor incidence, and this has been demonstrated to be due to an inhibitory action of green and black teas on nitrosamine-induced tumorigenesis (85). In addition, DAS affects the level of cytochromes P-450 by elevating the mRNA for P-450 2B1 (86), and this seems to show that DAS has two different biologic actions. It may be noted that another sulfur-containing compound, disulfiram, is also an inhibitor and inactivator of P-450 2E1 (87), which is in line with an earlier observation (88) that disulfiram inhibits aminopyrine demethylation in vivo and in vitro.

A second example is the action of a flavonoid from citrus fruits, naringin, the metabolism of which was studied by Booth *et al.* in 1958 (89). Naringenin, the aglycone of naringin, is readily formed in humans who have consumed grapefruit juice. This compound inhibits the oxidative metabolism of drugs (21) by inhibiting cytochrome P-450 3A4. Since this enzyme activates aflatoxin B1 to its ultimate carcinogen, we here have an example wherein a plant constituent inhibits the detrimental biologic action of a food contaminant (90).

This interaction of a constituent of grapefruit juice has been investigated for many diverse drugs; the calcium channel antagonist felodipine served as the first example. The interpretation of the interaction is more complex than a straightforward inhibition of the cytochrome P-450 3A4, (which seems to occur in the gut wall) but not of the hepatic monooxygenase; therefore, a more diverse mechanism of action will have to be envisaged, as pointed out by Bailey *et al*. (91). Besides a number of studies on dihydropyridine-type calcium channel blockers, about a dozen other non-related drugs have been shown to be altered in their pharmacokinetics (92), and this list is steadily increasing. A detailed study has shown that the CYP 3A4-mediated 7-hydroxylation of coumarin in a human is not inhibited by naringin when given in water instead of grapefruit juice, as measured by the urinary excretion of 7-hydroxcoumarin, while various amounts of naringin added to grapefruit juice show a dose-dependent inhibition. Thus, the inhibitory potency of grapefruit juice can be amplified by naringin. This experiment by Runkel *et al*. (93) shows that there is a clear synergism between grapefruit juice (or one of its components) and naringin, since grapefruit juice is necessary to elicit the inhibitory potency of naringin in vivo. It may be noted that the urinary excretion of naringin is greatly enhanced by grapefruit juice and occurs about 10 hours earlier (after 2 hours) compared to naringin in water. These studies by Legrum and coworkers (93) point to a more complicated mechanism and may also explain why naringin and orange juice are not effective. Current considerations also involve an inhibitory action of naringin or its hydrolysis product naringenin on the enterocytotic reverse transport of drugs into the intestinal lumen which may be facilitated by a phosphoglycoproein, similar to cellular multi drug resistance.

Another field of nutritional influence on drug metabolism has been opened by Henderson *et al*. (94), who have demonstrated very efficient in vitro inhibition of human P-450 enzymes by a prenylated flavonoid from hops. Hops contains a considerable number of flavonoids, among which xanthohumol seems to be the most efficient one in inhibiting 7-ethoxyresorufin deethylase activity of cDNA-expressed human P-450 CYP 1B1 at low micromolar concentrations. The inhibition is competitive in nature. The authors have investigated many more hops constituents with similar results. Of course, it remains to be verified whether the amount of ingested hops-related inhibitors may also be sufficient to exert in vivo effects in humans. As the metabolic activation of the carcinogen AFB1 in vitro also is inhibited by isoxanthohumol and 8-prenylnaringenin, the prevention of AFB1 induced human tumors might be possible by the therapeutic and at the same time pleasant regular consumption of beer.

CONCLUSIONS

The foregoing considerations explain the basic principles by which food components can interact with biologic functions in mammals. The cytochromes P-450, by no means, react uniformly to the biochemical challenge exerted by nutritional influences. This is a reflection of the fact that the cytochromes P-450 represent a very diverse family of enzymes. Furthermore, food deprivation may positively affect the ability to chemically convert toxicants. Also, the important role of lipids in the diet includes their positive effect on constitutive mixed-function oxidases as well as the induction of these enzymes by various substances by increasing their biosynthesis.

Attention has been called to the essential role of endoplasmic membranes as a modifying matrix for oxidative as well as conjugative enzymes. Finally, a few examples have been given to introduce the expanding field that deals with the effects of dietary plant constituents.

REFERENCES

1. Benet LZ, Mitchell JR, Sheiner LB. Pharmacokinetics, biotransformation of drugs. In: Gilman AG, Rall TW, Nies AS, Taylor P, eds. *Goodman and Gilman's the pharmacological basis of therapeutics*. New York, Oxford, Beijing, Frankfurt, Sao Paulo, Sydney, Tokyo, Toronto: Pergamon Press;1990; 13–20.
2. Waterman MR, Johnson EF, eds. *Cytochrome P-450*. New York: Academic Press; 1991 (*Methods in Enzymology; vol 206*).
3. Nebert DW, Nelson DR, Coon MJ, Estabrook RW, Feyereisen R, Fujii-Kuriyama Y, Gonzalez FJ, Guengerich FP, Gunsalus IC, Johnson EF, Loper JC, Sato R, Waterman MR., Waxman DJ. The P-450 superfamily: update on new sequences, gene mapping, and recommended nomenclature. *DNA Cell Biol* 1991; 10:1–14.
4. Whitlock JP Jr. The regulation of cytochrome P-450 gene expression. *Annu Rev Pharmacol Toxicol* 1986; 26:333–69.
5. Coon MJ, Ding X, Pernecky SJ, Vaz ADN. Cytochrome P-450: progress and predictions. *FASEB J* 1992; 6:669–73.
6. Gonzalez FJ. The molecular biology of cytochrome P-450s. *Pharmacol Rev* 1988; 40323–88.
7. Nebert DW, Gonzalez FJ. P-450 genes: structure, evolution, and regulation. *Annu Rev Biochem* 1987; 56:945–93.
8. Gonzalez FJ. Human cytochromes P-450: problems and prospects. *Trends Pharmacol Sci* 1992; 13:346–52.
9. Ioannides C. Nutritional modulation of cytochromes P450. In: Ioannides C, ed. *Nutritional and chemical toxicity*. Chichester: John Wiley & Sons; 1998; 115–59 (227 references).
10. Burchell B, Coughtrie MW. UDP-glucuronosyl transferases. *Pharmacol Ther* 1989; 43:261–89.
11. Tephly TR, Burchell B. UDP-glucuronosyl transferases: a family of detoxifying enzymes. *Trends Pharmacol Sci* 1990; 11:276–9.
12. Mulder GJ, Jakoby WB. Sulfation. In: Mulder GJ, ed. *Conjugation reactions in drug metabolism*. London, New York, Philadelphia: Taylor and Francis; 1990; 107–61.
13. Falany CN. Molecular enzymology of human liver cytosolic sulfotransferases. *Trends Pharmacol Sci* 1991; 12:255–9.

14. Netter KJ. Acetylation. In: Gorrod JW, Oelschläger H, Caldwell J, eds. *Metabolism of xenobiotics*. London, New York, Philadelphia: Taylor and Francis; 1988; 179–88.

15. Jakoby WB, Habig WH. Glutathione transferases. In: Jakoby WB, ed. *Enzymatic basis of detoxicazion, II*. New York, London: Academic Press; 1980; 63–94.

16. Pickett CB, Lu AY. Glutathione-S-transferases: Gene structure, regulation and biological function. *Annu Rev Biochem* 1989; 58:743–64.

17. Hahnemann B, Kuhn B, Heubel F, Legrum W. Selective induction of coumarin hydroxylase by N-containing heteroaromatic compounds. *Arch Toxicol Suppl* 1989; 13:297–301.

18. Legrum W, Frahseck J. Acceleration of 7-(^{14}C-methoxy)-coumarin-derived carbon dioxide exhalation by cobalt pretreatment in mice. *J Pharmacol Exp Ther* 1982; 221:790–4.

19. Netter KJ. Inhibition of oxidative drug metabolism in microsomes. *Pharmacol Ther* 1980; 10:515–35.

20. Testa B, Jenner P. Inhibitors of cytochrome P-450s and their mechanism of action. *Drug Metab Rev* 1981; 12:1–117.

21. Soons PA, Vogels BAPM, Roosemalen MCM, Shoemaker HC, Uchida E, Edgar B, Lundahl J, Cohen AF, Breimer DD. Grapefruit juice and cimetidine inhibit stereoselective metabolism of nitrendipine in humans. *Clin Pharmacol Ther* 1991; 50:394–403.

22. Campbell TC, Hayes JR. Role of nutrition in the drug metabolizing enzyme system. *Pharmacol Rev* 1974; 26:171–97.

23. Parke DV, Ioannides C. The role of nutrition in toxicology. *Annu Rev Nutr* 1981; 1:207–34.

24. Hathcock JN, Coon J, eds. *Nutrition and drug interrelations*. New York: Academic Press; 1978.

25. Hathcock JN, ed. Symposium. *Metabolic interactions of nutrition and drugs. Fed Proc* 1985; 44:123–4.

26. Netter KJ. Pharmacodietetics: perspectives in the nutritional manipulation of toxicity or drug effects. In: Olive G, ed. *Drug-action modifications, comparative pharmacology*. Oxford: Pergamon Press; 1979: 65–8 (*Advances in pharmacology and therapeutics: vol 8*).

27. Netter KJ. Toxicodietetics: dietary alterations of toxic action. In: Chambers PL, Gehring P, Sakal F, eds. *New concepts and developements in toxicology*. Amsterdam: Elsevier; 1986; 139–44.

28. Netter KJ. Nutrition and chemical toxicity. In: Parke DV, Ioannides C, Walker R, eds. *Food, nutrition and chemical toxicity*. London: Smith-Gordon; 1993; 17–26.

29. Rogers AE, Nields HM, Newberne PM. Nutritional and dietary influences on liver tumorigenesis in mice and rats. *Arch Toxicol, Suppl* 1987; 10:231–43.

30. Guengerich FP. Effects of nutritive factors on metabolic processes involving bioactivation and detoxication of chemicals. *Annu Rev Nutr* 1984; 4:207–31.

31. Bidlack WR, Brown RC, Mohan C. Nutritional parameters that alter hepatic drug metabolism, conjugation, and toxicity. *Fed Proc* 1986; 45:142–8.

32. Yang CS, Yoo J-SH. Dietary effects on drug metabolism by the mixed-function oxidase system. *Pharmacol Ther* 1988; 38:53–72.

33. Parke DV. Nutritional requirements for detoxication of environmental chemicals. *Food Addit Contam* 1991; 8:381–96.

34. Anderson KE, Kappas A. Dietary regulation of cytochrome P-450. *Annu R Nutr* 1991; 11:141–67.

35. Yang CS, Brady JF, Hong JY. Dietary effects on cytochromes P-450, xenobiotic metabolism, and toxicity. *FASEB J* 1992; 6:737–44.
36. Report of a WHO Study Group. Diet, nutrition, and the prevention of chronic diseases. *WHO Technical Report Series* 1990; 797.
37. Brüch M, Quintanilla ME, Legrum W, Koch J, Netter KJ, Fuhrmann GF. Effects of vanadate on intracellular reduction equivalents in mouse liver and the fate of vanadium in plasma, erythrocytes and liver. *Toxicology* 1984; 31:283–95.
38. Brüch M, Dietrich A, Netter KJ. A qualitative study on vanadate effects in the tetrahydofolate dependent formate transfer in vitro and in vivo in mice. *Naunyn-Schmiedeberg's Arch Pharmacol* 1987; 336:111–16.
39. Benevenga NJ, Steele RD. Adverse effects of excessive consumption of amino acids. *Annu Rev Nutr* 1984; 4:157–81.
40. Strubelt O, Dost-Kempf E, Siegers CP, Younes M, Völpel M, Preuss U, Dreckmann JG. The influence of fasting on the susceptibility of mice to hepatotoxic injury. *Toxicol Appl Pharmacol* 1981; 60:66–77.
41. Pessayre D, Dolder A, Artigon JY, Wandscheer JC, Descatoire V, Degott C, Benhamon JP. Effect of fasting on metabolite-mediated hepatotoxicity in the rat. *Gastroenterology* 1979; 77:264–71.
42. Hong JY, Pan JM, Gonzalez FJ, Gelboin HV, Yang CS. The induction of a specific form of cytochrome P-450 (P-450) by fasting. *Biochem Biophys Res Commun* 1987; 142(3):1077–83.
43. Imaoka S, Terano Y, Funae Y. Changes in the amount of cytochrome P-450s in rat hepatic microsomes with starvation. *Arch Biochem Biophys* 1990; 278:168–78.
44. Gil L, Orellana M, Vasquez H, Silva M. Nutrition-related alterations in liver microsomal testosterone hydroxylases. *Int J Androl* 1988; 11:339–48.
45. Ma Q, Dannan GA, Guengerich FP, Yang CS. Similarities and differences in the regulation of hepatic cytochrome P-450 enzymes by diabetes and fasting in male rats. *Biochem Pharmacol* 1989; 38:3179–84.
46. Knodell RG, Wood DG, Guengerich FP. Selective alteration of constitutive hepatic cytochrome P-450 enzymes in the rat during parenteral hyperalimentation. *Biochem Pharmacol* 1989; 38:3341–5.
47. Pohjanvirta R, Tuomisto J. 2,3,7,8-Tetrachlorodibenzo-*p*-dioxin enhances responsiveness to post-ingestive satiety signals. *Toxicology* 1990; 63:285–99.
48. Yoo JS, Ning SM, Pantuck CB, Pantuck EJ, Yang CS. Regulation of hepatic microsomal cytochrome P-45011E1 level by dietary lipids and carbohydrates in rats. *J Nutr* 1991; 121:959–65.
49. Yoo JS, Hong JY, Ning SM, Yang CS. Roles of dietary corn oil in the regulation of cytochromes P-450 and glutathion S-transferases in rat liver. *J Nutr* 1990; 120:1718–26.
50. Kwei GY, Zalesk J, Thurman RG, Kauffman FC. Enzyme activities associated with carcinogen metabolism in liver and nonhepatic tissues of rats maintained on high fat and food-restricted diets. *J Nutr* 1991; 121:131–7.
51. Marshall WJ, McLean AEM. A requirement for dietary lipids for induction of cytochrome P-450 by phenobarbitone in rat liver microsomal fraction. *Biochem J* 1971; 122:569–73.
52. Lambert L, Wills ED. The effect of dietary lipid peroxides, sterols and oxidised sterols on cytochrome P-450 and oxidative demethylation in the endoplasmatic reticulum. *Biochem Pharmacol* 1977; 26:1417–21.

53. Kim HJ, Choi ES, Wade AE. Effect of dietary fat on the induction of hepatic microsomal cytochrome P-450 isozymes by phenobarbital. *Biochem Pharmacol* 1990; 39: 1423–30.

54. Blanck A, Lindhe B, Porsch-Hallstrom I, Lindeskog P, Gustafsson JA. Influence of different levels of dietary casein on initiation of male rat liver carcinogenesis with a single dose of aflatoxin B1. *Carcinogenesis* 1992; 13:171–6.

55. Horiguchi M, Iwama M, Takahashi N, Iitoi Y, Sakamoto Y, Kanke Y. Long-term exposure of Wistar rats to high dietary sodium chloride level. II. Changes in hepatic drug-metabolizing and glutathione-related enzyme systems. *Int J Vitam Nutr Res* 1992; 62:54–9.

56. Kleman M, Overvik E, Mason G, Gustafsson JA. Effects of the food mutagens MeIQx and PhIP on the expression of cytochrome P-450IA proteins in various tissues of male and female rats. *Carcinogenesis* 1990; 11:2185–9.

57. Zaleski J, Kwei GY, Thurman RG, Kauffman FC. Suppression of benzo(a)pyrene metabolism by accumulation of triacylglycerols in rat hepatocytes: effect of high-fat and food-restricted diets. *Carcinogenesis* 1991; 12:2073–9.

58. Aster PL, Moon TW. Influence of fasting and diet on lipogenic enzymes in the american eel, *Anguilla rostrata LeSueur. J Nutr* 1981; 111:346–54.

59. Sachan DS, Das SK. Alterations of NADPH-generating and drug-metabolizing enzymes by feed restriction in male rats. *J Nutr* 1982; 112:2301–6.

60. Casazza JP, Veech RL. The content of pentose-cycle intermediates in liver in starved, fed ad libitum and meal-fed rats. *Biochem J* 1986; 236:635–41.

61. Inoue K, Sanada H, Wada O, Aoki N. Effects of dietary pyrazinamide, tryptophan, or nicotinic acid and gamma-ray irradiation on levels of NAD and NADP in various organs of mice. *J Nutr Sci Vitaminol Tokyo* 1989; 35:383–91.

62. Jaeschke H, Wendel A. Diurnal fluctuation and pharmacological alteration of mouse organ glutathione content. *Biochem Pharmacol* 1985; 34:1029–33.

63. Jaeschke H, Wendel A. Manipulation of mouse organ glutathione contents. I: Enhancement by oral administration of butylated hydroxyanisole and butylated hydroxytoluene. *Toxicology* 1985; 36:77–85.

64. Sastre J, Pallardo FV, Llopis J, Furukawa T, Vina JR, Vina J. Glutahione depletion by hyperphagia-induced obesity. *Life Sci* 1989; 45:183–7.

65. Reinke LA, Belinsky SA, Evans RK, Kauffman FC, Thurman RG. Conjugation of *p*-nitrophenol in the perfused rat liver: the effect of substrate concentration and carbohydrate reserves. *J Pharmacol Exp Ther* 1981; 217:863–70.

66. Catania VA, Carrillo MC. Intestinal phase II detoxification systems: effect of low-protein diet in weanling rats. *Toxicol Lett* 1990; 54:263–70.

67. Brown RC, Bidlack WR. Dietary magnesium depletion: p-nitroanisole metabolism and glucuronidation in rat hepatocytes and hepatic microsomal membranes. *Proc Soc Exp Biol Med* 1991; 197:85–90.

68. Pantuck EJ, Pantuck CB, Kappas A, Conney AH, Anderson KE. Effects of protein and carbohydrate content of diet on drug conjugation. *Clin Pharmacol Ther* 1991; 50:254–8.

69. Vang O, Jensen MB, Autrup H. Induction of cytochrome P4501A1 in rat colon and liver by indole-3-carbinol and 5,6-benzoflavone. *Carcinogenesis* 1990; 11:1259–63.

70. Song BJ, Gelboin HV, Park SS, Yang CS, Gonzalez FJ. Complementary DNA and protein sequences of ethanol-inducible rat and human rat enzyme. *J Biol Chem* 1986; 261:16689–97.

71. Amelizad Z, Narbonne JF, Wolf CR, Robertson LW, Oesch F. Effect of nutritional imbalances on cytochrome P-450 isozymes in rat liver. *Biochem Pharmacol* 1988; 37:3245–9.

72. Wade AE. Effects of dietary fat on drug metabolism. *J Environ Pathol Toxicol Oncol* 1986; 6:161–89.

73. Knodell RG, Whitmer DI, Holman RT. Influence of hyperalimentation on rat hepatic microsomal fluidity and function. *Gastroenterology* 1990; 98:1320–5.

74. Castuma CE, Brenner RR. The influence of fatty acid unsaturation and physical properties of microsomal membrane phospholipids on UDP-glucuronyl transferase activity. *Biochem J* 1989; 258:723–31.

75. Christon R, Fernandez Y, Cambon-Gros C, Periquet A, Deltour P, Leger CL, Mitjavila S. The effect of dietary essential fatty acid deficiency on the composition and properties of the liver microsomal membrane of rats. *J Nutr* 1988; 118:1311–18.

76. Saito M, Oh-Hashi A, Kubota M, Nishide E, Yamaguchi M. Mixed function oxidases in response to different types of dietary lipids in rats. *Br J Nutr* 1990; 63:249–57.

77. Wade AE, Bellows J, Dharwadkar S. Influence of dietary menhaden oil on the enzymes metabolizing drugs and carcinogens. *Drug Nutr interact* 1986; 4:339–47.

78. Dharwadkar SM, Bellows JT, Ramanathan R, Wade AE. In vitro activation and resultant binding of benzo(a)pyrene to DNA by microsomes from rats fed corn and menhaden oils. *Pharmacology* 1986; 33:190–8.

79. Yoo JSH, Hong JY, Ning SM, Yang CS. Roles of dietary corn oil in the regulation of cytochromes P-450 and glutathione-S-transferases in rat liver. *J Nutr* 1990; 120: 1718–26.

80. Yoo JSH, Ming SM, Pantuck CB, Pantuck EJ, Yang CS. Regulation of hepatic microsomal cytochrome P-450IIE1 level by dietary lipids and carbohydrates in rats. *J Nutr* 1991; 121:959–65.

81. Yoo JSH, Smith TJ, Ming SM, Lee Mi, Thomas PE, Yang CS. Modulation of the levels of cytochromes P-450 in rat liver and lung by dietary lipid. *Biochem Pharmacol* 1992; 43:2535–42.

82. Adoga GI. The mechanism of the hypolipidemic effect of garlic oil extract in rats fed on high sucrose and alcohol diets. *Biochem Biophys Res Commun* 1987; 142:1046–52.

83. Brady JF, Ishizaki H, Fukuto JM, Lin MC, Fadel A, Gapac JM, Yang CS. Inhibition of cytochrome P-450 2E1 by diallyl sulfide and its metabolites. *Chem Res Toxicol* 1991; 4:642–7.

84. Hong JY, Wang ZY, Smith TJ, Zhou S, Shi S, Pan J, Yang CS. Inhibitory effects of diallyl sulfide on the metabolism and tumorigenicity of the tobacco-specific carcinogen 4-(methylnitrosamino)-1-(3-pyridyl)-1-butanone (NNK) in A/J mouse lung. *Carcinogenesis* 1992; 13:901–4.

85. Wang ZY, Hong JY, Huang MT, Reuhi KR, Conney AH, Yang CS. Inhibition of N-nitrosodiethylamine- and 4-(methylnitrosamino)-1-(3-pyridyl)-1-butanone-induced tumorigenesis in A/J mice by green tea and black tea. *Cancer Res* 1992; 52:1943–7.

86. Brady JF, Wang MH, Hong JY, Xiao F, Li Y, Yoo JS, Ning SM, Lee MJ, Fukuto JM, Gapac JM, *et al*. Modulation of rat hepatic microsomal monooxygenase enzymes and cytotoxicity by diallyl sulfide. *Toxicol Appl Pharmacol* 1991; 108:342–54.

87. Brady JF, Xiao F, Wang MH, Li Y, Ming SM, Gapac JM, Yang CS. Effects of disulfiram on hepatic P-450IIE1, other microsomal enzymes and hepatotoxicity in rats. *Toxicol Appl Pharmacol* 1991; 108:366–73.

88. Honjo T, Netter KJ. Inhibition of drug demethylation by disulfiram in vivo and in vitro. *Biochem Pharmacol* 1969; 18:2681–3.

89. Booth AN, Jones FL, De Eds F. Metabolic and glucosuria studies on naringin and phloridzin. *J Biol Chem* 1958; 233:280–2.
90. Guengerich FP, Kim DH. In vitro inhibition of dihydropyridine oxidation and afla-toxin B1 activation in human liver microsomes by naringenin and other flavonoids. *Carcinogenesis* 1990; 11:2275–9.
91. Bailey DG, Malcolm J, Arnold O, Spence JD. Grapefruit juice–drug interactions. *Brit J Clin Pharmacol* 1998; 46:101–10.
92. Wunderer H. Wechselwirkungen: Nicht jeder Arzneistoff verträgt Grapefruitsaft. *Pharmazeut Zeit* 1998; 143:11–22.
93. Runkel M, Bourian M, Tegtmeier M, Legrum W. The character of inhibition of the metabolism of 1,2-benzopyrone (coumarin) by grapefruit juice in a human. *Eur J Clin Pharmacol* 1997; 53:265–9.
94. Henderson MC, Miranda CL, Stevens JF, Deinzer ML, Buhler DR. In vitro inhibition of human P450 enzymes by prenylated flavonoids from hops *Humulus lupulus*. *Xenobiotica* 2000; 30:235–51.

4

FOODS DERIVED FROM GENETICALLY MODIFIED CROP PLANTS

Roy L. Fuchs

INTRODUCTION

Food producing plants have been continuously genetically modified and improved for centuries. Initially these improvements were achieved by selecting seed from superior plants and reproducing them with continual selection and breeding. Traditional breeding methods have resulted in significant increases in productivity, with corn and wheat yields approximately doubling over the past 40–50 years. More recently, the ability to introduce DNA directly into crop plants has enabled the very specific and selective genetic modification of plants through techniques commonly referred to as plant biotechnology, genetic engineering or recombinant DNA methodology. Numerous traits are being assessed for their potential to yield products with the ability to: (a) protect plants against insect damage, fungal, viral or bacterial diseases; (b) provide selectivity to preferred herbicides; (c) directly enhance crop yields; (d) increase nutritional value; (e) reduce naturally occurring toxicants or allergens; (f) modify the ripening process and provide fruits or vegetables with superior flavor; (g) use plants as factories to make pharmaceutical products or to produce foods containing human vaccines; and many others.

The rapidly growing world population, the loss of top soil due to erosion and the loss of tropical forest, grasslands and other sources of biodiversity are all major and urgent challenges of tremendous global importance. For example, combining the projected doubling of the global population with a simultaneous increased demand for higher protein-containing foods results in tripling the amount of food that will be needed in the next 40–50 years relative to current needs. These numbers reflect the fact that even today an estimated 800 million people, including 200 million children, are chronically undernourished in the developing world and millions suffer from diseases associated with micronutrient deficiencies. To put this challenge into perspective, the world will have to produce more food in the next 40–50 years than has been produced in the history of our civilization.

Plant biotechnology provides a very important tool to help address many of these challenges more directly and effectively than was previously possible. Biotechnology and other technological innovations must be effectively integrated with the best current agricultural practices. It is only with effective global cooperation and utilization of the most productive and environmentally appropriate technologies that agricultural systems can meet these dramatic challenges.

This chapter will provide a brief overview of the methods used to genetically modify food crops, will describe some of the current products that have either been commercialized or are under development, and will provide a brief overview of the regulatory status and approaches used to assure the food, feed and environmental safety of these products.

PLANT TRANSFORMATION

The first genetically modified plants were produced in 1983 using the bacterium *Agrobacterium tumefaciens*. In nature, *A. tumefaciens* infects a plant and transfers a portion of it's DNA (called T-DNA for transfer DNA) into the infected plant, which results in the production of a localized disease on the recipient plant. To take advantage of these transfer capabilities, scientists deleted the disease-causing sequences from the *Agrobacterium* T-DNA and inserted, in its place, DNA that encodes for a specific trait of interest (e.g. insect protection). Appropriate regulatory signals were added to enable the gene to function optimally in the plant. Typically the engineering of the genes and regulatory signals is performed with autonomously replicating plasmid DNA in a laboratory strain of *Escherichia coli*. These plasmids replicate in both *E. coli* and *A. tumefaciens* so that the engineered plasmids can be transferred from *E. coli* to *A. tumefaciens* by a conjugation process. These plasmids typically contain two specific sequences, referred to as border sequences, that delineate the DNA that ultimately should be inserted in the plant genome. The desired gene(s) along with the regulatory sequences are contained within these border sequences. *A. tumefaciens* containing the engineered plasmid is incubated with appropriate tissue from the host plant under conditions which enable the gene(s) of interest to be transferred into the selected host plant. The DNA is stably inserted into the chromosome of the host plant. The inserted genes can then be transferred to new plant varieties using traditional breeding methods.

This system worked well initially with dicotyledonous plants (broad leaf plants like tomato, tobacco, cotton, soybean, etc.). Using more aggressive strains of *A. tumefaciens* with broader host ranges, this system has recently been expanded to function in the formerly recalcitrant cereal crops (e.g. corn, wheat, rice, etc.). In addition, techniques such as protoplast transformation and particle bombardment have been used to directly transfer DNA into cells and stably insert the DNA into the plant genome. These techniques have been particularly valuable in plants for which the *Agrobacterium* transformation method was not effective or efficient.

76

Table 1 Plant biotechnology products that have regulatory approval in at least one country

Company	Trait
AgrEvo	Phosphinothricin tolerant canola, corn, soybean, beet, rice
	Male sterile/phosphinothricin tolerant rapeseed, corn
	Phosphinothricin tolerant/lepidopteran resistant corn
	Fertility restorer/phosphinothricin tolerant rapeseed, corn
Agritope, Inc.	Modified fruit ripening tomato
Asgrow Seed Co.	CMV/WMV2/ZYMV resistant squash
BASF	Sethoxydim herbicide tolerant corn
Bejo-Baden	Male sterility chicory
Calgene Inc.	Fruit ripening tomato
	Bromoxynil tolerant cotton
	Bromoxynil tolerant/lepidopteran resistant cotton
	Laurate canola
China	Virus resistant tomato, tobacco
Ciba-Giegy	Lepidopteran resistant corn
Cornell U. /U. of Hawaii	PRSV virus resistant papaya
DeKalb Genetics Corp.	Phosphinothricin tolerant corn
	European corn borer resistant corn
	European corn borer resistant/phosphinothricin tolerant corn
DNA Plant Technology	Improved ripening tomato
DuPont	Sulfonylurea tolerant cotton
	High oleic acid soybean
Florigene	Carnations with increased vase life
	Carnations with modified flower color
Monsanto	Glyphosate tolerant soybean, cotton, canola, corn
	Improved ripening tomato
	Colorado potato beetle resistant potato
	Lepidopteran resistant cotton
	European corn borer resistant corn
	European corn borer resistant/glyphosate tolerant corn
	Colorado potato beetle resistant/PVY resistant potato
	Colorado potato beetle resistant/PLRV resistant potato
Mycogen	European corn borer resistant corn
Novartis	European corn borer resistant corn
	European corn borer resistant/phosphinothricin tolerant corn, sweet corn
Northrup King	European corn borer resistant corn
Rhone-Poulenc	Bromoxynil tolerant canola
Pioneer Hybrid	Male sterile/phosphinothricin tolerant corn
Seita	Bromoxynil tolerant tobacco
University of Saskatchewan	Sulfonylurea tolerant flax
Upjohn	WMV2/ZYMV virus resistant squash
Zeneca/Petoseed	Improved ripening tomato

Today, over a decade and a half since the initial reports of the first genetically modified plants, almost all economically important plants have been genetically modified. By the end of 1997, more than 70 different crop species had been transformed (1). Field tests have been conducted at more than 25,000 individual field sites, in at least 45 countries with at least 60 different crops. The number of genetically modified crops continues to grow rapidly, as indicated by numerous reports published routinely on the transformation of new plant species and on new traits being introduced.

Following extensive field testing, over 50 different products have been approved by the appropriate regulatory authorities for marketing in at least one country (Table 1). Approximately 3 million acres of genetically modified plant products were grown commercially in 1996, with over 30 million acres planted in 1997, over 65 million acres in 1998, over 98 million acres in 1999 and the number of acres are expected to be maintained or to increase again in 2000. Thirteen countries (US, Argentina, Canada, China, Australia, South Africa, Mexico, Spain, France, Portugal, Romania, Bulgaria and Ukraine) grew nine crops (soybean, corn, cotton, canola, potato, squash, papaya, tomato and tobacco) commercially in 1999 (2). Approximately 55% of the soybeans, 45% of the cotton and 35% of the corn grown in the United States in 1999 were genetically modified, while over 70% of the soybeans grown in Argentina were genetically modified. The details of some of these products will be briefly discussed within the appropriate categories below.

Genetically modified plant products have created significant economic value. In 1997 glyphosate-tolerance soybeans created an estimated global economic value of over $1 billion, and insect-protected cotton grown in the US created an economic value of approximately $240 million in 1996 and $190 million in 1997 (3,4). Farmers received the greatest portion of this value, with an estimated 76% of the value for glyphosate-tolerant soybeans in 1997, 59% of the value of insect-protected cotton in 1996 and 42% in 1997. Global sales of transgenic crops rose from approximately $75 million in 1995 to more than $2 billion in 1999.

IMPORTANT TRAITS

It is beyond the scope of this chapter to discuss all the traits that have been tested and the products that have been approved, so classes of the most advanced traits and important products will be briefly discussed. According to the Worldwide Web site *www.aphis.usda.gov/bbeg/bp/* provided by the US Animal and Plant Health Inspection Service (APHIS), the most frequently field tested traits are: insect protection, virus protection, fungal resistance, herbicide tolerance and food quality enhancements. In addition to these traits, several more recent, but important areas of research will be briefly summarized including the use of genetically modified plants for the production of biopolymers, pharmaceuticals and vaccines.

Insect protection

Several different approaches have been, or are being, evaluated to control a variety of insect pests that cause tremendous damage to the most agronomically important crops. These pests typically had been controlled by use of a variety of chemical insecticides. Insecticides and associated management practices cost approximately $10 billion annually worldwide. Yet 20% to 30% of total crop product is still lost due to insect pests. Based on the importance and value associated with effective insect control and the availability of microbial genes with insecticidal activity, genetically modified plants containing genes providing protection from insect damage were among the first field tested and marketed.

The bacterium *Bacillus thuringiensis* (*B.t.*) has been used as a microbial insecticide in farming products since 1961, establishing a long history of safe use. The insecticidal proteins produced by *B.t.* are specific and selective due to the receptor-mediated binding in the insect gut. However, the microbially-produced *B.t.* products account for less than 1% of all pesticides used in the United States. Disadvantages of these microbial-based products include: expensive production and application, need for repeated applications, inactivation by sunlight and being washed off by rain. These disadvantages have been overcome by introducing and expressing the genes encoding these proteins directly in the host plant. These genes were modified to more closely resemble plant genes to increase the level of expression of the insecticidal protein in the plant by as much as 100-fold compared to expression of protein from the native, bacterial gene and to levels within the crops sufficient to provide commercial protection against the targeted insect pests. The effectiveness of this delivery system is reflected in the successful launches of insect protected potato, cotton and corn products (5).

Approximately 1.8 million acres of genetically modified insect protected cotton varieties were grown in the US in 1996, the first year of introduction, accounting for approximately 13% of the US cotton crop. Approximately 2.2 million acres were planted in 1997, 2.6 million acres in 1998 and 4.0 million acres in 1999 (2). In addition, insect protected cotton varieties were planted in Australia, Mexico, China and South Africa. These plants provide protection against the three major insect pests in cotton: tobacco budworm (*Heliothis virescens*), pink bollworm (*Pentinophora gossypiella*) and cotton bollworm (*Helicoverpa zea*).

The US cotton growers planting these varieties significantly reduced their use of chemical insecticides (5). Increased effectiveness in control of these insect pests resulted in a significant yield increase in cotton lint production while also resulting in increased numbers of beneficial insects due to increased specificity of the pesticidal protein. Farmers have realized significant economic value, averaging on the order of $40 per acre increase in income. In spite of heavy infestations of bollworm, one of the targeted insect pests, 60% of the growers in the US needed no chemical insecticide applications to control these pests. Most growers who did apply insecticides to the genetically modified cotton varieties used only one application compared to an average of four to six applications for the nongenetically modified cotton varieties. Chemical insecticide use in the US was reduced by over 2 million pounds in 1998 due to the introduction of insect protected cotton varieties. Reductions of insecticide use were even more impressive in China, where more than 10 insecticide applications are typically used per growing season for cotton. Insecticide applications have been reduced by at least 80% by growing *B.t.* cotton (6).

Approximately 10,000 acres of genetically modified *B.t.* potato plants protected against Colorado potato beetle, the most damaging potato insect pest, were planted commercially in 1996. Growers planting the genetically protected potato plants used approximately 35% less insecticide to control the Colorado potato beetle compared to growers planting non-genetically protected potato varieties. Planted acres more than doubled in 1997 (25,000 acres) with another doubling in acres in 1998 (50,000 acres). In 1999, two new products were introduced which combined protection against the Colorado potato beetle with resistance to either potato leaf roll virus or to potato virus Y, two major viral diseases for potato.

Approximately 700,000 acres of genetically modified *B.t.* corn plants protected against corn borer were planted commercially in 1996. A number of additional insect protected, genetically modified corn products entered the market in 1997, which significantly increased the acreage to over 7 million acres. Acres doubled to over 14 million acres in 1998 and increased again in 1999 to approximately 18 million acres or approximately 25% of the planted corn acres in the US. One of the key advantages of *B.t.* corn varieties is that those varieties which provide effective protection of corn ears also drastically reduce the amount of secondary infection caused by the fungus *Fusarium*. The ability to control *Fusarium* results in up to 95% reduction of the level of the mycotoxin, fumonisin, under some conditions (7). Fumonisins are fungal toxins that produce morbidity and death in horses and swine (8) and have been linked in epidemiological studies to high rates of esophageal and liver cancer in African farmers (9). Fumonisin reductions have been demonstrated in the US, Italy and France with additional studies being conducted to verify reductions in other geographies where corn is grown globally (7,10).

An initial laboratory study conducted with *B.t.* corn (11) suggested that the pollen for some *B.t.* corn varieties may have the potential to harm monarch butterflies. Based on a considerable amount of subsequent field research on *B.t.* corn plants, assessment of the potential toxicity of corn pollen from *B.t.* corn plants to monarch butterflies and more importantly, the exposure of monarch butterflies to *B.t.*-containing corn pollen, the Environmental Protection Agency recently concluded that *B.t.* corn does not pose a significant risk to monarch butterflies (12).

Research continues to develop insect protection in many other crops, e.g. tomato, rice, peanuts, soybeans and chickpeas through insertion of *B.t.* proteins. To date, hundreds of different *B.t.* genes have been isolated and characterized from a broad variety of *Bacillus* species. With this diversity of genes, many additional insect pests will be controlled and crops protected. Of course, one of the primary focuses of those involved in the development of these products is to assure the long term durability and effectiveness of these products. Toward this end, numerous strategies have been developed and are being implemented and refined to assure that these products maintain their effectiveness long term. Products which contain two different *B.t.* proteins with different modes of action have been submitted for regulatory review. These products, combined with other insect resistance management strategies, should assure the long term durability. In contrast to many of the broad spectrum insecticides that are used, plants expressing *B.t.* proteins are selectively active against the targeted insect pests. This selectivity enables growers to maintain beneficial predatory insects in their fields which provide additional protection against insect pests and against some diseases that are transmitted by insects (e.g. viruses).

In addition to the genes from *B.t.*, many other sources have been, and continue to be, evaluated for insecticidal proteins (13). Among the proteins that have been identified are those that interfere with the nutritional requirements of the insect. For example, protease inhibitors or α-amylase inhibitors continue to be tested individually and in combination to provide insect protection. Chitinases and lectins also have been tested. Typically, these classes of proteins cause modest insect growth inhibition with little direct insect mortality, especially in comparison to the proteins produced by *B.t.* Several recent approaches that are likely to expand the range of insects that will be controlled by genetically modified plants as well as provide additional valuable mechanisms to enhance the durability of the *B.t.*-based products have been reported. These leads include a family of unique *B.t.* proteins with activity against corn root worm (*Diabrotica undecimpunctata*), a major corn pest, as well as a family of proteins called VIPs, isolated from vegetatively growing *Bacillus* species, which have activity against corn rootworm, as well as activity against lepidopteran insect pests. These various insecticidal leads likely will create a second generation of genetically modified plant products with built-in protection against a broader diversity of important insect pests.

Virus protection

Virus resistant genetically modified plants have been produced by using genes derived directly from the virus, so called pathogen-derived resistance (14). The first genetically modified plants, tobacco with increased resistance to tobacco mosaic virus (TMV), resulted from the expression of the TMV coat protein in tobacco plants (15). This coat protein-mediated resistance has been used extensively to confer viral resistance to numerous crops, including tobacco, tomato, squash, melon, papaya and potato. In 1987, TMV resistant tomatoes were the first transgenic tomato plants to be field tested (16).

Squash varieties resistant to zucchini yellow mosaic virus (ZYMV) and watermelon mosaic virus (WMV) were approved in the US in 1995 and marketed. These two viruses routinely reduce crop yields by 20–80% depending on production season and growing region. More recently, squash plants resistant to these two viruses plus a third destructive virus, cucumber mosaic virus (CMV), were developed (17). Marketing of these products will allow the more consistent production of squash varieties with significant reductions in the use of chemical insecticides that are typically required to control virus spread by insects. Coat protein genes have also been used to confer resistance to papaya ringspot virus (PRSV), which is being marketed in Hawaii (18) and to sweetpotato feathery mottle virus (SPFMV), which is being developed in Kenya (19). Resistance to PRSV has essentially saved papaya production in Hawaii, while resistance to SPFMV has the potential to increase sweetpotato yields in Kenya by up to 80%.

Several other viral-derived genes have been identified to provide effective control of viruses (20,21). For example, resistance to the most economically important virus disease in potato, potato leafroll virus (PLRV), was achieved by introducing a gene involved in viral replication (22,23). This virus typically causes decreases in potato tuber quality by causing net necrosis and reductions in potato production worldwide. The use of viral genes encoding dysfunctional movement proteins can inhibit viral spread, probably by inhibiting cell-to-cell movement of the virus (20). Combinations of virus resistance with resistance to key insect pests, as described above, will be critical, since insects often serve as vectors to transmit viruses. The best example of such combinations are potatoes engineered for resistance to PLRV and Colorado potato beetle (CPB) (22,23) or potatoes resistant to potato virus Y (PVY) and CPB (24). These potatoes are grown commercially in the USA and Canada. Chemical insecticide programs provide protection against both insect and viral disease. Combining these traits within transgenic plants leads to even more impressive reductions in dependence on traditional insecticide programs for crop production and results in the accompanying energy and resource savings associated with the production, distribution and application of synthetic chemical products. Transgenic plants also deliver higher yield and better quality food.

Fungal/bacterial disease resistance

The identification of anti-fungal genes and the production of genetically modified plants that effectively control fungal disease lag significantly behind the production of insect and virus resistant plants. Recent advances, however, indicate that this will only be a temporary situation; genetically modified plants that are effective in controlling important plant diseases are being produced and evaluated (25).

Initial efforts have focused on isolating and expressing in plants more effective hydrolytic enzymes (e.g. chitinases and glucanases) that degrade fungal cell walls, thereby increasing their fungal resistance. However, there remains a need for more potent enzymes to further enhance anti-fungal activity. Expression of plant defensins or other proteins induced upon fungal infection appears especially promising.

Identification of genes that may effectively control bacterial diseases in plants is also progressing rapidly. Commercial plants that provide resistance to fungal and bacterial diseases are still a few years away and must await more intensive assessment, especially at the whole plant level under field conditions.

Selectivity to herbicides

Weeds are one of the major agricultural pests that can devastate a crop if not managed properly. Weeds compete with crops and reduce yield, decrease harvest efficiency, decrease seed quality and serve as a reservoir for crop pests. Agricultural practices used prior to World War II were labor intensive and contributed to soil erosion. These processes have been replaced by the use of chemical herbicides that are the most effective, reliable and economic method of controlling weeds on large scale. Today herbicides are used on essentially 100% of the acreage of the major agronomic crops in developed countries.

Traditionally, herbicides have been selected based on the weeds to be controlled and the natural resistance of the crop to the herbicide. The use of this system has been tremendously successful in providing the grower with effective weed control. Biotechnology provides an opportunity to modify crops so they tolerate selected herbicides with preferred environmental properties, such as glyphosate. These herbicide tolerant crops allow the farmer to apply herbicide to planted fields, killing weeds but leaving the planted crop unaffected. This ability provides increased flexibility and cost savings to growers. In addition, farmers can move from using pre-emergent, soil incorporated herbicides to post-emergent herbicides that are applied on an "as needed" basis. This strategy can reduce the number and total amount of herbicides used and enable the application of herbicides that bind tightly to the soil and are less likely to enter the ground water, thereby providing significant environmental benefits.

Soybeans, canola, cotton and corn tolerant to glyphosate, oilseed rape and corn tolerant to phosphinotricin, and cotton tolerant to sulfonylureas are examples of genetically modified, herbicide tolerant crops that are entering the marketplace (26) (see Table 1). For example, glyphosate tolerant soybeans were planted on over 1.5% of the US soybean acres in 1996 by over 10,000 growers. Because of exceptional weed control and crop safety, these soybean varieties were rapidly adopted. Approximately 75% of the growers used only one herbicide application, providing a reduction in the amount and number of herbicides used compared to soybeans treated with conventional herbicide programs. In 1997, approximately 15% of the US soybean acres were planted with glyphosate tolerant soybeans. Over 25 million acres of glyphosate tolerant soybeans were planted in 1998 with over 37 million acres or greater than 50% of the US soybeans in 1999. Over 70% of the soybeans grown in Argentina in 1999 were tolerant to glyphosate.

It is important to recognize that improving crop yields is essential to meeting the increased food demands of an ever-increasing world population. Improving crop protection from pests and increasing crop yields through herbicide tolerance are two strategies that will help meet this need. Biotechnological enhancements of crops can assure that these changes occur quickly enough to meet urgent global food needs. At the same time, agricultural biotech practices have significant advantages for the environment in terms of reductions in soil erosion and tilling, and less agricultural chemical use and run-off. Furthermore, efforts underway will incorporate other approaches which should directly enhance crop yields.

FOOD QUALITY ENHANCEMENTS

Although many of the initial genetically modified plant products have improved agronomic traits, the first plant biotechnology product marketed in the US, the Flavr Savr™ tomato (Table 1), is a product with food quality enhancements. Many additional products with improved quality or nutritional properties are under development. A few of these will be briefly reviewed.

Texture, taste and aroma are important aspects for food products intended for direct consumer purchase. Five of the products shown in Table 1 are tomato products modified to alter the ripening and/or softening processes, allowing the fruit to ripen longer on the vine or have an increased shelf life. These enhancements have been achieved either by inhibiting the breakdown of cell wall components or by directly altering the synthesis of ethylene, the phytohormone that triggers the fruit ripening process.

Carbohydrates

Modifications of the carbohydrate content and specific carbohydrate composition in foods represent a significant opportunity to impact food quality and nutrition since these are the most important sources of energy in the diet. Among food crops, corn, rice, wheat, potato and cassava provide major contributions to the intake of dietary carbohydrate, in the form of starch and fiber.

The amount of starch has been increased in potatoes by introducing a gene encoding the non-regulated bacterial enzyme, ADP-glucose pyrophosphorylase, which is the key regulatory enzyme controlling starch biosynthesis in plants. Strategies to assure the uniformity of the increased starch content throughout the potato, which would lead to enhanced performance characteristics in potato chips and French fry products, are being assessed. The increase in starch content has been achieved at the expense of water content, resulting in both greater yield of processed potato products as well as reduced absorption of oil when frying. Consequently, chips and French fry products from these potatoes will have reduced fat and caloric content. In addition, tomatoes containing increased starch provide energy conservation and other economic benefits by reducing the amount of energy intensive processing required for the production of tomato sauce and tomato paste based products.

In addition to modifying the total starch content, modifications in the composition of starch also have been achieved. For example, amylose-free, amylopectin-enriched potatoes have been produced by inhibiting the production of granule-bound starch synthase to produce starch preferred for specific industrial applications. In addition, by redirecting the flow of carbon from starch biosynthesis, increases in sucrose content and production of new components like fructans or other sweeteners also can be realized.

84

Oils

Oils provide the second most important source of food energy for humans. Plant biotechnology efforts have focused largely on altering the fatty acid composition of canola, soybeans and sunflowers, all major sources of dietary oils, to provide either more stable sources of specific oils or oils with enhanced nutritional characteristics (27,28). The first modified oil product to be marketed was a high laurate canola oil (see Table 1) which serves as an alternative for oils with similar composition, such as tropical oils. A genetically modified soybean variety with increased levels of oleic acid, a mono-unsaturated fatty acid, and reduced levels of poly-unsaturated fatty acids, also has been produced. This product exhibits increased oxidative stability and significantly reduces the need for and costs of hydrogenation during oil processing. Elimination of chemical hydrogenation eliminates the production of *trans*-fatty acids, which have been linked to cardiovascular disease.

Proteins

Foods that provide sufficient quantities of the essential amino acids are critical to meet dietary needs, especially in developing countries where foods low in one or more of these amino acids are often consumed. Approaches to increase the amount of lysine in cereals and methionine in legumes are being assessed (29,30). These amino acids may be introduced in their free form (e.g. as lysine or methionine) or they may be added as part of proteins, particularly seed storage proteins that are present at relatively high levels in the grain and are rich in these essential amino acids. For example, total lysine content of both canola and soybean was increased two- and five-fold, respectively, by expressing non-feedback inhibited forms of two key enzymes in lysine biosynthesis. A two- to three-fold increase in the total lysine content of corn also was obtained using the same approach.

Vitamins/minerals

Numerous research efforts to increase the levels of key vitamins and minerals in genetically modified crops are on-going. In developing countries where malnutrition is a major concern, enhancing these components in staple foods may improve nutritional status and reduce the risk of disease. These projects include increasing the levels of β-carotene, the precursor of vitamin A, vitamin D, iron and other nutrients. The most exciting of these research programs is the production of a variety of rice with significantly increased levels of both β-carotene and iron (31). The World Health Organization estimates that vitamin A deficiency affects over 250 million children worldwide, causing night blindness and vulnerability to disease, and leading to over 200 million deaths per year. Plants genetically improved to contain increased amounts of β-carotene, the precursor of vitamin A, may play an important role to address this problem. This high β-carotene, high iron rice could also help address iron deficiency anemia, another widespread nutritional problem. Consumption of 300 g of this genetically modified rice, which contains increased levels of β-carotene and iron, could help satisfy the increased need for both vitamin A and iron. A canola variety with greatly increased levels of β-carotene in the oil also has been developed (32).

Eliminating food components

Plant biotechnology is also being used to reduce or eliminate a number of food components, particularly components which are undesirable. For example, the level of one of the major allergenic proteins in rice has been reduced by approximately 80% by researchers in Japan (33). Other efforts to modify the amino acid sequences which are responsible for eliciting the allergenic response also have proven successful. Efforts to reduce or eliminate other undesirable components of foods, e.g. glucosinolates in canola meal, protease inhibitors in beans, glyco-alkaloids in potatoes and mycotoxins in corn are being evaluated. Components such as caffeine from coffee beans can be eliminated or reduced to provide a coffee with no or very low caffeine level without using chemicals.

PLANTS AS PRODUCTION FACTORIES

Plants provide an unmatched capacity to produce products with a maximum efficiency and a minimum of energy – thus, in an extremely cost effective manner. As illustrated in the previously described examples of insect and disease control, plants can be bioengineered to produce their own insecticidal proteins using carbon dioxide, water and nitrogen, thereby reducing the need for expensive and energy intensive chemical insecticide manufacturing plants. The potato developed to contain genes that produce proteins that protect it from both the primary insect pest (Colorado potato beetle) and the major viral pest, potato leaf roll virus, is an excellent example. The combination of these two traits will enable potatoes to be grown without the application of chemical insecticides to control these two major pests and will result in significant environmental and economic savings. Only water, carbon dioxide and nitrogen are required for the genetically modified potato plant to protect itself from these pests. In contrast, extensive raw materials and energy in the form of oil are required to produce the approximately 5 million pounds of chemical insecticides that are currently used to control these pests annually in the US. Approximately 2.5 million pounds of waste are generated annually in the production process. Less than 5% of the applied insecticide actually reaches the target pest. Using the plant to produce these pesticides is therefore much more energy efficient and environmentally acceptable. Given these efficiencies in using plants as production factories, efforts are underway to produce protein-based pharmaceutical products and other compounds in plants. One of the most exciting applications of plant biotechnology is the ability to produce edible vaccines in plants. For example, transgenic potatoes containing the non-toxic binding subunit of the *Escherichia coli* enterotoxin (34) as well as trangenic potatoes which express the capsid protein of Norwalk virus (35) have been produced. The expressed proteins function as an oral immunogen in humans. Efforts are underway to introduce such vaccines into banana, which would be eaten raw and thus serve as an ideal source to deliver this technology to developing countries, where inexpensive vaccines are urgently needed.

FOOD, FEED AND ENVIRONMENTAL SAFETY CONSIDERATIONS

As plant biotechnology products have been developed, so have national and international regulatory oversight measures to assure the food, feed and environmental safety of these products. In the early to mid-1980s oversight in the US was provided by the National Institutes of Health for laboratory experimentation. In 1986, the US Office of Science and Technology published the "Coordinated Framework for Regulation of Biotechnology" which clearly delineated regulatory authority in the US to each of three existing regulatory agencies. The Food and Drug Administration (FDA) maintained authority to assure the food and feed safety of the products of plant biotechnology; the US Department of Agriculture (USDA), the authority to assure that these plants do not present any plant pest risks; and the Environmental Protection Agency (EPA), the authority to assess the safety of plants that produce a pesticidal substance (e.g. plants protected against insects, fungi, bacteria or viruses). EPA also maintained authority over herbicides used in conjunction with herbicide tolerant plants, as is the case with any herbicide.

The USDA regulates the movement and release of genetically modified plants under the Federal Plant Pest Act and the Plant Quarantine Act. The initial regulations, published in 1987, have been modified twice to reflect the experience gained. Permits or notifications (a simplified procedure whereby applicants notify the agency of the crop and trait to be tested and agree to follow prescriptive, crop-specific, containment procedures) are required for field testing. A determination that the genetically modified plant is not a plant pest (e.g. that the plant does not pose a risk to the environment or production agriculture) must be obtained prior to market introduction. Field tests at more than 14,000 individual sites in the US have been conducted, and determinations that genetically modified plant products are not plant pests have been made for over 50 different products, to date. These products are included in Table 1.

The FDA maintains its authority for the safety and wholesomeness of food and feed products, including those derived from genetically modified plants. The decision-tree approach used to assure the food and feed safety by developers of genetically modified plant products and the FDA is described in the 1992 FDA document, "Statement of Policy: Foods Derived from New Plant Varieties" (36). In this document, FDA concluded that foods derived from genetically modified plants do not pose any unique safety concerns. Therefore, FDA uses the same post-market food adulteration approach and regulations for these products is used for food and feed products derived from traditionally bred plant varieties. Consultations between developers and the FDA have been conducted for each of the products that have been marketed www.fda.gov. These products are included in Table 1.

The EPA has jurisdiction over plants with bioengineered pesticidal traits under the Federal Insecticide, Fungicide and Rotenticide Act (FIFRA). To date EPA has approved at least seven different products after assessing the safety of the introduced pesticidal trait. These products were also reviewed by FDA and USDA.

Other countries also have regulations in place for genetically engineered plant products. Health Canada regulates food safety, and Agriculture and Agri-Food Canada regulate feed and environmental safety as well as registration of specific plant varieties for certain crops. The Ministry of Heath and Welfare regulates food safety in Japan whereas the Ministry of Agriculture, Food and Fisheries regulates feed and environmental safety. At the European Union level, environmental assessments are conducted prior to placing a product on the market through the process described in the 90/220 EEC regulations. Food safety in the European Union is regulated through the newly implemented Novel Foods Regulation, replacing individual country food regulations that were in place in the United Kingdom, Denmark and The Netherlands. Feed safety is currently conducted within the 90/220 EEC process, however, a Novel Feed Regulation is under consideration. Many other countries also have, or are developing, regulations for these products.

Table 2 summarizes the types of information and data that are typically provided to the various regulatory authorities to assess the food, feed and environmental safety of genetically modified plants and plant products. Two recent reports from expert consultations sponsored by the United Nations Food and Agricultural Organization (FAO) and the World Health Organization (WHO) demonstrate that there is general consensus on the approach used and information appropriate to assess the food safety of genetically modified plant products (37,38). This approach is based on the concept of "substantial equivalence." From a food safety perspective, food products from genetically modified plants that are compositionally and nutritionally comparable (i.e. substantially equivalent) to the traditional counterparts can be introduced into the marketplace with the same degree of assurance as those products derived from their traditional counterparts. Using conventional standards such as foods from traditionally-bred food crops is a well accepted approach for food safety assessment. If there are newly expressed traits, such as insect protection conferred by the expression of a protein(s), the safety of this substance(s) is rigorously assessed using internationally accepted safety assessment approaches. These assessments address any ques-

Table 2 Data typically provided to regulatory agencies to support the food, feed and environmental safety of genetically modified plants and plant products

Food/feed safety assessment
Molecular characterization
   Gene source
   Transformation system
   Insert number
   Copy number
   Integrity of each inserted genetic element (promoter, coding sequence, termination sequence)
   Genetic stability
Protein safety assessment
   Source
   Host/processing
   Protein expression levels/consumption
   History of use of same/similar protein
   Safety to non-target organisms
      Function/specificity/mode-of-action
      Homology to known toxins/allergens
      Digestibility
      Potential toxicity testing (case-by-case)
   Allergenicity assessment
   Marker gene/protein safety
Nutritional equivalence
   Identification of key nutrients
   Levels of key nutrients relative to traditional counterpart
   Anticipated uses relative to historical uses
   Nutritional assessment of expressed trait
     (e.g. modified oil, carbohydrate, etc.)
   In vitro/in vivo nutritional studies (case-by-case)
Toxicological assessment
   Identification of key anti-nutrients/toxicants in host
     or organisms related to host
   Levels of key anti-nutrients relative to traditional counterpart
   Toxicological assessment of expressed trait

Environmental safety assessment
Molecular characterization (as above)
Outcrossing/gene flow (potential/impact)
Weediness
Competitiveness/survivability/dormancy
Morphological/phenotypic characteristics
Insect/disease susceptibility
Impact on non-target organisms
Agronomic performance
Resistance management (where appropriate)

tions concerning toxicity or allergenic potential of the newly expressed protein (39), as well as questions raised pertaining to the protein function and specificity. Gaining consensus and harmonization in international policies and regulations for food products derived from genetically modified plants is critical since many are commodity crops (e.g. corn, soybean, wheat, rice) destined for international commerce.

CONCLUSIONS

The benefits of genetically modified plant products are being realized as the first products establish their place in the market. Initial benefits were realized primarily by farmers who produce crops with increased yields and quality while using more environmentally friendly practices, such as decreased tilling and approaches which require less pesticide use. Although consumers may not see these benefits first-hand, nevertheless, they translate into a more stable and abundant food supply and an improved environment. Future generations of genetically modified plant products will include products with enhanced nutritional qualities that will directly improve the diets of people globally. Finally, the possibilities of using plants as clean, efficient factories for the production of medicines, biodegradable plastics and other important compounds are beginning to be explored.

ACKNOWLEDGMENTS

I thank Jim Astwood, Rick Goodman, Karen Marshall, Andrew Reed, Wojciech Kaniewski and Lisa Watson for critical review and constructive comments on this chapter.

REFERENCES

1. James C. *Global status of transgenic crops in 1997*. ISAAA Briefs No. 5. Ithaca, NY: ISAAA; 1997.
2. James C. *Preview: Global Review of Commercialized Transgenic Crops: 1999*. ISAAA Briefs No. 12. Ithaca, NY: ISAAA; 1999.
3. Gianessi LP, Carpenter JE. *Agricultural Biotechnology: Insect Control Benefits*. National Center for Food and Agricultural Policy; 1999.
4. Klotz-Ingram C, Jans S, Fernandez-Cornejo J, McBride W. Farm-level production effects related to the adoption of genetically modified cotton for pest management. *AgBioForum* 1999; 2(2):73–84.
5. Betz F, Hammond BG, Fuchs RL. Safety and advantages of *Bacillus thuringiensis* (*Bt*)-protected plants. *J Reg Tox Pharmacol* 2000; 32:156–73.
6. Xia JY, Cui JJ, Ma LH, Dong SX, Xui XF. The role of transgenic Bt cotton in integrated insect pest management. *Acta Gossypii Sinica* 1999; 11:57–64.
7. Munkvold GP, Hellmich RL, Rice, LG. Comparison of fumonisin concentrations in kernels of transgenic Bt maize hybrids and non-transgenic hybrids. *Plant Dis* 1999; 83:130–8.

8. Norred, WP. Fumonisins – mycotoxins produced. *J Tox Environ Health* 1993; 38:309–28.

9. Marasas WFO, Jaskiewicz K, Venter FS, van Schalkwyk DJ. *Fusarium moniliforme* contamination of maize in oesophageal cancer areas in the Transkei. *S Afr Med J* 1988; 74:110–14.

10. Masoero F, Moschini M, Rossi F, Prandini A, Fietri A. Nutritive value, mycotoxin contamination and in vitro rumen fermentation of normal and genetically modified corn (CryIA(B)) grown in northern Italy. *Maydica* 1999; 44:205–9.

11. Losey JE, Rayor LS, Carter ME. Transgenic pollen harms monarch larvae. *Nature* 1999; 3999:214.

12. Environmental Protection Agency (EPA). Biopesticides Registration Action Document. *Bacillus thuringiensis Plant-Pesticides* 2000; 57–8. http://www.epa.gov/scipoly/sap/2000/October/brad3/_enviroassessment.pdf

13. Estruch JJ, Carozzi N, Desai N, Duck NB, Warren GW, Koziel MG. Transgenic plants: an emerging approach to pest control. *Nature Biotechnology* 1997; 15:137–41.

14. Kaniewski WK, Lawson EC. Coat Protein and Replicase Mediated Resistance to Plant Viruses. In: Hadidi A, Khetarpal RK, Koganezawa H, eds. *Plant Virus Disease Control.* St. Paul, Minnesota: APS Press; 1998; 65–78.

15. Powell-Abel P, Nelson RS, De B, Hoffmann N, Rogers SG, Fraley RT, Beachy RN. Delay of disease development in transgenic plants that express the tobacco mosaic virus coat protein gene. *Science* 1986; 232:738–43.

16. Nelson RS, McCormick SM, Delannay X, Dube P, Layton J, Anders J, Kaniewska M, Proksch RK, Horsch RB, Rogers SG, Fraley RT, Beachy RN. Virus tolerance, plant growth and field performance of transgenic tomato plants expressing coat protein from tobacco mosaic virus. *Bio/Technology* 1988; 6:403–9.

17. Tricoli DM, Carney KJ, Russell PF, McMaster JR, Groff DW, Hadden KC, Himmel PT, Hubbard JR, Boeshore ML, Reynolds JF, Quemada HD. Field evaluation of transgenic squash containing single or multiple virus coat protein gene constructs for resistance to cucumber mosaic virus, watermelon mosaic virus 2 and zucchini yellow mosaic virus. *Bio/Technology* 1995; 13:1458–65.

18. Manshardt R. *The Development of Virus Resistant Papaya in Hawaii.* ISAAA Brief No. 11:9–11. Ithaca, NY: ISAAA; 1999.

19. Kaniewski W, Maingi D, Kaniewska M, Flasinski S, Lowe J, Kirubi D, Macharia C, Wambugu F, Horsch R, Hinchee M. *Engineered resistance to sweetpotato feathery mottle virus in sweetpotato.* Fifth Triennial Congress of the African Potato Association. Kampala, Uganda; 2000; 37.

20. Martin RR. Alternative strategies for engineering virus resistance in plants. In: Hadidi A, Khetarpal RK, Koganezawa H, eds. *Plant Virus Disease Control.* St. Paul, Minnesota: APS Press; 1998;121–8.

21. Miller ED, Hemenway C. History of coat protein-mediated protection. In: Foster GD, Taylor SC, eds. *Methods in Molecular Biology, Vol. 81: Plant Virology Protocols: From Virus Isolation to Transgenic Resistance.* Totowa, NJ: Humana Press Inc.; 1998; 25–38.

22. Lawson C, Weiss J, Thomas P, Kaniewski W. NewLeaf Plus Russet Burbank potatoes: Replicase-mediated resistance to potato leafroll virus. *Molecular Breeding* 2000 (Accepted).

23. Thomas P, Lawson C, Zalewski J, Reed G, Kaniewski W. Extreme resistance to potato leafroll virus in potato cv. Russet Burbank mediated by the viral replicase gene. *Virus Research* 2000; 71:49–62.

24. Perlak P, Kaniewski W, Lawson C, Vincent M, Feldman J. Genetically improved potatoes: Their potential role in integrated pest management. Environmental Biotic Factors in Integrated Plant Disease Control. *Proceed. 3rd EFPP Conference*; 1995; 451–4.

25. Shah DM, Rommens CMT, Beachy RN. Resistance to diseases and insects in transgenic plants: progress and applications to agriculture. *Trends in Biotechnology* 1995; 13:362–8.

26. Duke SO, ed. *Herbicide-Resistance Crops: Agricultural, Environmental, Economic, Regulatory and Technical Aspects*. Boca Raton, Florida: CRC Press, Inc.; 1996.

27. Budziszewski GJ, Croft KPC, Hildebrand DF. Uses of biotechnology in modifying plant lipids. *Lipids* 1996; 31:557–69.

28. Murphy DJ. Engineering oil production in rapeseed and other oil crops. *Trends in Biotechnology* 1996; 14:206–13.

29. Day PR. Genetic modification of proteins in food. *Critical Reviews in Food Science and Nutrition* 1996; S49–67.

30. Falco SC, Guida T, Locke M, Mauvais J, Sanders C, Ward RT, Webber P. Transgenic canola and soybean seeds with increased lysine. *Biotechnol* 1995; 577–82.

31. Gura T. Biotechnology: New genes boost rice nutrients. *Science* 1999; 285:994–5.

32. Shewmaker C, Sheehy JA, Daley M, Colburn S, Ke DY. Seed-specific overexpression of phytoene synthase: increase in carotenoids and other metabolic effects. *Plant J* 1999; 20(4):401–12.

33. Matsuda T. Application of transgenic techniques for hypo-allergenic rice. *Proceedings from the international symposium on novel foods regulation in the European Union – integrity of the process of safety evaluation*. Berlin, Germany; 1998; 311–14.

34. Tacket CO, Mason HS, Losonsky G, Clements JD, Levine MM, Arntzen CJ. Immunogenicity in humans of a recombinant bacterial antigen delivered in a transgenic potato. *Nat Med* 1998; 4:607–9.

35. Tacket CO, Mason HS, Losonsky G, Estes MK, Levine MM, Arntzen CJ. Human immune responses to a novel Norwalk virus delivered in transgenic potatoes. *J Infect Diseases* 2000; 182:302–5.

36. Food and Drug Administration. *Statement of policy: Foods derived from new plant varieties*: Notice. Federal Register 1992; 57(104):22984–3005.

37. FAO/WHO. *Biotechnology and food safety*. Report of a Joint FAO/WHO Joint Consultation. FAO, Rome; 1996.

38. FAO/WHO. *Safety aspects of genetically modified foods of plant origin*. Report of a Joint FAO/WHO Consultation, WHO, Geneva, Switzerland; 2000.

39. Metcalfe DD, Astwood JD, Townsend R, Sampson HA, Taylor SL, Fuchs RL. Assessment of the allergenic potential of foods derived from genetically engineered crop plants. *Critical Reviews in Food Science and Nutrition* 1996; S165–86.

5

ALLERGIC REACTIONS AND FOOD INTOLERANCES

Steve L. Taylor and Susan L. Hefle

INTRODUCTION

A *food sensitivity* is defined as an abnormal physiologic response to a particular food. This same food is safe for the vast majority of consumers to ingest. Although only a small percentage of the population is likely affected, food sensitivities are viewed by the public as a significant health concern (1,2).

Several different types of illnesses occur within the general category of food sensitivities. True food allergies and various types of food intolerances are included as forms of food sensitivity, because these illnesses affect only certain individuals within the population. These diseases are often grouped together under the general heading of "food allergy," although they can be distinguished from one another. The existence of several different types of individualistic adverse reactions to foods with varying symptomology, severity, prevalence, and causative factors is not recognized by some physicians. Consumers are even more likely to be confused regarding the definition and classification of the individualistic adverse reactions to foods. Consumers perceive that "food allergies" are quite common (1,2), but, in fact, many self-diagnosed cases of "food allergy" incorrectly associate foods with a particular malady or ascribe various mild forms of postprandial eating discomfort to this category of illness.

CLASSIFICATION

A classification scheme for the different types of illnesses that are known to occur in association with food ingestion and that only involve certain individuals in the population is provided in Table 1. Two major groups of food sensitivity are known: true food allergies and food intolerances. The true food allergies involve abnormal immunological mechanisms; food intolerances do not. The differences between *immunological* food allergies and *non-immunological* food intolerances are significant for the affected individual. Food intolerances can usually be managed

Table 1 Classification of individualistic adverse reactions to foods

---

True food allergies
Antibody-mediated food allergies
  IgE-mediated food allergies (peanut, cows' milk, etc.)
  Exercise-associated food allergies
Cell-mediated food allergies
  Celiac disease
  Other types of delayed hypersensitivity

Food intolerances
Anaphylactoid reactions
Metabolic food disorders
  Lactose intolerance
Idiosyncratic reactions
  Sulfite-induced asthma

---

by limiting the amount of the food or food ingredient that is eaten. In contrast, *total avoidance* of the offending food is typically necessary with true food allergies.

Food allergies are abnormal immunological responses to a particular food or food component, usually a naturally occurring protein (3,4). Two different types of immunological mechanisms are known to be involved with true food allergies. Immediate hypersensitivity reactions are antibody-mediated; IgE-mediated food allergies are the only well recognized form of antibody-mediated, immediate hypersensitivity associated with foods. Delayed hypersensitivity reactions are cell-mediated. The role of cell-mediated reactions in food allergies is far less well established. However, cellular mediation certainly plays a major role in celiac disease. Celiac disease will be the only type of delayed hypersensitivity discussed in this review.

In contrast, *food intolerances* do not involve abnormal responses of the immune system (5,6). Three major categories of food intolerances are recognized: anaphylactoid reactions, metabolic food disorders, and food idiosyncrasies.

In addition, *allergy-like intoxications* can often be confused with true food allergies and intolerances because the symptoms are identical (7). However, allergy-like intoxications can affect anyone in the population who eats a hazardous amount of the causative substance. Histamine poisoning is the primary example of an allergy-like intoxication.

IgE-MEDIATED FOOD ALLERGY

Mechanism

The mechanism involved in IgE-mediated, immediate hypersensitivity reactions to foods is depicted in Figure 1. This same type of IgE-mediated reaction is also responsible for allergic reactions to other environmental substances such as pol-

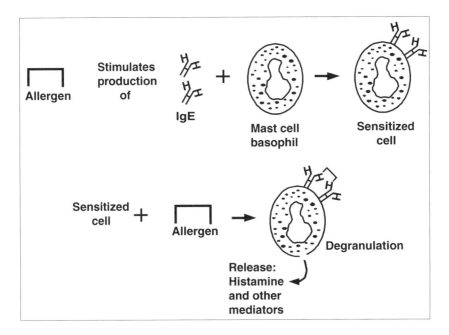

Figure 1 Mechanism of IgE-mediated food allergy.

lens, mold spores, animal danders, and bee venoms; only the source and structure of the allergen is different. In susceptible individuals, allergen-specific IgE antibodies are produced by B cells in response to the immunological stimulus created by exposure of the immune system to the allergen (3). Food allergens are usually naturally-occurring proteins present in the food (4), although only a few of the many thousands of proteins existing in the food supply are known to be allergens (8). In this sensitization phase of the allergic response, the allergen-specific IgE antibodies bind to the surfaces of mast cells in the tissues and basophils in the blood. No symptoms occur during this sensitization phase. However, upon subsequent exposure to the specific allergen, the allergen cross-links IgE-antibodies affixed to the surfaces of sensitized mast cells or basophils. This interaction causes the sensitized cells to degranulate and release a variety of potent, physiologically active mediators into the bloodstream and tissues. The granules within mast cells and basophils contain many of the important mediators of the allergic reaction. Other non-granule-associated mediators are also released simultaneously. The mediators are actually responsible for the symptoms of immediate hypersensitivity reactions. Histamine is perhaps the most important of the mediators released from mast cells and basophils during an allergic reaction. Histamine is responsible for many of the immediate effects encountered in IgE-mediated food allergies and can elicit inflammation, pruritis, and contraction of the smooth muscles in the blood vessels, gastrointestinal tract, and respiratory tract (9). Many

95

other important mediators have been identified including various leukotrienes and prostaglandins (10) This same type of IgE-mediated reaction is also responsible for allergic reactions to other environmental substances such as pollens, mold spores, animal danders, and bee venoms; only the source and structure of the allergen is different. The leukotrienes are associated with some of the symptoms that develop more slowly in IgE-mediated food allergies, such as late-phase asthmatic reactions.

During the sensitization phase, a susceptible individual forms allergen-specific IgE antibodies after exposure to a specific food protein. However, even among individuals predisposed to allergies, exposure to food proteins does not typically result in the formation of IgE antibodies. Exposure to a food protein in the gastrointestinal tract results in oral tolerance through either the formation of protein-specific IgG, IgM, or IgA antibodies or no immunological response whatsoever (clonal anergy) (11,12). Oral tolerance occurs to all dietary proteins in normal individuals and to the vast majority of dietary proteins even among individuals pre-disposed to development of food allergy. Heredity is probably the most important factor in predisposing individuals to the development of IgE-mediated allergies including food allergies (13). Approximately 65% of patients with clinically documented allergy have first degree relatives with allergic disease (13). Recently, studies with monozygotic and dizygotic twins confirm that genetics is an extremely important parameter and that identical twins may even inherit the likelihood of responding to the same allergenic food, e.g. peanuts (14,15). Conditions affecting the permeability of the small intestinal mucosa to proteins such as viral gastroenteritis, premature birth, and cystic fibrosis also seem to increase the risk of development of food allergy.

Symptoms

The onset time for IgE-mediated food allergies ranges from a few minutes to several hours after the consumption of the offending food. The symptoms of IgE-mediated food allergies range from mild and annoying to severe and life-threatening (Table 2). Only a few of these symptoms will be apparent in each allergic individual. The type of symptoms and their severity depend upon several factors including the individual, the amount of the offending food ingested, the tissue receptors that are affected, and the length of time since the last previous exposure.

Gastrointestinal and cutaneous symptoms (see Table 2) are among the more common manifestations of IgE-mediated food allergies. Respiratory symptoms are much less commonly encountered in food allergic reactions, but are more likely to be severe and life-threatening. Mild respiratory symptoms (rhinitis, rhinoconjuncitivitis) are much more likely to be encountered with exposure to environmental allergens such as pollens or animal danders that are airborne and inhaled directly into the respiratory tract. Those few food-allergic individuals who experience serious respiratory manifestations (asthma, laryngeal edema) in association with the inadvertent ingestion of the offending food are most likely to

Table 2 Symptoms associated with IgE-mediated food allergy

Gastrointestinal	Nausea
	Vomiting
	Diarrhea
	Abdominal cramping
Cutaneous	Urticaria
	Dermatitis or eczema
	Angioedema
	Pruritis
Respiratory	Rhinitis
	Asthma
	Laryngeal edema
Generalized	Anaphylactic shock

be at risk for life-threatening episodes (16). However, among the many symptoms involved in IgE-mediated food allergies, systemic anaphylaxis, also known as anaphylactic shock, is the most severe manifestation. Anaphylactic shock involves multiple organ systems (gastrointestinal, respiratory, cutaneous, and cardiovascular) and numerous symptoms. Death can occur from severe hypotension coupled with respiratory and cardiovascular complications. Anaphylactic shock is the most common cause of death in the occasional fatalaties associated with IgE-mediated food allergies (16,17). The number of deaths occurring from IgE-mediated food allergies is not recorded in most countries, but approximately 100 deaths are thought to occur in the US each year (18).

Mild symptoms are much more likely to occur with IgE-mediated food allergies than are severe symptoms such as anaphylactic shock. One of the more common and perhaps the most mild forms of IgE-mediated food allergy is the so-called oral allergy syndrome (OAS) (19). In OAS, symptoms are confined to the oropharyngeal area including pruritis, urticaria, and angioedema. OAS is most frequently associated with the ingestion of various fresh fruits and vegetables (19). OAS is an IgE-mediated reaction to specific proteins present in fresh fruits and vegetables, even though fresh fruits and vegetables contain comparatively low quantities of protein (19). The allergens in these fresh fruits and vegetables are apparently quite susceptible to digestive proteases in the gastrointestinal tract (20), so systemic reactions are rarely encountered to these foods. The allergens involved in OAS are also apparently heat-labile (20), since the heat-processed versions of these foods are not typically involved in initiation of OAS. In the case of OAS, affected individuals are initially sensitized to one or more pollens in the environment that cross-react with related proteins found in the fresh fruits and vegetables (20,22). Birch and mugwort pollens are frequently implicated. With OAS, sensitization to the pollen increases the likelihood of sensitization to specific foods.

In some cases of IgE-mediated food allergy, exercise must be done coincident with ingestion of the food for symptoms to occur (22). Exercise-induced food allergies are a subset of the immediate hypersensitivity reactions to foods. Several foods have been implicated in exercise-induced food allergies including shellfish (23), wheat (24), celery (25), and peach (26). The symptoms of this type of food allergy are individualistic and as variable as those involved in other food allergies. Exercise-induced allergies can also exist without any role for food intake (22). The mechanism of this illness is not well understood, although the involvement of IgE antibodies is apparent. With the recent emphasis on increased physical activity, reports of this condition could increase.

Prevalence

The overall prevalence of IgE-mediated food allergies is not precisely known but probably lies in the range of 2.0–2.5% for all age groups combined. Few epidemiological investigations involving clinical confirmation of IgE-mediated food allergies have been conducted using representative, unselected groups of adults. In a large-scale epidemiological investigation of the prevalence of IgE-mediated food allergy in the Netherlands (27), more than 10% of Dutch adults believed that they had adverse reactions to one or more specific foods, but food allergies were confirmed in approximately 2% with clinical histories and blinded food challenges. Among infants and young children, the prevalence of IgE-mediated food allergies is higher than for adults (28). Epidemiological investigations involving groups of unselected infants suggest that the prevalence of IgE-mediated food allergies is in the range of 4–8% in young infants (29,30). The results of the epidemiological investigations are supported by evidence obtained through consumer surveys. Surveys conducted in both the US and England indicate that the self-perceived prevalence of peanut allergy alone is 0.5–0.6% among all age groups (31,32). The US survey also included the self-perceived prevalence of tree nut allergies; the combined prevalence of peanut and tree nut allergies in this survey was 1.1% (31). Since peanut and tree nut allergies are often rather profound and easily diagnosed, any over-estimates associated with reliance on self-diagnosis among the consumers in these surveys are probably minimal. If the prevalence of peanut and tree nut allergies alone exceeds 1% and the prevalence of all food allergies among infants is 4–8%, then an overall estimate for IgE-mediated food allergies of 2.0–2.5% seems reasonable.

Persistence

Many food-allergic infants lose their food allergies within a few months to several years after the onset of the hypersensitivity (33–36). Allergies to certain foods, such as cows' milk (33), are more commonly outgrown than are allergies to certain other foods such as peanut (34). The mechanisms involved in the loss of sensitivity to specific foods are not precisely known, but the development of immunological tolerance is definitely involved (12).

Common allergenic foods and food ingredients

Eight foods or food groups, including cows' milk, eggs, fish, crustacea (shrimp, crab, lobster, etc.), peanuts, soybeans, tree nuts (walnuts, almonds, hazelnuts, etc.), and wheat are responsible for the vast majority of IgE-mediated food allergies on a worldwide basis (37). Some have suggested that sesame seeds should be added to this list of commonly allergenic foods (38–40), but, in 1999, the Codex Alimentarius Commission established an initial allergen list that incorporates only the first eight foods or food groups (sometimes referred to as the Big 8). Beyond the eight major foods or food groups, over 160 other foods have been documented to cause IgE-mediated food allergies (41). Any food that contains protein is likely to elicit allergic sensitization on at least rare occasions. The eight most commonly allergenic foods or food groups are foods that contain comparatively high amounts of protein and that are commonly consumed in the diet. However, several other commonly consumed foods with high protein contents, such as beef, pork, chicken, and turkey, are rarely allergenic.

Ingredients derived from the commonly allergenic foods will also be allergenic if they contain protein residues from the source material. Several categories of food ingredients, including edible oils, protein hydrolysates, lecithin, flavors, and gelatin can, on occasion, be derived from commonly allergenic sources.

Edible oils are not often a cause of allergic reactions. The refining process for edible oils should remove virtually all of the protein from the source material when hot solvent extraction is used. Highly refined peanut, soybean, and sunflower seed oils are documented to be safe for ingestion by individuals allergic to the source material (42–44). Oils from other sources such as sesame seed and tree nuts may receive less processing and contain allergenic residues (38,45). Oils produced by other procedures, such as cold-pressed oils, may also contain allergenic residues (46).

Protein hydrolysates can, on occasion, be a significant source of food allergens. Protein hydrolysates can be obtained from commonly allergenic sources such as soybean, wheat, milk, and peanuts. Several processes, including acid hydrolysis and enzymatic hydrolysis, are used to obtain these hydrolysates. The degree of hydrolysis of the proteins in hydrolysates is variable dependent upon the functional use, source, and method of hydrolysis. Partial hydrolysates, including many of those produced through enzymatic hydrolysis, are likely to retain their allergenicity. If the hydrolysates are extensively hydrolyzed, they may be safe for the vast majority of individuals allergic to the source material. The experiences obtained with hypoallergenic infant formula provide excellent insights into the safety of protein hydrolysates. Hypoallergenic infant formula based upon extensively hydrolyzed casein is safe for the vast majority of cows' milk-allergic infants (47). However, extensively hydrolyzed casein in hypoallergenic infant formula has been shown to trigger allergic reactions in some exquisitely sensitive cows' milk-allergic infants (48,49). However, infant formula based upon partial whey hydrolysates were even more likely to elicit allergic reactions in cows' milk-allergic infants (50,51).

The role of lecithin in food allergies is less certain, but probably not terribly significant. Lecithin is usually derived from soybean, but can also be derived from egg. Commercial soy lecithin contains trace residues of soy proteins. The soy protein residues in lecithin include IgE-binding proteins (52). However, the levels of soy allergens in soy lecithin may be insufficient to elicit allergic reactions in most soybean-allergic individuals. The allergenicity of soy and/or egg lecithin has not been documented through clinical challenge trials. Many soybean-allergic individuals do not avoid lecithin.

Flavors, both natural and artificial, can occasionally elicit allergic reactions. Only a few flavoring formulations contain protein and even fewer contain protein derived from allergenic sources (53). Allergic reactions have occurred to flavors especially in meat products where flavors can be used for dual-functional purposes (54–56).

Gelatin is considered a non-allergenic protein when ingested. Gelatin is most frequently derived from beef and pork. However, gelatin, especially kosher gelatin, can be derived from fish. The allergenicity of fish gelatin is unknown, although allergic reactions to fish gelatin among fish-allergic individuals has not been documented.

Food allergens

Allergens are almost always naturally occurring proteins found in foods (4). Only a comparatively small percentage of many proteins in foods are known to be allergens (8). Only a few of the known food allergens have been purified and characterized (4). Foods may contain both major and minor allergens. Major allergens are defined as those proteins that bind to serum IgE antibodies from more than 50% of patients with a specific food allergy. Multiple allergenic proteins exist in some commonly allergenic foods including peanuts, eggs, and cows' milk, while other foods such as Brazil nuts, codfish, and shrimp seem to have only one major allergen (4). For example, cows' milk contains three major allergens: casein, β-lactoglobulin, and α-lactalbumin (57), which also happen to be the major proteins in cows' milk. In addition, cows' milk contains several minor allergens including bovine serum albumin and lactoferrin (58). Peanuts contain at least three major allergens, namely *Ara h* 1, *Ara h* 2, and *Ara h* 3 (4). Like cows' milk, peanuts also contain a large number of minor allergens (59). In contrast, codfish, Brazil nut, and shrimp contain primarily one major allergenic protein: *Gad c* 1 (60), *Ber e* 1 (61), and *Pen a* 1 (62), respectively.

Diagnosis

Food hypersensitivities can be diagnosed in a number of ways. Self or parental diagnosis is a common practice but is often unreliable (29). Confirmation of an IgE-mediated food allergy requires careful medical attention. Initially, it must be determined that an adverse reaction occurs after the ingestion of a suspected food. If the number of foods suspected is confined to one or a limited number, and especially if the symptoms are very noteworthy, a food diary may be kept. This diary should record all foods consumed and any symptoms that occur coincident with ingestion of those foods. This will aid the physician in determining if an adverse reaction is occurring following the ingestion of a particular food.

However, the most reliable diagnostic procedure is the double-blind, placebo-controlled food challenge (DBPCFC) (63). The DBPCFC will unequivocally link ingestion of a specific food to elicitation of a specific set of symptoms. The DBPCFC should not be used when there is a history of life-threatening anaphylaxis to a suspected food (63). The DBPCFC is particularly useful when the role of a specific food or foods in a reaction is nebulous.

IgE-mediated allergic reactions may be confirmed by skin-prick test (SPT) or by the radioallergosorbent test (RAST) and similar immunoassays (64). The simplest procedure is the SPT (65). A small amount of food extract is applied to the skin, usually on the inside of the forearm or on the back, and the site is pricked with a needle to allow the extract (antigen) to enter. A wheal-and-flare developing at the site demonstrates that IgE in the skin has reacted with some protein in the extract. The RAST test is an in vitro test utilizing a sample of the patient's blood serum (66). The serum is reacted with an allergen that has been bound to some solid matrix. The degree of binding of allergen-specific IgE in the serum to the solid-phase allergen is assessed with radiolabeled antihuman IgE. The results of the RAST are equally as reliable as the SPT, but these tests are considerably more expensive. However, this may be the test of choice for patients with extreme sensitivities because SPTs may be hazardous in such patients (67).

Treatment

Once the offending food or foods has been identified, the major means of treatment for true food allergies is the specific avoidance diet (68). The patient must avoid the food or foods that cause the reaction. For example, if allergic to peanuts, simply avoid peanuts in all forms. Patients must have considerable knowledge of food composition. Dietitians may be helpful in teaching clients to interpret food labels to detect ingredients made from the offending food. Compliance with such avoidance diets is improved if the number of foods eliminated is kept to a minimum. Thus, accurate diagnosis is important. A few hypoallergenic foods are available for use by such patients. For example, most infants with cows' milk allergy can safely be fed alternative formulae such as casein hydrolysate formulae (47). However, there are case reports that have shown occasional adverse reactions to milk protein hydrolysates (48–51).

Individuals with IgE-mediated food allergies are often exquisitely sensitive to the offending food (68,69). Adverse reactions can occur following exposure to trace amounts of the offending food that might arise through various processing or preparation errors (70). Examples include the failure to adequately clean common equipment, the use of re-work (a common practice in certain segments of the industry involving the incorporation of left-over or misformulated quantities of a food product into subsequent batches of related products with identical or related formulations), and the inadvertent addition of an ingredient that is not supposed to be in the formulation (54,71–73). The existence of unlabeled foods especially in restaurant or other food service settings is another concern for the development of safe and effective avoidance diets. Many adverse reactions have occurred in such settings (16,17).

While, for practical purposes, complete avoidance must be maintained, threshold doses do exist below which allergic individuals will not experience adverse reactions. Hourihane *et al.* (74) evaluated the threshold doses in a group of peanut-allergic individuals who were challenged with low but varying doses of peanuts. In this group of 12 individuals, the most sensitive individual began to experience subjective symptoms when exposed to 100 µg of peanut protein and experienced mild, objective symptoms when exposed to 2 mg of peanut protein (74). However, four other peanut-allergic individuals in this same study with equally impressive histories of serious allergic reactions to peanuts had no reaction when exposed to the highest dose used in the trial, 50 mg of peanut protein (74). Thus, threshold doses seem to be low and variable among individuals allergic to any specific food. Also, the possibility exists that the threshold doses vary from one allergenic food to another.

In the construction of safe and effective avoidance diets, questions often arise regarding the need to avoid closely related foods. While cross-reactions can occur between closely related foods in the case of certain food groups, it does not seem to be possible to offer uniform advice on this particular aspect. For example, cross-reactions are known to occur among the various crustacean species (shrimp, crab, lobster, and crawfish) (75), different species of avian eggs (76), and cows' milk and goats' milk (77). In contrast, patients allergic to one or more species of fish can often consume other fish species without adverse reactions (78). However, the patterns of fish allergy are quite variable from one individual to another (78). Also, some peanut-allergic individuals are allergic to other legumes such as soybeans (79), but this is not a common occurrence (80). Clinical hypersensitivity to one legume, such as peanuts or soybeans, does not warrant exclusion of the entire legume family from the diet unless allergy to each individual legume is confirmed by clinical challenge trials (80).

Cross-reactions are also known to occur between certain types of pollens and foods especially with OAS. Examples would include ragweed pollen and melons, mugwort pollen and celery, mugwort pollen and hazelnut, and birch pollen and various foods including carrots, apples, hazelnuts, and potatoes (4).

Cross-reactions are also known to occur between allergies to natural rubber latex and certain foods such as banana, chestnut, and avocado (81).

Pharmacological approaches are available for the treatment of the symptoms that occur during an allergic reaction (82). Epinephrine (adrenaline) and antihistamines can be particularly useful in treating the symptoms of an allergic reaction. Those patients with a history of life-threatening reactions to foods are advised to carry an epinephrine-filled syringe with them at all times (83).

Prevention

The prevention of the development of IgE-mediated food allergies in infants requires early identification of high-risk infants, exclusion of commonly allergenic foods such as cows' milk, eggs and peanuts from the infant diet, breast-feeding for an extended period, possible use of hypoallergenic infant formulae, and the exclusion of commonly allergenic foods from the diet of the nursing mother (84,85). IgE-mediated food allergies are most likely to develop in high-risk infants, those infants born to parents with histories of allergic disease of any type (pollens, mold spores, animal danders, bee venoms, food, etc.). The restriction of the diet of the mother during pregnancy (excluding commonly allergenic foods such as peanuts) does not appear to be helpful in the prevention of allergy in the infant (84,85), which suggests that sensitization does not occur in utero. Breast-feeding for extended periods of time delays, but may not prevent, the development of IgE-mediated food allergies (86). However, infants can be sensitized to allergenic foods through exposure to the allergens in breast milk (87,88). Apparently, certain allergenic food proteins resist digestion, are absorbed at least to a small extent from the small intestine, and are secreted in breast milk leading to sensitization. The exclusion of certain commonly allergenic foods, such as peanuts, from the diet of the nursing mother will help to prevent sensitization through breast milk. While the elimination of peanuts from the diet of lactating women with high-risk infants is often recommended, the elimination of milk and eggs is not because these foods are usually considered to be too important nutritionally to exclude from the diets of lactating women. The use of probiotics during lactation may also help to lessen the likelihood of allergic sensitization (89). Hypoallegenic infant formula may also prevent the development of food allergies in high-risk infants (90), although these formulae are more often used to prevent reactions after sensitization has already occurred. The use of partial whey hydrolysate formula has been advocated for this purpose because the partial hydrolysate is more likely to prevent sensitization than a formula based on whole milk (91). High-risk infants may still develop food allergies once solid foods are introduced into the diet (87).

Effect of processing on allergenicity

Allergenic food proteins are remarkably stable to food processing conditions (20). Most allergenic food proteins are quite heat-stable; typical heat processing conditions do not alter the allergenicity of the resulting products (20). The allergens in some fruit and vegetable allergens are exceptions being quite heat-sensitive (92). And, the allergens present in some species of fish may be destroyed by canning processes, although other heating processes do not appear to affect these allergens (93). Food allergens also tend to be resistant to proteolysis allowing these proteins to survive digestive processes and arrive in the intestine in immunologically active form (20). However, the resistance to proteolysis means that these allergens may survive, in whole or in part, the acid and enzymatic hydrolysis methods used to prepare protein hydrolysates (20).

Impact of agricultural biotechnology

Concerns have arisen regarding the possible allergenicity of foods developed through agricultural biotechnology (8). In the development of transgenic foods one or more genes, coding for specific proteins, are transferred from one biological source to another. Since most food allergens are proteins, the theoretical possibility exists that these novel proteins could be or could become allergens. However, millions of proteins exist in nature including many within the food supply, and only a small percentage of those proteins are allergens (8). So, the probability of transferring a protein with allergenic potential into a transgenic food is quite small (8). Also, a decision-tree strategy has been developed to assess the potential allergenicity of novel proteins in genetically modified foods (94). The probability of transferring an allergen is enhanced if the gene is derived from a known allergenic source. The reliability of the decision tree assessment strategy was documented with the discovery that a gene from Brazil nuts cloned into soybeans to enhance the methionine content of soybeans coded for the heretofore unknown major allergen from Brazil nuts, *Ber e* 1 (61). Commercialization of that particular transgenic variety of soybeans was immediately halted.

DELAYED HYPERSENSITIVTY

Delayed hypersensitivities are cell-mediated involving tissue-bound T lymphocytes that are sensitized to a specific foodborne substance that triggers the reaction (95). These cell-mediated reactions result in tissue inflammation often localized to certain sites within the body. In the case of delayed hypersensitivity reactions, symptoms begin to appear 6–24 hours after consumption of the offending food.

CELIAC DISEASE

Pathogenesis and symptoms

Celiac disease, also known as celiac sprue, non-tropical sprue or gluten-sensitive enteropathy, is a malabsorption syndrome occurring in sensitive individuals upon the consumption of wheat, rye, barley, triticale, spelt, and kamut (96). Following the ingestion of these grains or protein-containing products derived from these grains, the absorptive epithelium of the small intestine becomes damaged. This results in a decreased number of epithelial cells that are critical for digestion and absorption. The mucosal enzymes necessary for digestion and absorption are also altered in the damaged cells. Thus, the absorptive cells are functionally compromised. This mucosal damage leads to nutrient malabsorption. The loss of absorptive function along with the ongoing inflammatory process results in a severe malabsorption syndrome characterized by diarrhea, bloating, weight loss, anemia, bone pain, chronic fatigue, weakness, muscle cramps, and, in children, failure to gain weight and growth retardation (96,97). Intraepithelial T cells in the small intestine may be involved in the inflammatory mechanism occurring with celiac disease (98,99). Thus, celiac disease may be considered as a form of cell-mediated food allergy (98).

In addition to the symptoms that occur within the first few days after ingestion of a product made from one of the offending grains, patients with celiac disease are at increased risk for certain chronic illnesses. For example, such individuals are at increased risk for development of T cell lymphoma (100). Celiac patients also seem to be more likely to have various autoimmune diseases including dermatitis herpetiformis, thyroid diseases, Addison's disease, pernicious anemia, autoimmune thrombocytopenia, sarcoidosis, insulin-dependent diabetes mellitus, and IgA nephropathy (101).

Prevalence

Celiac disease is an inherited trait, although its inheritance is complex (102). Celiac disease occurs in approximately 5% of first degree relatives of celiac patients and approximately 75% of monozygotic twin pairs are concordant for celiac disease (103). HLA class II genes are the major genes associated with celiac disease but concordance for celiac disease is only 25–40% in siblings who are identical for one or both HLA haplotypes. Thus, genes outside of the HLA locus likely play some as-yet undefined role in disease susceptibility.

The exact prevalence of celiac disease remains a subject of active debate. Statistics on the prevalence of celiac disease are complicated by use of different diagnostic approaches in different parts of the world. Celiac disease appears to be latent or asymptomatic in some individuals (104,105). The prevalence of celiac disease appears to be particularly high in certain European regions and in Australia (101,106). Celiac disease occurs in as many as 1 in every 250 people in some European regions (101). In the US, celiac disease seems much less common occurring in about 1 of every 3,000 individuals (107). However improved diagnosis may lead to a higher prevalence rate in the US (108). However, even within European populations, considerable variability is observed in the prevalence of celiac disease (101,106,109).

Causative factor

The ingestion of gliadin from wheat and related prolamin proteins from other grains is associated with the development of celiac disease (97,107). The prolamin fraction of wheat is known as gluten, so celiac disease is sometimes referred to as gluten-sensitive enteropathy. Apparently, a defect in mucosal processing of gliadin in celiac patients provokes the generation of toxic peptides that contribute to the abnormal T cell response and the subsequent inflammatory reaction (110). However, the mechanism involved in celiac disease and the exact role of gliadin remain to be determined.

Treatment

Like IgE-mediated food allergies, celiac disease is treated with use of an avoidance diet (111). Many celiac sufferers attempt to avoid all sources of wheat, rye, barley, and related grains including a wide variety of common food ingredients derived from these grains (111). The need to avoid ingredients that do not contain protein from the implicated grains is somewhat debatable, but widely practiced. Most of these individuals also avoid oats, although the role of oats in the elicitation of celiac disease has recently been refuted (112). However, oats are often contaminated with wheat commercially so some caution may still be necessary with respect to the ingestion of oats by celiac sufferers. Although evidence is scant, spelt and kamut, which are basically varieties of wheat, are likely to trigger celiac disease in susceptible individuals. The tolerance for wheat, rye, barley, and related grains among celiac sufferers is unknown. Most celiac sufferers attempt to practice complete avoidance. An enzyme-linked immunosorbent assay (ELISA) has been developed for the detection of gluten in foods (113). The lower limit of detection of wheat gluten in this ELISA is 0.016%. The availability of this ELISA has led some to conclude that foods containing gluten below 200 ppm are not hazardous for celiac patients. While this has not been conclusively proven, a few isolated studies have concluded that levels of 10 mg of gliadin per day will be tolerated by most patients with active celiac disease (114). Clearly, the threshold dose may vary from one individual to another since, in latent forms of celiac disease, normal dietary quantities of the offending grains seem to cause little problem.

FOOD INTOLERANCES

In contrast to true food allergies, food intolerances involve one of several non-immunological mechanisms. The distinction between true food allergies and food intolerances is important with respect to both mechanism and treatment. Individuals with various types of food intolerances are typically able to tolerate some amount of the offending food or food ingredient. In contrast, the threshold doses for the offending food with true food allergies is extraordinarily small. Thus, the dietary management of food intolerances is much easier. With a very few notable exceptions, little research has been conducted on food intolerances. In many cases, the cause–effect relationship between ingestion of the offending food or food ingredient and the adverse reaction has not even been carefully established.

ANAPHYLACTOID REACTIONS

Anaphylactoid reactions are caused by substances that bring about the non-immunologic release of chemical mediators from mast cells (69). The mediators are the same as in an IgE-mediated food allergy, however, the mechanism does not involve IgE mediation. Evidence for the existence of anaphylactoid reactions is largely circumstantial. None of the histamine-releasing substances in foods has ever been identified.

The best example of this type of reaction is the strawberry "allergy." The strawberry contains very little protein. No allergens have ever been identified, and there is no evidence of the existence of strawberry-specific IgE. However, strawberries have been known to cause allergy-like symptoms, frequently urticaria, in sensitive individuals. The most logical reason is an in vivo release of histamine and other mediators. However, a clear diagnosis of an anaphylactoid reaction is difficult to establish on the basis of such logic alone. The possibility that strawberry "allergy" is a form of OAS has not yet been ruled out either.

METABOLIC FOOD DISORDERS

Metabolic food disorders occur as the result of genetically determined metabolic deficiencies that either affect the host's ability to metabolize a specific substance in foods or heighten an individual's sensitivity to a particular foodborne chemical (69). Lactose intolerance and favism are examples of metabolic food disorders.

Lactose intolerance

Lactose intolerance results from an inherited deficiency of the enzyme, lactase or β-galactosidase, in the intestinal mucosa (115). As a result of this deficiency, lactose, the primary sugar in milk and milk products, cannot be metabolized into the respective monosaccharides, galactose and glucose. The undigested lactose cannot be absorbed by the small intestine and passes into the colon, where bacteria metabolize the lactose into CO_2, and H_2O. The symptoms characteristic of lactose intolerance include bloating, flatulence, abdominal cramping, and frothy diarrhea (115).

Lactose intolerance affects a large number of people worldwide. The frequency is high in black Americans, Native Americans, Hispanics, Asians, Jews, and Arabs, affecting as many as 60–90% (116). In contrast, the prevalence among North American Caucasians in about 6–12% (117). The symptoms of lactose intolerance may rarely be present from birth, but more commonly become apparent later in life as the activity of intestinal lactase diminishes.

The diagnosis of lactose intolerance is made following the administration of a lactose tolerance test (LTT). A fasting patient is challenged orally with a 50-g lactose load. Blood glucose levels are monitored. An increase of blood glucose of

25 mg/dl or more is considered normal. The patient is also monitored for the development of characteristic symptoms.

A diagnosis based on the LTT results may be exaggerated. A 50-g dose of lactose is equivalent to the amount of lactose in a liter or more of milk. This would not be considered a "normal" serving for one individual for one meal. It is possible that some of these individuals may tolerate lactose in smaller quantities. Many individuals with a positive LTT with 50 g of lactose can consume an 8-oz serving of milk with no adverse effects (118,119). The frequency and severity of symptoms increase as the lactose dose exceeds 12 g, the amount found in a cup of milk (120,121).

The usual treatment for lactose intolerance is the avoidance of dairy products containing lactose. Lactose-intolerant individuals can usually tolerate some lactose in their diets (118,119). Additionally, some dairy products, such as yogurt and acidophilus milk are better tolerated than other dairy products, apparently because they contain bacteria with β-galactosidase (122,123). Thus, lactose intolerance is a rather manageable condition.

Favism

Favism is an intolerance to the consumption of fava beans or the inhalation of pollen from the *Vicia faba* plant. As a result, the individual suffers from acute hemolytic anemia (124). The symptoms characteristic of favism include pallor, fatigue, dyspnea, nausea, abdominal and/or back pain, fever, and chills. In rare cases, hemoglobinuria, jaundice, and renal failure may occur. The onset time is quite rapid, usually occurring in 5–24 hours after ingestion. Favism is a self-limited disease with a prompt and spontaneous recovery assuming no further exposure. Favism is most prevalent when the *Vicia faba* plant is in bloom, causing elevated levels of airborne pollen, and also when the bean is available in the market. Fava beans are not commonly consumed in the United States.

Favism affects individuals with an inherited deficiency of erythrocyte glucose-6-phosphate dehydrogenase (G6PDH) (124). G6PDH is a critical enzyme in erythrocytes, where it is essential to maintain adequate levels of the reduced form of glutathione (GSH) and nicotinamide adenine dinucleotide phosphate (NADPH), which prevent oxidation. The red blood cells of individuals with G6PDH deficiency are susceptible to oxidative damage. Fava beans contain several naturally occurring oxidants, including vicine and convicine, which are able to damage the erythrocytes of G6PDH-deficient individuals. G6PDH deficiency is the most common inherited enzymatic defect in humans on a worldwide basis affecting 100 million people (124). The greatest prevalence of G6PDH deficiency occurs among Oriental Jewish communities in Israel, Sardinians, Cypriot Greeks, American blacks, and certain African populations. The trait is virtually non-existent in northern European nations, North American Indians, and Eskimos. Favism occurs primarily in the Mediterranean area, the Middle East, China, and Bulgaria, where the genetic trait is fairly prevalent and where the fava bean is frequently consumed. The diagnosis of G6PDH deficiency is made through an assay for enzymatic activity in isolated red blood cells. The treatment of favism is to avoid the ingestion of fava beans and/or the inhalation of the plant pollen.

IDIOSYNCRATIC REACTIONS

Idiosyncratic reactions refer to those adverse reactions to food experienced by certain individuals for which the mechanism is unknown. An immense number of mechanisms may be involved in these reactions, allowing a wide variety of ill-nesses with symptoms that can range from very minor to life-threatening. The role of specific foods or food ingredients has not been well established in many of these illnesses. The best example of an idiosyncratic reaction is sulfite-induced asthma (125). In the case of sulfite-induced asthma, the cause–effect relationship is very well established. However, the cause–effect relationship between a specific food or food component and a particular adverse reaction has often not been well established in other types of food idiosyncrasies. Psychosomatic illnesses are also included in this category.

Sulfite-induced asthma

Sulfites are used as additives in a variety of foods with residual levels ranging from <10 ppm in a variety of foods to >2,000 ppm in dried fruits (126). They can also occur naturally in foods, especially fermented foods such as wine, but rather low levels are typically formed during food fermentation (126). They are commonly used in the food processing industry to control enzymatic and non-enzymatic browning, to prevent undesirable bacterial growth, to condition doughs, to provide antioxidant protection, and as a bleaching agent in the production of maraschino cherries and hominy (126). Sulfites are also added to many pharmaceutical products as an antioxidant and for the prevention of non-enzymatic browning.

Asthma is the most prominent symptom involved in sulfite sensitivity (127–129). Only a small percentage of asthmatics are sulfite-sensitive (129). Severe asthmatics who are dependent on steroids for control of their asthma symptoms are the primary risk group for sulfite sensitivity (129). However, only about 5% of these severe asthmatics are sulfite-sensitive (129). However, sulfite-induced asthma can be quite severe; deaths have been documented (126,130).

The only effective technique for diagnosing sulfite sensitivity is through a sulfite challenge test in a clinical setting (131). The double-blind challenge may be conducted with capsules or acidic solutions of sulfite. Due to the release of SO_2 vapor from acidic beverages, asthmatics may be more sensitive to the acidic beverages than to capsules (132). The inhalation of SO_2 vapors during ingestion of acidic beverages may be a factor (132).

Sulfites are very reactive chemicals in foods (126). They can react with a number of food components including reducing sugars, proteins, vitamins, ketones, and aldehydes. Very little free sulfite remains in most foods.

Sulfite-sensitive asthmatics must avoid certain sulfited foods and beverages in their diets (66). Sulfites added to foods must be declared on product labels so avoidance diets are reasonably easy to develop (125). Sulfite-sensitive asthmatics can tolerate the ingestion of small quantities of sulfites, especially when the sulfites are incorporated in certain types of foods (125,127,133). The threshold for sulfite does vary from one patient to another. However, reactions have not been observed to food products containing <10 ppm SO_2 residues which is the limit for labeling purposes. Thus, while sulfite-induced asthma poses a considerable risk to sensitive individuals, this condition is manageable once it is recognized.

Tartrazine sensitivity

The azo dye, tartrazine (FD&C Yellow #5), is one of the dyes approved for use in foods and beverages as a coloring agent. Controversy exists over the existence of tartrazine sensitivity (131,134). In 1959, Lockey (135) conducted unblinded oral challenges with tartrazine, claiming three patients presented with a rash after ingesting the dilute tartrazine solutions. Numerous other studies, including double-blind studies, reported tartrazine involvement in chronic urticaria (131). In many of the studies reviewed, antihistamines were withheld or the information concerning pretreatment with antihistamines was not provided (131,134). The urticarial reactions observed in the challenge trials may have resulted from the withdrawal of antihistamines or other medications rather than the challenge with tartrazine (131, 134). In a double-blind, placebo-controlled challenge trial with tartrazine in chronic urticaria in which antihistamines were not withheld, only 1 of 24 patients developed urticaria and this reaction required a rather large provoking dose of 50 mg (134).

Many studies have been done to link tartrazine to asthma (131,134). These studies also often involved the withholding of medications or failed to provide information concerning the administration of antihistamines and bronchodilators during the course of the study. In reviews of these studies, it was concluded that the results did not establish the existence of tartrazine-induced asthma but that the change in FEV_1 (forced expiratory volume in 1 second as measured by spirometry) values reflected an exacerbation of asthma symptoms due to withholding bronchodilators (131,134,136). In studies in which medications were not witheld, tartrazine did not induce asthma (131,134,136).

Monosodium glutamate sensitivity

Monosodium glutamate (MSG) is a widely used food additive. MSG is used in commercial preparation of manufactured and restaurant foods to enhance flavors. MSG also occurs naturally in virtually all foods. MSG is the sodium salt of one of the most common amino acids in the human body.

MSG ingestion has been linked primarily with Chinese restaurant syndrome (CRS), now more appropriately called MSG-symptom complex. MSG-symptom complex is a mild, subjective, and transient syndrome characterized by burning and flushing, pressure and tightness, tingling, and numbness restricted to the face, neck, upper chest, and upper arms (137). The symptoms involved in this syndrome are highly subjective and cannot be measured objectively during challenge studies. However, the results of many challenge studies do not provide any compelling evidence for the existence of MSG-symptom complex (137). Most subjects experienced no symptoms or there were no differences in frequency of reactions to MSG vs. placeboes. A few individuals may react with the mild, subjective symptoms described above when exposed to doses of MSG exceeding 3 g in a single meal (137). Although it is possible to ingest 3 g of MSG in a single meal, it would not be a common occurrence. Thus, the MSG-symptom complex has not been firmly established and occurs, if it occurs at all, only at very high levels of exposure in a very small group of susceptible individuals (138).

Asthma provoked by MSG has also been documented by several challenge studies (139). However, there is only one report of MSG-induced asthma that was confirmed by DBPCFC (140). The remainder of the cases are anecdotal reports or reports of single-blind challenges performed in patients with a history of unstable asthma and conducted with placeboes administered first in the challenge sequence. Some positive reactions to MSG occurred 6–12 hours after challenge (141,142); these are likely to be false-positive responses. Any positive responses tended to occur after challenge with rather large MSG doses of 2 g or more. Although MSG may possibly provoke asthma in a few individuals, more careful studies including DBPCFCs will be necessary to confirm this association and determine its prevalence.

Other idiosyncratic reactions

Many of the other idiosyncratic reactions also involve various food additives. Examples would include the role of other food colors in asthma and/or chronic urticaria, the role of butylated hydroxyanisole (BHA) and butylated hydroxytoluene (BHT) in chronic urticaria, and the role of aspartame in migraine headache and/or urticaria (131). As previously noted, the cause–effect relationship has not been firmly established for most of these conditions. In the case of aspartame and its alleged role in migraine headaches and/or urticaria, the relationships have been disproven by subsequent clinical studies (143–145). For the other food additive-associated idiosyncratic reactions, the associations remain unproven and unlikely.

ALLERGY-LIKE INTOXICATIONS

Histamine poisoning, also known as scombroid fish poisoning, is the sole example of an allergy-like intoxication (146). Although histamine poisoning is not a food sensitivity, the symptoms are often confused with an allergic reaction. Histamine poisoning is the result of consuming foods containing high levels of histamine. All consumers are susceptible to histamine poisoning.

Mechanism

Histamine is formed in foods primarily from the growth of bacteria possessing the enzyme, histidine decarboxylase, which converts the naturally occurring amino acid, histidine, to histamine (146). Histamine poisoning is precipitated by the ingestion of foods containing high levels of histamine. Such foods do not always appear to be spoiled. If sufficient levels of histamine are ingested, the body's ability to metabolize and detoxify the histamine is overwhelmed (146). Excess histamine is absorbed, enters the bloodstream, and encounters histamine receptors in various tissues. The interaction of this histamine with the tissue-based receptors produces symptoms that are very similar to those encountered in IgE-mediated allergic reactions.

Incidence and implicated foods

Histamine poisoning is frequently associated with the ingestion of spoiled fish, both scombroid fish (tuna, mackerel, bonito) and non-scombroid fish (mahi-mahi, bluefish, sardines) (146). The only other food associated with histamine poisoning is cheese, especially Swiss cheese (147).

Most of the cases of histamine poisoning are not reported or correctly diagnosed. It is a very mild and self-limited illness in most cases, and a physician is probably seldom consulted. Thus, the true prevalence is difficult to estimate, although several dozen cases are reported by the Centers for Disease Control in the US each year. Other countries reporting cases of histamine poisoning include England, Japan, Australia, and other European countries. The potential exists wherever fish is consumed on a regular basis.

Symptoms

The symptoms that manifest from histamine poisoning are very similar in nature to those seen with IgE-mediated food allergies. Common symptoms include rash, pruritis, urticaria, gastrointestinal disturbances, oral burning sensation, flushing, and hypotension. The onset time for histamine poisoning is relatively short, ranging from a few minutes to a few hours. The mild symptoms usually resolve within a few hours. Most affected individuals will experience only a few symptoms. The severity of symptoms experienced may be dependent upon the amount of histamine consumed.

Histamine poisoning can be distinguished from IgE-mediated food allergy on the following bases: (a) the patient does not have a previous history of allergy to the food in question, (b) the patient has consumed one of the foods commonly implicated in histamine poisoning immediately prior to onset of symptoms, and (c) other consumers of the implicated food are afflicted with the same symptoms (146). The attack rate in group outbreaks of histamine poisoning is often 50–100% (146). While the diagnosis of histamine poisoning may be inferred from the patient history, foods consumed, symptoms experienced, group attack rate, and the onset time of the illness, the only method to confirm the diagnosis of histamine poisoning is to analyze the implicated food for the presence of histamine (146). When cases of histamine poisoning are suspected, the local health authorities should be notified so that they can obtain the samples necessary for analysis. Notification of local health authorities will also lead to the removal of contaminated fish or other foods from the marketplace to prevent further cases if suspicions are confirmed by analysis.

Control measures

Proper handling and storage temperatures are necessary to control histamine formation in fish (146). Raw fish should be kept cold at all times. Histamine is heat-stable, so it survives heat processing and preparation procedures.

The treatment of histamine poisoning is the administration of antihistamines (148). Even without treatment, the duration of symptoms is short-lived. Most symptoms will pass within a few hours, some lasting as long as 24–48 hours. Treatment with antihistamines is recommended, as it will shorten the duration and lesson the discomfort associated with the illness (148).

CONCLUSIONS

Food allergies and intolerances affect only a small percentage of the population. However, these reactions can be quite severe and even life-threatening in some of these individuals. The primary means of management of these illnesses is the avoidance of the offending food(s). However, the avoidance of a specific food can be difficult, and complete success is unlikely.

IgE-mediated food allergies are well understood and relatively easy to diagnose even though management can be difficult. Although many food allergens have been identified, further progress is needed in the identification and characterization of food allergens, methods for their detection in foods, the removal or inactivation of these allergens to provide hypoallergenic foods, and the development of improved treatment modalities.

The existence of other types of allergic mechanisms in food allergies remains to be established. However, celiac disease may be an example of an important illness that involves such mechanisms.

The role of foods in certain food intolerances such as lactose intolerance and sulfite-induced asthma is well established. However, for the vast majority of these illnesses, the relationship to foods remains to be firmly established, and the mechanism of the illness remains undetermined. Obviously, progress in the understanding of these illnesses will be necessary if there is to be any hope for improved treatment of such illnesses.

REFERENCES

1. Sloan AE, Powers ME. A perspective on popular perceptions of adverse reactions to foods. *J Allergy Clin Immunol* 1986; 78:127–32.
2. Altman DR, Chiaramonte LT. Public perception of food allergy. *Environ Toxicol Pharmacol* 1997; 4:95–9.
3. Mekori YA. Introduction to allergic disease. *Crit Rev Food Sci Nutr* 1996; 36: S1–S18.
4. Bush RK, Hefle SL. Food allergens. *Crit Rev Food Sci Nutr* 1996; 36:S119–S163.
5. Anderson JA. The establishment of common language concerning adverse reactions to foods and food additives. *J Allergy Clin Immunol* 1986; 78:140–3.
6. Taylor SL. Allergic and sensitivity reactions to food components. In: Hathcock JN, ed. *Nutritional toxicology*, vol II. New York: Academic Press; 1987; 173–98.
7. Taylor SL, Hefle SL. Allergylike intoxications from foods. In: Frieri M, Kettelhut B, eds. *Food hypersensitivity and adverse reactions – a practical guide for diagnosis and management*. New York: Marcel Dekker; 1999; 141–53.
8. Taylor SL. Genetically-engineered foods: commercial potential, safety, and allergenicity. *Food Australia* 1996; 48:308–11.
9. Bush RK, Taylor SL. Histamine. In: Macrae R, Robinson RK, Sadler M, eds. *Encyclopedia of Food Science, Food Technology and Nutriton*, Vol. 4. London: Academic Press; 1993; 2367–71.
10. Johnston SL, Holgate ST. Cellular and chemical mediators: their roles in allergic diseases. *Curr Opinion Immunol* 1990; 513–24.
11. Sicherer SH, Sampson HA. Cows' milk protein-specific IgE concentrations in two age groups of milk-allergic children and in children achieving clinical tolerance. *Clin Exp Allergy* 1999; 29:507–12.
12. Strobel S. Oral tolerance: immune responses to food antigens. In: Metcalfe DD, Sampson HA, Simon RA, eds. *Food allergy – adverse reactions to foods and food additives*, 2nd edn. Boston: Blackwell Science; 1997; 107–35.
13. Chandra RK. Food allergy: setting the theme. In: Chandra RK, ed. *Food allergy*. St. John's, Newfoundland: Nutrition Research Education Foundation; 1987; 3–5.
14. Lack G, Fox DES, Golding J. The role of the uterine environment in the pathogenesis of peanut allergy. *J Allergy Clin Immunol* 1999; 103:S95.
15. Sicherer SH, Furlong TJ, Gelb BD, Desnick RJ, Sampson HA. Peanut allergy in twins. *J Allergy Clin Immunol* 2000; 105:S181.
16. Sampson HA, Mendelson L, Rosen J. Fatal and near-fatal anaphylactic reactions to foods in children and adolescents. *N Engl J Med* 1992; 327:380–4.
17. Yuninger JW, Sweeney KG, Sturner WQ, Giannandrea LA, Teigland JD, Bray M, Benson PA, York J, Biedrzycki L, Squillace D, Helm R. Fatal food-induced anaphylaxis. *J Am Med Assoc* 1988; 260:1450–2.
18. Miller R, Ghatek A, Rothman P, Neugut A. Anaphylaxis in the United States: an investigation into its epidemiology. *J Allergy Clin Immunol* 2000; 105:S349.

19. Pastorello E, Ortolani C. Oral allergy syndrome. In: Metcalfe DD, Sampson HA, Simon RA, eds. *Food allergy – adverse reactions to foods and food additives*, 2nd edn. Boston: Blackwell Science; 1997; 221–33.

20. Taylor SL, Lehrer SB. Principles and characteristics of food allergens. *Crit Rev Food Sci Nutr* 1996; 36:S91–S118.

21. Calkhoven PG, Aalbers M, Koshte VL, Pos O, Oei HD, Aalberse, RC. Cross-reactivity among birch pollen, vegetables and fruits as detected by IgE antibodies is due to at least three distinct cross-reactive structures. *Allergy* 1987; 42:382–90.

22. Tilles SA, Schocket AL. Exercise- and pressure-induced syndromes. In: Metcalfe DD, Sampson HA, Simon RA, eds. *Food allergy – adverse reactions to foods and food additives*, 2nd edn. Boston: Blackwell Science; 1997; 303–9.

23. Maulitz RM, Pratt DS, Schocket AL. Exercise-induced anaphylactic reaction to shell-fish. *J Allergy Clin Immunol* 1979; 63:433–4.

24. Palosuo K, Alenius H, Varjonen E, Koivuluhta M, Mikkola J, Keskinen H, Kalkkinen N, Reunala T. A novel wheat gliadin as a cause of exercise-induced anaphylaxis. *J Allergy Clin Immunol* 1999; 103:912–17.

25. Kidd JM, Cohen SH, Sosman AJ, Fink JN. Food-dependent exercise induced anaphylaxis. *J Allergy Clin Immunol* 1983; 71:407–11.

26. Buchbinder EM, Bloch KJ, Moss KJ, Guiney TE. Food-dependent, exercise-induced anaphylaxis. *J Am Med Assoc* 1983; 250:2973–4.

27. Neistijl Jansen JJ, Kardinaal AFM, Huijbers G, Vlieg-Boestra BJ, Martens BPM, Ock-huizen T. Prevalence of food allergy and intolerance in the adult Dutch population. *J Allergy Clin Immunol* 1994; 93:446–56.

28. Sampson HA. Food allergy. *Curr Opinion Immunol* 1990; 2:542–7.

29. Bock SA, Lee WY, Remigio LK, May CD. Studies of hyper-sensitivity reactions to foods in infants and children. *J Allergy Clin Immunol* 1978; 62:327–34.

30. Halpern SR, Sellars WA, Johnson RB, Anderson DW, Saperstein S, Reisch JS. Development of childhood allergy in infants fed breast, soy or cow milk. *J Allergy Clin Immunol* 1973; 51:139–51.

31. Sicherer SH, Munoz-Furlong A, Burks AW, Sampson HA. Prevalence of peanut and tree nut allergy in the U.S. determined by a random digit dial telephone survey. *J Allergy Clin Immunol* 1999; 103:559–62.

32. Emmett SE, Angus FJ, Fry JS, Lee PN. Perceived prevalence of peanut allergy in Great Britain and its association with other atopic conditions and with peanut allergy in other household members. *Allergy* 1999; 54:380–5.

33. Hill DJ, Hosking CS. Patterns of clinical disease associated with cow milk allergy in childhood. *Nutr Res* 1992; 12:109–21.

34. Bock SA, Atkins FM. The natural history of peanut allergy. *J Allergy Clin Immunol* 1989; 83:900–4.

35. Ford RPK, Taylor B. Natural history of egg hypersensitivity. *Arch Dis Child* 1982; 57:649–52.

36. Sampson HA, Scanlon SM. Natural history of food hypersensitivity in children with atopic dermatitis. *J Pediatr* 1989; 115:23–7.

37. Food and Agricultural Organization of the United Nations, *Report of the FAO Technical Consultation on Food Allergies*, Rome, Italy, November 13–14, 1995.

38. Kanny G, de Hauteclocque C, Moneret-Vautrin DA. Sesame seed and sesame seed oil contain masked allergens of growing importance. *Allergy* 1996; 51:952–7.

39. Sporik R, Hill D. Allergy to peanuts, nuts, and sesame seed in Australian children. *Br Med J* 1996; 313:1477–8.

40. Bruijnzeel-Koomen C, Ortolani C, Aas K, Bindslev-Jensen C, Bjorksten B, Moneret-Vautrin D, Wuthrich B. Adverse reactions to food. *Allergy* 1995; 50:623–35.

41. Hefle SL, Nordlee JA, Taylor SL. Allergenic foods. *Crit Rev Food Sci Nutr* 1996; 36:S69–S89.

42. Hourihane JO'B, Bedwani SJ, Dean TP, Warner JO. Randomised, double-blind, crossover challenge study of allergenicity of peanut oils in subjects allergic to peanut. *Br Med J* 1997; 314:1084–8.

43. Bush RK, Taylor SL, Nordlee JA, Busse WW. Soybean oil is not allergenic to soybean-sensitive individuals. *J Allergy Clin Immunol* 1985; 76:242–5.

44. Halsey AB, Martin ME, Ruff ME, Jacobs FO, Jacobs RL. Sunflower oil is not allergenic to sunflower seed-sensitive patients. *J Allergy Clin Immunol* 1986; 78:408–10.

45. Teuber SS, Brown RL, Haapanen LAD. Allergenicity of gourmet nut oils processed by different methods. *J Allergy Clin Immunol* 1997; 99:502–7.

46. Hoffman DR, Collins-Williams C. Cold-pressed peanut oils may contain peanut allergen. *J Allergy Clin Immunol* 1994; 93:801–2.

47. Hill DJ, Hosking CS. The management and prevention of food allergy. In: Frieri M, Kettelhut B, eds. *Food hypersensitivity and adverse reactions – a practical guide to diagnosis and management.* New York: Marcel Dekker; 1999; 423–47.

48. Saylor JD, Bahna SL. Anaphylaxis to casein hydrolysate formula. *J Pediatr* 1991; 118:71–4.

49. Rosenthal E, Schlesinger Y, Birnbaum Y, Goldstein R, Benderly A, Freier S. Intolerance to casein hydrolysate formula. *Acta Paediatr Scand* 1991; 80:958–60.

50. Businco L, Cantani A, Longhi M, Giampietro PG. Anaphylactic reactions to a cow's milk whey protein hydrolysate (Alfa-Re Nestle) in infants with cow's milk allergy. *Ann Allergy* 1989; 62:333–5.

51. Tarim O, Anderson VM, Lifshitz F. Fatal anaphylaxis in a very young infant possibly due to a partially hydrolyzed whey formula. *Arch Pediatr Adolescent Med* 1994; 148:1224–8.

52. Muller U, Weber W, Hoffmann A, Franke S, Lange R, Vieths S. Commercial soybean lecithins: a source of hidden allergens? *Z Lebensm Unter Forsch* 1998; 207:341–51.

53. Taylor SL, Dormedy ES. The role of flavoring substances in food allergy and intolerance. *Adv Food Nutr Res* 1998; 42:1–44.

54. Gern JE, Yang E, Evrard HM, Sampson HA. Allergic reactions to milk-contaminated "non-dairy" products. *N Engl J Med* 1991; 324:976–9.

55. McKenna C, Klontz KC. Systemic allergic reaction following the ingestion of undeclared peanut flour in a peanut-sensitive woman. *Ann Allergy Asthma Immunol* 1997; 79:234–6.

56. St. Vincent JCM, Watson WTA. Unsuspected source of cow's milk protein in food. *J Allergy Clin Immunol* 1994; 93:209.

57. Taylor SL. Immunologic and allergic properties of cows' milk proteins in humans. *J Food Prot* 1986; 49:239–50.

58. Baldo BA. Milk allergies. *Aust J Dairy Technol* 1984; 39:120–8.

59. Barnett D, Baldo BA, Howden MEH. Multiplicity of allergens in peanuts. *J Allergy Clin Immunol* 1983; 72:61–8.

60. Elsayed S, Bennich H. The primary structure of allergen M from cod. *Scand J Immunol* 1975; 4:203–8.

61. Nordlee JA, Taylor SL, Townsend JA, Thomas LA, Bush RK. Identification of Brazil nut allergen in transgenic soybeans. *N Engl J Med* 1996; 334:688–92.

62. Daul CB, Slattery M, Reese G, Lehrer SB. Identification of the major brown shrimp (*Penaeus aztecus*) allergen as the muscle protein tropomyosin. *Int Arch Allergy Immunol* 1994; 105:49–55.

63. Bock SA, Sampson HA, Atkins FM, Zeiger RS, Lehrer S, Sachs M, Bush RK, Metcalfe DD. Double-blind, placebo-controlled food challenge (DBPCFC) as an office procedure: a manual. *J Allergy Clin Immunol* 1988; 82:986–97.

64. Van Arsdel PP, Larson EB. Diagnostic tests for patients with suspected allergic disease. Utility and limitations. *Ann Int Med* 1989; 110:304–12.

65. Dreborg S. Skin test in diagnosis of food allergy. *Allergy Proc* 1991; 12:251–4.

66. Adolphson CR, Yunginger JW, Gleich GJ. Standardization of allergens. In: Rose NR, Friedman H, eds. *Manual of clinical immunology*, 2nd edn. Washington DC: American Soc Microbiology; 1980; 778–88.

67. Metcalfe DD. Diagnostic procedures for immunologically-mediated food sensitivity. *Nutr Rev* 1984; 42:92–7.

68. Taylor SL, Bush RK, Busse WW. Avoidance diets – how selective should we be? *N Engl Reg Allergy Proc* 1986; 7:527–32.

69. Taylor SL, Nordlee JA, Rupnow JH. Food allergies and sensitivities. In: Taylor SL, Scanlan RA, eds. *Food toxicology – a perspective on the relative risks*. New York: Marcel Dekker; 1989; 255–95.

70. Taylor SL. Elimination diets in the diagnosis of atopic dermatitis. *Allergy* 1989; 44(Suppl 9):97–100.

71. Taylor SL. Chemistry and detection of food allergens. *Food Technol* 1992; 46(5): 146–52.

72. Yunginger JW, Gauerke MB, Jones RT, Dahlberg MJE, Ackerman SJ. Use of radio-immunoassay to determine the nature, quantity and source of allergenic contamination of sunflower butter. *J Food Prot* 1983; 46:625–8.

73. Laoprasert N, Wallen ND, Jones RT, Hefle SL, Taylor SL, Yunginger JW. Anaphylaxis in a milk-allergic child following ingestion of lemon sorbet containing trace quantities of milk. *J Food Prot* 1998; 61:1522–4.

74. Hourihane JO'B, Kilburn SA, Nordlee JA, Hefle SL, Taylor SL, Warner JO. An evaluation of the sensitivity of subjects with peanut allergy to very low doses of peanut: a randomized, double-blind, placebo-controlled food challenge study. *J Allergy Clin Immunol* 1997; 100:596–600.

75. Daul CB, Morgan JE, Lehrer SB. Hypersensitivity reactions to crustacea and mollusks. *Clin Rev Allergy* 1993; 11:201–22.

76. Langeland T. A clinical and immunological study of allergy to hen's egg white. VI. Occurrence of proteins cross-reacting with allergens in hen's egg white as studied in egg white from turkey, duck, goose, seagull, and in hen egg yolk, and hen and chicken sera and flesh. *Allergy* 1983; 38:399–412.

77. Bernard H, Creminon C, Negroni L, Peltre G, Wal JM. IgE cross-reactivity with caseins from different species in humans allergic to cows' milk. *Food Agric Immunol* 1999; 11:101–11.

78. Bernhisel-Broadbent J, Scanlon SM, Sampson HA. Fish hypersensitivity. I. In vitro and oral challenge results in fish allergic patients. *J Allergy Clin Immunol* 1992; 89:730–7.

79. Herian AM, Taylor SL, Bush RK. Identification of soybean allergens by immunoblotting with sera from soy-allergic adults. *Int Arch Allergy Appl Immunol* 1990; 92:193–8.

80. Bernhisel-Broadbent J, Sampson HA. Cross-allergenicity in the legume botanical family in children with food hypersensitivity. *J Allergy Clin Immunol* 1989; 83:435–40.

81. Blanco C, Carrillo T, Castillo R, Quiralte J, Cuevas M. Latex allergy: clinical features and cross-reactivity with fruits. *Ann Allergy* 1994; 73:309–14.

82. Furukawa CT. Nondietary management of food allergy. In: Chiaramonte LT, Schneider AT, Lifshitz F, eds. *Food allergy – a practical approach to diagnosis and management.* New York: Marcel Dekker; 1988; 365–75.

83. Atkins FM. The basis of immediate hypersensitivity reactions to foods. *Nutr Rev* 1983; 41:229–34.

84. Kjellman NI, Bjorksten B. Natural history and prevention of food hypersensitivity. In: Metcalfe DD, Sampson HA, Simon RA, eds. *Food allergy – adverse reactions to foods and food additives*, 2nd edn. Boston: Blackwell Scientific; 1997; 445–59.

85. Zeiger RS, Heller S, Mellon MH, Forsythe AB, O'Connor RD, Hamburger RN, Schatz, M. Effect of combined maternal and infant food-allergen avoidance on development of atopy in early infancy: a randomized study. *J Allergy Clin Immunol* 1989; 84:72–89.

86. Zeiger RS, Heller S. The development and prediction of atopy in high-risk children: follow-up at seven years in a prospective randomized study of combined maternal and infant food allergy avoidance. *J Allergy Clin Immunol* 1995; 95:1179–90.

87. Van Asperen PP, Kemp AS, Mellis CM. Immediate food hypersensitivity reactions on the first known exposure to food. *Arch Dis Child* 1983; 58:253–6.

88. Gerrard JW. Allergy in breast fed babies to ingredients in breast milk. *Ann Allergy* 1979; 42:69–72.

89. Kirjavainen PV, Apostolou E, Salminen SJ, Isolauri E. New aspects of probiotics – a novel approach in the management of food allergy. *Allergy* 1999; 54:909–15.

90. Businco L, Dreborg S, Einarsson R, Giampietro PG, Host A, Keller RM, Strobel S, Wahn U. Hydrolysed cow's milk formulae. Allergenicity and use in treatment and prevention. An ESPACI position paper. *Pediatr Allergy Immunol* 1993; 4:101–11.

91. Vandenplas Y, Hauser B, Van den Borre C, Clybouw C, Mahler T, Hachimi-Idrissi S, Deraeve L, Malfroot A, Dab I. The long-term effect of a partial whey hydrolysate formula on the prophylaxis of atopic disease. *Eur J Pediatr* 1995; 154:488–94.

92. Jankiewicz A, Baltes W, Bogl K, Werner K, Dehne LI, Jamin A, Hoffmann A, Haustein D, Vieths S. Influence of food processing on the immunochemical stability of celery allergens. *J Sci Food Agric* 1997; 75:357–70.

93. Bernhisel-Broadbent J, Strause D, Sampson HA. Fish hypersensitivity. II. Clinical relevance of altered fish allergenicity caused by various preparation methods. *J Allergy Clin Immunol* 1992; 90:622–9.

94. Metcalfe DD, Astwood JD, Townsend R, Sampson HA, Taylor SL, Fuchs RL. Assessment of the allergenic potential of foods derived from genetically engineered crop plants. *Crit Rev Food Sci Nutr* 1996; 36:S165–86.

95. Kniker WT. Delayed and non-IgE-mediated reactions. In: Frieri M, Kettelhut B, eds. *Food hypersensitivity and adverse reactions – a practical guide to diagnosis and management.* New York: Marcel Dekker; 1999; 165–217.

96. Ferguson, A. Gluten-sensitive enteropathy (celiac disease). In: Metcalfe DD, Sampson HA, Simon RA, eds. *Food allergy – adverse reactions to foods and food additives*, 2nd edn. Boston: Blackwell Science; 1997; 287–301.

97. Skerritt JH, Devery JM, Hill AS. Gluten intolerance: chemistry, celiac-toxicity, and detection of prolamins in foods. *Cereal Foods World* 1990; 35:638–44.

98. Strober W. Gluten-sensitive enteropathy: a nonallergic immune hypersensitivity of the gastrointestinal tract. *J Allergy Clin Immunol* 1986; 78:202–11.
99. Logan RFA, Rifkind EA, Turner ID, Ferguson A. Mortality in celiac disease. *Gastroenterol* 1989; 97:265–71.
100. Logan RFA. Descriptive epidemiology of celiac disease. In: Branksi D, Rozen P, Kagnoff MF, eds. *Gluten-sensitive enteropathy, Frontiers in Gastrointestinal Research*, vol 19. Basel: Karger; 1992; 1–14.
101. Kagnoff MF. Immunobiology of coeliac disease. In Domschke W, Stoll R, Brasitus TA, Kagnoff MF, eds. *Intestinal mucosa and its disease*. London: Kluwer Academic Publishers; 1998; 313–22.
102. Greco L, Corazza MCB, Clot F, Fulchignoni-Lataud MC, Percopo S, Zavattari P, Bouguerra F, Dib C, Tosi R, Troncone A, Ventura A, Mantavoni W, Maguzza G, Gatti R, Lazzari A, Guinta A, Perri F, Iaconon G, Cardi E, de Virgiliis S, Cataldo F, de Angelis G, Musumeci S, Ferrari R, Balli F, Bardella MT, Volta U, Catassi C, Torre G, Eliaou JF, Serre JL, Clerget-Darpoux F. Genome search in celiac disease. *Am J Hum Genet* 1998; 62:669–75.
103. Holtmeier W, Rowell DL, Nyberg A, Kagnoff MF. Distinct δ T cell receptor repertoires in monozygotic twins concordant for coeliac disease. *Clin Exp Immunol* 1997; 107:148–57.
104. Duggan JM. Recent developments in our understanding of adult coeliac disease. *Med J Australia* 1997; 166:312–5.
105. Troncone R. Latent coeliac disease in Italy. *Acta Pediatr* 1995; 54:1252–7.
106. Troncone R, Greco L, Auricchio S. Gluten-sensitive enteropathy. *Pediatr Clin North Am* 1996; 43:355–73.
107. Kasarda DD. The relationship of wheat protein to celiac disease. *Cereal Foods World* 1978; 23:240–4, 262.
108. Fasano A. Where have all the American celiacs gone? *Acta Pediatr* 1996; Suppl 412:20–4.
109. George EK, Mearin ML, van der Velde EA, Houwen RHJ, Bouquet J, Gijsbers CFM, Vandenbroucke JP. Low incidence of childhood celiac disease in the Netherlands. *Pediatr Res* 1995; 37:213–18.
110. Cornell HJ. Coeliac disease: a review of the causative agents and their possible mechanisms of action. *Amino Acids* 1996; 10:1–19.
111. Hartsook EI. Celiac sprue: sensitivity to gliadin. *Cereal Foods World* 1984; 29: 157–8.
112. Janatuinen EK, Pikkarainen PH, Kemppainen TA, Kosma VM, Jarvinen MK, Uusitupa MIJ, Julkunen RJK. A comparison of diets with and without oats in adults with celiac disease. *N Engl J Med* 1995; 333:1033–7.
113. Skerritt JH, Hill AS. Enzyme immunoassay for determination of gluten in foods: collaborative study. *J Assoc Off Anal Chem* 1991; 74:257–64.
114. Hekkens WTJM, van Twist de Graaf M. What is gluten-free – levels and tolerances in the gluten-free diet. *Nahrung* 1990; 34:483–7.
115. Suarez FL, Savaiano DA. Diet, genetics, and lactose intolerance. *Food Technol* 1997; 51(3):74–6.
116. Kocian J. Lactose intolerance. *Int J Biochem* 1988; 20:1–5.
117. Sandine WE, Daly M. Milk intolerance. *J Food Prot* 1979; 42:435–7.
118. Welsh D. Diet therapy in adult lactose malabsorption: present practices. *Am J Clin Nutr* 1978; 31:592–6.

119. Suarez FL, Savaiano DA, Levitt MD. A comparison of symptoms after the consumption of milk or lactose-hydrolyzed milk by people with self-reported severe lactose intolerance. *N Engl J Med* 1995; 333:1–4.

120. Johnson AO, Semenya JG, Buchowski MS, Enwonwu CO, Scrimshaw NS. Correlation of lactose maldigestion, lactose intolerance and milk intolerance. *Am J Clin Nutr* 1993; 57:399–401.

121. Reasoner J, Maculan TP, Rand AG, Thayer WR. Clinical studies with low-lactose milk. *Am J Clin Nutr* 1981; 34:54–60.

122. Gallagher CR, Molleson AL, Caldwell JH. Lactose intolerance and fermented dairy products. *Cult Dairy Prod J* 1975; 10(1):22–4.

123. Kolars JC, Levitt MD, Aouji M, Savaiano DA. Yogurt: an autodigesting source of lactose. *N Engl J Med* 1984; 310:1–3.

124. Mager J, Chevion M, Glaser G. Favism. In: Liener IE, ed. *Toxic constituents of plant foodstuffs*, 2nd edn. New York: Academic Press; 1980; 265–94.

125. Taylor SL, Bush RK, Nordlee JA. Sulfites. In: Metcalfe DD, Sampson HA, Simon RA, eds. *Food allergies – adverse reactions to foods and food additives*, 2nd edn. Boston: Blackwell Scientific; 1997; 339–57.

126. Taylor SL, Higley NA, Bush RK. Sulfites in foods: uses, analytical methods, residues, fate, exposure assessment, metabolism, toxicity, and hypersensitivity. *Adv Food Res* 1986; 30:1–76.

127. Stevenson DD, Simon RA. Sensitivity to ingested metabi-sulfites in asthmatic subjects. *J Allergy Clin Immunol* 1981; 68:26–32.

128. Baker GJ, Collett P, Allen DH. Bronchospasm induced by metabisulfite-containing foods and drugs. *Med J Australia* 1981; 2:614–17.

129. Bush RK, Taylor SL, Holden K, Nordlee JA, Busse WW. The prevalence of sensitivity to sulfiting agents in asthmatics. *Am J Med* 1986; 81:816–20.

130. Yang WH, Purchase ECR, Rivington RN. Positive skin tests and Prausnitz-Kustner reactions in metabisulfite-sensitive subjects. *J Allergy Clin Immunol* 1986; 78:443–9.

131. Bush RK, Taylor SL. Adverse reactions to food and drug additives. In: Middleton E, Reed CE, Ellis EF, Adkinson NF, Yunginger JW, Busse WW, eds. *Allergy – Principles and Practice*, vol II, 5th edn. St. Louis: Mosby; 1998; 1183–98.

132. Delohery J, Simmul R, Castle WD, Allen D. The relationship of inhaled sulfur dioxide reactivity to ingested metabisulfite sensitivity in patients with asthma. *Am Rev Resp Dis* 1984; 130:1027–32.

133. Taylor SL, Bush RK, Selner JC, Nordlee JA, Weiner MC, Holden K, Koepke JW, Busse WW. Sensitivity to sulfited foods among sulfite-sensitive asthmatics. *J Allergy Clin Immunol* 1988; 81:1159–67.

134. Stevenson DD, Simon RA, Lumry WR, Mathison DA. Adverse reactions to tartrazine. *J Allergy Clin Immunol* 1986; 78:182–91.

135. Lockey SD. Allergic reactions to F.D. and C. yellow #5 tartrazine, an aniline dye as a coloring agent in various steroids. *Ann Allergy* 1959; 17:719–21.

136. Stevenson DD, Simon RA, Lumry WR, Mathison DA. Pulmonary reactions to tartrazine. *Pediatr Allergy Immunol* 1992; 3:222–7.

137. Life Sciences Research Office, Federation of American Societies for Experimental Biology. *Analysis of adverse reactions to monosodium glutamate (MSG)*. Bethesda MD: FASEB; 1995.

138. Filer LJ, Stegink LD, Applebaum A, Chiaramontc L, Fcrnstrom J, Schiffman S, Taylor S. A report on the proceedings of an MSG workshop held in August 1991. *Crit Rev Food Sci Nutr* 1994; 34:159–74.
139. Schwartzstein RM. Pulmonary reactions to monosodium glutamate. *Pediatr Allergy Immunol* 1992; 3:228–32.
140. Schwartzstein RM, Kelleher AB, Weinberger SE, Weiss JW, Drazen JM. Airway effects of monosodium glutamate in subjects with chronic stable asthma. *J Asthma* 1987; 24:167–72.
141. Allen DH, Delohery J, Baker G. Monosodium L-glutamate-induced asthma. *J Allergy Clin Immunol* 1987; 80:530–7.
142. Moneret-Vautrin DA. Monosodium glutamate-induced asthma: study of the potential risk of 30 asthmatics and a review of the literature. *Allergy Immunol (Paris)* 1987; 19(1):29–35.
143. Schiffman SS, Buckley CE, Sampson HA, Massey EW, Baraniuk JN, Follett JV, Warwick ZS. Aspartame and susceptibility to headache. *N Engl J Med* 1987; 317:1181–5.
144. Garriga MM, Berkebile C, Metcalfe DD. A combined single-blind, double-blind, placebo-controlled study to determine the reproducibility of hypersensitivity reactions to aspartame. *J Allergy Clin Immunol* 1991; 87:821–7.
145. Geha R, Buckley CE, Greenberger P, Patterson R, Polmar S, Saxon A, Rohr A, Yang W, Drouin M. Aspartame is no more likely than placebo to cause urticaria/ angioedema: results of a multicenter, randomized, double-blind, placebo-controlled, crossover study. *J Allergy Clin Immunol* 1993; 92:513–20.
146. Taylor SL. Histamine food poisoning: toxicology and clinical aspects. *Crit Rev Toxicol* 1986; 17:91–128.
147. Stratton JE, Hutkins RW, Taylor SL. Biogenic amines in cheese and other fermented foods: a review. *J Food Prot* 1991; 54:460–9.
148. Taylor SL, Stratton JE, Nordlee JA. Histamine poisoning (scombroid fish poisoning): an allergy-like intoxication. *J Toxicol Clin Toxicol* 1989; 27:225–40.

6

ETHANOL TOXICITY AND NUTRITIONAL STATUS

Helmut K. Seitz and Paolo M. Suter

Without a doubt alcohol (in this text used interchangably with the term ethanol) remains the favorite mood-altering drug worldwide, especially in industrialized regions such as Europe and the United States, and alcoholism retains its status as a progressive, sometimes fatal disease. Nutritional and hepatic disorders remain at the forefront of the medical problems associated with alcohol abuse. However, alcohol is not only a psychoactive drug, but also a food, since it is rich in energy, and many societies have alcoholic beverages integrated into their daily food supply. In contrast, in other societies alcohol is mainly consumed for its mood-altering effects. Under both circumstances a large intake of alcohol has profound effects on nutritional status.

Based on national food consumption data, alcohol contributes about 4.5% of total energy intake of the American diet (1,2). The figure is higher in the alcohol-consuming population as illustrated by one study of male business executives and professionals drinking alcohol on one or more days of the week. In one-fourth of the subjects, ethanol accounted for 5–10% of total calories, and in one-fourth of the subjects accounted for even more than 10% of the daily energy intake (3). The heavy drinker, however, may derive half or more of his or her daily energy from ethanol. In this situation primary malnutrition occurs when other nutrients in the diet are displaced by the high energy content of the alcoholic beverages or because of associated medical and socioeconomic disorders. Furthermore, alcoholic beverages are usually devoid of minerals, vitamins, and protein. Because large quantities of alcohol supply much of the daily caloric requirement and decreased intake of other nutrients may result, the nutrient density of the diet decreases. Secondary malnutrition may result from maldigestion and/or malabsorption of nutrients caused by gastrointestinal complications (especially in the pancreas and the small intestine) due to alcoholism. Apart from dietary intake and nutritional imbalances, the overall alcohol-mediated nutritional changes are determined by associated organ damage. Alcoholic gastritis causes anorexia, retching, and vomiting. Intestinal alterations lead to malabsorption of specific nutrients such as vitamins and trace elements. Pancreatic disease results in steatorrhea with malabsorption of essential fatty acids, fat-soluble vitamins, trace minerals (such as zinc) and calcium. Finally, alcoholic liver disease leads to a failure of protein biosynthesis, amino acid imbalance, reduced zinc storage, and enhanced metabolism and failure to store vitamins A and B6. Thus, there are many reasons why chronic alcoholics may become malnourished (4–6). The potential to develop malnutrition in chronic alcoholism is great, however, it has to be stressed that alcoholism does not always lead to malnutrition. In assessing the nutritional alterations in alcohol consumers it is important to distinguish between socially adequate working alcoholic subjects and the so-called derelict type, who is often physically ill. In the latter group clinical signs of malnutrition and specific deficiencies are very common (7).

There is no doubt that ethanol exerts its toxicity not only through malnutrition, but also by various other mechanisms mostly depending on the metabolism and susceptibility of the target organ. These mechanisms include the physicochemical effects of ethanol on biologic membranes; its effect on the intermediary metabolism of carbohydrates, lipids, and proteins; its oxidation to acetaldehyde, a highly reactive and toxic intermediate; and its interactions with the metabolism of xenobiotics, including drugs, environmental toxins, and procarcinogens via cytochrome P-450 2E1 (5,6), to mention only a few.

It is impossible to discuss all aspects of ethanol-associated toxicity in this chapter. And it would also be far beyond the scope of this overview to discuss in detail all nutritional losses due to ethanol. The focus of this summary will be on the most important and often also less-recognized nutritional disorders such as alterations of nutrient activation, utilization, and degradation, which will also engender secondary malnutrition through altered bioavailability. In addition, major emphasis will be placed on new aspects of interactions of ethanol in energy metabolism and mechanisms of primary and secondary malnutrition due to heavy alcohol intake.

MALNUTRITION AS A FACTOR OF ETHANOL RELATED TOXICITY IN VARIOUS DISEASES

Ethanol-induced malnutrition is of great clinical significance. Thus, various deficiencies of vitamins and trace elements that occur in chronic alcoholics lead to certain diseases. Among these, alcoholic liver disease, alcohol-induced central nervous system disorders, and alcohol-associated cancers are the most important.

Although traditionally the disorders affecting the liver have been attributed exclusively to the different nutritional deficiencies accompanying alcoholism (8), studies carried out over the last three decades indicate that in addition to the role of dietary deficiencies, alcohol *per se* can be incriminated as a direct etiologic factor in the production of alcoholic liver disease. Indeed, even in the absence of dietary deficiencies alcohol can lead to the development of fatty liver in humans (9). A variety of studies, however, revealed that at the biochemical cellular level both toxic and nutritional effects of ethanol are often intertwined. A central mechanism due to chronic ethanol consumption is the induction of the activity of the microsomal ethanol oxidizing system (MEOS) (10) involving a microsomal cytochrome P-450 2E1. Cross induction of other cytochromes P-450 enhances vitamin A degradation and contributes to vitamin A depletion (11). Conversely, nutritional deficiencies may seriously impair detoxification processes. For example, deficiencies in vitamin E and glutathione, both important factors in free radical trapping and thus in detoxification of intermediates occurring from the metabolism of drugs and xenobiotics, may lead to lipid peroxidation in the liver, a situation believed to be of pathogenic importance with respect to alcoholic liver disease (12). Thus, at the hepatic cellular level, the distinction between nutritional and toxic effects of alcohol is now fading and the dichotomy between toxicity and nutritional factors has been bridged.

With respect to ethanol-induced central nervous system disorders, thiamin (13–16) seems to be of considerable importance. Thiamin deficiency leads to Wernicke–Korsakoff syndrome, to cardiac beriberi, and possibly to polyneuropathy (17–20). It is difficult to separate the contribution of alcohol and thiamin depletion to neuropsychiatric abnormalities in the alcoholic. Depletion of thiamin leads to reduced cerebral oxidative metabolism (21). Such alterations may account for many of the features encountered in acute alcohol intoxication and may precipitate alcohol withdrawal syndromes, including delirium tremens. In patients, thiamin therapy causes corrections of mental abnormalities that persist after recovery from acute alcohol intoxication or delirium tremens. In alcoholics with neuropathy, 86% showed a low level of blood thiamin (22–24), and 45% of alcoholics without neuropathy have lowered levels of this vitamin, many of whom have clinical evidence of reduced neural transmission.

A concomitant decrease in circulating riboflavin, nicotinic acid, pantothenic acid, biotin, vitamin B6, folic acid, and/or vitamin B12 was present in 90% of this group (23). The classic cardiac manifestation of thiamin deficiency, beriberi heart disease with a high cardiac output, is still encountered, but more frequently, alcoholics with thiamin depletion have cardiomyopathy (25).

Nutritional deficiencies have also been incriminated in alcohol-associated carcinogenesis of the upper alimentary and respiratory tracts (26). A variety of factors may be involved in esophageal carcinogenesis. Zinc deficiency, often observed in alcoholics, leads to an enhanced activation of nitrosamines via cytochrome P-450 in the mucosa cells of the esophageal epithelium (27). Most of the hypotheses relating a specific type of deficiency to cancer development are based on animal experiments only. For example, zinc deficiency enhanced tumor occurrence induced by nitroso-methylbenzylamine, and this was further increased by chronic ethanol administration (28). Observations of low zinc and vitamin A plasma levels in patients with squamous cell carcinoma of the esophagus (29,30), and of an inverse relationship between the intake of vitamins A and C and esophageal cancer (20), or head and neck cancer patients (31–33) added supportive data for humans.

124

Other important clinical consequences of ethanol-associated nutrient deficiencies include night blindness and testicular atrophy associated with decreased spermatogenesis due to vitamin A deficiency (34), and megaloblastic anemia due to folate deficiency (35). Folate deficiency is a good example to show that ethanol induced nutrient deficiency leads by itself to an enhancement of the functional impairment caused by alcohol, since folate deficiency also results in malabsorption of other vitamins and nutrients, including folate itself.

INTERACTIONS OF ETHANOL WITH SPECIFIC NUTRIENTS

One characteristic feature of ethanol metabolism is its metabolic priority position, i.e. the metabolic oxidation of ethanol has an absolute priority in intermediary metabolism. In contrast to other metabolic pathways, no feedback control system exists for ethanol metabolism. Thus, all ingested ethanol has to be oxidized continuously by the liver and other organs such as the gastrointestinal tract. Ethanol is toxic to the biochemical and physiologic integrity of the human body, so different systems for effective, complete, and fast ethanol elimination exist. However, all of these pathways of ethanol elimination interfere at several central positions with other biochemical pathways involving all energy substrates and all essential nutrients. Therefore it is not surprising that, due to this priority position of ethanol metabolism as well as the production of several intermediate metabolites of ethanol (see p. 128), most nutrients may be affected when large amounts of ethanol are ingested chronically. It is important to remember that there is no threshold level at which the negative alcohol effects on nutriture start, although there is probably a large interindividual variation in this threshold level. The acute ingestion of ethanol leads to important biochemical changes, however, the potential for damage is much smaller due to the single-exposure situation.

For many people ethanol is a component of the everyday diet, and ethanol is used often as a relatively cheap energy source. In this setting ethanol has to be viewed not only as an energy and potential nutrient source, but also as a drug and/or toxin because of its effect on nearly all biochemical and physiologic systems. Excessive ethanol consumption – if continued long enough – always leads to primary and secondary malnutrition, and malnutrition due to ethanol abuse itself also induces alterations of multiple organ functions, such as morphologic and functional changes in the small intestine (Figure 1). The high incidence of malnutrition among heavy alcohol consumers has a multifactorial origin such as poor intake; decreased/altered absorption; increased losses; increased metabolic requirements; and altered activation, utilization, and degradation of nutrients (5).

Not only are micronutrients (vitamins and trace elements) affected by ethanol, but macronutrients such as protein, carbohydrates, and lipids are also. In alcoholic patients consuming at least 1 g of ethanol/kg body weight per day, more calories are derived from carbohydrates (and also alcohol) than in non-drinking

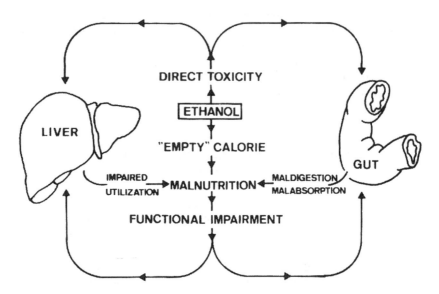

Figure 1 Interaction of direct toxicity of ethanol with malnutrition due to primary or secondary deficiencies. Secondary malnutrition may be caused by either maldigestion and malabsorption or impaired utilization (decreased activation and/or increased inactivation) of nutrients. Both "direct toxicity" of ethanol and malnutrition (whether primary or secondary) may affect function and structure of liver and gut. (With permission from Lieber C.S. (224))

normal controls, so that the percentage of the energy intake obtained from protein and fat are lower than in the non-consumers. When the energy contribution derived from ethanol is removed, the fat and protein contribution are similar to those in controls (7). Such a dietary imbalance may increase the demand for specific nutrients, such as thiamin. In addition, ethanol ingestion enhances nitrogen loss in urine of both rats and humans, thereby possibly increasing protein requirements.

Nutritional losses may also be due to ethanol associated malabsorption. Ethanol ingestion results in morphologic and biochemical alterations of the mucosal cell (36–39), including changes in motility and an inhibition of the Na-K-ATPase enzyme located in the basolateral membrane of the mucosal cell, which is involved in various active transport systems (38). Ethanol inhibits the intestinal absorption of glucose (40–42). Of clinical interest is the fact that 18–74% of alcoholics exhibit malabsorption for xylose, which returns to normal within 2 weeks of abstinence (43). Alcoholism also results in an enhancement of lactose intolerance associated with a decrease in mucosal lactase activity in blacks but far less in white Americans (44). Ethanol also inhibits the absorption of various amino acids (39), which may explain why many alcoholics have an elevated fecal nitrogen excretion. In addition, ethanol inhibits the activity of membrane-bound peptidases in vitro (42), and of cytoplasmic peptidase in vivo (45). These enzymes are important prerequisites for amino acid absorption. Another major reason for maldigestion is alcoholic pancreatic insufficiency, which cannot be discussed here in detail, and therefore the reader is referred to a recent review article (46). Besides the effects of alcohol on carbohydrate absorption, alcohol may enhance the development of diabetes mellitus. In this context it should be recognized that diabetic alcoholic subjects are not necessarily obese (for a clinical review (47)).

Effects of ethanol on energy metabolism

The energy content of 1 g of ethanol as determined in a bomb calorimeter is 30 kJ (i.e., 7.1 kcal). Thus ethanol's caloric content is between that of carbohydrates (16.7 kJ/g or 4 kcal/g) and fat (37.6 kJ/g or 9 kcal/g). The caloric content of alcohol is high and should not be disregarded in the overall energy balance, however, as will be discussed later, alcohol calories count differently from one individual to the other. Ethanol accounts for up to 5.6% of total energy intake in the United States (1). However, in regular ethanol consumers up to 10% of the daily caloric intake may be furnished by ethanol, and in heavy alcohol consumers this figure may rise up to 50% of the daily energy intake (3). The biologic utilization of fat and carbohydrates is nearly 100% (48). Despite the apparent importance of ethanol calories as an energy source in the overall energy intake, it is not clear what proportion of the energy derived from ethanol can be used by the body for ATP (adenosine triphosphate) production.

Quantity and *frequency* of ethanol intake are the two major modulators of the effects of ethanol on nutritional status. In the moderate drinker, ethanol can be added to the usual food intake, or the usual energy sources (i.e. carbohydrates and lipids) may be replaced by ethanol calories. Both patterns are found in healthy moderate ethanol consumers. However, with increasing ethanol intake and frequency of consumption more and more food is substituted or replaced by ethanol, and concomitantly the risk of developing primary and/or secondary malnutrition is increased proportionally (Figure 2). As mentioned above the distribution of the different energy sources is affected by the ingested alcohol. Since most alcoholic beverages contain virtually no essential nutrients, only "empty" calories are supplied to the body, and the nutrient density of the ingested foods decreases, and malnutrition develops.

Weight loss is a common feature among chronic alcohol consumers. The isocaloric substitution of ethanol for up to 50% of dietary carbohydrates in an otherwise balanced diet resulted in weight loss. When ethanol was added to the usual diet, weight gain occurred, however, this weight gain was less than the one achieved by the isocaloric addition of carbohydrate or fat (49,50). In view of the difficulty of accurately assessing ethanol intake in free-living subjects, there is some controversy as to what extent ethanol intake in moderate drinkers contributes to the variation (mainly increases) of body weight. Ethanol ingestion usually leads to increased oxygen consumption, corresponding to an increased energy expenditure. Therefore one calorie provided by ethanol cannot be counted as

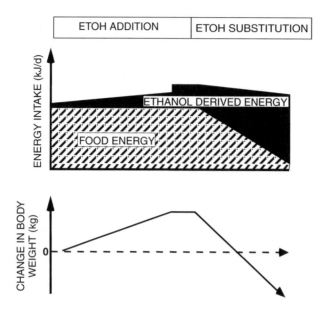

Figure 2 Dynamics in ethanol (ETOH) intake and consequences on food and nutritional status: A model (according to Suter PM *et al.* (53))

1 cal provided by fat or carbohydrates. This increased metabolic rate after the ingestion of food and also ethanol is known as *thermogenesis*, or the thermic effect of food (51). The thermic effect of ethanol depends on the pathway of ethanol metabolism. In the case of ethanol metabolism in the MEOS (i.e. when large doses of ethanol are consumed), oxidation without phosphorylation (i.e. oxidation without ATP production) occurs, and energy is wasted. When small doses of ethanol are ingested, ethanol is oxidized in the alcohol dehydrogenase (ADH) pathway, where the reducing equivalents in the form of NADH can be used for ATP generation (52,53). Other mechanisms of energy wastage (such as the creation of a hyperthyroid state) also are proposed (54,55).

One basic interaction between ethanol and overall nutriture occurs at the intake level (i.e. primary malnutrition). At very high intakes alcohol is one of the most powerful anorectic substrates. At lower intake levels alcohol has just the opposite effect. Most light to moderate alcohol consumers know the appetite enhancing effect of small amounts of alcohol. In several recent studies alcohol has been shown to enhance the development of a positive energy balance due to the enhancement of fat intake (56–58). At moderate consumption levels alcohol has no inhibitory effect on food intake, even when large amounts of food are consumed. Further the addition of alcohol to a meal enhanced food intake in the corresponding meal (mainly fat intake) and also the food intake in a consecutive meal (56,59). These data suggest that the combination of a diet rich in fat with alcohol enhances over-feeding and obesity.

Many different metabolic interactions of ethanol are known, however, the suppression of lipid oxidation is probably the one with the most important patho-physiologic consequences (e.g. hepatic steatosis). Since ethanol suppresses lipid oxidation at the organ level, it has to be assumed that ethanol suppresses lipid oxidation at the whole-body level as well. Recently the effect of ethanol ingestion on the 24-hour oxidation of different substrates at the whole-body level was assessed by indirect calorimetry (60). Eight healthy light ethanol consumers were studied during two 24-hour indirect calorimetry sessions. They consumed normal food, and the energy content of their diet corresponded to their calculated requirements. Ethanol (96 ± 4 g) was either added to their diet or substituted for other foods (isocalorically for fat and carbohydrates). Both ethanol consumption patterns led to an increase in 24-hour energy expenditure of up to 7% of the con-trol day's energy expenditure. Both ethanol consumption patterns did *not* affect carbohydrate oxidation rates, however, the addition and the substitution of ethanol reduced 24-hour lipid oxidation rates by about 45 g/day. This corresponds to about one third of the control day's lipid oxidation rate. The suppressive effect of ethanol on lipid oxidation was only found during the period in which ethanol was ingested and metabolized in the body. Thus, in moderate drinkers ethanol slows lipid oxidation, which leads to an increased storage of lipids in the adipose tissue. This is probably the reason why people become overweight when they drink moderate amounts of ethanol if this effect is not counterbalanced.

This suppressive effect on lipid oxidation eventually may be even more important for the ethanol effect on body weight than the absolute caloric value of ethanol itself. Habitual ethanol consumption in excess of energy needs can thus be judged as a risk factor for obesity since it favors lipid storage. In the heavy drinker this metabolic feature can be overridden by the increased metabolism in the MEOS and the increased substitution of other energy sources by ethanol. Reducing fat intake in moderate and light alcohol drinkers thus helps to avoid the potential weight gain due to ethanol. It might be hypothesized that a large frac-tion of the weight gain occurring during the period between 20 and 45 years might be caused or at least promoted by the ingestion of ethanol in addition to energy needs. It seems reasonable to include alcohol intake in the strategies to control the pandemic of obesity (61,62) and thus the control of chronic diseases.

In the present context it has to be mentioned that alcohol may affect the fat distribution pattern (63). The impact of overweight and obesity as cardiovascular risk factors is also determined by the fat distribution pattern (64); individuals with an increased abdominal fat mass do have a higher overall cardiovascular risk independent from other risk factors (65). In unpublished data from a cross-sectional study we did not find a difference of the effect of different alcoholic beverages on fat distribution pattern (Suter *et al.* unpublished data). In a study by Duncan *et al.* (66) alcohol from beer and liquor seems to be of major importance in the enhancement of the abdominal obesity, however, not alcohol consumed in the form of wine. The metabolic effects of alcoholic beverages on lipid oxidation are caused by the alcohol and accordingly the different effects of alcohol according to the beverage type on fat distribution pattern seems to be surprising and is most likely due to other characteristics of the consumer with certain beverage preferences. The latter would be in agreement with the differences of different alcoholic beverages upon cardiovascular risk (67,68). The pathophysiologic impact of alcohol induced abdominal obesity is not exactly known. In view of the increased chronic disease risk of abdominal fat, any factor promoting the abdominal accumulation of fat should be controlled, including (moderate) alcohol consumption. The major modifiers of the abdominal fat mass are any caloric excess, physical inactivity, psychological stress (including depression), probably the menopause and weight cycling. Some of these conditions are associated with a tendency for higher alcohol intakes. In addition it has been postulated that the rise in blood pressure and/or alcohol induced hypertension may be caused in part by the alcohol induced abdominal obesity (69).

Vitamin A

Patients with alcoholic liver disease have very low levels of vitamin A in their livers at all stages of their disease (70). Various animal experiments have clearly shown that chronic ethanol administration resulted in a depression of hepatic vitamin A that could not be attributed to insufficient vitamin A intake or malabsorption (71). Furthermore there were no significant changes of serum vitamin A or retinol binding protein (72). Subsequently, when dietary vitamin A was virtually eliminated, the depletion rate of vitamin A from hepatic storage was more than two times faster in ethanol-fed rats than in controls (71). All this suggests that at least in part some mechanisms other than malabsorption are responsible for this effect.

Increased mobilization of vitamin A from the liver to other tissues occurs after an acute dose of ethanol, which significantly decreases hepatic vitamin A. This is associated with an increase in serum lipoprotein-bound retinyl esters (73). Conceivably, the effect seen after a single dose of ethanol could be sustained during prolonged intake (74). In addition, enhanced catabolism of vitamin A occurs in the liver after chronic ethanol ingestion. Glucuronides of retinoic acid and retinol are excreted in the bile. Retinoic acid can also be metabolized through a microsomal cytochrome P-450-dependent enzyme system (75,76) and this metabolism can be induced by chronic ethanol consumption as well as by massive dietary vitamin A supplementation (77). In addition, a microsomal system was discovered that converts retinol to various polar metabolites (78). This microsomal system was found to be inducible not only by ethanol, but also by other drugs such as phenobarbital, and the intake of both alcohol and barbiturate results in potentiating the depletion of hepatic vitamin A (79). Since alcohol abuse is often associated clinically with overuse of other drugs, this potentiation may be meaningful in terms of vitamin A depletion in a much larger section of the population. Indeed, some clinical results indicate that drug use may also be associated with severe hepatic vitamin A depletion (80). Another NAD^+-dependent microsomal system converts retinol to retinal (81). This microsomal retinal dehydrogenase is inducible by chronic alcohol intake and may also contribute to hepatic vitamin A depletion, especially in view of the insensitivity of this enzyme to inhibition by ethanol.

The decrease in hepatic vitamin A after alcohol consumption may have a number of adverse consequences, including liver alterations and enhanced carcinogenesis (26). The question, therefore, must be raised whether vitamin A should be supplemented in these patients or not, and what the correct dose of vitamin A supplementation is, especially in view of the well-known hepatotoxicity of vitamin A. Chronic supplementation of vitamin A in amounts as low as 5,000 IU/day results in increased fasting plasma retinyl esters, particularly in the elderly (82), and this was associated with biochemical evidence of liver damage. In addition, concomitant alcohol intake has been shown to potentiate the hepatotoxicity of vitamin A (83,84). In the absence of signs of early vitamin A deficiency, great caution should be exercised in the selection of the dose of vitamin A used for the treatment of malnutrition-associated alcoholism. This may be especially relevant in advanced age, since in animal experiments it has been shown that chronic ethanol consumption strikingly decreases the amount of hepatic retinol binding protein (74).

In addition, not only vitamin A supplementation, but also supplementation with β-carotene may be toxic with ethanol (85). It has been shown in studies that the administration of β-carotene to prevent lung cancer showed an opposite effect. Individuals who received β-carotene exhibited a higher incidence of lung cancer as compared to those on placebo (86). In a reevaluation of the data it could be concluded that only those individuals who had an additional ethanol intake of more than 11 g/day developed cancer with β-carotene (87). Thus, the combination of β-carotene and alcohol potentiates not only hepatic toxicity leading to increased fibrogenesis but also carcinogenesis (88).

One explanation for this observation is the fact that retinol and ethanol both are metabolized primarily by alcohol dehydrogenase, and that a competitive inhibition of the retinol metabolism may occur during ethanol consumption. Furthermore, retinoic acid (RA) is also degraded by cytochrome P-450 2E1 (89) which is enhanced following chronic alcohol intake. As a result RA levels are extremely low, at least in the liver (90), but probably also in the epithelium of the gastrointestinal tract. The low RA levels are associated with a functional down-regulation of RA receptors and with an up-regulation of AP-I gene expression (89). This leads to an enhanced cell proliferation and a loss of cell differentiation, important factors enforcing malignant cell transformation due to ethanol.

Vitamin D

Receptors for vitamin D have been found in most cell systems of the human body (91,92,93). Thus vitamin D and its metabolites are of great importance not only for the maintenance of calcium homeostasis, but also for overall health. Vitamin D can be obtained from the diet or from endogenous synthesis in the skin upon ultraviolet light exposure with photoactivation of the vitamin D precursor 7-dehydrocholesterol. Dietary vitamin D sources are in general of no importance for the maintenance of an adequate vitamin D nutriture, except in countries where vitamin D is added to different food sources. Vitamin D metabolism involves several steps of activation by hydroxylation, one of which occurs in the liver (94,95). Chronic excessive ethanol consumption may result in osteopenia, osteoporosis, and an increased frequency of bone fractures (96). Ethanol can interfere at several levels of the vitamin D endocrine system, and thereby affect calcium homeostasis. Factors contributing to the development of alcohol-associated bone disease may include nutritional deficiencies of calcium and vitamin D, liver disease (impaired activation of vitamin D and/or increased degradation/inactivation of vitamin D by microsomal drug-metabolizing enzymes), hypogonadism, hypercortisolism, and other factors (97). The interactions of ethanol with the vitamin D endocrine system and bone metabolism are very complex and also depend on the pattern of ethanol consumption. Acute ethanol ingestion does not affect vitamin D metabolism, however, alcohol-induced transitory hypoparathyroidism leads to hypercalciuria and hypocalcemia and probably also hypermagnesiuria (97). Acute ingestion of ethanol leads further to a decrease of the γ-carboxylated bone protein osteocalcin (97), indicating altered bone formation.

Chronic ethanol ingestion leads to multiple and important alterations in the vitamin D endocrine system. Alcoholics have low levels of different vitamin D metabolites, including 1,25-dihydroxyvitamin D (98). The low levels are caused by several factors, including low dietary intake, malabsorption of vitamin D, increased biliary excretion of vitamin D, and increased degradation of vitamin D and its metabolites by the ethanol-induced cytochrome P-450 enzyme system (99). The synthesis of vitamin D in the skin is the major source of vitamin D, and reduced sunlight exposure may be an important cause of the impaired vitamin D status in heavy alcohol consumers. Low serum 1,25-dihydroxyvitamin D is found independently of liver disease, and the capacity of the 25-hydroxylation seems to be maintained even in the presence of liver cirrhosis (100–102).

Ethanol ingestion leads to suppression of osteoblastic function, which is also reflected in the low serum osteocalcin levels after acute and chronic ethanol intake (60,67). The interaction between vitamin D, bone metabolism and alcohol and the risk of osteoporosis is of great importance in women regularly consuming alcohol. Nevertheless a protective effect of small amounts of alcohol on bone mass has been reported repeatedly (103). It is possible that alcohol may enhance the synthesis of estrogens from precursors by the stimulation of adipose tissue aromatase. Although the latter effects may be of clinical significance it is possible that low levels of alcohol consumption are just a proxy for other protective behaviors in relation to bone mass.

In view of the multiple functions of vitamin D aside from its effects on calcium homeostasis, an impaired vitamin D status may be disadvantageous in the regeneration of liver tissue. In addition it was shown that vitamin D depletion was accompanied by hyporesponsiveness of the priming of the liver for compensatory growth upon a challenge (104).

Vitamin E

Ethanol also produces increased breakdown of other lipid-soluble vitamins such as hepatic α-tocopherol (vitamin E), possibly secondary to a marked increase in the formation of hepatic α-tocopherol quinone, a metabolite of α-tocopherol, by free radical reaction (12). Since chronic ethanol consumption has been associated with compromised antioxidant defense system status and an increase in lipid peroxidation events, the significant decrease of vitamin E, the principal lipid-soluble antioxidant, may contribute at least in part to the enhanced hepatic lipid peroxidation seen in alcoholics. Besides vitamin E, other antioxidants, such as vitamin C, glutathione, and selenium, are also strikingly decreased following chronic ingestion of alcohol, and this will be discussed in the appropriate sections of this chapter. It is also interesting that chronic alcohol consumption significantly alters the distribution of α-tocopherol and γ-tocopherol not only in hepatic tissue, but in extrahepatic tissues as well (105).

Supplementation of vitamin E in rats chronically fed ethanol and given an esophageal carcinogen resulted in a reduction of esophageal tumors as compared to the administration of ethanol alone (106). It was therefore suggested that free radicals generated during ethanol metabolism (107) may be responsible for the enhancement of carcinogenesis and that the antioxidative action of vitamin E resulted in decreased carcinogenesis. It was also observed that vitamin E supplementation leads to a normalization of colorectal cell regeneration which is stimulated by chronic alcohol ingestion (108). Thus, vitamin E supplementation in the alcoholic may be helpful in cancer prevention. However, clinical trials in patients with alcoholic liver disease did not show a therapeutic effect of vitamin E in the liver (109).

Effects of ethanol on water-soluble vitamins

B-complex vitamins

Several water-soluble vitamins are members of the so-called vitamin-B-complex group. The solubility of a vitamin influences its metabolic action and importance as well as metabolic fate. The B-vitamins are basically of importance in all body cells since they function as important cofactors in many different reactions of intermediary metabolism. In the following paragraphs the major aspects of ethanol interaction with these B-vitamins will be briefly discussed. The data on the incidence of vitamin B deficiency may vary widely for the same vitamin, since the populations studied were very heterogeneous, and different techniques of assessment were used.

Thiamin (*vitamin B₁*) and its different metabolites are important for oxidative decarboxylation reactions and thus energy production. Alcoholism is probably the most important cause of thiamin deficiency in all age groups in acculturated societies (110–112). Thiamin deficiency has been found in up to 80% of alcoholics, regardless of the presence of liver disease. In up to 25% of alcoholic patients admitted to general hospitals, some degree of thiamin deficiency can be diagnosed. The etiology of thiamin deficiency in the alcoholic subject is multicausal. Excessive ethanol consumption with reduced vitamin B_1 intake is one important cause of the vitamin B_1 deficiency. However, thiamin absorption is very sensitive to ethanol, and even in healthy nonalcoholic subjects thiamin absorption is inhibited by the coingestion of ethanol (113,114). In the heavy alcohol consumer other mechanisms for malabsorption (e.g. folate deficiency, see p. 137) may accentuate vitamin B_1 deficiency. Alcohol and its metabolites may further interfere with the intermediary metabolism of the vitamin, leading to an increased requirement. In the presence of liver disease there may also be a deficiency of the thiamin pyrophosphate apoenzyme, and because of the impaired liver function, the vitamin cannot be converted into its active form, thiamin pyrophosphate. Vitamin B_1 is mainly stored in the liver and muscle in the form of thiamin pyrophosphate, however, in the alcoholic subject with impaired liver function and reduced muscle mass, storage capacity is reduced considerably. The alcohol induced magnesium deficiency in addition may enhance thiamin deficiency, since an adequate magnesium nutriture is required for the activation of vitamin B_1 (115). This is also in agreement with the observation that magnesium may improve alcohol related pathologies in the brain (116). Urinary excretion of thiamin is also accentuated by the inestion of ethanol (117). In addition, ethanol ingestion directly affects the brain thiamin metabolism, leading to alcoholic encephalopathy (Wernicke–Korsakoff syndrome) (118).

The water-soluble *vitamin B₂* (*riboflavin*) functions mainly in two coenzyme forms (flavin mononucleotide (FMN) and flavin adenine dinucleotide (FAD)) in several oxidation-reduction reactions. Up to 50% of alcoholics have been reported to have a riboflavin deficiency independent of the degree of liver disease (119). The deficiency is mainly caused by poor intake, however, reduced bioavailability may contribute considerably. Recently it was shown that ethanol reduces the bioavailability of riboflavin by a direct impairment of the intraluminal hydrolysis of FAD (the predominant dietary form of the vitamin) and that ethanol inhibited the activities of different FAD- or FMN-dependent enzymes (120). Ethanol also affects tissue transformation of the riboflavin molecule and thus might modify different vitamin B_2 dependent reactions (121). Isolated riboflavin deficiency occurs rarely, so that other symptoms related to deficiencies of several B-complex vitamins predominate in the clinical picture of the ethanol abuser. It is thought that riboflavin deficiency may contribute to esophageal cancer seen in the alcoholic (26).

Niacin (*vitamin B₃*) includes the two compounds, nicotinic acid and nicotinamide. Niacin can be obtained either preformed from the diet or in part by conversion of dietary tryptophan to niacin. Biochemical niacin deficiency (usually in combination with other vitamin deficiencies) can be encountered in heavy alcohol consumers. The biochemical deficiency is accompanied by a reduction in hepatic niacin content, and the severity of the reduction is related to the degree of liver damage. Poor niacin intake, decreased hepatic storage, increased requirements, and also a decreased hepatic conversion to the different active coenzyme forms (NAD and NADP) (122) contribute to the niacin deficiency in alcoholics. The classic clinical symptom of niacin deficiency is pellagra, which is characterized by a "three D" symptomatic (dermatitis, diarrhea, and dementia) and can still be found in alcoholics.

One important source of niacin is the interconversion of tryptophan to niacin. Thus the nutritional requirements are expressed as niacin equivalents (15), where 60 mg of tryptophan are required to give the same response as 1 mg of niacin. In healthy subjects under normal conditions about 1.5% of dietary tryptophan is converted to niacin. The efficiency of this interconversion depends, however, on different endocrine and nutritional factors. Nutrient deficiencies frequently encountered in heavy alcohol consumers – such as vitamin B_6, riboflavin, and zinc deficiency – may all lead to an impairment of this interconversion and thus to niacin deficiency. The interconversion mainly takes place in the liver, thus being more altered in the presence of liver cirrhosis or other alcohol induced pathologies. Tryptophan metabolism is, however, also affected by alcohol in other organs, mainly the central nervous system (123). Further, the essential amino acid, tryptophan, is one of the limiting amino acids in the diet, and the lack of this niacin precursor in the diet of the alcoholic subject is often encountered. In addition evidence suggests that an altered tryptophan metabolism may be of importance in determining alcohol preference and also addiction (124).

The term *vitamin B_6* stands for three interrelated compounds – pyridoxine, pyridoxamine, and pyridoxal – and their phosphorylated derivatives. Pyridoxal-5-phosphate (PLP) is the coenzyme form of the vitamin and is involved in more than 100 enzymatic reactions in intermediary metabolism. PLP is derived more or less solely from synthesis in the liver, and there is only a very small circulating storage pool of vitamin B_6 in the body. Therefore it is not surprising that in view of the central role of the liver in vitamin B_6 metabolism, any impairment of liver function will immediately lead to an alteration of vitamin B_6 metabolism (125).

Chronic ingestion of ethanol leads to decreased circulating levels of vitamin B_6 (126). Low plasma PLP levels may be found in up to 50% of heavy ethanol consumers without liver disease, and this number increases to 100% in subjects with alcoholic cirrhosis (127). In heavy alcohol consumers (with and without liver disease) the liver tissue content of vitamin B_6 (mainly in the form of PLP) is decreased (128). The impaired vitamin B_6 status in alcoholism is caused by different mechanisms. Low intake of vitamin B_6 is important, however, vitamin B_6 absorption is probably not directly affected by ethanol, although the vitamin may show a reduced bioavailability due to an impaired liberation of the vitamin from different (protein) binding sites in food (129,130). Besides the low PLP levels in alcoholic subjects, the metabolic activation of pyridoxine to pyridoxal is reduced after the oral and/or intravenous application of the vitamin and may even be completely lacking (131). This may be caused by an increased total plasma clearance of the coenzyme as a consequence of an increased degradation of free PLP in the liver (132). But ethanol itself also reduces the activation of the vitamin, and acetaldehyde accelerates the degradation of intracellular pyridoxal phosphate (133). Furthermore, the urinary excretion of certain vitamin B_6 metabolites may be increased (132) and may contribute to an increased loss of the vitamin. In view of these multiple ethanol-induced alterations of vitamin B_6 metabolism, it is not surprising that normalization of vitamin B_6 nutriture by supplements in the alcoholic patient is not always successful. Because of the considerable toxicity of vitamin B_6 (134), the dosage should not be too high, i.e. not exceeding 50 mg/day if given over long periods of time.

Vitamin B_{12} (cyanocobalamin) serves as a cofactor in only two enzymes: methionine synthetase and L-methylmalonyl-CoA mutase. Vitamin B_{12} is absorbed by an active, intrinsic factor mediated transport in the terminal ileum. After absorption the vitamin is transported to, and stored in, the liver. Due to the large liver stores and the reserve capacity for sufficient vitamin B_{12} absorption, the deficiency of this vitamin is only rarely encountered in chronic alcoholic subjects. The absorption of cyanocobalamin is, however, impaired in chronic alcohol consumers (135). This malabsorption is probably caused by free radicals generated during the gastric metabolism of ethanol which may alter the intrinsic factor binding of vitamin B_{12} (136).

Lower plasma/serum *biotin* levels have been found in alcoholic subjects (22). Said *et al.* (137) showed recently that chronic ethanol feeding in rats led to a selective inhibition of biotin absorption at physiologic concentrations. Furthermore, acetaldehyde inhibited the transport of biotin to a much higher degree than ethanol itself.

It is not clear to what extent heavy ethanol consumption impairs *pantothenic acid* nutriture. By using urinary excretion testing and/or plasma pantothenic acid levels, an impaired status was found in alcohol abusers as compared to nonabusers. Liver pantothenate levels were also reduced as a function of alcoholic liver damage. Ethanol impairs the conversion of pantothenate to coenzyme A (CoA) in liver and other tissues (138).

Homocysteine (see also paragraph on folate and methyl groups in this chapter) has been identified as an potential independent cardiovascular risk factor (139–142). As an important intermediate metabolite in the transmethylation and transsulfuration pathways, an accumulation of homocysteine may have deleterious effects as demonstrated in homocysteinemia, a disease of the group of inborn errors of metabolism (142,143). Several vitamins may be of importance in the modulation of homocysteine concentration. Vitamin B_{12} and folate are important cofactors in the remethylation of homocysteine to methionine (see also these vitamin sections in this chapter (139,144,145). During the postprandial phase homocysteine is condensed with the amino acid serine in the transsulfuration pathway, which depends on an adequate supply of vitamin B_6 (139). As discussed above all three vitamins involved in the modulation of homocysteine may be affected at several levels by the ingestion of alcohol. Accordingly, it is not surprising that alcoholics have elevated levels of homocysteine (146–148). To what extent these elevated homocysteine levels are important in the development of classical alcohol related pathologies is not known. It seems that the atherogenic potential of this amino acid can be counterbalanced by other antiatherosclerotic effects of alcohol (149,150).

Alcohol may further directly decrease the activity of the methionine synthase (151). To what extent the alcohol effects on plasma homocysteine levels can be counterbalanced by vitamin supplements is not exactly known. From the present evidence it can be concluded that the pathophysiologic potential of homocysteine in the alcoholic is not comparable to the risk in non-alcohol consumers. Nevertheless it is important to remember that elevated homocysteine levels can be controlled by nutritional factors, even in situations of a genetic predisposition such as the presence of thermolabile variants of the methylenetetrahydrofolate reductase (MTHFR).

Folate

Folate deficiency is the most commonly found impairment of vitamin nutriture in malnourished alcoholics (152). In a group of unselected alcoholics, 38% had low serum folate and 18% had low red blood cell folate levels (153). Since beer contains considerably more folic acid than other alcoholic beverages, higher serum folate concentrations are found in alcoholics drinking beer than in those drinking predominantly spirits or wine (154).

Folate deficiency may cause megaloblastic anemia. It also causes malabsorption of thiamin and vitamin B_{12} as well as of folate itself (155). Megaloblastic anemia is only found in individuals with both a small folate pool and a diet low in folate. If the body has an adequate folate pool, anemia will not occur immediately if the diet is lacking folate, not even in combination with alcohol. However, a prolonged inadequate diet may deplete folate stores. Chronic alcohol consumption impairs folate coenzymes, increases hematologic indices of megaloblastic anemia, and may cause malabsorption of enterohepatic circulating folate in folate deficiency even when other essential nutrients are provided (156).

The low folate levels in the alcoholic result partly from primary and secondary malnutrition (poor intake and decreased absorption). Acute alcohol administration to rats produced a marked increase in the urinary excretion of folate compounds, which led to a decrease in plasma folate levels (157). In monkeys given 50% of energy as ethanol, 2 years of ethanol feedings resulted in mild hepatic injury, with a significant decrease in hepatic folate levels, but the processes of reduction, methylation, and formylation of reduced folate and the synthesis of polyglutamyl folates were not affected (158). Thus, decreased folate levels after alcohol are due either to a decreased ability to retain folate in the liver or to increased breakdown of folate. Superoxide may contribute to folate cleavage, and this may be relevant to the alcoholic in view of the evidence for enhanced free radical generation in association with purine degradation (159). In vitro, metabolism of acetaldehyde via xanthine oxidase (160) or aldehyde oxidase (161) may generate free radicals, but the concentrations of acetaldehyde required are rather high to be of significance in vivo, at least in the liver. However, in the rectum considerable amounts of acetaldehyde are produced by bacteria, which could decrease local folate concentrations and therefore could contribute to ethanol-associated rectal carcinogenesis (162). The megaloblastic anemia due to folate deficiency seen in the alcoholic is accentuated by ethanol, presumably by the inhibition of the thymidilate synthetase (163).

Indeed, low folate levels have been reported recently in the colorectal mucosa in rats chronically fed ethanol (164). It seems likely that these low folate concentrations may be responsible at least in part for the observed hypomethylation of colonic DNA in ethanol consuming animals, an important factor in colorectal carcinogenesis (165). In these studies a specific hypomethylation of the P52 gene could not be found (165). Low vitamin B_6 levels may also contribute to folate deficiency (148).

Vitamin C

Vitamin C (ascorbic acid) is widely distributed in many different food sources, and dietary intake is the most important determinant for adequate vitamin C nutriture. The classic vitamin C deficiency disease is scurvy, which is characterized by anemia and alterations in protein and collagen metabolism leading to the clinical signs of ecchymoses and hemorrhages in different body tissues (166). It is widely believed that there is no more scurvy in our modern societies, however, there are many at-risk population strata including heavy alcohol consumers (167–169), and the condition may even go unrecognized for a long period of time (170). Marginal deficiency due to low intake is rather common in certain substrata of the population, such as heavy alcohol drinkers, and thus scurvy can be seen even nowadays when vitamin C is readily available.

In general ethanol may contribute to and/or potentiate ascorbic acid deficiency. Alcohol drinkers have lower plasma and lower leukocyte ascorbic acid levels compared with non- or only light consumers. Acute ingestion of ethanol enhances urinary losses (171,172). However, reduced ascorbic acid status in heavy alcohol consumers is basically due to low intake, since dietary intake of vitamin C is the major determinant of vitamin C status. The inadequate vitamin C nutriture may contribute to some of the ethanol induced pathologies because of reduced antioxidative defense. Recently an ascorbate-dependent ethanol oxidation was described (173), and in vitamin C-deficient guinea-pigs ethanol metabolism was reduced (174). It is, however, controversial whether ascorbic acid affects the rate of ethanol elimination (173,175). In addition, vitamin C may protect certain body tissues (such as the heart muscle) from the deleterious effects of acetaldehyde (176).

In view of newly discovered functions of vitamin C and its potential importance in disease modulation, higher intake recommendations have been suggested (177). It remains to be proven whether an increased intake at the population level will result in an improved health, however at the level of the heavy alcohol consumer there is no doubt about potentially large health effects of an increased vitamin C intake. Alcohol metabolism is always associated with an increased production of free radicals and thus also a higher potential for free radical damage. Even in non alcoholic subjects the acute ingestion of alcohol is associated with an increased lipid peroxidation as assessed by the measurement of urinary parameters of oxidative stress (178). It is plausible that an optimal vitamin C status may counteract some of these adverse effects.

The cytotoxic effects of ethanol may be counteracted also by vitamin C (179). Recently higher recommended dietary intakes of vitamin C for the general population have been recommended (177) and animal experiments support the higher requirement in regular alcohol consumers (180). Of course other antioxidants are also of importance and the protection from alcohol related damage may be mediated independently by vitamin C and Vitamin E (181). Despite these favorable effects of antioxidants it should be remembered that the control of alcohol intake remains the cornerstone for the prevention of alcohol related damage.

Alcohol and methyl group metabolism

Chronic ethanol consumption leads both to methyl deficiency and to increased requirement for methyl groups. Methyl groups play an important role in the regulation of many cell functions. The methylation of DNA is the only postsynthetic DNA modification that occurs in mammalian cells (182) and represents one central mechanism in the regulation of gene expression. The active methyl donor is S-adenosylmethionine (S-AME). S-AME transfers methyl groups primarily to cysteine bases of DNA, to uridylate to generate thymidine or to polyenylphosphatidylethanol to produce polyenylphosphatidylcholine (PPC), which is an important constituent of cell membranes. Indeed, it has been shown that chronic ethanol consumption in rats and baboons results in a decrease of PPC associated with severe membrane injury, and that supplementation of PPC prevents the progression of alcoholic liver disease (183,184). Besides its membrane stabilizing effect, PPC has antioxidative, metabolic and antifibrotic effects (185–187).

The cause of decreased methyl transfer following ethanol ingestion is complex and includes folate, vitamin B_{12} and vitamin B_6 deficiencies and an inhibition of methyl transferase activity by acetaldehyde. In addition, ethanol also inhibits phosphatidylethanol methyl transferase (PEMT), the enzyme responsible for the generation of PPC (188).

It has been claimed that the ethanol mediated alteration of methyl transfer is a major factor in the pathogenesis of alcoholic liver disease, therefore, attempts to correct for this deficiency have been made in clinical trials. Most recently it has been reported that the administration of S-AME to patients with alcoholic liver cirrhosis improved their survival significantly (189). Another multicenter study with PPC in the United States is currently being performed.

Effects of ethanol on electrolytes and trace elements

Magnesium

Magnesium plays an important role as a cofactor of enzymes involved in carbohydrate, lipid, and amino acid metabolism. Magnesium deficiency as assessed by measurement of plasma/serum magnesium levels is commonly found in individuals with excessive ethanol consumption (190). This hypomagnesemia and intracellular deficit of magnesium (191) is caused by decreased dietary intake, malabsorption, increased urinary excretion (due to reduced renal reabsorption) (192), secondary hyperaldosteronism, diarrhea, and also eventually vomiting (193). Alcohol-induced reductions in tissue magnesium levels are well known, especially in the cardiac muscle, leading to the classic signs of rhythm disturbances and contributing to coronary insufficiency (194,195). Lower levels of magnesium alone may have adverse health effects, such as contributing to the development of hypertension (196).

Magnesium is involved in the maintenance of membrane integrity and thus also may contribute to the protection of liver membranes from alcohol- and non-alcohol related damage (197). The maintenance of magnesium nutriture is very efficiently regulated, and magnesium is widely distributed in many different food sources. Magnesium deficiency symptoms do occur in alcoholic subjects, however, other nutrient deficiencies seem to be of greater importance. One important interaction should be mentioned in the present context. The active form of vitamin B_1 is thiamin diphosphate which has to occur at the level of the liver and which is dependent on an adequate magnesium nutriture. Alcohol interferes at different levels in the metabolism of thiamin, and a vitamin B_1 deficiency is widespread in alcohol consumers (198,199). If vitamin B_1 supplements are given the adequacy of the magnesium nutriture should be monitored. This interaction is especially of importance in subjects with cardiac failure (18) (in part caused by alcoholic cardiomyopathy) and concomitant diuretic therapy (200). Diuretics lead to an increased excretion of thiamin (200) and magnesium thus exacerbating the cardiac symptoms. Magnesium deficiency may play an important modulatory role in the pathogenesis of stroke (201–203), which is more common in alcohol consumers (204,205). Different mechanisms for the stroke protective effects of magnesium have been proposed, one of them being the reduction of vascular spasms (116). The clinical significance of an adequate magnesium nutriture in alcoholics may be also of importance regarding carcinogenesis (206).

Zinc

Zinc is one of the most abundant trace elements in the human body, and thus it is not surprising that excessive ethanol ingestion affects zinc metabolism at several levels.

Low plasma and serum zinc levels in chronic alcohol consumers were recognized more than 30 years ago (207) and have been confirmed in many consecutive studies. Low cell and/or tissue zinc levels have been reported in the liver and leukocytes (polymorphonuclear cells) (208,209). It is important to recognize that the lower zinc liver levels occur not only in advanced alcoholic liver disease (liver cirrhosis) but in less advanced alcoholic liver disease (209). The reason for the low plasma zinc concentration in advanced alcoholic liver disease is the altered distribution of the zinc bound to different protein carriers. In cirrhosis, less zinc is bound to the albumin fraction, which is reduced due to the impaired liver function (210). The low blood and tissue levels may be caused by different mechanisms. The anorectic effect of large doses of ethanol reduces dietary zinc intake. Further, zinc absorption has been shown to be impaired by ethanol-induced gastrointestinal damage, however, increased zinc absorption has been reported in certain studies. One important reason for the impaired zinc absorption is the exocrine pancreatic insufficiency (211). The increased absorption may be a reflection of the depleted body zinc stores, since zinc absorption is also regulated by the zinc status of the whole organism. Malabsorption of zinc could be caused by decreased levels of zinc and copper binding protein, metallothionein. Another mechanism for the impairment of zinc status is the increased excretion of zinc in urine. This may be caused by the diuretic effect of ethanol and/or the decreased tissue uptake (mainly in the liver) of zinc, thus increasing the blood levels leading to a relative increase in urinary zinc excretin.

Zinc plays an important role in many metabolic reactions, and thus it is not surprising that many alcoholics show multiple clinical signs of zinc deficiency. Patients with alcoholic liver disease thus may exhibit abnormalities of taste and smell, hypogonadism, and infertility, as well as perturbations in dark adaptation. The latter symptom is caused by zinc-induced alterations in vitamin A metabolism, since zinc is of importance in the conversion of retinol to retinal as well as in the synthesis and secretion of retinol binding protein in the liver. Zinc deficiency can increase ethanol toxicity, since ADH, the rate-limiting enzyme in ethanol metabolism that converts ethanol to acetaldehyde, is a zinc metalloenzyme containing four zinc atoms. Impaired zinc nutriture thus slows down ethanol elimination and increases ethanol toxicity (212).

Many other enzyme systems are impaired by zinc deficiency in alcoholic subjects. One classical example is the altered activity of the zinc–copper enzyme, superoxide dismutase, which plays an important role in the protection of oxidative tissue damage. Due to altered zinc status with altered superoxide dismutase activity the hepatocytes may become more vulnerable to oxidative damage (213). In addition, zinc deficiency may also contribute to ethanol-associated esophageal cancer (214). Zinc supplementation in alcoholics has been reported to be successful in improving zinc-dependent biochemical reactions and also the overall clinical well-being of the patients.

Zinc plays an important role in immunomodulation, and in head and neck cancer patients zinc deficiency was associated with an increased tumor size and impaired cell mediated immune dysfunction (31).

Zinc may also play an important role in the metabolism of alcohol. A recent study with Wistar rats reported an improved gastric ADH activity upon zinc supplementation and an improvement of the first pass metabolism of alcohol (215). An improvement of the first pass metabolism may lessen the risk for the development of alcoholic hepatic lesions (215).

Selenium

The biologic role of selenium is related to its presence at the active site of the enzyme glutathione peroxidase, which catalyzes the hydrolysis of different organic peroxides (216,217). Glutathione protects hepatocytes from oxidative damage by lipid peroxides, and this protective function also is partially mediated by the selenium-containing enzyme glutathione peroxidase. Serum, plasma, erythrocyte, and leukocyte selenium are lowered in individuals with excessive ethanol consumption (218–221). The decrease in selenium concentration in blood (plasma/serum) and the different blood cells depends on the degree of liver damage, and the abnormality of selenium status correlates with different laboratory parameters of liver function (220–222). The liver also plays a central role in the metabolism of trace elements. In alcoholic patients liver selenium stores were much lower than in patients dying from other causes (220). Twenty-four-hour urinary selenium excretion as an index of selenium nutriture was reported to be reduced in subjects consuming excessive amounts of ethanol (223). Dietary deficiency of selenium may lead to liver cell necrosis, and increased hepatic lipid peroxidation may be an important contributor to alcoholic liver injury.

It has been suggested that hepatic selenium deficiency may contribute directly to the pathogenesis of alcoholic liver disease (220). The mechanism of selenium deficiency is not yet clear, however, inadequate dietary intake is of primary importance. This is also supported by the observation by Dutta *et al.* (219) that plasma selenium levels increased during alcohol detoxification after admission to a hospital. Assessment of dietary intake of selenium is very unreliable because of the wide variation of selenium in food and the large variability of selenium bioavailability from food.

SUMMARY, CONCLUSIONS, AND RECOMMENDATIONS

Alcohol may be both a nutrient and a toxin depending on various factors, including amount and frequency of consumption, age, gender, cultural drinking habits, concomitant diseases, and drug use. However, it is clear that alcoholism and/or alcohol-associated organ injury has become one of the major health problems worldwide.

One of the important mechanisms of alcohol toxicity is malnutrition. However, it is well established that alcohol may also exert its toxicity by other mechanisms, such as alteration of intermediary metabolism, acetaldeyde accumulation, and physicochemical alterations of cell membranes. Nutritional toxicity of ethanol may be mediated directly by nutritional deficiencies (primary malnutrition) or by alterations of metabolic pathways due to ethanol-induced organ damage (secondary malnutrition). Alcohol-associated malnutrition is especially relevant in alcoholic liver disease, central nervous system disorders, immunosuppression, and alcohol-associated carcinogenesis.

The caloric value of ethanol (1 g ethanol = 7.1 cal) as a potential risk factor for weight gain is well known in the moderate drinker. However, with increased intake of ethanol those calories cannot be used fully for biologically utilizable energy. The result is weight loss and wasting. Alcohol interferes at various levels of nutrient handling and metabolism (intake, digestion, absorption, activation, excretion). Of considerable importance is the interaction of ethanol with several vitamins and trace elements, leading to biochemical and clinical deficiencies.

In summary, it is concluded that chronic alcoholism exerts its toxicity at least in part through nutritional deficiencies and interaction with the metabolism of nutrients. Some of the data dealing with the effect of ethanol on nutrient activation, utilization, and degradation have been collected recently. Thus the effect of ethanol, especially on vitamin A, folate, vitamin E, selenium, and zinc, may not only be of relevance with respect to liver disease, but also may play a role in ethanol-associated carcinogenesis. However, further work is needed to establish such a role on a biochemical and molecular biologic level.

The best strategy to avoid any nutritional deficiencies in heavy drinkers is nevertheless the reduction of excessive ethanol consumption. In subjects consuming light to moderate amounts of alcohol the overall diet should be balanced according to the present recommendations and an overall healthy life style should be pursued.

REFERENCES

1. Block G, Dresser CM, Hartman AM, Carroll MD. Nutrient sources in the American diet: Quantitative data from the NHANES II survey. *Am J Epidemiol* 1985; 122:27–40.
2. Scheig R. Effects of ethanol on the liver. *Am J Clin Nutr* 1970; 23:467–73.
3. Bebb HT, Houser HB, Witschi JC, Littell AS, Fuller RK. Calorie and nutrient contribution of alcoholic beverages to the usual diets of 155 adults. *Am J Clin Nutr* 1971; 24:1042–52.
4. Seitz HK, Simanowski UA. Metabolic and nutritional effects of ethanol. In: Hathcock JN, ed. *Nutritional Toxicology*, vol II. New York: Academic Press; 1987; 63–103.
5. Lieber CS. Alcohol, liver and nutrition. *J Am Coll Nutr* 1991; 10:602–32.
6. Lieber CS. *Medical and Nutritional Complications of Alcoholism.* New York: Plenum Publishing Corporation; 1992.
7. Sherlock S. Nutrition and the alcoholic. *Lancet* 1984; i:436–9.
8. Best CH, Hartroft WS, Lucas CC, Rideout JH. Liver damage produced by feeding alcohol or sugar and its prevention by choline. *Br Med J* 1949; 2:1001–6.

9. Lieber CS, Jones DP, DeCarli LM. Effects of prolonged ethanol intake: production of fatty liver despite adequate diets. *J Clin Invest* 1965; 44:1009–21.

10. Lieber CS, DeCarli LM. Hepatic microsomal ethanol-oxidizing system: in vitro characteristics and adaptive properties in vivo. *J Biol Chem* 1970; 245:2505–12.

11. Leo MA, Lieber CS. Hypervitaminosis A: a liver's lover lament. *Hepatology* 1988; 8:412–17.

12. Kawase T, Kato S, Lieber CS. Lipid peroxidation and antioxidant defense systems in rat liver after chronic ethanol feeding. *Hepatology* 1989; 10:815–21.

13. Holzbach E. Thiamin absorption in alcoholic delirium patients. *J Stud Alcohol* 1996; 57:581–4.

14. Harper CG, Sheedy DL, Lara AI, Garrick TM, Hilton JM, Raisanen J. Prevalence of Wernicke–Korsakoff Syndrome in Australia: has thiamine fortification made a difference? *Med J Aust* 1998; 168:542–5.

15. National Academy of Sciences. Standing Committee on the Scientific Evaluation of Dietary Reference Intakes -F-a-N-B. Dietary Reference Intakes. Thiamin, Riboflavin, Niacin, Vitamin B6, Folate, Vitamin B12, Pantothenic Acid, Biotin, and Choline. Washington DC (USA): National Academy Press; 1998.

16. Charness ME, Simon RP, Greenberg DA. Ethanol and the nervous system. *N Engl J Med* 1989; 321:442–54.

17. Sokoll LJ, Morrow FD. Thiamin. In: Hartz SC, Rosenberg IH, Russell RM, eds. *Nutrition in the Elderly. The Boston Nutritional Status Survey*. London, UK: Smith-Gordon & Co Ltd; 1992; 111–17.

18. Kwok T, Falconer-Smith JF, Potter JF, Ives DR. Thiamine status of elderly patients with cardiac failure. *Age Aging* 1992; 21:67–71.

19. Ben-Hur T, Wolff E, River Y. Thiamin deficieny is common in Israel. *Harefuah* 1992; 123:382–4.

20. Cook CC, Thomson AD. B-complex vitamins in the prophylaxis and treatment of Wernicke-Korsakoff syndrome. *Br J Hosp Med* 1997; 57:461–5.

21. Leevy CM, Zetterman RK. Malnutrition and alcoholism: an overview. In: Rothschild MA, Oratz M, Schreiber SS, eds. *Alcohol and abnormal protein synthesis, biochemical and clinical aspects*. New York: Pergamon Press; 1975; 3–15.

22. Fennelly J, Frank O, Baker H, Leevy CM. Peripheral neuropathy of the alcoholics: I. Aetiological role of aneurin and other B-complex vitamins. *Br Med J* 1969; 2:1290–2.

23. Fennelly J, Frank O, Baker H, Leevy CM. Effect of alcohol on excretion of some water soluble vitamins. *Br J Med* 1964; 2:1290–2.

24. Fennelly JO, Frank O, Baker H, Leevy CM. Peripheral neuropathy of the alcoholic. Etiologic role of thiamin and other B-complex vitamins. *Br Med J* 1964; 2:1290–2.

25. Webster MWI, Ikram H. Myocardial function in alcoholic cardiac beriberi. *Int J Cardiol* 1987; 17:213–16.

26. Seitz HK, Pöschl G, Simanowski UA. Alcohol and Cancer. In: Galanter M, ed. *Recent Advances in Alcoholism*. New York: Plenum Press; 1998; 67–134.

27. Barch DH, Kuemmerle SC, Hollenberg PF, Iannaccone PM. Esophageal microsomal metabolism of N-nitrosomethylbenzylamine in the zinc deficient rat. *Cancer Res* 1984; 44:5629–33.

28. Gabrial GN, Schrager TF, Newberne PM. Zinc deficiency, alcohol, and retinoid: association with esophageal cancer in rats. *J Natl Cancer Inst* 1982; 68:785–9.

29. Mellow MH, Layne EA, Lipman TO, Kaushik M, Hostetler C. Plasma zinc and vitamin A in human squamous carcinoma of the esophagus. *Cancer* 1983; 51:1615–20.

30. Willet WC, McMahon B. Diet and cancer. *N Engl J Med* 1984; 310:633–8.
31. Prasad AS, Beck FW, Doerr TD, *et al*. Nutritional and zinc status of head and neck cancer patients: an interpretive review. *J Am Coll Nutr* 1998; 17:409–18.
32. Doerr TD, Marks SC, Shamsa FH, Mathog RH, Prasad AS. Effects of zinc and nutritional status on clinical outcomes in head and neck cancer. *Nutrition* 1998; 14: 489–95.
33. Doerr TD, Prasad AS, Marks SC, *et al*. Zinc deficiency in head and neck cancer patients. *J Am Coll Nutr* 1997; 16:418–22.
34. Roe DA. Drug induced nutritional deficiencies. Westport: AVI Publ Company; 1976.
35. Lindenbaum J, Roman MR. Nutritional anaemia in alcoholism. *Am J Clin Nutr* 1980; 33:2727–35.
36. Persson J, Berg NO, Sjolund K, Stenling R, Magnusson PH. Morphologic changes in the small intestine after chronic alcohol consumption. *Scan J Gastroenterol* 1990; 25:173–84.
37. Seitz HK, Velasquez D, Waldherr R, Veith S, Czygan P, Weber E. Duodenal gamma-glutamyl transferase activity in human biopsies: effect of chronic ethanol consumption and duodenal morphology. *Europ J Clin Invest* 1985; 15:192–6.
38. Seitz HK, Simanowski UA. Alkohol und Intestinum. *Verdauungskrankheiten* 1986; 4:54–60.
39. Mezey E. Effect of ethanol on intestinal morphology, metabolism, and function. In: Seitz HK, Kommerell B, eds. *Alcohol related diseases in gastroenterology*. New York: Springer Publishing Company; 1985; 342–60.
40. Dinda PK, Beck IT. On the mechanism of the inhibitory effect of ethanol on intestinal glucose and water absorption. *Am J Dig Dis* 1977; 22:529–33.
41. Dinda PK, Beck IT. Ethanol induced inhibition of glucose transport across the isolated brush border membrane of hamster jejunum. *Dig Dis Sci* 1981; 26:23–32.
42. Dinda PK, Beck IT. Effect of ethanol on peptidases of hamster jejunal brush border membranes. *Am J Physiol* 1982; 242:6442–50.
43. Mezey E. Metabolic effects of ethanol. *Federation Proc* 1985; 44:134–8.
44. Perlow W, Baraona E, Lieber CS. Symptomatic intestinal disaccharidase deficiency in alcoholics. *Gastroenterology* 1977; 77:680–4.
45. Dinda PK, Beck IT. Effects of ethanol on cytoplasmic peptidases of the jejunal epithelial cell of the hamster. *Dig Dis Sci* 1984; 29:46–53.
46. Niebergall-Roth E, Harder H, Singer M. A review: acute and chronic effects of ethanol and alcoholic beverages on the pancreatic exocrine secretion in vivo and in vitro. *Alcohol Clin Exp Res* 1998; 22:1570–83.
47. Greenhouse L, Lardinois C. Alcohol-associated diabetes mellitus. A review of the impact of alcohol consumption on carbohydrate metabolism. *Arch-Fam-Med* 1996; 5:229–33.
48. James WPT, Schofield EC. *Human energy requirements. A manual for planners and nutritionists*. Oxford, GB: Oxford University Press; 1990.
49. Pirola RC, Lieber CS. The energy cost of the metabolism of drugs, including ethanol. *Pharmacology* 1972; 7:185–96.
50. Pirola RC, Lieber CS. Energy wastage in rats given drugs that induce microsomal enzymes. *J Nutr* 1975; 105:1544–8.
51. Jequier E, Tappy L. Regulation of body weight in humans. *Physiol Rev* 1999; 79: 451–80.
52. Lieber C. Perspectives: do alcohol calories count? *Am J Clin Nutr* 1991; 54:976–82.

53. Suter PM, Häsler E, Vetter W. Effects of alcohol on energy metabolism and body weight regulation: Is alcohol a risk factor for obesity? *Nutr Reviews* 1997; 55.

54. Israel Y, Bernstein J. Liver hypermetabolic state after chronic ethanol consumption: hormonal interrelations and pathogenic implications. *Fed Proc* 1975; 34:2052–9.

55. Israel Y, Macdonald VA, Bernstein J. Metabolic alterations produced in the liver by chronic ethanol feeding. *Biochem J* 1973; 134:523–9.

56. Tremblay A, St. Pierre S. The hyperphagic effect of a high-fat diet and alcohol intake persists after control for energy density. *Am J Clin Nutr* 1996; 63:479–82.

57. Tremblay A, Buemann B, Thériault G, Bouchard C. Body fatness in active individuals reporting low lipid and alcohol intake. *Eur J Clin Nutr* 1995; 49:824–31.

58. Tremblay A, Wouters E, Wenker M, St. Pierre S, Bouchard C, Després JP. Alcohol and high-fat diet: a combination favoring overfeeding. *Am J Clin Nutr* 1995; 62:639–44.

59. Westerterp-Plantenga MS, Verwegen CRT. The appetizing effect of an apéritif in overweight and normal-weight humans. *Am J Clin Nutr* 1999; 69:205–12.

60. Suter PM, Schutz Y, Jéquier E. The effect of ethanol on fat storage in healthy subjects. *N Engl J Med* 1992; 326:983–7.

61. Must A, Spadano J, Coakley EH, Field AE, Colditz G, Dietz WH. The Disease Burden Associated with Overweight and Obesity. *JAMA* 1999; 282:1523–9.

62. Mokdad AH, Serdula MK, Dietz WH, Bowman BA, Marks JSl, Koplan JP. The Spread of the Obesity Epidemic in the United States, 1991–1998. *JAMA* 1999; 282: 1519–22.

63. Suter PM, Maire R, Vetter W. Alcohol consumption: a risk factor for abdominal fat accumulation in men. *Add Biol* 1997; 2:101–3.

64. Björntorp P. How should obesity be defined? *J Int Med* 1990; 227:147–9.

65. Walker SP, Rimm EB, Ascherio A, Kawachi I, Stampfer MJ, Willett WC. Body size and fat distribution as predictors of stroke among US men. *Am J Epidemiol* 1996; 144: 1143–50.

66. Duncan BB, Chambless LE, Schmidt MI, *et al.* Association of the waist-to-hip ratio is different with wine than beer or hard liquor. *Am J Epidemiol* 1995; 142:1034–8.

67. Klatsky AL. Is it the drink or the drinker? Circumstantial evidence only raises a probability. *Am J Clin Nutr* 1999; 69:2–3.

68. Klatsky AL, Friedman GD, Armstrong MA. The relationship between alcoholic beverage use and other traits to blood pressure: a new Kaiser Permanente study. *Circulation* 1986; 73:628–36.

69. Suter PM, Maire R, Vetter W. Is an increased waist:hip ratio the cause of alcohol-induced hypertension. The AIR94 study. *J Hypertens* 1995; 13:1857–62.

70. Leo MA, Lieber CS. Hepatic vitamin A depletion in alcoholic liver injury. *N Engl J Med* 1982; 307:597–601.

71. Sato M, Lieber CS. Hepatic vitamin A depletion after chronic ethanol consumption. *J Nutr* 1981; 111:2015–23.

72. Vahlquist A, Sjoelund A, Norden A, Peterson PA, Stigmark G, Johansson B. Plasma vitamin A transport and visual dark adaptation in diseases of the intestine and liver. *Scand J Clin Lab Invest* 1978; 38:301–8.

73. Sato M, Lieber CS. Changes in vitamin A status after acute ethanol administration in the rat. *J Nutr* 1982; 112:1188–96.

74. Mobarhan S, Seitz HK, Russell RM, *et al.* Age related effects of chronic ethanol intake on vitamin A status in Fisher 344 rats. *J Nutr* 1991; 121:510–17.

75. Roberts AB, Nichols MD, Newton DL, Sporn MB. In vitro metabolism of retinoic acid in hamster intestine and liver. *J Biol Chem* 1979; 254:6296–303.

76. Roberts AB, Lamb LC, Sporn MB. Metabolism of all-trans-retinoic acid in hamster liver microsomes: oxidation of 4-hydroxy- to 4-keto-retinoic acid. *Arch Biochem Biophys* 1980; 199:374–83.

77. Leo MA, Iida S, Lieber CS. Retinoic acid metabolism by a system reconstituted with cytochrome P-450. *Arch Biochem Biophys* 1984; 234:305–12.

78. Leo MA, Lieber CS. New pathway for retinol metabolism in liver microsomes. *J Biol Chem* 1985; 260:5228–31.

79. Leo MA, Lowe N, Lieber CS. Potentiation of ethanol induced hepatic vitamin A depletion by phenobarbital and butylated hydroxytoluene. *J Nutr* 1987; 117:70–6.

80. Leo MA, Lowe N, Lieber CS. Decreased hepatic vitamin A after drug administration in men and in rats. *Am J Clin Nutr* 1984; 40:1131–6.

81. Leo MA, Kim CI, Lieber CS. NAD$^+$-dependent retinol dehydrogenase in liver microsomes. *Arch Biochem Biophys* 1987; 259:241–9.

82. Krasinski SD, Russell RM, Otradovec CL, *et al.* Relationship of vitamin A and vitamin E intake to fasting plasma retinol, retinol binding protein, retinyl esters, carotene, alpha-tocopherol, and cholesterol among elderly people and young adults: increased plasma retinyl esters among vitamin A-supplement users. *Am J Clin Nutr* 1989; 49: 112–20.

83. Leo MA, Arai M, Sato M, Lieber CS. Hepatotoxicity of vitamin A and ethanol in the rat. *Gastroenterology* 1982; 82:194–205.

84. Leo MA, Lieber CS. Hepatic fibrosis after long term administration of ethanol and moderate vitamin A supplementation in the rat. *Hepatology* 1983; 2:1–11.

85. Leo MA, Cho-Il K, Lowe N, Lieber CS. Interaction of ethanol with β-carotene: delayed blood clearance and enhanced hepatotoxicity. *Hepatology* 1992; 15:883–91.

86. The Alpha-Tocopherol -B-C-C-P-S-G. The effect of vitamin E and beta-carotene on the incidence of lung cancer and other cancers in male smokers. *N Engl J Med* 1994; 330:1029–35.

87. Albanes D, Heinonen OP, Taylor PR, *et al.* α-Tocopherol and β-carotene supplements and lung cancer incidence in the Alpha-Tocopherol, Beta-Carotene Cancer Prevention Study: Effects of Base-line Characteristics and Study Compliance. *J Natl Cancer Inst* 1996; 88:1560–70.

88. Leo MA, Aleynik SI, Aleynik MK, Lieber CS. β-Carotene beadlets potentiate hepatotoxicity of alcohol. *Am J Clin Nutr* 1997; 66:1461–9.

89. Liu C, Russell RM, Seitz HK, Wang XD. Ethanol enhances retinoic acid metabolism into polar metabolites in rat liver via induction of cytochrome P450 2E1. *Gastroenterology* 2001; 120:179–89.

90. Wang XD, Liu C, Chung J, Stickel F, Seitz HK, Russell RM. Chronic alcohol intake reduces retinoic acid concentration and enhances AP-1 (c-jun and c-fos) expression in rat liver. *Hepatology* 1998; 28:744–50.

91. Wood RJ, Fleet JC. The genetics of osteoporosis: vitamin D receptor polymorphisms. *Annu Rev Nutr* 1998; 18:233–58.

92. Segaert S, Bouillon R. Vitamin D and regulation of gene expression. *Curr Opin Clin Nutr Metab Care* 1998; 347–54.

93. Berg JP, Haug E. Vitamin D: a hormonal regulator of the cAMP signaling pathway. *Crit Rev Biochem Mol Biol* 1999; 1999:315–23.

94. Holick MF. McCollum Award Lecture 1994: Vitamin D – new horizons for the 21st century. *Am J Clin Nutr* 1994; 60:619–30.

95. Vieth R. Vitamin D supplementation, 25-hydroxyvitamin D concentrations, and safety. *Am J Clin Nutr* 1999; 69:842–56.

96. Von Moreau KB, Mueller B, Drisch D, Osswald B, Seitz HK. Alcohol and trauma. *Alcoholism Clin Exp Res* 1992; 16:141–3.

97. Laitinen K, Lamberg-Allard C, Tunninen R, *et al*. Transient hypoparathyroidism during acute ethanol intoxication. *N Engl J Med* 1991; 324:721–7.

98. Lalor BC, France MW, Powell D, Adams PH, Counihan TB. Bone and mineral metabolism and chronic alcohol abuse. *Q J Med* 1986; 59:497–511.

99. Jung RT, Davis M, Hunter JO, Chalmer TM, Lawson DE. Abnormal vitamin D metabolism in cirrhosis. *Gut* 1978; 19:290–3.

100. Mobarhan SA, Russell RM, Recker RR, Posner DB, Iber FL, Miller P. Metabolic bone disease in alcoholic cirrhosis: A comparison of the effect of vitamin D_2, 25-hydroxyvitamin D, or supportive treatment. *Hepatology* 1984; 4:266–73.

101. Compston JE. Hepatic osteodystrophy: vitamin D metabolism in patients with liver disease. *Gut* 1986; 27:1073–90.

102. Gascon-Barre M. Influence of chronic ethanol consumption on the metabolism and action of vitamin D. *J Am Coll Nutr* 1985; 4:565–74.

103. Feskanich D, Korrick SA, Greenspan SL, Rosen HN, Colditz GA. Moderate alcohol consumption and bone density among postmenopausal women. *J Womens Health* 1999; 8:65–73.

104. Ethier C, Goupil D, Gascon-Barre M. Influence of the calcium and vitamin D endocrine system on the "priming" of the liver for compensatory growth. *Endocrine Research* 1991; 17:421–36.

105. Meydani M, Seitz HK, Blumberg JB, Russell RM. Effect of chronic ethanol feeding on hepatic and extrahepatic distribution of vitamin E in rats. *Alcoholism Clin Exp Res* 1991; 15:771–4.

106. Eskelson CD, Odeleye OE, Watson RR, Earnest D, Mufti SI. Modulation of cancer growth by vitamin E and alcohol. *Alkohol & Alcoholism* 1993; 28:117–26.

107. Albano E, Clat P. Free radicals and ethanol toxicity. In: Preedy VR, Watson RR, eds. *Alcohol and the gastrointestinal tract*. Boca Raton (FL, USA): CRC Press; 1995; 57–68.

108. Vincon P, Wunderer J, Simanowski UA, *et al*. Inhibition of alcohol-associated hyperregeneration in the rat colon by α-tocopherol. *Alcohol Clin Exp Res* (abstract) 2000; 24:213A.

109. Wenzel G, Kulinski B, Ruhlmann C, Ehrhardt D. Alkoholische Hepatitis – eine "freie Radikale" assoziierte Erkrankung. Letalitätssenkung durch adjuvante Antioxidantientherapie (Alcoholic toxic hepatitis – a free radical associated disease: decreased mortality by adjuvant antioxidant therapy). *Z Gesamte Inn Med* 1993; 48:490–6.

110. Bovet P, Larue D, Fayol V, Paccaud F. Blood thiamin status and determinants in the population of Seychelles (Indian Ocean). *J Epidemiol Community Health* 1998; 52:237–42.

111. Jamieson CP, Obeid OA, Powell-Tuck J. The thiamin, riboflavin and pyridoxine status of patients on emergency admission to hospital. *Clin Nutr* 1999; 18:87–91.

112. Iber FL, Blass JP, Brin M, Leevy CM. Thiamin in the elderly – relation to alcoholism and to neurological degenerative disease. *Am J Clin Nutr* 1982; 36:1067–82.

113. Hoyumpa AM, Breen KJ, Schenker S, Wilson FA. Thiamine transport across the rat intestine. II. Effect of ethanol. *Lab Clin Med* 1975; 86:803–16.

114. Tomasulo PA, Kater RMH, Iber FL. Impairment of thiamine absorption in alcoholism. *Am J Clin Nutr* 1968; 21:1341–4.

115. Zieve L. Influence of magensium deficiency on the utilization of thiamine. *Ann NY Acad Sci* 1969; 162:732–43.

116. Ema M, Gebrewold A, Altura BT, Zhang A, Altura BM. Alcohol-induced vascular damage of brain is ameliorated by administration of magensium. *Alcohol* 1998; 15: 95–103.

117. Neville JN, Eagles JA, Samson G, Olson RE. Nutritional status of alcoholics. *Am J Clin Nutr* 1968; 21:1329–40.

118. Butterworth RF, D'Amour M, Bruneau J, Heroux M, Brissette S. Role of thiamine deficiency in the pathogenesis of alcoholic peripheral neuropathy and the Wernicke–Korsakoff Syndrome: An update. In: Palmer TN, ed. *Alcoholism: A molecular perspective*. New York: Plenum Press; 1991; 269–73.

119. Baines M. Detection and incidence of B and C vitamin deficiency in alcohol related illness. *Ann Clin Biochem* 1978; 15:307–18.

120. Pinto J, Huang YP, Rivlin RS. Mechanisms underlying the differential effects of ethanol on the bioavailability of riboflavin and flavin adenine dinucleotide. *J Clin Invest* 1987; 79:1343–8.

121. Ono S, Takahashi H, Hirano H. Ethanol enhances the esterification of riboflavin in rat organ tissue. *Int J Vitam Nutr Res* 1987; 57:335.

122. Magni G, Amici A, Emanuelli M, Raffaelli N, Ruggieri S. Enzymology of NAD$^+$ synthesis. *Adv Enzymol Relat Areas Mol Biol* 1999; 73:135–82.

123. Badawy AA, Rommelspacher H, Morgan CJ, *et al.* Tryptophan metabolism in alcoholism. Tryptophan but not excitatory amino acid availability to the brain is increased before the appearance of the alcohol-withdrawal syndrome in men. *Alcohol Alcohol* 1998; 33:616–25.

124. Bano S, Morgan CJ, Badawy AA, *et al.* Tryptophan metabolism in male Sardinian alcohol-preferring (sP) and -non-preferring (sNP) rats. *Alcohol Alcohol* 1998; 33:220–5.

125. Driskell JA. Vitamin B6. In: Machlin LJ, ed. *Handbook of vitamins. Nutritional, biochemical, and clinical aspects*. New York: Marcel Dekker Inc; 1984:379–401.

126. Lumeng L, Ting-Kai L. Vitamin B6 metabolism in chronic alcohol abuse. *J Clin Invest* 1974; 53:693–704.

127. Davis RE, Smith KK. Pyridoxal and folate deficiency in alcoholics. *Med J Aust* 1974; 2:357–60.

128. Baker H, Frank O, Ziffer H, Goldfarb S, Leevy CM, Sobotka H. Effect of hepatic diseases on liver B-complex vitamin titers. *Am J Clin Nutr* 1964; 14:1–8.

129. Middleton HM, Mills III, Singh M. Effect of ethanol on the uptake of pyridoxine-HCL in rat jejunum. *J Nutr* 1984; 39:54–61.

130. Bonjour JP. Vitamins and alcoholism: III. Vitamin B6. *Internat J Vit Nutr Res* 1980; 50:216–30.

131. Mitchell D, Wagner C, Stone WJ, Wilkinson GR, Schenker S. Abnormal regulation of plasma pyridoxal-5'-phosphate in patients with liver disease. *Gastroenterology* 1976; 71:1043–9.

132. Henderson JM, Codner MA, Hollins B, Kutner MH, Merrill AH. The fasting B6 vitamer profile and response to a pyridoxine load in normal and cirrhotic subjects. *Hepatology* 1986; 6:464–71.

133. Veitch RL, Lumeng L, Li TK. Vitamin B6 metabolism in chronic alcohol abuse. The effect of ethanol oxidation on hepatic pyridoxal 5'-phosphate metabolism. *J Clin Invest* 1975; 55:1026–32.

134. Schaumberg H, Kaplan J, Windebank A, Vick N, Rasmus S, *et al.* Sensory neuropathy from pyridoxine abuse: a new megavitamin syndrome. *New Engl J Med* 1983; 309:445–7.

135. Lindenbaum J, Lieber C. Alcohol induced malabsorption of vitamin B12 in man. *Nature* 1969; 224:806.

136. Shaw S, Herbert V, Colman N, Jayatilleke E. Effect of ethanol-generated free radicals on gastric intrinsic factor and glutathione. *Alcohol* 1990; 7:153–7.

137. Said HM, Sharifian A, Bagherzadeh A, Mock D. Chronic ethanol feeding and actue ethanol exposure in vitro: effect on intestinal transport of biotin. *Am J Clin Nutr* 1990; 52:1083–6.

138. Israel BC, Smith CM. Effects of acute and chronic ethanol ingestion on pantothenate and CoA status of rats. *J Nutr* 1987; 117:443–51.

139. Guba SC, Fonseca V, Fink LM. Hyperhomocysteinemia and thrombosis. *Semin Thromb Hemost* 1999; 25:291–309.

140. Refsum H, Ueland P, Nygard O, Vollset SE. Homocysteine and cardiovascular disease. *Ann Review Med* 1998; 49:31–62.

141. Blom HJ. Determinants of plasma homocysteine. *Am J Clin Nutr* 1998; 67:188–9.

142. McCully KS. Homocysteine and Vascular Disease. *Nature Medicine* 1996; 2: 386–9.

143. McCully KS. Vascular pathology of homocysteinemia: implications for the pathogenesis of arteriosclerosis. *Am J Pathol* 1969; 56:111–23.

144. McCully KS, Olszewski AJ, Vezeridis MP. Homocysteine and lipid metabolism in atherogenesis: effect of the homocysteine thiolactonyl derivatives, thioretinaco and thioretinamide. *Atherosclerosis* 1990; 83:197–206.

145. Lindenbaum J, Savage DG, Stabler SP, Allen RH. Diagnosis of cobalamin deficiency: II. Relative sensitivities of serum cobalamin, methylmalonic acid, and total homocysteine concentrations. *Am J Haematol* 1990; 34:99–107.

146. Nygard O, Refsum H, Ueland PM, Vollset SE. Major lifestyle deterimants of plasma total homocysteine distribution: the Hordaland Homocysteine Study. *Am J Clin Nutr* 1998; 67:263–70.

147. Cravo ML, Gloria LM, Selhub J, *et al.* Hyperhomocysteinemia in chronic alcoholism: correlation with folate, vitamin B12, and vitamin B6 status. *Am J Clin Nutr* 1996; 63:220–4.

148. Stickel F, Choi SW, Kim YI, *et al.* Effect of chronic alcohol consumption on total plasma homocystein levels in rats. *Alcoholism Clin Exp Res* 2000; 24:259–64.

149. Rehm J, Bondy S. Alcohol and all-cause mortality: an overview. In: Chadwick DJ, Goode JA, eds. *Alcohol and Cardiovascular Diseasees*. Chichester: John Wiley & Sons; 1998; 68–85.

150. Chadwick DJ, Goode JA, eds. *Alcohol and Cardiovascular Diseases*. Chichester: John Wiley & Sons (Novartis Foundation Symposium 216); 1998.

151. Kenyon SH, Nicolaou A, Gibbons WA. The effect of ethanol and its metabolites upon methionine synthase activity in vitro. *Alcohol* 1999; 15:305–9.

152. Baker HA. A vitamin profile of alcoholism. In: Hank A, ed. *Vitamins in medicine: recent therapeutic aspects*. Bern (Switzerland): Hans Huber Publisher; 1983; 179–84.

153. World MJ, Ryle PR, Jones D, Shaw GK, Thomson AD. Differential effect of chronic alcohol intake and poor nutrition on body weight and fat stores. *Alcohol* 1984; 19:281–90.

154. Bonjour JP. Vitamins and alcoholism II: folate and vitamin B12. *Int J Vitam Nutr Res* 1980; 50:96–121.

155. Green PH. Alcohol, nutrition, and malabsorption. *Clin Gastroenterol* 1983; 12:563–74.

156. Blocker DE, Thenen SW. Intestinal absorption, liver uptake, and excretion of 3H-folic acid in folic acid deficient, alcohol consuming nonhuman primates. *Am J Clin Nutr* 1987; 46:503–10.

157. McMartin KE. Increased urinary folate excretion and decreased plasma folate levels in rat after acute ethanol treatment. *Alcoholism Clin Exp Res* 1984; 8:172–8.

158. Tamura T, Romero JJ, Watson JE, Gong EJ, Halsted CH. Hepatic folate metabolism in chronic alcoholic monkey. *J Lab Clin Med* 1981; 97:654–61.

159. Kato S, Kawase T, Alderman J, Inatomi N, Lieber CS. Role of xanthin oxidase in ethanol induced lipid peroxidation in rats. *Gastroenterology* 1990; 98:203–10.

160. Shaw S, Jayatilleke E, Herbert V, Colman N. Cleavage of folates during ethanol metabolism: role of acetaldehyde/xanthine oxidase generated superoxide. *Biochem J* 1989; 257:277–80.

161. Shaw S, Jayatilleke E. The role of aldehyde oxidase in ethanol induced hepatic lipid peroxidation in the rat. *Biochem J* 1990; 268:579–83.

162. Seitz HK, Simanowski UA, Garzon FT, *et al.* Possible role of acetaldehyde in ethanol-related rectal cocarcinogenesis in the rat. *Gastroenterology* 1990; 98:406–13.

163. Sullivan LW, Herbert V. Suppression of hematopoesis by ethanol. *J Clin Invest* 1964; 43:2048–62.

164. Homann N, Tillonen J, Salaspuro M. Microbially produced acetaldehyde from ethanol may increase the risk of colon cancer via folate deficiency. *Int J Cancer* 2000 (in press).

165. Choi SW, Stickel F, Baik HW, Kim YI, Seitz HK, Mason J. Chronic alcohol consumption induces genomic, but not p53-specific, DNA hypomethylation in the colon of the rat. *J Nutr* 1999; 129:1945–50.

166. Jaffe GM. Vitamin C. In: Machlin LJ, ed. *Handbook of vitamins. Nutritional, biochemical, and clinical aspects.* New York: Marcel Dekker Inc; 1984; 199–244.

167. Hirschmann JV, Raugi GJS. Adult scurvy. *J Am Acad Dermatol* 1999; 41:895–906.

168. Sasaki G, Satoh T, Yokozeki H, Katayama I, Nishioka K. Confluent ecchymoses on the lower extremities of a malnourished patient. *J Dermatol* 1999; 26:399–401.

169. Sloan B, Kulwin DR, Kersten RC. Scurvy causing bilateral orbital hemorrhage. *Arch Ophthalmol* 1999; 117:842–3.

170. Gonzalez-Gay MA, Garcia-Porrua C, Lueiro M, *et al.* Scurvy can mimic cutaneous vasculitis. Three case reports [letter]. *Rev Rhum Engl Ed* 1999; 66:360–1.

171. Mochizuki S, Yoshida A. Effects of dietary ethanol on ascorbic acid and lipid metabolism, and liver drug-metabolizing enzymes in rats. *J Nutr Sci Vitaminol* (Tokyo) 1989; 35:431–40.

172. World MJ, Ryle PR, Thomson AD. Alcoholic malnutrition and the small intestine. *Alcohol Alcoholism* 1985; 20:89–124.

173. Susick RL, Zannoni VG. Ascorbic acid and ethanol oxidation. *Biochem Pharmacol* 1984; 33:3963–9.

174. Dow J, Goldberg A. Ethanol metabolism in the vitamin C deficient guinea-pig. *Biochem Pharmacol* 1975; 24:863–6.

175. Lieber C. Metabolic effects of ethanol and its interaction with other drugs, hepatotoxic agents, vitamins, and carcinogens: A 1988 update. *Sem Liver Dis* 1988; 8:47–68.

176. Sprince H, Parker CM, Smith GG, Gonzales LJ. Protection action of ascorbic acid and sulfur compounds against acetaldehyde toxicity: implications in alcoholism and smoking. *Agents Actions* 1975; 5:164–72.

177. Carr AC, Frei B. Toward a new recommended dietary allowance for vitamin C based on antioxidant and health effects in humans. *Am J Clin Nutr* 1999; 69:1086–107.

178. Meagher EA, Barry OP, Burke A, *et al*. Alcohol-induced generation of lipid peroxidation products in humans. *J Clin Invest* 1999; 104:805–13.

179. Wickramasinghe SH, Hasan R. In vitro effects of vitamin C on the cytotoxicity of post-ethanol serum. *Biochem Pharmacol* 1994; 48:621–4.

180. Zloch Z, Ginter E. Moderate alcohol consumption and vitamin C status in the Guinea-Pig and the rat. *Physiol Res* 1995; 44:173–8.

181. Glascott PA, Gilfor E, Serroni A, Farber JL. Independent antioxidant action of vitamin E and C in cultured rat hepatocytes intoxicated with allyl alcohol. *Niochem Pharmacol* 1996; 52:1245–52.

182. Richardson B, Yung R. Role of DNA methylation in the regulation of cell function. *J Lab Clin Med* 1999; 134:333–40.

183. Lieber CS, Robins SJ, Li J, *et al*. Phosphatidylcholine protects against fibrosis and cirrhosis in the baboon. *Gastroenterology* 1994; 106:152–9.

184. Lieber CS, De-Carli LM, Mak KM, Kim CI, Leo MA. Attenuation of alcohol-induced hepatic fibrosis by polyunsaturated lecithin. *Hepatology* 1990; 12: 1390–8.

185. Li J, Kim CI, Leo MA, Mak KM, Rojkind M, Lieber CS. Polyunsaturated lecithin prevents acetaldehyde-mediated hepatic collagen accumulation by stimulating collagenase activity in cultured lipocytes. *Hepatology* 1992; 15:373–81.

186. Aleynik SI, Leo MA, Ma X, Aleynik MK, Lieber CS. Polyenylphosphatidylcholine (PPC) prevents hepatic lipid peroxidation and attenuates associated injury induced by CCl4 in rats and by alcohol in baboons. *Gastroenterology* 1995; 108:A1025.

187. Navder KP, Baraona E, Lieber CS. Polyenylphosphatidylcholine attenuates alcohol-induced fatty liver and hyperlipemia in rats. *J Nutr* 1997; 127:1800–6.

188. Lieber CS, Robins SJ, Leo MA. Hepatic phosphatidylethanolamine methyltransferase activity is decreased by ethanol and increased by phosphatidylcholine. *Alcoholism Clin Exp Res* 1994; 18:592–5.

189. Mato JM, Cámara J, Fernández-de-Paz J, *et al*. S-Adenosylmethionine in alcoholic liver cirrhosis: a randomized, placebo-controlled, double-blind, multicentre clinical trial. *J Hepatol* 1999; 30:1081–9.

190. Petroianu A, Barquete J, Plentz d-A, Bastos C, Maia DJ. Acute effects of alcohol ingestion on human serum concentration of calcium and magnesium. *J Int Med Res* 1991; 19:410–13.

191. Wacker WEC, Parisi AF. Magnesium metabolism. *N Engl J Med* 1968; 278:772–6.

192. Kalbfleisch JM, Linderman RD, Ginn HE, Smith WO. Effects of ethanol administration on urinary excretion of magnesium and other electrolytes in alcoholic and normal subjects. *J Clin Invest* 1963; 42:1471–5.

193. Flink EB. Magnesium deficiency in alcoholism. *Alcohol Clin Exp Res* 1986; 10:590–4.

194. Greenspon AJ, Schaal SF. The "holiday heart": Electrophysiological studies of alcohol effects in alcoholics. *Ann Int Med* 1983; 98:135–9.

195. Iseri LT. Magnesium and cardiac arrhythmias. *Magnesium* 1986; 5:111–26.

196. Witteman JCM, Willett WC, Stampfer MJ, *et al*. Relation of moderate alcohol consumption and risk of systemic hypertension in women. *Am J Cardiol* 1990; 65:633–7.

197. Gullestad L, Dolva LO, Soyland E, *et al*. Effect of oral magnesium treatment in chronic alcoholics. In: Lasserre B, Durlach J, eds. *Magnesium: A relevant Ion.* London, UK: John Libbey & Company Ltd; 1991; 405–9.

198. Baum RA, Iber FL. Thiamin: The interaction of aging, alcoholism and malabsorption in various populations. *World Rev Nutr Diet* 1984; 44:85–116.

199. Basel HG. Alcohol intake and thiamin. *Ernaehrung/Nutrition* 1987; 11:329–33.

200. Lubetsky A, Winaver J, Seligmann H, *et al*. Urinary thiamine excretion in the rat: Effects of furosemide, other diuretics, and volume load. *J Lab Clin Med* 1999; 134:232–7.

201. Altura BM, Gebrewold A, Zhang A, Altura BT, Gupta RK. Magnesium deficiency exacerbates brain injury and stroke mortality induced by alcohol: a 31P-NMR in vivo study. *Alcohol* 1998; 15:181–3.

202. Suter PM. The effect of potassium, magnesium, calcium, and fiber on risk of stroke. *Nut Reviews* 1999; 57:84–91.

203. Durlach J, Bac P, Bara M, Guiet-Bara A. Cardioprotective foods and nutrients: possible importance of magnesium intake. *Magnes Res* 1999; 12:57–61.

204. Altura BM, Altura BT, Gebrewold A. Alcohol-induced spasms of cerebral blood vessels: Relation to cerebrovascular accidents and sudden death. *Science* 1983; 220: 331–3.

205. Altura BM, Altura BT. Magnesium and cardiovascular biology: an important link between cardiovascular risk factors and atherosclerosis. *Cell Mol Biol Res* 1995; 41:347–59.

206. Rivlin RS. Magnesium deficiency and alcohol intake: mechanisms, clinical significance and possible relation to cancer development. *J Am Coll Nutr* 1994; 416–23.

207. Vallee BL, Wacker WEC, Bartholomay AF, Robin ED. Zinc metabolism in hepatic dysfunction. I. Serum zinc concentrations in Laennec's cirrhosis and their validation by sequential analysis. *N Engl J Med* 1956; 255:403–8.

208. Goode HF, Kelleher J, Walker BE, Hall RI, Guillou PJ. Cellular and muscle zinc in surgical patients with and without gastrointestinal cancer. *Clin Sci* 1990; 247–52.

209. Bode JC, Hanisch P, Henning H, Koenig W, Richter FW, Bode C. Hepatic zinc content in patients with various stages of alcoholic liver disease and in patients with chronic active and chronic persistent hepatitis. *Hepatology* 1988; 8:1605–9.

210. Boyett JD, Sullivan JF. Distribution of protein-bound zinc in normal and cirrhotic serum. *Metabolism* 1970; 19:148–57.

211. Ijuin H. Evaluation of pancreatic exocrine function and zinc absorption in alcoholism. *Kurume Med J* 1998; 45:1–5.

212. Milne DB, Canfield WK, Gallagher SK, Hunt JR, Klevay LM. Ethanol metabolism in postmenopausal women fed a diet marginal in zinc. *Am J Clin Nutr* 1987; 46:688–93.

213. Ritland S, Aaseth J. Trace elments and the liver. Journal of Hepatology 1987; 5:118–22.

214. Seitz HK, Simanowski UA. Alcohol and carcinogenesis. *Annu Rev Nutr* 1988; 8: 99–119.

215. Caballeria J, Gimenez A, Andreu H, *et al*. Zinc administration improves gastric alcohol dehydrogenase activity and first-pass metabolism of ethanol in alcohol fed rats. *Alcohol Clin Exp Res* 1997; 21:1619–22.

216. Levander OA. A global view of human selenium nutrition. *Ann Rev Nutr* 1987; 7:227–50.

217. Harris ED. Regulation of antioxidant enzymes. *FASEB J* 1992; 6:2675–83.
218. Aaseth J, Thomassen Y, Norheim G. Decreased serum selenium in alcoholic cirrhosis. *N Engl J Med* 1980; 303:944–5.
219. Dutta SK, Miller PA, Greenberg LB, Levander OA. Selenium and acute alcoholism. *Am J Clin Nutr* 1983; 38:713–18.
220. Dworkin BM, Rosenthal WS, Stahl RE, Panesar NK. Decreased hepatic selenium content in alcoholic cirrhosis. *Dig Dis Sci* 1988; 33:1213–17.
221. Dworkin B, Rosenthal WS, Jankowski RH, Gordon GG, Haldea D. Low blood selenium levels in alcoholics with and without advanced liver disease. *Dig Dis Sci* 1985; 30:838–44.
222. Korpela H, Kumpulainen J, Luoma PV, Arranto AJ, Sotaniemi EA. Decreased serum selenium in alcoholics as related to liver structure and function. *Am J Clin Nutr* 1985; 42:147–51.
223. Rissanen A, Sarlio-Lahteenkorva S, Alfthan G, Gref CG, Keso L, Salaspuro M. Employed problem drinkers: a nutritional risk group? *Am J Clin Nutr* 1987; 45: 456–61.
224. Lieber CS. The influence of alcohol on nutritional status. *Nut Reviews* 1988; 46: 241–54.

7

INTERACTIONS BETWEEN FOODS, NUTRITIONAL SUPPLEMENTS AND DRUGS

John A. Thomas and Richard Cotter

I. INTRODUCTION

Food–drug interactions can be the result of different mechanisms and may lead to a decrease in drug efficacy or increases in toxicity. There have been a number of very informative recent reviews on food–drug interactions (15,27,42,51,52,56, 60,62). In addition, several recent reviews pertaining to food–drug interactions have focused on specialized conditions such as in the elderly (53), in the critically ill (54), in hospitalized patients (18) and in patients receiving specialized nutritional support (35). Unfortunately, despite efforts to clearly define the actual incidence of food–drug interactions, it has been difficult to define the magnitude of the problem, which will most likely expand as the population ages. It is estimated that two-thirds of patients aged 65 years and older use one or more drugs daily (50). Furthermore, older patients experience adverse drug reactions more frequently than do younger adults (2). Not only has the incidence of drug–food interactions increased in older patients, but generally drugs have become more potent. Along with the increased use of specialized nutritional support in patients requiring multiple medications, there is a concern about the possibility for pharmacologic–nutritional interactions, and the consequences of therapeutic outcomes (35,42). It is known that the frequency of drug interactions is higher in the critically ill patient simply because of the larger number of drugs they receive (45).

This chapter will re-examine the role of physiologic factors that affect a drug's absorption and the importance of bioavailability. Further, certain disease states necessitate additional vigilance in predicting food–drug interactions. Finally, this chapter will describe such interactions that may be manifest during nutritional supplementation either through enteral or parenteral feeding.

II. PHYSIOLOGIC FACTORS AFFECTING
ABSORPTION

The presence of food in the gastrointestinal (G-I) tract can significantly affect the oral bioavailability of drugs (12,53,56). Drugs are absorbed after oral administration as a consequence of a complex set of interactions between the drug, its particular formulation and the G-I tract. Food can affect transit profiles, pH and its solubilization capacity. Food can modify the rate and/or the extent of absorption. It can also modify pre-systemic metabolism and systemic drug clearance (12,15, 54). Such modifications caused by food can produce variation in the drug's efficacy and toxicity.

Historically, the presence of food in the G-I tract was regarded as a simple barrier to a drug's absorption and that oftentimes a drug should be taken on a so-called empty stomach. Increasingly, however, potential drug–food interactions should be considered individually. Indeed, the nature and magnitude of the effect of food on bioavailability is related to the drug, its formulation and dosage, as well as the amount and composition of the food itself (60).

Physical and chemical interactions can occur between drugs and various components of food (15). There are interactions involving drug release, drug dissolution and drug degradation. Generally, increasing the dissolution rate increases absorption. Changes in postprandial absorption are often a function of the changes related to progressing from a fasted to a fed state. These changes in G-I absorption may be reflected in:

- changes in secretion of gastric acid, bile and pancreatic fluids
- changes in gastric and intestinal motility, and
- modifications in visceral blood and lymph flow.

Most of the clinically significant nutrient–drug interactions involve the absorptive process and are profoundly affected by the adrenergic nervous system within the G-I tract. Adrenergic mechanisms have a significant modulating action upon gut motility, blood flow and mucosal transport (14). Both α and β subtypes of adrenoreceptors have been recognized at different levels of the gastrointestinal tract and are involved in the regulation of motility and secretion; thus, they may influence drug–nutrient interactions. α^1-adrenoreceptors are located post-junctionally on mostly smooth muscle cells, while α^2-adrenoreceptors can be found both at pre- and post-synaptic sites. Several agents can act selectively as agonists as well as antagonists (Table 1).

Excluding ethanol, few drugs are absorbed to any significant degree in the stomach; most acidic or basic drugs are absorbed in the small intestine. Hence, gastric function has a major effect upon both the rate and degree of drug absorption, as changes in gastric motility can affect the residence time of the food and/ or drug in the G-I tract. The composition of the diet and the timing of meals can also influence drug absorption; for example, delays in gastric emptying time may

Table 1 Receptors modulating gastrointestinal motility and secretion

	Receptor type				
	α_1	α_2	β_1	β_2	β_3
Distribution in the gut	Smooth muscle	Adrenergic neurons (autoreceptors)	Smooth muscle	Smooth muscle	Smooth muscle
	Gastric neurons	Myenteric neurons (heteroceptors)	Enteric neurons (?)	Peripheral noradrenergic neurons (?)	
Functional response	Smooth muscle contraction or relaxation	Smooth muscle presynaptic-inhibition	Smooth muscle relaxation	Smooth muscle relaxation	Smooth muscle relaxation
	Neuronal depolarization (gastric neurons)	Smooth muscle contraction		Facilitation of transmitter release	
Selective agonists	Phenylephrine, methoxamine, cirazoline	Clonidine, azepexole	Xamoterol, prenalterol	Terbutaline, salbutamol, nitrodrine	
Selective antagonists	Prazosin, corynanthine	Yohimbine, idazoxan, rauwolscine	Betaxolol, atenolol		

Source: Modified from DePonti *et al.* (1996) (14).

be caused by fatty foodstuffs and thus affect a drug's absorption. In addition, foods can enhance drug absorption (Table 2) (24,39), can delay drug absorption (Tables 2 and 3) (17,39,46), or can decrease drug absorption (Table 4) (17,27,46). Several mechanisms whereby foodstuffs and drugs can interact result in an altered pharmacological response. Such mechanisms involve physiological changes in blood levels of the drug as a result of the food either increasing or decreasing its rate of absorption.

Because lipids often constitute a major constituent of food, they often influence drug absorption. Ingested lipids decrease gastric motility. Further, the presence of lipid digestion products in the upper small intestine induce the secretion of biliary and pancreatic fluids which, in turn, significantly alter the luminal environment. The presence of bile enhances the bioavailability of many poorly water soluble drugs by facilitating the rate of dissolution and/or solubility (12). The effect of bile salt solubilization has been examined for many drugs (e.g. griseofulvin, digoxin, diazepam, etc.) and typically, solubility may increase upwards of two-fold (12). However, not all interactions of poorly water-soluble drugs with bile salts lead to increased solubilization (e.g. tubocurarine, neomycin, kanamycin and various other large molecular weight antibiotics).

The interaction between food and drugs may involve several physiological factors whereby a drug affects processes related to eating, sensory appreciation of the food, swallowing, digestion, gastric emptying, nutrient absorption, nutrient

Table 2 Drug–nutrient interactions causing enhanced absorption[a]

Drug	Mechanism
Atovaquone	High fat meal facilitates absorption
Carbamazepine	Increased bile production; enhanced dissolution and absorption
Cyclosporin	Increased bioavailability
Diazepam	Food enhances enterohepatic recycling of drug; increased dissolution secondary to gastric acid secretion
Dicumerol	Increased bile flow; delayed gastric emptying permits dissolution and absorption
Erythromycin	Unknown
Griseofulvin	Drug is lipid-soluble, enhanced absorption
Hydralazine	Food reduces first-pass extraction and metabolism, blocks enzymic transformation in gastrointestinal tract
Hydrochlorothiazide	Delayed gastric emptying enhances absorption from small bowel
Itraconazole	Bioavailability enhanced by the co-administration of a cola drink[b]
Labetalol	Food may reduce first-pass extraction and metabolism
Lithium citrate	Purgative action decreases absorption
Lovastatin	Unknown
Metoprolol	Food may reduce first-pass extraction and metabolism
Misoprostol	Food decreases its side effects
Nitrofurantoin	Delayed gastric emptying permits dissolution and increased absorption
Phenytoin	Delayed gastric emptying and increased bile production improves dissolution and absorption
Propoxphene	Delayed gastric emptying improves dissolution and absorption
Propranolol	Food may reduce first-pass extraction and metabolism
Spironolactone	Delayed gastric emptying permits dissolution and absorption; bile may solubilize

Sources:
a Modified from Anderson (1998) (4); Katz and Dejean (1985) (25).
b Lange *et al.* 1997 (31).
  See also Thomas *et al.* (1998) (52).

metabolism or renal excretion of nutrients. The mechanisms of food–drug interactions have not been fully understood, but they most likely involve both direct as well as indirect factors (Table 5). Although the exact number of drugs which affect G-I absorption is not known, there are probably about 100 to 150 separate chemical entities that elicit such actions.

Physiological changes can occur at the lumen of the G-I tract or at the membrane (15). Luminal factors include delayed gastric emptying, increased acid secretion, motility and bile secretions. Membrane-associated factors include region-dependent absorptive activity, active absorptive processes, intestinal elimination and splanchnic blood flow.

The oral administration of drugs represents a convenient, non-invasive route, and dosing can be associated with daily routines, such as meal times, which tends to increase patient compliance. Nevertheless, this association can result in an increased incidence of nutrient–drug interactions, with certain foods able to

Table 3 Drugs–nutrient interactions causing delayed absorption

Drug	Mechanism
Paracetamol (acetaminophen)	High pectin foods act as absorbent and protectant
Ampicillin	Reduction in stomach fluid volume
Amoxicillin	Reduction in stomach fluid volume
Aspirin	Direct interference; change in gastric pH
Atenolol	Mechanism unknown, possibly physical barrier
Cephalosporins	Mechanism unknown
Cimetidine	Mechanism unknown
Digoxin	High-fiber, high-pectin foods bind drug
Frusemide	Mechanism unknown
Glipizide	Unknown
Metronidazole	Mechanism unknown
Piroxicam	Mechanism unknown
Quinidine	Possibly protein binding
sulphonamides	Mechanism unknown; may be physical barrier
Valproic acids	Mechanism unknown

Sources: Modified from Katz and Dejean (1985) (25); Kessler *et al.* (1982) (25); Kirk (1995) (27). See also Thomas *et al.* (1998) (52).

decrease, delay, or increase the absorption of drugs, hence altering their bioavailability, solubility of gastric fluid and gastric emptying time (55). Delayed drug absorption does not necessarily imply that less total drug is actually absorbed, but that peak blood levels of the drug may be achieved over a longer period of time. Drugs that bind or complex to nutrients are often unavailable for absorption.

Food (and food components) can affect the bioavailability of drugs by binding directly to the drug, by changing luminal pH, gastric emptying, intestinal transit, mucosal absorption and splanchnic-hepatic blood flow (4,5). Food-induced changes in the bioavailability of some drugs may be partially dependent upon hepatic biotransformation, as evidenced by absorbed nutrients competing with drugs for first-pass metabolism in the intestine or the liver. Many drugs undergo metabolic transformation by enteric micro-organisms that may, in turn, be affected by nutrients. Hence, the drug's metabolism may also be changed.

III. BIOAVAILABILITY AND METABOLISM

Bioavailability defines that fraction of the drug's dosage that reaches the systemic circulation metabolically unchanged (61). There are numerous factors that can modify a drug's disposition such as genetic traits, age, sex, pregnancy and, of course, diet itself (21). Indeed, bioavailability can be modified by foods and leads to changes in the drug's pharmacokinetics and/or pharmacodynamics.

Some foods may enhance a drug's bioavailability (e.g. pleconaril, triclabendazole), may reduce bioavailability (e.g. eplastigmine) or have no effect (e.g. zolmitriptan).

Table 4 Drugs–nutrient interactions causing decreased absorption[a]

Drug	Mechanism
Ampicillin	Reduction in stomach fluid volume
Atenolol	Mechanism unknown; possibly physical barrier
Azithromycin	Reduces bioavailability
Captopril	Mechanism unknown
Chlorpromazine	Drug undergoes first-pass metabolism in gut; delayed gastric emptying affects bioavailability
Desmopressin	Administering drug in a fasting state prolongs its duration of action[b]
Eptastigmine	Reduce bioavailability[c]
Erythromycin	Mechanism unknown
Fluoxetine	Inhibits amino acid transport of leucine[d]
Isoniazid	Food raises pH preventing dissolution and absorption; also delayed gastric emptying
Levodopa	Drug competes with amino acids for absorption and transport
Lincomycin	Mechanism unknown
Methyldopa	Competitive absorption
Nafcillin	Mechanism unknown; may be alteration of gastric fluid on pH
Penicillamine	May form chelates with calcium or iron
Penicillin G	Delayed gastric emptying, gastric acid degradation; impaired dissolution
Propantheline	Mechanism unknown
Ropinirole	Reduces rate of absorption[e]
Tetracyclines	Binds with calcium ions or iron salts forming insoluble chelates
Zidovudine	Food decreases concentration of drug

Sources:
a Modified from Anderson (1988) (4); Katz and Dejean (1985) (25); Kessier *et al.* (1992) (26); Kirk (1995) (27).
b Rittig *et al.* (1998) (40).
c Bjornsson *et al.* (1998) (8).
d Urdaneta *et al.* (1998) (57).
e Biefelelal (1998) (9).
  See also Thomas *et al.* (1998) (52).

Table 5 Drug–food interactions affecting absorptive processes

Indirect mechanisms
- Drug-induced alterations in gastrointestinal motility (e.g. anticholinergics)
- Drug-induced malabsorption syndromes (e.g. Neomycin)

Direct mechanisms
- Drug-induced pH alterations in gastrointestinal tract (e.g. antacids)
- Drug-induced changes in bioavailability (e.g. absorption to drug – kaolin/pectin)
- Drug-induced retardation of absorption (e.g. charcoal)
- Drug-binding/chelation (e.g. anionic exchange resins – Cholestyramine; metal ions – iron, calcium)

Sources: Modified from Ama *et al.* (1986) (3). See also Thomas *et al.* (1998) (52).

For example, to enhance the oral bioavailability of pleconaril, a broad spectrum antipocornaviral agent, it should be co-administered with a fat-containing meal (1). Food can likewise improve the systemic availability of triclabendazole (32). Conversely, other drugs such as eplastigmine have their bioavailability significantly reduced by food causing a reduction in efficacy (8). Some agents (e.g. zolmitriptan) exhibit good oral bioavailability at therapeutic doses and have dose-proportional pharmacokinetics that are not affected by food (44). The importance of biliary salts and their actions upon solubilizing the drug can also influence bioavailability (see Section II). Thus, the effect of food upon bioavailability tends to vary according to the particular drug and is related to such factors as its water solubility, action by biliary secretions and other physicochemical interactions occurring in the G-I tract.

A serendipitous observation made almost a decade ago revealed that grapefruit juice increases the bioavailability, as well as efficacy and toxicity, of the calcium channel blocker felodipine (6). This example constituted one of the first observed pharmacokinetic interactions between a citrus juice and a drug. Since many medications are ingested with breakfast, this interaction became even more important. Further, such interactions may lead to increases in blood drug levels from about 200 to 900% (48). It is now recognized that grapefruit juice (fresh or frozen) interferes with cytochrome P-450 isoenzymes (30). The effect appears to be mediated mainly by suppression of the cytochrome P-450 enzyme CYP 3A4 in the small intestine wall. Grapefruit juice may also activate P-glycoprotein-mediated drug transport systems (47). These actions lead to a diminished first pass metabolism and increased maximal plasma concentrations of substrates of this enzyme (16).

Grapefruit juice has achieved notoriety as a potent inhibitor of several P-450 enzymes (CYP1A2, 2A6, and 3A4), but its predominant effect is on 3A4 located in the gut wall (23). The inhibitory substance is not naringenin, but a furanocoumarin compound widely found in nature (namely 6,7-dihydroxybergamottin) (30). It is noteworthy that several selective serotonin reuptake inhibitors (SSRIs) (e.g. citalopram, fluoxetine, paroxetine and sertraline), all drugs used in the treatment of obsessive-compulsive disorders can inhibit cytochrome P-450 enzymes. Cyclosporine (10), terfenadine (13), lovastatin (24) and the protease inhibitor saquinavir (28) can interact with grapefruit juice, generally resulting in increased bioavailability of these agents. Table 6 depicts a more extensive list of drug interactions with grapefruit juice (16). Felodipine, another calcium antagonist, appears to have a propensity to interact with grapefruit juice, leading to increased bioavailability and toxicity. Cyclosporin is metabolized by the same enzymes as calcium antagonists.

Most of the drugs that interact with grapefruit juice are subject to first pass degradation, and are known to be mediated by cytochrome P-450 (3A4/5). Also, most of these drugs undergo Phase I metabolism (16). The antihistamines torfenadine and astemizole undergo extensive first pass metabolism. Grapefruit juice has the potential to modify steroid metabolism (e.g. estradiol and cortisol). Benzodiazepines (e.g. midazolam, triazolam) can have their CNS effects prolonged by grapefruit juice presumably acting through enhanced bioavailability.

Table 6 Relative relationship between bioavailability
of selected drugs due to grapefruit juice

Drug	Relative bioavailability[a]
Felodipine	+++++
Nisoldipine	+++++
Saquinavir	++++
Nitrendipine	++++
Nifedipine	++++
Nimodipine	++++
Midazolan	+++
Trizolan	+++
Cyclosporin	++
Ethinylestradiol	++
Amiodipine	++
Diltiazem	++
Quinidine	++

Source: Modified from Fuhr (1998) (16).

Notes

a Using an arbitrary scale (1 to 5 + with 5 being the most
enhanced bioavailability), and based upon an increase in
AUC (area under the curve).

Mechanistically, the components of grapefruit juice which most probably cause the interaction with the involved drugs are psoralen derivatives. Regardless of the molecular mechanism, patients should refrain from drinking grapefruit juice when ingesting drugs that are known to undergo extensive metabolism.

It is recognized that drugs may be metabolized by two basic reactions, i.e. Phase I and Phase II. Phase I reactions include oxidation, hydroxylation, reduction or hydrolysis and result in changes to a functional group on the drug molecule. The mixed function oxidase system (MFOS) is an inducible enzyme system that catalyzes the oxidation of a wide variety of drugs. The MFOS is found primarily in the endoplasmic reticulum of the liver and other tissues. Phase II reactions include conjugation to glucuronate or glutathione, and acetylation or sulphonation to functional group(s) on the drug molecule. Modifications of functional groups frequently render the drug more water-soluble (or polar) and thus more readily excreted by the kidney. Conjugation enzymes are present in the endoplasmic reticulum of the cytoplasm. Several oxidized products of the MFOS are substrates for conjugating enzymes.

The metabolism of drugs by Phase I and Phase II reactions is catalyzed by various enzymes, and the formation of metabolites necessitates that other substances be provided by the body through nutrition. Several nutrients and micronutrients (e.g. vitamins) exert significant roles in Phase I oxidation reactions (21). Phase II reactions involving conjugation depend upon the body to provide carbohydrates, amino acids, fats and proteins.

IV. INTERACTION EFFECTS ON DRUG ABSORPTION

Food can influence drug absorption through physicochemical and physiological interactions (see Sections II & III). Many of the studies pertaining to such interactions involve meals of mixed composition (59). Meals high in fat are recommended by the US FDA for studies of a drug's oral bioavailability and bioequivalence in order to fully assess food interactions. Further, specific meal composition can aid in more accurately predicting particular mechanistic aspects of food–drug interactions that affect drug absorption (15,33,37).

Factors that increase the risk of clinically significant interactions include a narrow therapeutic index, complexity of the pharmacokinetics (e.g. absorption, hepatic metabolism, etc.) and elimination by the kidneys. Generally, food does not affect most drugs which are absorbed evenly throughout the intestinal tract. Conversely, drugs that are primarily absorbed in the upper intestine exhibit diminished absorption when administered with meals.

The actual composition of the meal itself may affect the potential interaction through different mechanisms (Table 7). Meals rich in carbohydrates, including sucrose and starch, may delay gastric emptying, thus leading to increased absorption. Protein-rich diets and those containing high fiber content generally reduce the rate of absorption. Lipid-rich meals, acting through enhanced solubilization by biliary and pancreatic secretions, may facilitate the absorptive process.

Insight into a drug's pharmacokinetic properties and its formulation may aid in predicting potential interactions. A biopharmaceutical classification has been devised based on rate-limiting factors for absorption as a function of drug properties and G-I physiology (15). This classification includes:

Class I   High Solubility/High Permeability (e.g. acetaminophen)
Class II   Low Solubility/High Permeability (e.g. digoxin)
Class III High Solubility/Low Permeability (e.g. furosemide)
Class IV   Low Solubility/Low Permeability (e.g. amphotericin B)

Table 7 Meal composition and drug absorption

Meal	Interaction
Carbohydrate-rich	Delayed gastric emptying
Protein-rich	Increased gastric pH; binding
Fat-rich	Increased lumenal fluid volume; increased biliary and pancreatic secretions; solubilization of lipophilic drugs by bile salts
Artificial fat (e.g. Olestra)	Increase high lipophilic compounds partitioning in the non-absorbable fat
Dietary fiber	Binding; decreased emptying and decrease diffusion rate

Source: Modified from Fleisher *et al.* (1999) (15).

Limiting factors are related to drug dissolution and drug permeability. For example, Class I drugs exhibit both rapid dissolution and high membrane permeability while Class IV drugs are poorly water-soluble and have low membrane permeability. The usefulness of this biopharmaceutical classification is in defining screening procedures for drug dosage forms to predict the limitations of oral bioavailability (15). It also aids in predicting drug interactions that affect oral drug absorption. Finally, this classification system may be useful in those drug absorption interactions where G-I residence time has been affected by controlled release products.

V. DISEASES AND DRUG INTERACTIONS – ABSORPTION

Intestinal malabsorption can occur in aging as well in many disease states where there is impairment of one or more physiological processes involved in digestion and absorption of nutrients (Table 8). Malabsorption can be further affected by the presence of different drugs. Both pathogenic organisms and a variety of drugs are associated with provoking diarrhea, leading to further modifications of the residence time of a given drug within the G-I tract. The mechanism(s) of disease-related changes in G-I absorptive processes differs and ranges from effects on gastric acid to a loss of mucosal surface area (Table 8).

While the aging process is physiological, and not pathological, there are a number of changes that can affect drug absorption in the elderly (20,49,53). In addition, aging can lead to changes in drug distribution, metabolism and excretion (50) (Table 9). Aging is occasionally associated with an increase in gastric pH, a decrease in intestinal blood flow and increased gastric emptying time (Table 9). In contrast to what was commonly believed, nearly 90% of elderly people exhibit normal acid secretory activity (22). With increasing gastric emptying time, the drug may be in extended contact with the gastric mucosa, including

Table 8 Effect of aging and selected disease states on G-I absorptive processes[a]

Condition/State	G-I absorptive processes
Aging	Increased gastric pH; reduced gastric acid secretions
AIDs	Increased gastric pH; decreased acidity
Celiac sprue (gluten-sensitive enteropath)	Reduced absorptive surface; many factors involved, including genetics
Cystic fibrosis	Decreased intestinal pH
Idiopathic Inflammatory Bowel Disease (IBD) (e.g. ulcerative colitis; Crohn's Disease)	Increased paracellular transport; decreased carrier-mediated transport

Sources: Modified from Fleisher et al. (1999) (15); Thomas and Burns (1998) (53) and Thomas et al. (2000) (54).

Table 9 Aging, and its effect on pharmacokinetics[a]

Pharmacokinetic parameter	Age-related changes
Absorption	*G-I*
	Increased gastric pH
	Decreased G-I tract blood flow
	Increased gastric emptying time
Distribution	*Body composition*
	Decreased lean body mass
	Increased fatty tissue
	Decreased total body water
	Decreased plasma protein (albumin)
Metabolism	*Hepatic*
	Decreased liver size
	Decreased microsomal enzyme activity
	Decreased hepatic blood flow
	Increased incidence of drug-induced hepatitis
Excretion	*Renal*
	Decreased creatinine clearance
	Decreased renal size
	Decreased renal blood flow
	Decreased glomerular filtration
	Decreased numbers of functioning glomeruli

Sources: Stein (1984) (50). See also Thomas and Burns 1998 (53).

regional blood supply, and may reduce gastric intestinal blood flow. Age-related changes in body composition, including increases in body fat, decreases in lean body mass and decreases in total body water, can affect drug disposition in several ways. Hypoalbuminemia often seen in malnourished elderly also may affect plasma protein binding.

Hepatic metabolism can be reduced in the elderly. Drug metabolism in the geriatric patient is affected by a decrease in liver size, a diminution in hepatic microsomal enzyme systems, a reduction in regional blood flow and by an increased incidence of drug-induced hepatitis. In older patients, hepatic blood flow can be half that observed in younger individuals. Hepatic microsomal enzyme systems responsible for Phase I (e.g. oxidation reduction and hydrolysis) and Phase II (e.g. conjugation) metabolic transformations can be greatly reduced in the elderly.

Aging leads to a decrease in the size of the kidney, less renal blood flow and a diminished glomerular filtration rate (GFR) (Table 9). The actual number of normal or physiologically functioning nephrons may be decreased in the elderly, which is correlated with the ability of the kidney to effectively eliminate drugs. In the elderly, these excretory mechanisms may not be functioning optimally or they may have undergone pathological changes.

The elderly are more at risk of adverse effects and clinically important outcomes of drug–nutrient interactions (41). These increased risks are likely to be caused by multiple drug usage, age-related modifications in drug disposition, geriatric pathologies which might impair drug clearance or simply because subgroups of the elderly may be malnourished. Loss of appetite in the elderly is not uncommon.

The elderly often take more drugs because of various physiological deteriorations. Frequently, the elderly consume over-the-counter (OTC) drug products such as cathartics, vitamin/minerals and antacids. It has been estimated that by the year 2000, approximately 50% of all long term care drugs will be OTC products (11). A common form of drug–nutrient interaction in the elderly is mineral deficiency caused by the frequent use of diuretics leading to potassium and magnesium loss (29,41).

Nutritional modulation of lifespan in the elderly also indicates that alterations in the micronutrient and macronutrient constituents of the diet can be affected by gene expression (34). Nutritional problems in the elderly are related not only to multiple drug use, but to the consumption of specialized diets for one or more chronic illnesses. Adverse effects such as nausea, vomiting and diarrhea can also affect drug–nutrient interactions. Cytotoxic drugs can damage mucosal cells in the G-I tract while enteric microflora can be suppressed by certain antibiotics.

Diseases often associated with the elderly include hypertension, cardiac failure osteoporosis, and renal insufficiency, and raise particular concerns about selected classes of drugs and to what extent they affect nutrition. Drug-induced adverse effects can complicate the therapy by affecting nutritional status. Digoxin, while effective for congestive heart disease, possesses inherent anorexic properties including nausea and vomiting. Loop diuretics not only facilitate the loss of sodium, but also potassium, magnesium, calcium and thiamin. Osteoporosis can be exacerbated in the elderly postmenopausal patient undergoing therapy with loop diuretics. Osteoporosis is commonly associated with the malabsorption of calcium (36).

VI. MEDICAL NUTRITION

Enteral and parenteral nutrition depend upon several physiological processes including both metabolism and excretion. Enteral nutrition also utilizes absorptive processes. Other mechanisms that can affect drug–nutrient interactions may include protein binding, altered solubility and chelation.

Total parenteral nutrition (TPN) is provided to patients intravenously when the G-I tract is not functioning and most of the energy requirements can be provided by isotonic fat solutions. In patients receiving TPN, over three-fourths of the drugs used lead to some degree of altered nutritional status (43). Furthermore, TPN is more apt to be associated with infectious, metabolic, and fluid complications than enteral nutrition. Prolonged TPN may lead to atrophy of the G-I tract and to hepatic complications (19). Thus, careful nutritional support of these patients is imperative to improve nutritional status, facilitate wound healing and restore immune status.

Enteral nutrition rather than TPN should be used whenever possible (38). There are at least six categories of Medicare (USA) approved enteral products

Table 10 Effect of enteral formula and other diets on dietary components on G-I absorption[a]

Enteral formula	Liquid phenytoin levels may be decreased
Milk and dairy products	Tetracycline levels may be decreased
Grapefruit juice[b]	Nifedipine levels may be decreased
Meat and high protein foods	Levodopa levels may be decreased
Food or milk	*Increased levels* *Decreased levels*
	Nitrofurantoin Rifampin
	Hydralazine Isoniazid
	Propanolol Captopril
	Atenolol
	Dadolol
	Timolol

Sources:
a Modified from Varella *et al.* (1997) (58).
b See Section III for further details.

ranging from the so-called standard nutrition to calorie-dense products and from disease specific to those containing specialized nutrients.

Enteral formulas can affect G-I absorption (Table 10). Formulas that are either low or high in fat, protein or carbohydrates can affect the absorption of certain drugs. For example, the pharmacokinetics of theophylline can be profoundly affected by the composition of the diet (35,42,58). Despite the benefits of enteral nutrition, there are associated drug-related problems. Diarrhea is perhaps the most common adverse effect of enteral feeding. Oftentimes in order to meet dosage requirements, intravenous drug formulations use sorbitol as a diluent. Sorbitol can cause diarrhea. Several oral liquid formulations contain sorbitol (Table 12). The etiology of diarrhea in tube-fed patients may be multifactorial since many other drugs can contribute to the increased frequency of watery stools.

Pharmacotherapy can significantly affect fluid and electrolyte balance, vitamin homeostasis and acid–base balance. For example, when administering either enteral or parenteral formulas in conjunction with certain medications, it is necessary to recognize that many drugs, particularly antibiotics, may increase sodium levels and hence affect electrolyte balance (Table 11). Sodium levels in these drug formulations may range from 12 mEq (e.g. ampicillin) to 120 mEq (e.g. carbicillin), a 10-fold difference. Thus, a patient treated with ticarcillin would need considerably less sodium added to the TPN solution.

In addition to antibiotics, and of course diuretics, other drugs affect sodium homeostasis. Cisplatin can cause renal loss of sodium. Sodium loading can prevent renal toxicity caused by amphotericin B (42). Conversely, dietary sodium restriction can potentiate the blood pressure effects of antihypertensive agents (7).

Measuring sodium levels in parenteral and enteral solutions is important. Likewise, potassium homeostasis must be monitored since many drugs and hormones (e.g. insulin) can produce hypokalemia. Hence, the failure to add sufficient potassium to IV solutions can lead to electrolyte imbalances (42). Conversely, ACE

167

Table 11 Sodium levels present in selected antibiotics

Drug	Sodium content (mEq/g)	Usual daily dose (g)	Sodium delivery (mEq)
Carbenicillin	5.0	24	120.0
Ticarcillin	5.2	18	93.6
Azlocillin	2.2	18	39.6
Ceftizoximine	2.6	12	31.2
Metronidazole (ready-to-use vial)	14.0	2	28.0
Cephradine	6.0	4	24.0
Piperacillin	1.9	12	22.8
Cefamandole	3.3	6	19.8
Cefoxitin	2.3	8	18.4
Nafcillin	2.9	6	17.4
Mezlocillin	1.9	8	15.2
Moxalactam	3.8	4	15.2
Cefotetan	3.5	4	14.0
Ampicillin	3.0	4	12.0

Sources: From Mowatt-Larssen and Brown (1994) (35); Sacks and Brown (1994) (42).

inhibitors and potassium-sparing diuretics tend to promote potassium retention. Heparin, as well as non-steroidal anti-inflammatory drugs (NSAIDs), can promote hyperkalemia.

It should be evident that the concomitant use of drugs during medical nutrition regimens can cause nutritional–pharmacologic interactions. A knowledge of these potential interactions is important for successful therapeutic outcomes. Specific clinical conditions, particularly in the critically ill, necessitate a complete understanding of such potential drug–nutrient interactions. Optimal pharmacologic therapy, coupled with complete diet intake, must be compatible for the well-

Table 12 Selected liquid drug formulations and their content of sorbitol

Drugs	Dosage	Sorbitol (g)
Propranolol solution	80 mg × 3 doses	37.8
Acetaminophen elixir	650 mg × 4 doses	28.4
Acetaminophen elixir	650 mg × 4 doses	<16.3
Mylanta suspension	30 ml × 4 doses	16.0
Cimetidine solution	300 mg × 4 doses	11.5
Codeine phosphate solution	30 mg × 4 doses	11.2
Ranitidine syrup	150 mg × 2 doses	4.0
Bactrim suspension	20 ml × 2 doses	2.8
Calcium carbonate suspension	1,250 mg × 2 doses	2.8
Furosemide solution	40 mg × 1 dose	2.0

Source: See Varella *et al.* (1997) (58).

being and recovery of the patient in need of either parenteral or enteral nutritional support.

VII. CONCLUDING REMARKS

Several physiological and pathological factors can influence the incidence of drug–nutrient interactions. Of the various physiologic processes involved, the G-I absorptive process appears to play a significant role in a drug's eventual efficacy and toxicity. Several diseases, along with the aging process, can affect the magnitude and incidence of drug–nutrient interactions. Finally, medical nutrition, more specifically enteral and parenteral, can involve drug–nutritional interactions. An understanding not only of the physiological processes involved (and their possible impairment or compromised state), but also of the impact of drugs and dietary components, can lead to successful therapeutic outcomes.

REFERENCES

1. Abdel-Rahman SM, Kearns GL. Single-dose pharmacokinetics of a pleconaril (VP63843) oral solution and effect of food. *Antimicrobial Agents and Chemotherapy* 1998; (42):2706–9.
2. Arhonheim J. Practical pharmacology for older patients: avoiding adverse drug effects. *Mt. Sinai J Med* 1993; 60:497–501.
3. Ama PFM, Poehlman ET, Simoneau JA, Boulay MR, Theriault G, Tromblay A, Bouchard C. Fat distribution and adipose tissue metabolism in non-obese male black African and Caucasian subjects. *International Journal of Obesity* 1986; 10: 503–14.
4. Anderson KE. Influences of diet and nutrition on clinical pharmacokinetics. *Clinical Pharmacokinetics* 1988; 14:325–46.
5. Anderson KE, Kappas A. How diet affects drug metabolism. *Hospital Therapy* 1987; 4:93–102.
6. Bailey DG, Spence JD, Munoz C, Arnold JMO. In action of citrus juices with felodipine and nifedipine. *Lancet* 1991; 337:268–9.
7. Bennett WM. Drug interactions and consequences of sodium restriction. *American Journal of Clinical Nutrition* 1997; 65:6785–815.
8. Bjornsson TD, Troetel WM, Imbimbo BP. Effect of food on the absorption of eptastigmine. *Eur J Clin Pharmacol* 1998; 54:243–7.
9. Brefel C, Thalamas C, Rayet S, Lopez-Gil A, Fitzpatrick K, Bullman S, Citerone DR, Taylor AC, Montastruc JL, Rascol O. Drug interactions and consequences of sodium restriction. *Br J Clin Pharmacol* 1998; 45:412–15.
10. Brunner LJ, Munar MY, Vallian J, Wolfson M, Stennett DJ, Meyer MM, Bennett WM. Interactions between cyclosporine and grapefruit juice requires long-term ingestion in stable renal transplant recipients. *Pharmacotherapy* 1998; 18: 23–9.
11. Cardinale V. Stemming the tide of polymedicine. *Drug Topics* 1998; 132:36–41.

12. Charman WN, Porter CJH, Mithani S, Dressman JB. Physicochemical and physiological mechanisms for the effects of food on drug absorption: the role of lipids and pH. *J Pharm Sci* 1997; 88:269–82.

13. Clifford CP, Adams DA, Murray S, Taylor GW, Wilkins MR, Boobis AR, Davies DS. The cardiac effects of terfenadine after inhibition of its metabolism by grapefruit juice. *Eur J Clin Pharmacol* 1997; 52:311–15.

14. DePonti F, Giaroni C, Consentino M, Lecchini S, Frigo G. Adrenergic mechanisms in the control of gastrointestinal motility: from basic science to clinical applications. *Pharmacology and Therapeutics* 1996; 69:59–78.

15. Fleisher D, Li C, Zhou Y, Pao L-H, Karim A. Drug, meal and formulation interactions influencing drug absorption after oral administration: clinical implications. *Clin Pharmaokinet* 1999; 36:233–54.

16. Fuhr U. Drug interactions with grapefruit juice: extent, probable mechanism and clinical relevance. *Drug Safety* 1998; 18:251–72.

17. Garabedian-Ruffalo SM, Syrja-Farber M, Lanius PM, Plucinski A. Monitoring of drug–drug and drug–food interactions. *American Journal of Hospital Pharmacy* 1998; 45:1530–4.

18. Gauthier I, Malone M. Drug–food interactions in hospitalised patients: methods of prevention. *Drug Safety* 1998; 18:363–93.

19. Gottschlich MM, Matarse LEA, Shronts EP. *Nutrition Support Dietetics Core Curriculum* 2nd edn. Silver Spring, MD: Aspen Co; 1992.

20. Hämmerlein A, Derendorf H, Lowenthal D. Pharmacokinetic and pharmacodynamic changes in the elderly: clinical implications. *Clin Pharmacokinet* 1998; 35:49–64.

21. Hoyumpa A, Schenker S. Major drug interactions: effect of liver disease, alcohol, and malnutrition. *Annual Reviews of Medicine* 1982; 33:113–49.

22. Hurwitz A, Brady DA, Schaal SE, *et al*. Gastric acidity in older adults. *JAMA* 1997; 278:659–82.

23. Jefferson JW. Drug and diet interactions: avoiding therapeutic paralysis. *J Clin Psychiatry* 1998; 59(suppl 16):31–9.

24. Kantola T, Kivistö KT, Neuvonen PJ. Grapefruit juice greatly increases serum concentrations of lovastatin and lovastatin acid. *Clin Pharmacol Ther* 1998; 63: 397–402.

25. Katz NL, Dejean A. Interrelationships between drugs and nutrients. *Pharm Index* 1985; 12:9–15.

26. Kessler DA, Taylor MR, Maryanski JH, Flamm EL, Kahl LS. The safety of foods developed by biotechnology. *Science* 1992; 256:1747–832.

27. Kirk JK. Significant drug–nutrient interactions. *Am Fam Physician* 1995; 51: 1175–82.

28. Kupferschmidt HH, Fattinger KE, Ha HR, Follath F, Krähenbühl S. Grapefruit juice enhances the bioavailability of the HIV protease inhibitor saquinavir in man. *Br J Clin Pharmacol* 1998; 45:355–9.

29. Lamy PP. Effects of diet and nutrition on drug therapy. *Journal of the American Geriatric Society* 1982; 30:S99–112.

30. Landrum-Michalets E. Update: clinically significant cytochrome P-450 drug interactions. *Pharmacotherapy* 1998; 18:84–112.

31. Lange D, Pavao JH, Jacqmin P, Woestenborghs R, Ding C, Klausner M. The effect of co-administration of a cola beverage on the bioavailability of intraconazole in patients with acquired immunodeficiency syndrome. *Curr Therap Res* 1997; 58:202–12.

32. Lecaillon JB, Godbillon J, Campestrini J, Naquira C, Miranda L, Pacheco R, Mull R, Poltera AA. Effects of food on the bioavailability of triclabendazole in patients with fascioliasis. *Br J Clin Pharmacol* 1998; 45:601–4.

33. Li C, Pao L-H, Li L, *et al*. Regional intestinal absorption and meal effects on LY303366 in canine model. *Pharm Sci* 1998; (supp) S10.

34. Mooradian AD. Nutritional modulation of life span and gene expression. *Annals of Internal Medicine* 1988; 109:891–2.

35. Mowatt-Larssen C, Brown R. Drug–nutrient interactions. In: Zaloga GP, ed. *Nutrition in Critical Care*. St. Louis, MO: Mosby Co; 1994; 487–503.

36. Nordin B. Calcium and osteoporosis. *Nutrition* 1997; 13:664–86.

37. Pao LH, Zhou SY, Cook C, *et al*. Reduced systemic availability of an antiarrhythmic drug, bidisomide, with meal co-administration: relationship with region-dependent intestinal absorption. *Pharm Res* 1998; 15:221–7.

38. Parrish CR, McCray SF, Wolf AMD. When is nutrition support appropriate? *Hospital Medicine* 1999; 50–5.

39. Randle NW. Food or nutrient effects on drug absorption: a review. *Hospital Pharmacy* 1987; 22:694–7, 718.

40. Rittig S, Jensen AR, Jensen KT, Pedersen EB. Effect of food intake on the pharmacokinetics and antidiuretic activity of oral desmopressin (DDAVP) in hydrated normal subjects. *Clinical Endocrinology* 1998; 48:235–41.

41. Roe DA. Therapeutic significance of drug–nutrient interactions in the elderly. *Pharmacological Reviews* 1984; 36:1095–225.

42. Sacks GS, Brown RO. Drug–nutrient interactions in patients receiving nutritional support. *Drug Therapy* 1994; 3:35–42.

43. Schneider PJ, Mirtallo J. Medication profiles in TPN patients. *Nutr Supp Serv* 1983; 3:40–6.

44. Seaber EJ, Peck RW, Smith DA, Allanson J, Hefting NR, van Lier JJ, Sollie FAE, Wemer J, Jonkman JHG. The absolute bioavailability and effect of food on the pharmacokinetics of zolmitriptan in healthy volunteers. *Br J Clin Pharmacol* 1998; 46:433–9.

45. Sierra P, Castillo J, Gomez M, Sorribes V, Monterde J, Castano J. Servicio de anestesioilogia, Fundacion Puigvert, Barcelona. *Rev Esp Anestesiol Reanim* (Spain) 1997; 44:383–7.

46. Smith CH, Bidlack WR. Dietary concerns associated with the use of medications. *Journal of the American Diet Association* 1984; 84:901–14.

47. Soldner A, Christians U, Susanto M, Wacher VJ, Silverman JA, Benet LZ. Grapefruit juice activates P-glycoprotein-mediated drug transport, *Pharmaceutical Research* 1999; 16:478–85.

48. Spence JD. Drug interactions with grapefruit: whose responsibility is it to warn the public? *Clin Pharmacol Ther* 1997; 61:395–400.

49. Steffens DC, Krishnan KRR. Metabolism, bioavailability, and drug interactions, *Clinics in Geriatric Medicine* 1998; 14:17–31.

50. Stein BE. Avoiding drug reactions: seven steps to writing safe prescriptions. *Geriatrics* 1984; 49:28–36.

51. Thomas JA, Tschanz C. Nutrient–drug interactions. In: Kotsonis FN, Mackey M, Hjelle JJ, eds. *Nutritional Toxicology*, Target Organ Toxicity Series. New York: Raven Press; 1994; 139–48.

52. Thomas JA, Stargel WW, Tschanz C. Interactions between drugs and diet. In: Ioannides C, ed. *Nutrition and Chemical Toxicity*. J Wiley & Sons Ltd; 1998; 161–82.

53. Thomas JA, Burns RA. Important drug–nutrient interactions in the elderly. *Drugs & Aging* 1998; 13:199–209.

54. Thomas JA, Stargel WW, Cotter R. Drug–nutrient interactions in the critically ill. In: Kenneth A Kudsk, MD, ed. *Update in Intensive Care and Emergency Medicine.* Springer-Verlag Publ; 2000; 34:151–65.

55. Trovato A, Nuhlicek DN, Midtling JE. Drug–nutrient interactions. *American Family Physician* 1991; 44:1651–8.

56. Tschanz C, Stargel WW, Thomas JA. Interactions between drugs and nutrients. *Advances in Pharmacology* 1996; 35:1–26.

57. Urdaneta E, Idoate I, Larralde J. Drug–nutrient interactions: inhibition of amino acid intestinal absorption by fluoxetine. *Br J Nutr* 1998; 79:439–46.

58. Varella L, Jones E, Meguid M. Drug–nutrient interactions in enteral feeding: a primary care focus. *The Nurse Practitioner* 1997; 6:98.

59. Welling PG. Effects of food on drug absorption. *Pharmacol Ther* 1989; 43:425–41.

60. Welling PG. Effects of food on drug absorption. *Annual Reviews of Nutrition* 1996; 16:383–415.

61. Winstanley PA, Orme ML'E. The effects of food on drug bioavailability. *British Journal of Clinical Pharmacology* 1989; 28:621–8.

62. Yamreudeewong W, Henann NE, Fazio A, *et al.* Drug–food interactions in clinical practice. *J Fam Pract* 1995; 40:376–84.

8

PHYTOSTEROLS IN HUMAN
HEALTH

David Kritchevsky

INTRODUCTION

Plant sterols resemble cholesterol structurally in that they all have a steroid nucleus 3-β-hydroxyl group, and double bond between carbon atoms 5 and 6. Bean (1) has described 44 sterols present in plants. The major ones resemble cholesterol but have substitutions and/or unsaturation in the side chain. The principal plant sterols are β-sitosterol (24-α-ethylcholesterol), campesterol (24-α-methylcholesterol), and stigmasterol (Δ^{22}, 24-α-ethylcholesterol). Itoh (2) and Weihrauch and Gardner (3) summarized the sterol composition of a number of foods of plant origin. β-sitosterol makes up 45–95% of the total sterol of edible plant oils. Campesterol comprises up to 30% of the total sterols of seed oils, and stigmasterol may account for up to 25% of the sterols of some seed oils. Another phytosterol, ergosterol ($\Delta^{7,22}$, 24-α-methylcholesterol) is the principal sterol (90–100%) of yeast and is present in significant amounts in corn, cotton seed, peanut, and linseed oils. Pine oil also yields a mixture of phytosterols, mainly β-sitosterol, which are called tall oil sterols (Tables 1 and 2).

EARLY ANIMAL STUDIES

The first report of the unabsorbability of phytosterols was by Ellis and Gardner (4) in 1912. Several decades later, Schoenheimer and his colleagues demonstrated that sitosterol, stigmasterol, ergosterol, and brassicasterol were not absorbed by rabbits or rodents (5–8). Schoenheimer reviewed their studies in 1931 (9). The first evidence that these sterols could interfere with absorption of cholesterol was provided by Peterson (10) who fed chickens (starting cholesterol level 91 ± 14 mg/dl) a basal diet containing 4% cottonseed oil (cholesterol level 85 ± 8 mg/dl) or the same diet plus 1% cholesterol which led to serum cholesterol levels of 553 ± 95 mg/dl. Addition of 1% soy sterols or 1% sitosterol to the cholesterol-containing diet brought blood cholesterol levels down to the

Table 1 Sterol content of selected vegetable oils

Oil	% of total sterols					
	% Non-sapifiable in oil	β-sito-sterol	Stigma-sterol	Campe-sterol	Brassica-sterol	Avena-sterol
Castor	0.5	44	2	10	—	21
Cocoa Butter	0.4	59	26	9	tr	3
Coconut	0.4	58	13	8	tr	14
Corn	1.3	66	6	23	tr	1
Cotton-seed	0.6	93	1	4	tr	2
Linseed	0.9	46	9	29	tr	13
Olive	0.8	91	1	2	—	2
Palm	0.4	74	8	14	tr	2
Peanut	0.4	64	9	15	tr	8
Rice Bran	4.2	49	15	28	tr	5
Safflower	0.6	52	9	13	tr	1
Soybean	0.4	53	20	20	tr	3
Sunflower	0.7	60	8	8	—	4
Wheat Germ	3.2	67	tr	22	tr	6

Source: After Itoh *et al.* (2).

normal range (109 ± 19 mg/dl). In a subsequent experiment, Peterson *et al.* (11) demonstrated that addition of 1.3% soy sterol to a diet containing 1% cholesterol totally inhibited atherogenesis. Liver cholesterol levels which had been elevated eight-fold by cholesterol feeding were only twice normal on the cholesterol-soy sterol regimen. Peterson *et al.* (12) also carried out a dose-response study in chicks in which a diet containing 1% cholesterol was mixed with 0.25, 0.50, 1, 2, or 3% soy sterol (Table 3). There was a normalization of cholesterol levels with increasing levels of soy sterol. Addition of 0.25, 0.50, or 1% soy sterol reduced serum cholesterol levels by 36, 72, and 87%, respectively. When the diet contained 2 or 3% soy sterol, cholesterol levels were in the normal range. When the cholesterolemic diet contained 1% ergosterol or stigmasterol, cholesterol levels were reduced by 84 and 79%, respectively. Hernandez *et al.* (13,14), using cholesterol labeled with carbon 14, confirmed that simultaneous feeding with mixed soy sterols or β-sitosterol inhibited absorption of cholesterol.

Although early experiments suggested that plant sterols were totally unabsorbed, the availability of radiolabeled materials permitted more accurate assessment of absorbability. Gould *et al.* (15) fed tritium labeled sitosterol to rats and estimated absorption to be about 30–40% of an equivalent dose of cholesterol. Studies of direct lymphatic absorption of β-sitosterol in the rat indicate that about 3–6% of this sterol is absorbed (13,16,17).

Table 2 Major sterols of selected foods (mg/100 g)

Food	Total sterol	β-sitosterol	Campesterol	Stigmasterol
Vegetables				
Brussels sprouts	24	17	6	—
Cabbage	11	7	2	—
Cauliflower	18	12	6	2
Onion	15	12	1	—
Fruits				
Apple	12	11	1	—
Banana	16	11	2	3
Cherry	12	12	tr	—
Orange	24	17	1	3
Peach	10	6	1	3
Pear	8	7	—	—
Nuts				
Almond	143	122	5	3
Cashew	158	130	13	tr
Pecan	108	88	4	2
Pistachio	108	90	6	2
Walnut	108	87	6	—
Legumes				
Broad bean	124	95	8	9
Kidney bean	127	91	3	31
Pea	135	106	10	10

Source: After Weirauch and Gardner (3).

Table 3 Dose response to soy Sterol of chick plasma and liver cholesterol (Male chicks on diet at age 3 weeks)

Addition	Plasma cholesterol (mg/dl)		Liver cholesterol % dry weight
	3 weeks	6 weeks	6 weeks
4% CSO[a]	100	83	1.00
4% CSO + 1% Cholesterol (CSOC)	752	1226	6.88
CSOC + 0.25% Soy sterol	481	787	5.97
CSOC + 0.5% Soy sterol	229	349	4.61
CSOC + 1.0% Soy sterol	153	163	2.38
CSOC + 2% Soy sterol	100	91	1.37
CSOC + 3% Soy sterol	95	108	1.17

Source: After Peterson *et al.* (12).

Note
a CSO = Cottonseed Oil.

Vahouny and co-workers (18) compared the lymphatic absorption of sitosterol, stigmasterol, and fucosterol (24-ethylidine cholesterol) and found them to be absorbed at levels of 3.1, 4.3, and 3.7%, respectively. Under the same conditions, cholesterol absorption measured 46.8 ± 5.0%. When the three plant sterols were

Table 4 Absorption of plant sterols by various
          animal species

Species	No. of studies	% Absorption
Rat	4	4
Rabbit	2	0
Dog	1	15
Fowl	1	2
Guinea pig	1	<5
Man	5	6

Source: Summarized from Pollak and Kritchevsky (28).

fed (individually) with cholesterol, both sitosterol and stigmasterol inhibited absorption by 54%, but fucosterol had no effect. A later study (19) showed that fucosterol did, indeed, inhibit cholesterol absorption, but not as effectively as sitosterol. Ikeda and Sugano (20) suggested that restriction of the micellar solubility of cholesterol was the major determinant of inhibition of cholesterol absorption by sitosterol. Phytosterols seemed to displace cholesterol from micelles, but other sites of discrimination were also present (19–21). Mattson *et al.* (22) demonstrated that phytosterol esters inhibited absorption of dietary cholesterol as effectively as free phytosterols.

Salen *et al.* (23) studied sitosterol absorption and metabolism in man. When 240–320 mg of sitosterol were fed, absorption was about 1.5–5% and acidic metabolites were recovered from the feces. Boberg *et al.* (24) have reported that while the rat can convert plant sterols into C_{21} bile acids, conversion of sitosterol to C_{24} bile acids does not occur (25). Vervet monkeys fed 300 mg of β-sitosterol daily for 5 days excreted acidic steroids adduced to be 27-carboxysitosterol and 27-carboxysitostanol (26). Using an intestinal perfusion technique, Heinemann *et al.* (27) compared absorption of several phytosterols in 10 healthy human subjects. Mean hourly absorption (%) was: cholesterol, 31.2 ± 9.1; campestanol, 12.5 ± 4.8; campesterol, 9.6 ± 13.8; stigmasterol, 4.8 ± 6.5; and sitosterol, 4.2 ± 4.2. A summary of data on phytosterol absorption by different animal species is presented in Table 4 (28).

EARLY HUMAN STUDIES

Pollak (29) was the first to demonstrate that a plant sterol could reduce plasma cholesterol levels in humans. His cohort of 26 healthy men had an average cholesterol level of 267 ± 14 mg/dl at the outset of the study. They were given 5.7 or 10 g of plant sterols daily (average dose 8.1 g) for 13 ± 1 days. Average cholesterol level for the group fell to 193 ± 7 mg/dl, a reduction of 27.7%. For the next few years following Pollak's observation, there was a virtual flood of papers attesting, for the most part, to the hypocholesterolemic effects of soy sterol preparations. A wide variety of preparations containing soy or phytosterols from a number of sources were fed to subjects who ranged from healthy to ones with coronary disease. A few of the papers will be cited here, but they are all presented and discussed by Pollak and Kritchevsky (28). Beveridge, who was a pioneer in work on diet and hypercholesterolemia, concluded that the hypocholesterolemic action of corn oil could be due, largely, to its phytosterol content (30–32). Best *et al.* (33) treated nine subjects with β-sitosterol (5–6 gm before eating) and observed a 20% reduction in total serum cholesterol. They observed no toxic or undesirable side effects. Joyner and Kuo (24) gave seven subjects a 13% β-sitosterol suspension (daily dosage 6–15 g/d) for 4 weeks and observed an average fall in serum cholesterol of 10%. Four other subjects were given 10, 12, or 15 g of a suspension of β sitosterol in water for 4 weeks and their average serum cholesterol fell by 25%. Barber and Grant (35) fed sitosterol which was incorporated into biscuits. Each biscuit contained 1.5 g of sitosterol and their 26 subjects ingested two biscuits before every meal. Overall, a 7% reduction in mean serum cholesterol was observed. Sachs and Weston (36) gave normal subjects 9 g of a sitosterol mixture daily and, over a 4-week period, observed an average reduction in serum cholesterol of 12.7%. Farquhar *et al.* (37) administered a total of 12–18 g/d of a β-sitosterol preparation immediately before meals. They observed a significant reduction of 19.5% in serum cholesterol levels. Lees and Lees (38) confirmed earlier studies of the hypocholesterolemic action of sitosterol preparations. When the commercial suspension of sitosterol was used in seven subjects given 15 g/d of sterol, a reduction of 14% in serum cholesterol was seen. A purified β-sitosterol preparation made by spray drying (93% sitosterol) was effective at lower dosages (3–6 g). In five patients fed the commercial preparation of sitosterol (Cytellin), the authors found plasma concentrations of β-sitosterol and campesterol of 0.5 ± 0.07 and 16 ± 3.2 mg/dl, respectively. The average serum β-sitosterol concentration in 13 subjects fed 3 g/d of the purified preparation was 1.16 ± 0.13 mg/dl. Concern was expressed that the observed levels of serum campesterol might be atherogenic. It would seem that if campesterol were atherogenic, lesions would have been observed in susceptible animal species such as rabbits, but, to date, there have been no reports of aortic plaques in rabbits fed non-atherosclerotic regimens.

In a multicenter study (39), subjects were fed different phytosterol preparations. In the first study, 12 adults were fed 18 g/d (three equal doses) of the commercial soy sterol preparation that contained a sitosterol/campesterol ratio of about 2:1. Cholesterol levels fell by 12.2%. In the second study (six adult males), a spray-dried dispersible sterol preparation was fed at 18 g/d. The observed reduction in cholesterol levels was 12.2%. The first two experiments covered about 300 days. The third study involved use of a tall oil (pine) sterol powder containing 5% campesterol. In nine adults fed 3 g/d of powder for 216 days, serum cholesterol was reduced by 12.3%. The fourth study was carried out in two phases and included children and adolescents as well as adults. On 3 g/d of tall oil sterols fed for about 6 months, cholesterol levels in 38 probands fell by 8%; 6 g/d

Table 5 Influence of phytosterols in human cholesterolemia
(52 studies)

Average number of subjects	17 ± 3
Average phytosterol dose	13 ± 1.1 g
Average duration of study	27 ± 4 weeks
Average cholesterol reduction	$20 \pm 1.5\%$

Source: Summarized from Pollak and Kritchevsky (28).

of tall oil sterol suspension fed to 29 subjects for about 4.5 months led to a 6% reduction in cholesterol. The study showed that the form in which sitosterol was presented could affect the extent of its hypocholesterolemic activity. Table 5 presents a summary of results obtained in 52 human studies.

SITOSTANOL STUDIES

Sugano and his collaborators first investigated the lipid-lowering effects of phytostanols. They compared the effects of mixed phytosterols and phytostanols on serum lipids in rats fed diets containing no cholesterol or 1% cholesterol (40). Phytostanols appeared more hypolipidemic than phytosterols, and a dose–response effect was noted. The plant sterols and stanols also lowered liver cholesterol levels. Campesterol, campestanol, sitosterol, and sitostanol were recovered from the plasma.

A second study (41) offered a direct comparison of the hypocholesterolemic effects of β-sitosterol and β-sitostanol in rats. The rats were fed varying levels of safflower oil, corn oil, and lard and 0 or 0.5% cholesterol, β-sitosterol or β-sitostanol. There were no deleterious effects on growth. Sitostanol was consistently more hypocholesterolemic (Table 6). There was no effect on serum triglycerides. Comparison of the metabolism of radio-labeled β-sitosterol and β-sitostanol showed that the latter was turned over more rapidly and excreted in feces to a greater extent (97% vs. 88% recovery). The authors suggested that β-sitostanol might be a more effective hypocholesterolemic agent than β-sitosterol (42). A study in cholesterol-fed rabbits showed β-sitostanol to be more hypocholesterolemic and anti-atherogenic than β-sitosterol (43). Sitostanol was also shown to reduce the solubility of cholesterol in bile salt micelles in vivo more effectively than sitosterol (44). Ntanios *et al.* (45) have also shown dietary sitosterol to reduce severity of atherosclerosis in cholesterol-fed rabbits. Sitostanol has also been shown to significantly lower plasma cholesterol levels in cholesterol-fed hamsters (46).

Heinemann *et al.* (47) were the first investigators to study the effect of sitostanol on serum cholesterol levels in human subjects. They fed six hypercholesterolemic subjects a sitostanol preparation (500 mg/d) for 4 weeks. The dosage form was capsules containing 250 mg of sitostanol partly dissolved and partly dispersed in 368 mg of sunflower oil. After 4 weeks of therapy, plasma cholesterol

Table 6 Effects of sitosterol and sitostanol on cholesterol metabolism in rats

Group	Sterols added (%)			Plasma cholesterol (mg/dl)	Liver cholesterol (mg/g)
	Cholesterol	Sitosterol	Sitostanol		
1	0	0	0	120 ± 11	2.52 ± 0.08
2	0	0.5	0	176 ± 6	2.25 ± 0.06
3	0	0	0.5	115 ± 5	2.45 ± 0.10
4	0	0	0	99 ± 5	3.15 ± 0.20
5	0.5	0	0	141 ± 4	22.5 ± 2.0
6	0.5	0.5	0	117 ± 4	4.00 ± 0.24
7	0.5	0	0.5	96 ± 5	4.09 ± 0.21

Source: After Sugano *et al.* (41).

Notes
Diets contained 5% safflower oil.
6 rats/group fed 23 days.

and triglycerides were reduced by 15 and 20%, respectively. LDL and HDL cholesterol levels were 15 and 6% lower, respectively, and the LDL/HDL cholesterol ratio was 10% lower.

Miettinen *et al.* (48) partially overcame the sitostanol solubility problem by interesterifying the stanol with rapeseed oil and incorporating the esterified product into margarine. They studied 153 mildly hypercholesterolemic subjects for one year. One-third of the subjects were given sitostanol-free margarine and the others ingested either 1.8 or 2.6 g of sitostanol daily. Lipid levels in the control group were virtually unchanged, but total and LDL-cholesterol were significantly reduced in the test subjects (Table 7). Further studies showed that the sitostanol was itself unabsorbable and interfered with cholesterol absorption as well as reducing levels of circulating sitosterol and campesterol (49). Miettinen and his co-workers have tested the effects of sitostanol esters in a number of clinical conditions, and other investigators have also been quick to investigate stanol ester effects on hyperlipidemia. Jones *et al.* (50) have reviewed the literature pertaining

Table 7 Percent change of serum lipids of subjects (51/gp) fed a control diet or 1.8 or 2.6 gm sitostanol ester daily

Parameter	Control	Group	
		1.8 g/d	2.6 g/d
Total cholesterol	+0.85	−7.76	−10.26
LDL-cholesterol	−1.26	−9.80	−16.25
HDL-cholesterol	−1.89	+1.75	0
LDL/HDL cholesterol	−3.00	−11.19	−16.56
Triglycerides	+5.83	+2.75	−1.80

Source: After Miettinen *et al.* (48).

to phytosterol (sitosterol and sitostanol) effects on hypercholesterolemia. Nguyen (51) limited his review to studies of plant stanol esters. He cites 16 studies in which an average of 26 ± 6 subjects were fed an average of 2.3 ± 0.25 g/d of sitostanol ester. Over a period of 12.3 ± 3.9 weeks, serum or plasma cholesterol was reduced by $8.3 \pm 0.5\%$ and LDL-cholesterol by $12.1 \pm 0.7\%$. Hallikainen *et al.* (52) have confirmed that a dose of 1.6 g stanol is sufficient to effect significant reduction of total and LDL cholesterol, and increasing the dose by 50 or 100% (2.4 or 3.2 g/d) provides no significant additional effect.

There are two major sources of sitosterol: soybean oil or pine oil. The two differ in the spectrum of sterols/stanols accompanying the sitosterol. Hallikainen and Uusitupa (53) compared the effects of margarines containing wood stanol and vegetable oil stanol on 55 hypercholesterolemic subjects who were given a high fat diet for 4 weeks, then given a low fat diet containing wood or vegetable stanol or no stanol (control). The results are summarized in Table 8. The wood-derived stanol margarine was a somewhat more effective hypocholesterolemic agent. Plat and Mensink (54) carried out a similar study in subjects who had not been primed with a high fat diet and concluded that, despite different sitostanol/campestanol ratios, the wood and vegetable oil-derived phytosterol mixtures had similar total- and LDL-cholesterol lowering effects. Westrate and Meijer (55) compared the effects of sterol or stanol ester-containing margarines in a large group of subjects and found them to be of equal efficacy in cholesterol lowering (Table 9). Small bowel absorption of cholesterol, tested in ileostomy subjects, was 56% in the control period, 38% when the subjects were provided with a sterol ester margarine, and 39% when a stanol ester margarine was fed (56).

Table 8 Effects of margarines containing wood (W) or vegetable (V) stanol esters fed for 8 weeks as part of an NCEP step II diet (values $\pm$ SD)

Number	Group		
	W 18	*V 20*	*Control 17*
Starting values (mg/dl)			
Cholesterol	253 ± 30	237 ± 31	234 ± 21
LDL cholesterol	176 ± 28	164 ± 33	165 ± 23
LDL/HDL cholesterol	3.15	3.01	3.14
Eight week value (mg/dl)			
Cholesterol	206 ± 29	199 ± 30	215 ± 19
LDL cholesterol	135 ± 30	133 ± 29	148 ± 22
LDL/HDL cholesterol	2.47	2.54	2.79
Percent change			
Cholesterol	18.6	16.0	8.1
LDL-cholesterol	23.3	18.9	10.3
LDL/HDL-cholesterol	21.6	15.6	11.1

Source: After Hallikainen and Uusitupa (53).

Table 9 Serum lipids of 95 volunteers consuming different margarines

Margarine	Total cholesterol	Serum lipid (mg/dl)		
		HDL-C	LDL-C	LDL/HDL-C
Vegetable	200	48	130	2.71
Sitostanol ester	186	48	114	2.38
Sitosterol ester	184	48	113	2.35
Rice bran	198	48	128	2.67
Shea nut	199	48	129	2.68

Source: After Westrate and Meijer (55).

PHYTOSTEROLS AND CANCER

Although the major interest in phytosterols has been their effect on cholesterolemia and atherosclerosis, there are some reports concerning effects in cancer that are notable. Raicht *et al.* (57) studied effects of a diet containing 0.2% β-sitosterol on experimental colon carcinogenesis in rats. The rats were treated with the direct acting carcinogen, N-methyl-N-nitrosourea. After 28 weeks, tumor incidence in the sitosterol-fed rats was significantly lower than that seen in the control group (33 vs. 54%), and there were 38% fewer tumors per tumor-bearing animal. Awad *et al.* (58) have found β-sitosterol to inhibit the growth in vitro of HT-29 human colon cancer cells. One possible mode of action is reduction of testosterone metabolism by inhibiting action of testosterone-metabolizing enzymes (50). In a review of plant sterol effects in atherosclerosis, Pollak (60) discusses the prostate carcinostatic effects of various sitosterol preparations, but presents no specific literature references. The dosage (in capsule form) is rather modest (30 mg/d to 1.76 g/d).

TOXICOLOGY

Shipley *et al.* (61) found no evidence of toxicity of either β-sitosterols (tall oil) or soy sterols in rats, rabbits, and dogs fed large doses of the test compounds for periods as long as 2 years. There was no evidence of sterol deposition in any tissue and no increase in sterol content of either the liver or aorta. Malini and Vanithakumari (62) isolated β-sitosterol from the leaves of *Anacardium occidentale L.* and administered it subcutaneously to male and female rats for 60 days in doses of 250, 500, or 1,000 mg per 100 g body weight/d. They found no lesions in liver or kidney or in a number of blood/serum parameters. The higher intravenous doses led to significant reductions in serum cholesterol. In a later paper (63), they reported antifertility effects in male rats, but data derived from subcutaneous administration are not germane to feeding studies.

There are now available exhaustive studies relating to the safety of sterol (64–68) and stanol (69–73) esters. Insofar as sterol esters are concerned, there was no evidence of estrogenicity in either in vivo or in vitro assays (64). A subchronic 90-day oral toxicity study carried out in male and female Wistar rats fed 0, 0.16, 1.6, 3.2, or 8% sterol ester showed virtually no significant changes in weight gain, hematology, blood chemistry, or organ weights (65). A two-generation reproduction study showed no differences in a number of reproduction parameters including male or female fertility, litter size, fecundity, and duration of gestation. There were no effects on sexual maturation parameters (66). The authors concluded that there were no adverse effects at the doses tested.

Several safety studies were also carried out in human volunteers (12 males, 12 females) fed a control vegetable margarine or one containing 8.6 g of phytosterols (46% sitosterol, 26% campesterol, 20% stigmasterol) for 21 days (males) or 28 days (females). Fecal excretion (mg/dry weight) of neutral steroids was increased by about 4.5–5.0-fold. The major excretion products were β-sitosterol, campesterol, stigmasterol, and cholesterol. An appreciable increase in sitostanol, campestanol, and stigmastanol was also observed. Excretion of total bile acids was decreased by about 27% as was that of secondary bile acids (67). These findings may reflect the faster transit of cholesterol. Ingestion of the phytosterol ester-enriched margarine by male and female volunteers produced no differences in fecal short chain fatty acids, fecal bacterial enzyme activity, or fecal bacterial count. Hormonal assay in the female volunteers showed no differences in follicle-stimulating hormone, luteinizing hormone, progesterone, estradiol, estrone, or sex hormone binding globulin (68).

After a two-generation study of phytostanol esters in rats, Whittaker *et al.* (69) concluded that there were no adverse effects on reproduction, pup mortality, or pup growth. Developmental toxicity was studied using vegetable oil-derived stanol esters (70). No statistical differences were seen in uterine or placental weights, number of fetuses, fetal weight, or fetal sex ratios. The authors concluded that there were no adverse effects on either reproduction or development. Turnbull *et al.* (71) conducted an oral toxicity study in which rats were fed stanol esters prepared from either wood- or vegetable oil-derived stanols. No adverse effects were seen in either red cell and coagulation parameters or in white cell blood values. Clinical chemistry values were also within the normal range with either stanol ester source. Plasma levels of vitamins E, K, and D were reduced on diets containing 5% of stanol ester. Total plasma plant sterol content was significantly reduced in male and female rats fed either wood- or vegetable oil-derived stanol esters at 0.2, 1.0, or 5% of the diet. Levels of plasma campesterol fell with increasing amounts of dietary stanol ester; levels of campesterol rose, however, more on the vegetable oil-derived material than on the wood-derived material. Neither free phytostanols or phytostanol esters showed evidence of estrogenic or uterotrophic activity (72). The plant stanol esters (either wood- or vegetable oil-derived) were also found to be non-genotoxic using either bacterial or mammalian gene mutation assays (73).

The toxicologic data show that phytosterol or phytostanol esters are safe dietary ingredients. However, the metabolic effect of specific phytosterols have come under question. Lees *et al.* (39) observed increased levels of serum campesterol in subjects on a commercial sitosterol preparation and suggested that this particular sterol might have atherogenic potential. In 1943, Hardegger *et al.* (74) analyzed a large amount of aortic tissue and found mostly cholesterol plus some oxidized derivatives. There are only a few examples in the literature in which sterols other than cholesterol have been fed and then aortic tissue analyzed. In 1954, Cook *et al.* (75) maintained New Zealand white rabbits for 23–75 days on diets containing 16.6% olive oil and 1% cholesterol, lathosterol, 7-dehydrocholesterol, or cholestanol. There were only a few rabbits in each dietary group. All the sterols caused lipidemia. Cholesterol feeding led to well-defined arterial plaques in the males and less severe plaques in the females. Both lathosterol and 7-dehydrocholesterol yielded plaques just visible to the naked eye. Feeding of cholestanol crystallized from methanol led to visible lesions in one rabbit and no lesions in another. Cholestanol crystallized from petroleum ether led to visible plaques in one animal, and pure cholestanol supplied by a pharmaceutical source gave barely visible lesions in two rabbits. Clearly, the method of sterol preparation had an effect. Curran and Costello (76) fed two rabbits 9 g of dehydrocholesterol for 4 weeks and observed increases in digitonin-precipitable sterols of serum and liver. They did not report atherosclerosis. They also reported recovery of soybean sterols from livers of rabbits fed 9 g of soy sterols weekly. Shipley *et al.* (61) could not reproduce the Curran and Costello finding. The purity of the dehydrocholesterol may have affected the findings.

Triparanol is a drug which inhibits cholesterol synthesis late in the synthetic cycle with concomitant accumulation of desmosterol (24-dehydrocholesterol). When rabbits were fed normal or atherogenic diets containing 0.2% triparanol, desmosterol was recovered from their aortas. Desmosterol was present in both the uninvolved and atherosclerotic areas of the aorta. The ratio of cholesterol to desmosterol in the plaques was about 24:1. Small amounts of cholestanol and coprostanol were also isolated from the plaques. The ratio of cholesterol to desmosterol in the aortas of control rabbits was 1.5:1.0 (77).

The foregoing suggest that only sterols with the intact eight carbon atom side chain of cholesterol are taken up by arterial tissue. The possibility that campesterol or any other phytosterol can be atherogenic is amenable to experimental verification, and experiments in that direction should be carried out.

Sanders *et al.* (78) have compared absorption of radioactive carbon or tritium labeled phytosterols in male and female rats. The labeled compounds were administered by gavage and the animals were necropsied 4 days later. Radioactivity was measured in expired air, excreta, soft tissues, and carcass (Table 10). Adrenal glands, ovaries, and intestinal epithelia showed the highest levels and longest retention of radioactivity. Absorption in female rats was higher than in male rats (Table 11).

The data obtained from analysis of the carbon and tritium labeled sterols and stanols are consistent. That carbon and tritium labeled cholesterol track together

Table 10 Recovery of ^{14}C and ^{3}H from rats gavaged with labeled sterols or stanols

Substance	No. of Rats	Gender[a]	% Recovery	
			Feces	Total
[4-^{14}C]-β-sitosterol	4	M	96.54 ± 7.85	100.00 ± 7.67
[4-^{14}C]-β-sitosterol	4	F	85.23 ± 3.84	97.85 ± 6.20
[4-^{14}C]-β-sitostanol	4	M	87.93 ± 4.00	95.96 ± 3.43
[4-^{14}C]-β-sitostanol	4	F	88.09 ± 3.61	99.47 ± 2.98
[5,6-^{3}H]-β-sitostanol	4	F	75.07 ± 15.86	95.03 ± 2.24
[3α-^{3}H]-campesterol	3	F	76.45 ± 2.88	94.17 ± 3.14
[5,6-^{3}H]-campestanol	3	F	90.39 ± 1.05	97.97 ± 2.46
[3-^{3}H]-stigmasterol	3	F	84.65 ± 2.64	95.03 ± 0.33
[4-^{14}C]-cholesterol	3	M	51.32 ± 17.02	81.23 ± 12.85
[4-^{14}C]-cholesterol	3	F	58.70 ± 5.59	91.55 ± 1.15

Source: After Sanders *et al*. (78).

Note
a M = Male; F = Female.

Table 11 Apparent absorbed dose (%) of labeled phytosterols
24 after dosing

Substance	% Absorption	
	Males (4)	Females (4)
[4-^{14}C]-cholesterol	29.4 ± 3.3	27.1 ± 5.2
[4-^{14}C]-β-sitosterol	1.9 ± 1.1	4.3 ± 1.3
[4-^{14}C]-β-sitostanol	0.5 ± 0.3	1.2 ± 0.8

Source: After Sanders *et al*. (78).

in the rat was confirmed a half century ago (79). However, there is the possibility of tritium transfer from steroids to fatty acids and other lipids (80). Whole body radioautographs of rats made 24 hours after a single gavage dose of carbon labeled cholesterol, β-sitosterol, or β-sitostanol reflect the level of uptake of these substances. The least deposition was observed with β-sitostanol. Uptake of labeled β-sitosterol was found in the ovary, spleen, adrenal, kidney, and bone marrow, among other tissues. Labeled sterols, after injection into the chick yolk sac, appear in many tissues including the brain (81).

β-sitosterolemia is a rare lipid storage disease that can lead to xanthomatosis and atherosclerosis (82). Glueck and his collaborators (83,84) have shown that subjects with premature coronary disease and their first-degree relatives are hypercholesterolemic and also exhibit elevated levels of phytosterols. The authors seem to imply that phytosterolemia may have contributed to their condition. The condition can be ameliorated when the subjects are placed on a low cholesterol, low phytosterol, low fat diet. While the possibility exists that phytosterolemia may be found in previously asymptomatic subjects placed on phytosterol therapy,

Table 12 Mechanisms of plasma cholesterol reduction
by sitosterol

---

1. Mixed crystal formation
2. Adsorption: Aggregation
3. Decrease in solubility of cholesterol micelles
4. Enhanced cholesterol excretion
5. Interference with cholesterol synthesis
6. Competitive esterification
7. Competition for "acceptor" sites
8. Effects on membrane permeability

Source: Summarized from Pollak and Kritchevsky (28).

the chances are that this is a reality for only a few of the exposed subjects. If phytosterols were a general danger, one would expect to see more coronary disease in vegetarian populations. However, we should be alert to the possibility that phytosterolemia may occur even if the odds of this happening are slight. It is apparent from the foregoing data and other experiments (23,24,26,78,81,85) that phytosterols can be metabolized by mammalian enzymes when exposed to them. The barrier is their low absorbability. The data suggest diligence in monitoring treatment, but not panic.

Plant sterols which were shown to be anti-atherogenic and hypocholesterolemic almost a half century ago present a relatively simple and safe way for lowering blood lipids and the risk of coronary heart disease. Their precise method of action is still moot, but a number of possible mechanisms of action have been proposed (28). These are summarized in Table 12.

REFERENCES

1. Bean GA. Phytosterols. *Adv Lipid Res* 1973; 11:193–218.
2. Itoh T, Tamura T, Matsumoto T. Sterol composition of 19 vegetable oils. *J Am Oil Chem Soc* 1973; 50:122–5.
3. Weihrauch JL, Gardner JM. Sterol content of foods of plant origin. *J Am Dietetic Assoc* 1978; 73:39–47.
4. Ellis GW, Gardner JA. The origin and destiny of cholesterol in the organism VIII. On the cholesterol content of the liver of rabbits under various diets and during inanition. *Proc Roy Soc (Lond)* 1912; B84:461–70.
5. Schönheimer R. Versuch einer bilanz am kaninchen bei verfutterung mit sitosterin. *Hoppe-Seyler's Z Physiol Chem* 1929; 180:24–32.
6. Schönheimer R. Über die sterine des kaninchenkotes. *Hoppe-Seyler's Z Physiol Chem* 1929; 180:32–7.
7. Schönheimer R, von Behring H, Hummel R. Über die spezifität der resorption von sterinen, abhangig von ihrer konstitution. *Hoppe-Seyler's Z Physiol Chem* 1930; 192:117–24.
8. Schönheimer R. Die spezifität der cholesterin resorption und ihre biologische bedeutung. *Klin Wochschr* 1932; 11:1793–6.

185

9. Schönheimer R. New contributions in sterol metabolism. *Science* 1931; 74:579–84.

10. Peterson DE. Effect of soybean sterols in the diet on plasma and liver cholesterol in chicks. *Proc Soc Exp Biol Med* 1951; 78:143–7.

11. Peterson DW, Nichols CW Jr, Shneour EA. Some relationships among dietary sterols, plasma and liver cholesterol levels, and atherosclerosis in chicks. *J Nutr* 1952; 47:57–65.

12. Peterson DW, Shneour EA, Peek NF, Gaffey HW. Dietary constituents affecting plasma and liver cholesterol in cholesterol-fed chicks. *J Nutr* 1953; 50:191–201.

13. Hernandez HH, Peterson DW, Chaikoff IL, Dauben WG. Absorption of cholesterol-4-C^{14} in rats fed mixed soybean sterols and sitosterol. *Proc Soc Exp Biol Med* 1953; 83:498–9.

14. Hernandez HH, Chaikoff IL. Do soy sterols interfere with absorption of cholesterol? *Proc Soc Exp Biol Med* 1954; 84:541–4.

15. Gould RG. Absorbability of β-sitosterol. *Trans NY Acad Sci* 1955; 18:129–34.

16. Swell L, Trout EC Jr, Field H Jr, Treadwell CR. Absorption of ^{3}H-β-sitosterol in the lymph fistula rat. *Proc Soc Exp Biol Med* 1959; 100:140–2.

17. Sylven C, Borgstrom B. Absorption and lymphatic transport of cholesterol and sitosterol in the rat. *J Lipid Res* 1964; 10:179–82.

18. Vahouny GV, Connor WE, Subramaniam S, Lin DS, Gallo LL. Comparative lymphatic absorption of sitosterol, stigmasterol, and fucosterol and differential inhibition of cholesterol absorption. *Am J Clin Nutr* 1983; 37:805–9.

19. Ikeda I, Tanaka K, Sugano M, Vahouny GV, Gallo LL. Inhibition of cholesterol absorption by plant sterols. *J Lipid Res* 1988; 29:1573–82.

20. Ikeda I, Sugano M. Some aspects of mechanism of inhibition of cholesterol absorption by β-sitosterol. *Biochem Biophys Acta* 1983; 732:651–8.

21. Ikeda I, Tanaka K, Sugano M, Vahouny GV, Gallo LL. Discrimination between cholesterol and sitosterol for absorption in rats. *J Lipid Res* 1988; 29:1583–91.

22. Mattson FH, Volpenhein RA, Erickson BA. Effect of plant sterol esters on the absorption of dietary cholesterol. *J Nutr* 1977; 107:1139–46.

23. Salen G, Ahrens EH Jr, Grundy SM. Metabolism of β-sitosterol in man. *J Clin Invest* 1979; 49:952–67.

24. Boberg KM, Lund E, Olund J, Bjorkheim I. Formation of C_{21} bile acids from plant sterols in the rat. *J Biol Chem* 1990; 265:7967–75.

25. Boberg KM, Einarsson K, Bjorkheim I. Apparent lack of conversion of sitosterol into C_{24}-bile acids in humans. *J Lipid Res* 1990; 31:1083–8.

26. Kritchevsky D, Davidson LM, Mosbach EM, Cohen BL. Identification of acidic steroids in feces of monkeys fed β-sitosterol. *Lipids* 1981; 16:77–8.

27. Heinemann T, Axtmann G, von Bergman K. Comparison of intestinal absorption of cholesterol with different plant sterols in man. *Eur J Clin Invest* 1993; 23:827–31.

28. Pollak OJ, Kritchevsky D. *β-sitosterol*. Basel: S. Karger; 1981.

29. Pollak OJ. Reduction of blood cholesterol in man. *Circulation* 1953; 7:702–6.

30. Beveridge JMR, Connell WF, Mayer GA, Firstbrook JB, DeWolfe MS. The effect of certain vegetable and animal fats on the plasma lipids of humans. *J Nutr* 1955; 56: 311–20.

31. Beveridge JMR, Connell WF, Mayer GA. Dietary factors affecting the level of plasma cholesterol in humans: the role of fat. *Can J Biochem Physiol* 1956; 34:441–55.

32. Beveridge JMR, Connell WF, Mayer GA, Haust HL. Plant sterols, degree of unsaturation, and hypocholesterolemic action of certain fats. *Can J Biochem Physiol* 1958; 36:895–911.

33. Best MM, Duncan CH, Van Loon EJ, Wathen JD. Lowering of serum cholesterol by the administration of a plant sterol. *Circulation* 1954; 10:201–6.

34. Joyner C Jr, Kuo PT. The effect of sitosterol administration upon the serum cholesterol level and lipoprotein pattern. *Am J Med Sci* 1955; 230:636–47.

35. Barber JM, Grant AP. The serum cholesterol and other lipids after administration of sitosterol. *Brit Heart J* 1955; 17:296–8.

36. Sachs BA, Weston RE. Sitosterol administration in normal and hypercholesterolemic subjects. *Arch Int Med* 1956; 97:738–52.

37. Farquhar JM, Smith RE, Dempsey ME. The effect of beta sitosterol on the serum lipids of young men with arteriosclerotic heart disease. *Circulation* 1956; 14:77–82.

38. Lees RS, Lees AM. Effects of sitosterol therapy on plasma lipid and lipoprotein concentrations. In: Greten H, ed. *Lipoprotein Metabolism.* Berlin: Springer-Verlag; 1976; 119–24.

39. Lees AM, Mok HYI, Lees RS, McCluskey MA, Grundy SM. Plant sterols as cholesterol lowering agents: clinical trials in patients with hypercholesterolemia and studies of sterol balance. *Atheroclerosis* 1977; 28:325–38.

40. Sugano M, Kamo F, Ikeda I, Morioka H. Lipid-lowering activity of phytostanols in rats. *Atherosclerosis* 1976; 24:301–9.

41. Sugano M, Morioka H, Ikeda I. A comparison of hypocholesterolemic activity of β-sitosterol and β-sitostanol in rats. *J Nutr* 1977; 107:2011–19.

42. Ikeda I, Sugano M. Comparison of absorption and metabolism of β-sitosterol and β-sitostanol in rats. *Atherosclerosis* 1978; 30:227–37.

43. Ikeda I, Morioka H, Sugano M. The effect of dietary β-sitosterol and β-sitostanol on the metabolism of cholesterol in rats. *Agric Biol Chem* 1979; 43:1927–33.

44. Ikeda I, Kawasaki A, Samezima K, Sugano M. Antihypercholesterolemic activity of β-sitostanol in rats. *J Nutr Sci Vitaminol* 1981; 27:243–51.

45. Ntanios FY, Jones PJH, Frohlich JJ. Dietary sitostanol reduces plaque formation but not lecithin-cholesterol acyl transferase activity in rabbits. *Atherosclerosis* 1998; 138:101–10.

46. Ntanios FY, Jones PJH. Effects of variable dietary sitostanol concentrations on plasma lipid profile and phytosterol metabolism in hamsters. *Biochem Biophys Acta* 1998; 1390:237–44.

47. Heinemann T, Leiss O, von Bergmann K. Effect of low-dose sitostanol on serum cholesterol in patients with hypercholesterolemia. *Atherosclerosis* 1986; 61:219–23.

48. Miettinen TA, Puska P, Gylling H, Vanhanen H, Vartainen E. Reduction of serum cholesterol with sitostanol-ester margarine in a mildly hypercholesterolemic population. *N Engl J Med* 1995; 333:1308–12.

49. Vanhanen HT, Kajander J, Lehtovirta H, Miettinen TA. Serum levels, absorption efficiency, faecal elimination and synthesis of cholesterol during increasing doses of dietary sitostanol esters in hypercholesterolaemic subjects. *Clin Sci* 1994; 87: 61–7.

50. Jones PJH, MacDougall DE, Ntanios F, Vanstone CA. Dietary phytosterols as cholesterol-lowering agents in humans. *Can J Physiol Pharmacol* 1997; 75:217–27.

51. Nguyen T. The cholesterol-lowering action of plant stanol esters. *J Nutr* 1999; 129: 2109–12.

52. Hallikainen MA, Sarkkinen ES, Uusitupa MIJ. Plant stanol esters affect serum cholesterol concentrations of hypercholesterolemic men and women in a dose-dependent manner. *J Nutr* 2000; 130:767–76.

53. Hallikainen MA, Uusitupa MIJ. Effects of 2 low-fat stanol ester-containing margarines on serum cholesterol concentrations as part of a low-fat diet in hypercholesterolemic subjects. *Am J Clin Nutr* 1999; 69:403–10.

54. Plat J, Mensink RP. Vegetable oil based versus wood based stanol ester mixtures: effects on serum lipids and hemostatic factors in non-hypercholesterolemic subjects. *Atherosclerosis* 2000; 148:101–12.

55. Westrate JA, Meijer GW. Plant sterol-enriched margarines and reduction of plasma total- and LDL-cholesterol concentrations in normocholesterolemic and mildly hyper-cholesterolemic subjects. *Eur J Clin Nutr* 1998; 52:334–43.

56. Normen L, Dutta P, Lia A, Andersson H. Soy sterol esters and β-sitostanol ester as inhibitors of cholesterol absorption in human small bowel. *Am J Clin Nutr* 2000; 71:908–13.

57. Raicht RF, Cohen BI, Fazzini EP, Sarwal, AN, Takahashi M. Protective effect of plant sterols against chemically induced colon tumors in rats. *Cancer Res* 1980; 40:403–5.

58. Awad AB, Von Holtz RL, Cone JP, Fink CS, Chen Y-C. β-sitosterol inhibits the growth of HT-29 human colon cancer cells by activating the sphingomyelin cycle. *Anticancer Res* 1998; 18:471–9.

59. Awad AB, Hartati MS, Fink CS. Phytosterol feeding induces alterations in testosterone metabolism in rat tissues. *J Nutr Biochem* 1998; 9:712–17.

60. Pollak OJ. Effect of plant sterols on serum lipids and atherosclerosis. *Pharmacol Therap* 1985; 31:177–208.

61. Shipley RE, Pfieffer RR, Marsh MM, Anderson RC. Sitosterol feeding. Chronic animal and clinical toxicology and tissue analysis. *Circulation* 1958; 6:373–82.

62. Malini T, Vanithakumari G. Rat toxicity studies with β-sitosterol. *J Ethnopharmacol* 1990; 28:221–34.

63. Malini T, Vanithakumari G. Antifertility effects of β-sitosterol in male albino rats. *J Ethnopharmacol* 1991; 35:149–53.

64. Baker VA, Hepburn PA, Kennedy SJ, Jones PA, Lea LJ, Sumpter JP, Ashby J. Safety evaluation of phytosterol ester. Part 1. Assessment of oestrogenicity using a combination of *in vivo* and *in vitro* assays. *Food Chem Toxicol* 1999; 37:13–22.

65. Hepburn PA, Horner SA, Smith M. Safety evaluation of phytosterol esters. Part 2. Subchronic 90-day oral toxicity study on phytosterol esters – a novel functional food. *Food Chem Toxicol* 1999; 37:521–32.

66. Waalkens-Berendsen DH, Wolterbeek APM, Wijnands MVW, Richold M, Hepburn PA. Safety evaluation of phytosterol esters. Part 3. Two generation reproduction study in rats with phytosterol esters – a novel functional food. *Food Chem Toxicol* 1999; 37:683–96.

67. Westrate JA, Ayesha R, Bauer-Plank C, Drewitt PN. Safety evaluation of phytosterol esters. Part 4. Faecal concentrations of bile acids and neutral sterols in healthy normo-lipidemic volunteers consuming a controlled diet with or without a phytosterol ester-enriched margarine. *Food Chem Toxicol* 1999; 37:1063–71.

68. Ayesh R, Westrate JA, Drewitt PN, Hepburn PA. Safety evaluation of phytosterol esters. Part 5. Faecal short chain fatty acid and microflora content, faecal bacterial enzyme activity and serum female sex hormones in healthy normolipidaemic volunteers consuming a controlled diet either with or without a phytosterol ester-enriched margarine. *Food Chem Toxicol* 1999; 37:1127–38.

69. Whittaker MH, Frankos VH, Wolterbeek APM, Waalkens-Berendsen DH. Two generation reproductive toxicity study of plant stanol esters in rats. *Reg Toxicol Pharmacol* 1999; 29:196–204.

70. Slesinski RS, Turnbull D, Frankos VH, Wolterbeek APM, Waalkens-Berendsen DH. Developmental toxicity study of vegetable oil-derived stanol fatty acid esters. *Reg Toxicol Pharmacol* 1999; 29:227–33.

71. Turnbull D, Whittaker MH, Frankos VH, Jonker D. 13-week oral Toxicity study with stanol esters in rats. *Reg Toxicol Pharmacol* 1999; 29:216–26.

72. Turnbull D, Frankos VH, Leeman WR, Jonker D. Short term tests of estrogenic potential of plant stanols and plant stanol esters. *Reg Toxicol Pharmacol* 1999; 29:211–15.

73. Turnbull D, Frankos VH, van Delft JHM, De Vogel N. Genotoxicity evaluation of wood-derived and vegetable oil-derived stanol esters. *Reg Toxicol Pharmacol* 1999; 29:205–210.

74. Hardegger E, Ruzicka L, Tagmann E. Untersuchungen über organextrakte. Zur kenntnis der unverseifbaren lipoide aus arteriosklerotischen aorten. *Helv Chim Acta* 1943; 30:2205–21.

75. Cook RP, Kliman A, Fieser LF. The absorption and metabolism of cholesterol and its main companions in the rabbit – with observations on the atherogenic nature of the sterols. *Arch Biochem* 1954; 52:439–50.

76. Curran GL, Costello RL. Effect of dehydrocholesterol and soybean sterols on cholesterol metabolism in rabbit and rat. *Proc Soc Exp Biol Med* 1956; 91:52–6.

77. Kritchevsky D, Fumagalli R, Cattabeni F, Tepper SA. Effect of triparanol on sterol composition in normal and atherosclerotic rabbit aorta. *Rivista di Farmacologia e Terapia* 1970; 1:455–63.

78. Sanders DJ, Minter HJ, Howes D, Hepburn PA. The safety evaluation of phytosterols esters. Part 6. The comparative absorption and tissue distribution of phytosterols in the rat. *Chem Toxicol* 2000; 38:485–91.

79. Biggs MW, Kritchevsky D, Kirk MR. Assay of samples doubly labeled with radioactive hydrogen and carbon. *Anal Chem* 1952; 24:223–4.

80. Kritchevsky D, Werthessen NT, Shapiro IL, Nair PP, Turner DA. Transfer of radioactivity of cholesterol-7-α^3H to fatty acids of tissue lipids *in vivo*. *Nature* 1965; 207:194–5.

81. Kritchevsky D, Defendi V. Deposition of tritium labeled sterols (cholesterol, sitosterol, lanosterol) in brain and other organs of the growing chicken. *J Neurochem* 1962; 9:421–5.

82. Bhattacharyya AK, Connor WE. β-sitosterolemia xanthamatosis. A newly described lipid storage disease in two sisters. *J Clin Invest* 1974; 53:1033–43.

83. Glueck CJ, Speers J, Tracy T, Streicher P, Illig E, Vandergrift J. Relationship of serum plant sterols (phytosterols) and cholesterol in 595 hypercholesterolemic subjects, and familial aggregation of phytosterols, cholesterol, and premature coronary disease in hyperphytosterolemic probands and their first degree relatives. *Metabolism* 1991; 40:842–8.

84. Glueck CJ, Streicher P, Illig E. Serum and dietary phytosterols, and coronary heart disease in hyperphytosterolemic probands. *Clin Biochem* 1992; 25:331–4.

85. Kritchevsky D, Staple E, Whitehouse MW. Oxidation of ergosterol by rat and mouse liver mitochondria. *Proc Soc Exp Biol Med* 1961; 106:704–8.

9

THRESHOLD OF REGULATION: A UNIFYING CONCEPT IN FOOD SAFETY ASSESSMENT

W. Gary Flamm, Frank N. Kotsonis and Jerry J. Hjelle

The concept of a threshold of regulation has likely always been a part of risk control strategies because the common-sense notion that many risks are too small to warrant attention and/or remediation is ingrained in all systems used for controlling risk (1). A threshold of regulation concept is not rooted in a callous disregard for small risks, but instead, is based on the understanding that a framework for setting priorities is necessary if risks are to be effectively controlled. In no area is the case for threshold of regulation more persuasive than in food. Given that foods contain hundreds of thousands of natural chemical substances and that thousands more are produced when foods are stored or cooked, or when new varieties are introduced, the need for a threshold of regulation concept is compelling (2).

Prior to World War II, the concept of threshold of regulation was widely applied by industry and government in decision-making to control food-borne risks. The term, threshold of regulation, was probably not used as such but the concept was likely a critical factor when deciding which risks could be dismissed as inconsequential and which needed to be addressed.

A threshold of regulation for food ingredients is evident in the legislative history of the 1958 Food Additives Amendment of the Federal Food, Drug, and Cosmetic Act (FD&C), when Congress accepted a "reasonable certainty of no harm" as the applicable standard to be applied to food additives on the grounds that proof of safety to an absolute certainty is unattainable. Congress, therefore, accepted the principle that some uncertainty about the risk posed by food additives must, as a practical matter, be accepted. An example of this principle in practice was the use of a 100-fold safety factor. Food & Drug Administration (FDA) scientists such as Lehman and Fitzhugh (3) understood that the use of the 100-fold safety factor did not guarantee absolute safety, but in their judgment represented a reasonable certainty of no harm standard when based on appropriately designed and properly conducted studies. Application of the 100-fold safety factor in the early 1960s by the Joint FAO/WHO Expert Committee on Food

Additives led to the use of the concept of an acceptable daily intake (4,5). The design and conduct of the toxicologic studies used to establish safety also require the implicit consideration of a threshold of risk as practical decisions must be made that ultimately affect the sensitivity of the tests (6).

While lawmakers and scientists tend to hold many different views on food safety, history shows they agree that, in matters pertaining to food-borne risk, the only absolute certainty is that there will always be some uncertainty. However, public acceptance of uncertainty may ultimately require judgments based only on publicly accepted policies. In the present paper, we will discuss the general paradigms used or considered by the FDA for regulating drugs used in food-producing animals, direct food and color additives, migrants from food packaging, constituents in additives, pesticide residues, traditional foods, and foods derived from new plant varieties. Drugs used in food-producing animals bear the distinction of being the first product category for which the FDA developed formal procedures in 1977 for judging whether the intended use of a substance (as an animal drug) is above or below a regulatory threshold, food itself being the last. In 1992, the FDA developed a policy to determine when it is necessary to consult with or seek FDA approval for new foods derived from new plant varieties (7).

DRUGS USED IN FOOD-PRODUCING ANIMALS

The Delaney Anticancer Clause in Section 512 of the FD&C Act, states that animal drugs intended for use in food-producing animals shall not be approved for use if such drug has been found to induce cancer when ingested by humans or animals or, after tests that are appropriate for the evaluation of the safety of such drug, induces cancer in humans or animals, except under conditions of use in which the animal is not adversely affected and no residue of the additive is found in any edible portion of such animal after slaughter or in any food derived from the living animal. The term "no residue" as used in the statutory language of Section 512 requires definition. In theory some drug residue will persist after use though perhaps at levels far below the sensitivity of current analytical methods designed to detect the residue. Were the term "no residue" taken simply at face value, its definition would be entirely dependent on methodologic sensitivity. As the intent of the law is to protect the public health, it did not make sense to allow no residue to be defined solely by the sensitivity of the analytical method. Consequently, the FDA needed to develop, by regulation a definition for no residue that took into account the purpose and intent of the law. The basic philosophy employed by the FDA was to define "no residue" in terms of the cancer risk posed to consumers. Assessing the cancer risk posed to consumers is done by conducting cancer studies in animals and extrapolating the results to humans (8,9). The FDA has concluded that upper-limit estimates of lifetime risk no greater than one occurrence of cancer per million individuals is an acceptable level of risk and is consistent with the meaning of "no residue" as intended by Congress (10).

In order to comply with the congressional intent regarding the use of drugs in food-producing animals, as required in the no residue provision of the Delaney Clause, FDA began to build a system for conducting risk assessment of carcinogens in the early 1970s (10). In the course of developing a policy/regulatory definition for "no residue," the FDA was also compelled to address the thorny issue of residues of metabolites of the animal drug known to induce cancer in humans or animals. As the number of metabolites may range into the hundreds, it became apparent that, as a practical matter, not every metabolite could be tested with the same thoroughness as the parent animal drug. This forced the FDA to consider for the first time threshold assessment. Threshold assessment combines information on the structure and in vitro biologic activity of a metabolite for the purpose of determining whether carcinogenicity testing is necessary. This decision-making approach likely represents the first time the FDA formalized a threshold concept for making judgments about the need for requiring carcinogenicity testing.

DIRECT FOOD AND COLOR ADDITIVES

For the most part, ingredients that are added to food are either generally recognized as safe (GRAS), food additives, or color additives. Under Section 201(s) of the FD&C Act, a food additive is "any substance the intended use of which results or may reasonably be expected to result, directly or indirectly, in its becoming a component or otherwise affecting the characteristics of any food" if the substance is not GRAS. Direct food additives are added as such to food. A color additive, as defined in Section 201(t), is "a dye, pigment, or other substance...(synthesized)...extracted, isolated or otherwise derived...from a vegetable, animal, or mineral or other source and...when added or applied to a food...is capable of imparting color thereto."

In the early 1980s the FDA published guidelines on "Toxicological Principles for the Safety Assessment of Direct Food Additives and Color Additives Used in Food" (11), which, because of its red cover, is often referred to as the "Redbook." The guidelines set out a tiered system of information requirements for additives in food. In addition to providing guidance regarding toxicological testing, it sets out a concept of priority that ranks compounds according to the level of health concern based on the extent of human exposure and probable toxicologic effects. To determine the extent of testing required for approval of a new food or color additives, it combines information about chemical structure with anticipated exposure from intended use in order to assign a "concern level" (Table 1).

Substances that are classified as having structure A, defined as those chemical structures of the least toxicologic concern, fall into concern level I provided anticipated exposure is below a dietary level of 0.05 ppm. Between 0.05 and 1.0 ppm, substances classified as structure A are placed in concern level II. Exposures anticipated to be above 1.0 ppm come under level III and require the greatest amount of toxicologic testing. The other structure classes, B and C, are handled

Table 1 Assignment of concern level

Structure category A	Structure category B	Structure category C	Concern level
<0.05 ppm in the total diet (<0.0012 mg/kg/day) or	<0.025 ppm in the total diet (<0.00063 mg/kg/day) or	<0.0125 ppm in the total diet (<0.00031 mg/kg/day) or	I
≥0.05 ppm in the total diet (≥0.0012 mg/kg/day) or	≥0.05 ppm in the total diet (≥0.00063 mg/kg/day) or	≥0.0125 ppm in the total diet (≥0.00031 mg/kg/day) or	II
≥1 ppm in the total diet (≥0.025 mg/kg/day)	≥0.5 ppm in the total diet (≥0.0125 mg/kg/day)	≥0.25 ppm in the total diet (≥0.0063 mg/kg/day)	III

in a similar manner, except that structures with higher probable toxicity invoke higher concern levels at lower dietary concentrations. The FDA developed this approach in order to codify its decision-making regarding the type of testing required (Table 2) for different intended uses of different food or color additives, recognizing that not all potential risks need to be addressed in the same way or to the same extent. For substances classified as A structures with dietary concentrations below 1 ppm, no carcinogenicity testing is required unless there is evidence from either the battery of short-term tests or the short-term feeding study to indicate the substance may be a carcinogen. In essence, this approach is equivalent to a

Table 2 Tests for each concern level

Concern level	Tests required
I	Short-term feeding study (at least 28 days in duration) Short-term tests for carcinogenic potential that can be used for conduction of lifetime carcinogenicity bioassays and may assist in the evaluation of results from such bioassays, if conducted
II	Subchronic feeding study (at least 90 days in duration) in a rodent species Subchronic feeding study (at least 90 days in duration) in a non-rodent species Multigeneration reproduction study (minimum of two generations with a teratology phase) in a rodent species Short-term tests for carcinogenic potential
III	Carcinogenicity studies in two rodent species A chronic feeding study at least 1 year in duration in a rodent species (may be combined with a carcinogenicity study) Long-term (at least 1 year in duration) feeding study in a non-rodent species Multigenerational reproduction study (minimum of two generations) with a teratology phase in a rodent species Short-term tests for carcinogenic potential

threshold assessment process as it accepts the concept that not all substances need to be tested for carcinogenicity.

MIGRANTS FROM FOOD PACKAGING MATERIAL

Under Section 201(s) of the FD&C Act, food additives are defined as "...any substance the intended use of which results or may reasonably be expected to result directly or indirectly, in its becoming a component or otherwise affecting the characteristics of any food including any substance intended for use in producing, manufacturing, packaging, processing, preparing, treating, packaging, transporting, or holding food..." Because this broad definition includes food packaging and components of food packaging that might migrate into food and because of the myriad number and complexity of food packaging (thousands of substances are listed under title 21 of the Code of Federal Regulations (CFR) as food additives and thousands more may migrate to food at levels below detection and are not listed or considered as food additives), the FDA has tried to develop a threshold of regulation scheme to encompass migrants from food packaging material. In fact, over the years the FDA has been forced to make threshold of regulation decisions on a case-by-case basis. That such decisions are permitted by law is supported by the United States Circuit Court of Appeals of the District of Columbia decision in the Case of Monsanto vs. Kennedy, 1979, in which the Court ruled (12) that the FDA has discretion in deciding whether a small migratory amount of a substance, in this case acrylonitrile, should be considered a food additive within the meaning of Section 201(s). The Court cited the legal principle, *de minimis non curat lex*, which loosely translated means "the law does not concern itself with trifles," and ruled that the FDA is not required under the law to consider every substance that might migrate at extremely low levels into food from food packaging material a food additive and that the common-sense legal principle of *de minimis* could be applied instead. This led to the development of the term *de minimis* risk.

De minimis risk has not been viewed by the FDA as identical or equivalent to acceptable risk. Acceptable risks are those risks that are controlled, such as the restrictions on the use of food ingredients or exposure to lead as an impurity in a food additive. *De minimis* risk has been viewed by the FDA as a risk that is so small and self-limited that it does not require the same type of identification, control, and monitoring that attends the risk from an impurity identified in the specifications of a food ingredient. Under the *de minimis* concept, a substance that migrates into food need not be considered a food additive or even a constituent of a food additive if its level of migration is so small that identification, control, and monitoring of it as either an additive or a component of an additive would be unwarranted.

A scheme for establishing a threshold of regulation for food contact substances (i.e. components of indirect additives) has been proposed and discussed by Rulis

(13,14) and Flamm *et al.* (15). The scheme relies on an evaluation of the potency of over 300 carcinogens in order to estimate levels of migration that would represent risk sufficiently small to be unworthy of any further regulatory consideration. The details of such a scheme are still under review, but in the meantime, practical demands require that such decisions be made on a case-by-case basis. Rulis (14) has used distribution frequencies of carcinogen potency (16–18) to determine a dietary concentration (0.5–1.0 ppb) that would have a median upper-bound presumptive risk of carcinogenesis from lifetime exposure of 1×10^{-6}. In this approach, all components of indirect additives are assumed to have potency similar to the 300 known animal carcinogens evaluated. In this manner, it was estimated that the median lifetime risk of causing cancer is one in a million when dietary concentrations are in the 0.5–1.0 ppb concentration range for known carcinogens. Others have reviewed these findings and have concluded, based on the conservative methodologic assumptions used, that a 1-ppb dietary concentration is a reasonable threshold for components of indirect additives (19).

THE CONSTITUENTS POLICY AND CHEMICAL SPECIFICATIONS

All substances, including food and color additives, have a myriad of impurities (20). Such impurities may come from the starting material or reaction products, or be adventitious components. In any case, impurities will exist, some at levels that can be readily detected and others at undetectable levels. Decisions concerning the safety of impurities and the development of appropriate specifications for food or color additives must take into consideration the concept of threshold of regulation. Recognition that no substance is entirely pure and that many non-carcinogenic additives may contain carcinogens forced the FDA to recognize the need for and to propose and apply its impurities or constituents policy (20).

In 1982, the FDA proposed, in an advance notice of proposed rule-making, that banning food and color additives simply because they were found to contain a trace level of a known animal carcinogen was not reasonable since many additives may contain carcinogenic impurities (20). The FDA asserted that the mere fact that the additive contains an impurity known to be carcinogenic should not automatically cause the FDA to ban the additive under the Delaney Anticancer Clause (FD&C Act, Sections 409 and 706). Instead, the FDA proposed to take regulatory action on the estimated cancer risk posed by the impurity, based on its concentration and intended condition of use. Thus, the FDA adopted a risk assessment approach and a level of risk equivalent to that used previously for residues of animal drugs. Again, the benchmark for risk was a one in a million lifetime risk of cancer.

The need for the proposal was clear; increased analytical sensitivity could lead to the possible ban of food and color additives for which the risk of cancer was vanishingly low. In its advance notice of proposed rule-making, the FDA concluded that the 1979 decision by the United States Court of Appeals for the District of Columbia in Monsanto vs. Kennedy was supportive of this new policy (12). The agency applied its new impurities or constituents policy in allowing the continued use of the color additive Green No. 6, which contain *p*-toluidine, a mouse carcinogen. A subsequent court decision, Scott vs. Young, ruled in favor of the agency (21). In addition, in the decision of Public Citizen vs. Department of Health and Human Services (22), the court held that "application of a *de minimis* exception for constituents of a color additive, however, seems to us materially different from use of such a doctrine for the color additive itself."

In the case of the constituents policy, an individual impurity of a food or color additive may have, by itself, undergone substantial toxicologic evaluation, including carcinogenicity studies. More often, however, impurity safety is established through testing the food or color additive *per se*. In order to establish appropriate specifications for a new food or color additive, attention must be given to the potential of impurities to cause adverse health effects under conditions of intended use. To properly assess the potential for impurities to occur at levels that may be harmful, and to set specifications intended to exclude this possibility, it is necessary to carefully consider the level and nature of impurities in the final product. As a practical matter and because of time and cost considerations, established specifications must be relatively simple and straightforward. In fact, established specifications provide reasonable assurance that the ingredient is of suitable purity for its intended conditions of use providing the production process is conducted under good manufacturing practices (23). It is generally neither necessary nor practical for established specifications to require extensive analysis of individual impurities in order to certify each production lot as being of suitable purity.

It should be emphasized that for specifications to perform their intended purpose of assuring suitable purity and thus safety of the product, it is necessary that the synthesis and production of the product be strictly controlled. Such controls include consistency in the quality of the starting materials, in the method of synthesis, in the purification process, in the equipment used, and other relevant quality control measures. A loss of consistency in these areas can result in the presence of new impurities that may escape detection by the analyses required to establish compliance of the product with existing specifications. For this reason, any significant change in the synthetic process, or any significant change in starting materials, must invoke consideration of whether existing specifications are adequate. In instances in which it has been concluded that the existing specifications are not adequate, further analytical research is necessary to provide qualitative and quantitative information about impurities that will permit an evaluation of the safety of the final product and establishment of new specifications if necessary.

There is a critical linkage between the thoroughly tested toxicologic sample and its impurities and the safety of new lots or batches that may have different impurities. If there are significant changes in the type and quantity of impurities from those present in the toxicologic sample, assurance of safety cannot be guaranteed without further consideration and/or toxicity testing. This concept is the inherent underlying principle governing the certification of FD&C or D&C colors by the FDA. Both FD&C and D&C colors are subject to certification under section 706(c). Some color additives, such as natural colors (e.g. grape skin extract), mineral-based colors (e.g. calcium carbonate, iron oxides), and contact lens colors are exempt from certification on the grounds that certifications are not necessary for the protection of the public health as provided under Section 706(c) of the FD&C Act. Certification means that the FDA must chemically analyze every batch to ensure it is in compliance with specifications for that color additive and does not significantly depart in composition from the batch of color additive (i.e. the toxicologic sample) on which safety testing was conducted.

In the end, there must be a rational basis for decisions pertaining to impurities in the final product not demonstrated to be safe from testing of the toxicologic sample. For impurities present at levels below a projected dietary level of approximately 1 ppb, the concept of negligible or *de minimis* risk as described above (13,19) is considered appropriate. For impurities whose concentrations in the diet are greater than 1 ppb, but less than 50 ppb, acute toxicity testing, in accordance with FDA principles and guidelines, is considered adequate for chemicals whose structures do not carry a suspicion of carcinogenicity (24). The concept that all risks are not worthy of formal regulatory control and thus subject to threshold of regulation considerations is implicit in the principles that guide the process both whereby the safety of impurities is assessed and the way in which chemical specifications for food and color additives are established.

PESTICIDE RESIDUES

A pesticide is defined under the FD&C Act as any substance used to control pests within the meaning of the Federal Insecticide, Fungicide and Rodenticide Act (FIFRA), in the production, storage, or transportation of raw agricultural commodities [Section 201(q)]. The Pesticide amendments of 1956 to the FD&C Act (Section 408) were the first amendments to the FD&C Act requiring premarket clearance evaluations of the safety of chemicals added to food. Currently, the US Environmental Protection Agency (EPA) is responsible for evaluating the safety of pesticides before issuing tolerances.

A major part of the registration process governed by FIFRA involves tolerance setting. Pesticides intended for use on food crops must be granted tolerances or exempted from tolerances under the FD&C Act. Tolerances for raw agriculture commodities were established under Section 408 of the Federal Food Drug and Cosmetic Act. If the pesticide chemical was found to concentrate in any of the processed fractions from the studies done on the food crops, the residues in the processed fraction(s) were considered to be intentional food additives and were required to be assigned "Food Additive Tolerances" under Section 409 of the Act. However, under the new Food Quality Protection Act (FQPA) of 1996,

Section 201(s) of the FD&C Act excludes pesticides from the definition of "food additive" – even in the case of concentration of residues in processed fractions. Consequently, the Delaney Clause is no longer applicable *for pesticides*. The Delaney Clause has *not* been repealed from Section 409 and continues to apply to intentional food additives *other than pesticides*.

The FQPA requires that an additional 10-fold safety factor: "...shall be applied for infants and children to take into account potential pre- and post-natal toxicity and completeness of the data with respect to exposure and toxicity to infants and children." Therefore, the "default" assumption is that the additional 10-fold safety factor will be applied to the chemical safety assessment resulting in a 1000-fold safety factor. The additional 10-fold safety factor may be reduced depending on factors including adequacy of exposure assessment, adequacy of the toxicological database and the nature and severity of any adverse effects observed in the safety studies.

The role of the threshold of regulation concept for pesticides is perhaps most clear with respect to the process of determining if a tolerance needs to be established. In those cases in which it is determined that no tolerance needs to be set for the application of a specific pesticide, use of the threshold of regulation concept is evident.

TRADITIONAL FOODS

The FDA's approach to assessing the safety of traditional food is markedly different from the safety assessments described above, but like those above, an implicit threshold of regulation is clearly evident. Traditional foods are treated differently from food ingredients by the FDA in assessing safety, because the FD&C Act mandates different treatment in Section 402(a)(1). This section of the FD&C Act prohibits sale of any food that contains an "added" poisonous or deleterious substance that "may render" the food injurious to health. However, if the poisonous or deleterious substance is a normal component of that food and is not an "added" substance, the food is prohibited from sale only if the quantity of such substance in the food "ordinarily renders the food injurious to health." In other words, Section 402(a)(1) provides two different safety standards depending on whether the poisonous or deleterious substance is added or is a normal component of food.

If the substance is added, the mere possibility that it may be injurious could result in the FDA prohibiting its sale. But for a non-added component of food, the mere possibility that it may be injurious would not suffice under the FD&C Act to prohibit the sale of the affected food. To prohibit the sale of such affected food, the FDA would have to show that the poisonous or deleterious substance was present at levels that rendered the food ordinarily injurious to health. The FDA has regarded the language of 402(a)(1), "...ordinarily render it injurious to health...," to mean that the FDA must demonstrate that consumers have become ill or have otherwise suffered an impairment of health as a direct consequence of consuming the affected food. The standard of "ordinarily render injurious to health" for normal food components recognizes that traditional foods may contain low levels of toxic components, such as solanine in potatoes. However, only if concentrations of such components are sufficiently high so as to ordinarily render the food to be injurious, is the food deemed unsafe and adulterated.

It is also well recognized that a small percentage of consumers can develop allergic reactions to proteins in traditional foods. Allergic reactions are rarely severe; however, a very small percentage of consumers have severe reactions to the proteins in such common staples as milk and eggs. These foods are not ordinarily injurious and, furthermore, the law requires the labeling of food ingredients in finished foods so that consumers with food allergies can manage their diet.

Section 402(a)(1) of the FD&C Act was signed into law in 1938 and has its origins in the original Food & Drugs Act of 1906. The dichotomous treatment described above in which an added substance is subjected to a more risk averse standard than a normal component of food, which is not considered to be added, reflects a basic difference in how Congress viewed the benefits of food additives vs. the benefits of food itself. Congress enacted legislation, the Food Additives Amendment of 1958, which served to eliminate much of the uncertainty pertaining to the safety of food additive substances by requiring appropriate safety evaluation. On the other hand, traditional food was presumed to be safe, and for it to be found unsafe, it must have demonstrably affected the health of consumers of that food.

While neither Congress nor the FDA has explicitly referred to the 402(a)(1) safety standard for traditional food as coming under the threshold of regulation concept, it is nevertheless apparent that the construction Congress passed into law established the equivalent of a threshold of regulation for traditional foods and their normal non-added components. The threshold for traditional food is "ordinarily injurious to health." In other words, neither the Congress nor the FDA is concerned with possible risks from traditional food unless or until it meets or exceeds the threshold of "ordinarily injurious."

FOODS DERIVED FROM NEW PLANT VARIETIES

Given that substances added to food under the meaning of 402(a)(1) are subject to a different safety standard than components of traditional food, the FDA has had to decide whether food derived from new plant varieties, including plants altered through genetic engineering, would be considered a food additive, color additive, GRAS ingredient, or food. The FDA has recently addressed this issue in the "Statement of Policy: Foods Derived from New Plant Varieties" (7).

The FDA takes a common-sense approach to deciding the above question by apparently concluding that the unaltered portion of the food should be considered traditional food and subject to the "ordinarily injurious to health" safety standard. Exceptions to this may be invoked if the normal food component were present at levels that significantly exceed the range found in the traditional food. In that case such component could be regarded as an "added" substance subject to the "may render injurious" standard. In addition, components that are new to the altered food are to be considered added substances. This means that such substances would have to be either subject to pre-market approval as a regulated food additive or color additive or determined to be GRAS by qualified experts. In either case, assurance of safety would need to be provided to establish that the food is not adulterated because it contains a substance that may render the food injurious to health.

In most cases the added substance, which is the result of genetic modification, will be the same or substantially equivalent to substances commonly found in food such as proteins, fats and oils, and carbohydrates. The FDA has determined that new substances should be subject to food additive regulations only in those cases where they raise questions of safety sufficient to warrant formal pre-market review and approval by the FDA. The criteria that would trigger regulation of such a substance as a food additive are discussed in the *Federal Register* announcement (7). A rather elaborate and detailed decision-tree scheme is provided to offer guidance regarding when it is necessary to consult with the FDA and when the new component of an altered food requires formal FDA pre-market approval and clearance. For example, the FDA indicated in its policy announcement that nucleic acids, which are present in the cells of every living organism, including every plant and animal used for food, do not raise a safety concern as a component of food, and as such, are presumed to be GRAS. The FDA further indicated that the expression product or products present in altered foods will typically be proteins or substances produced by the action of enzymes, such as carbohydrates and fats and oils. As these substances are already present in food, the FDA reasons they are unlikely to raise a safety question sufficient to challenge their presumed GRAS status.

CONCLUSIONS

The concept of threshold of regulation is likely as old as regulations themselves. In the case of health risks, threshold of regulation can mean different things to different people. For some, the threshold of regulation means establishing a safe level that permits use of the substance under conditions of intended use. To others, the threshold of regulation means that substances under the threshold are not legally food or color additives; while to still others, it means that the risk from the substance is so small that it can be totally disregarded. The concept and any resulting actions are likely to trigger spirited debate from those who don't understand or don't agree with the decisions being made. In the past and to this day, many threshold of regulation decisions have been made on a case-by-case basis. However, without guidelines or a well-formulated framework such decisions have the appearance of being post facto and are subject to the accusation of being inconsistent with policy and other decisions made in accord with existing policy.

REFERENCES

1. Scheuplein R, Flamm WG. A historical perspective on FDA's use of risk assessment. In: Middlekauff RD, Shubik P, eds. *International food regulation handbook*. New York: Marcel Dekker Inc; 1989; 27–52.

2. Kotsonis FN, Burdock GA, Flamm WG. Food Toxicology. In: Klaassen CD, ed. Casarett and Doull's Toxicology: The Basic Science of Poisons, 6th edn. New York: McGraw-Hill; 2001.

3. Lehman A, Fitzhugh G. 100-fold margin of safety. *Q Bull Assoc Food Drug Off* 1954; 33–5.

4. WHO. Evaluation of the toxicity of a number of antimicrobials and antioxidants. *WHO Tech Rep Ser* 1962; 228.

5. Lu FC. Acceptable daily intake: inception, evolution, and application. *Regul Toxicol Pharmacol* 1988; 8:45–60.

6. Scheuplein R. Risk assessment and food safety: a scientist's and a regulator's view. *Food Drug Cosmet Law J* 1987; 42:237–50.

7. FDA. Statement of policy: foods derived from new plant varieties. *Fed Regist* 1992; 57:22984–3005.

8. Flamm WG. Critical assessment of carcinogenic risk policy. *Regul Toxicol Pharmaco* 1989; 9:216–24.

9. Flamm WG, Lorentzen R. Quantitative risk assessment (QRA): a special problem in the approval of new products. In: Cothern CR, Mehlman MA, Marcus WL, eds. *Risk assessment and risk management of industrial environ chemicals*. Princeton, NJ: Princeton Scientific Publishing Co Inc; 1988; 91–108.

10. FDA. Food-producing animals, criteria and procedures for evaluating assays for carcinogenic residues. *Fed Regist* 1977; 42.

11. FDA. *Toxicological principles for the safety assessment of direct food additives and color additives used in food*. Washington, DC: Bureau of Foods; 1982.

12. DC Circuit Monsanto vs. Kennedy. 613 *Fed Reporter* 2nd Series; 1979; 947.

13. Rulis A. *De minimis* risk and the threshold of regulation. In: Felix CW, ed. *Food protection technology*. Chelsea, Michigan: Lewis Pub Inc; 1987; 29–38.

14. Rulis AM. Threshold of regulation. Options for Handling Minimal Risk Situations. In: Finley JW, Robinson SF, Armstrong DJ, eds. *Food safety assessment*. Washington, DC: American Chemical Society; 1992; 132–9. (*ACS Symposium Series* 484.)

15. Flamm WG, Lake LR, Lorentzen RJ, Rulis AM, Schwartz PS, Troxell TC. Carcinogenic potencies and establishment of a threshold of regulation for food contact substances. In: Whipple C, ed. *De minimis risk*. New York/London: Plenum; 1987; 87–92.

16. Gold LS, Sawyer CB, Magaw R, *et al*. A carcinogenic potency database of the standardized results of animal bioassays. *Environ Health Perspect* 1984; 58:9–319.

17. Gold LS, de Veciana M, Backman GM, *et al*. Chronological supplement to the Carcinogenic Potency Database: standardized results of animal bioassays published through December 1982. *Environ Health Perspect* 1986; 67:161–200.

18. Gold LS, Slone TH, Backman GM, *et al*. Second chronological supplement to the Carcinogenic Potency Database: standardized results of animal bioassays published through December 1984 and by the National Toxicology Program through May 1986. *Environ Health Perspect* 1987; 74:237–329.

19. Munro IC, Rapporteur. Safety assessment procedures for indirect food additives: an overview. *Regul Toxicol Pharmacol* 1990; 12:2–12.

20. FDA. Policy for regulating carcinogenic chemicals in food and color additives. *Fed Regist* 1982; 47:14464–70.
21. United States Court of Appeals, Sixth Circuit; Scott vs. Food and Drug Administration. 728 *Fed Reporter*, 2nd Series, 1984; 322.
22. United States Court of Appeals for the DC Circuit; Public Citizen vs. Dept of Health and Human Services; 1987; No. 86–5150.
23. Modderman JP. Specifications. In: Middlekauff RD, Shubik P, eds. *International food regulation handbook*. New York: Marcel Dekker Inc; 1989; 283–307.
24. Kokoski DJ, Flamm WG. Establishment of acceptable limits of intake. In: *Proceedings of the second national conference for food safety*. Washington, DC: Dept of Health and Human Services; 1984; 61–72.

10

BIOTECHNOLOGY-DERIVED AND NOVEL FOODS: SAFETY APPROACHES AND REGULATIONS

Gerrit J.A. Speijers, M. Younes and J.E.N. Bergmans

INTRODUCTION

In the history of mankind, the attainment of quantities of food sufficient to ensure a healthy and productive life has always been a major concern. Major milestones in society's effort to ensure adequate food supplies have been the development of agricultural methods and the domestication of livestock in most parts of the world. During the course of these developments, new food items have continuously been selected and introduced as part of the human diet, the assessment of their safety being a consequence of "trial and error." This holds true for edible plants (e.g. potatoes), food products of animal origin, and microorganisms (e.g. those involved in the preparation of fermented foods). Seen in this context, the dietary inclusion of "novel" foods has been and still is a continuing process. The aspect of novelty may easily be expanded to include biotechnology-derived food items. Seen in a broader sense of selection and introduction of genetic variations, biotechnology also has a long history of use in food production and processing. In fact, traditional breeding and selection techniques have a long tradition in the development of new varieties of microbial, plant or animal origin for such uses. The process of selective breeding, however, by crossing and selection, is slow and its possibilities are limited because of the limited genetic diversity of the parent organisms, and the results are often unpredictable. By way of contrast, the "novel" methods of biotechnology (genetic engineering; cell fusion) permit the introduction of rapid and more precisely targeted genetic changes. The aspect of novelty thus may apply to food items or ingredients that have not been used for human consumption before as well as to those that have been produced using new breeding or processing methods (1). In fact, there is still a lack of consensus on the definition of the term " novel food," although it generally covers both foods made by new processing techniques, including those based on methods of molecular biology, and foods that have not been previously used for human consumption (which often is true only for a certain country or region) (1–4).

The EU novel food regulation (4) has divided novel foods in six subcategories:

- Foods and food ingredients containing or consisting of genetically modified organisms within the meaning of EC Directive 90/220/EEC (4);
- Foods and food ingredients produced from, but not containing, genetically modified organisms;
- Foods and food ingredients with a new or intentionally modified primary molecular structure;
- Foods and food ingredients consisting of or isolated from micro-organisms, fungi or algae;
- Foods and food ingredients consisting of or isolated from plants, and food ingredients isolated from animals, except for foods and food ingredients obtained by traditional propagating or breeding practices and which have a history of safe use; and
- Foods and food ingredients to which have been applied a production process not currently used, where that process gives rise to significant changes in the composition or structure of the foods or food ingredients, which affect their nutritional value, metabolism, or level of undesirable substances.

Before the introduction of any novel food to the market, both safety and nutritional evaluations have to be carried out, with special focus on specifications as well as production and processing methods employed. Because of the diversity of products that (might) fall into the category of novel foods, procedures to be followed for safety evaluation are not easy to set up and it is neither possible nor desirable to draw up guidelines or rigid protocols for any part of it. It is generally accepted, however, that safety evaluation of novel foods requires an integrated multidisciplinary approach covering analytical, nutritional, and toxicologic aspects.

APPLICATIONS OF BIOTECHNOLOGY AND TYPES OF NOVEL FOODS

According to the broad definition mentioned above, novel foods cover products developed for use as food constituents or in processing, including certain metabolites and food additives, like amino acids, citric acid, vitamins, enzymes, and polysaccharides, as well as biomass, including whole foods and other complex mixtures. Novel foods may be derived from microorganisms, higher plants, or animals. Although many of the principles involved in the applications of new methodologies as well as in safety evaluation are similar in all three groups, each group will be discussed separately.

Microorganisms

Microorganisms have a long history of safe use in food production (e.g. in the preparation of beer, wine and bread), although their role was not recognized until Pasteur's discovery. In fact fermented foods are major constituents of the human diet and include various products derived from plants (alcoholic beverages, tea, coffee, bread), as well as fermented fish, meat and milk products (e.g. cheese) (5,6). Besides contributing to the diversification of the diet, fermentation may also have significant contributions in increasing palatability, acceptability, nutritional value, and shelf life of foods (2,7).

Microorganisms employed for fermentation purposes may be bacteria, yeast, or fungi. They may become an integral part of the food, but are in many instances removed after having produced the desired metabolic effects, so that only their soluble products are consumed. Besides these traditional applications, microbial biomass, mostly derived from yeast, algae, or fungi has been considered as a potential food source. Both the isolated cell protein and the total cell material are termed "single-cell protein" (5,6,8,9).

Traditional selection and breeding techniques, such as crossbreeding, hybridization, and mutation, have been employed to improve the efficacy of a number of microorganisms used in food processing and production (2,5). Techniques based on molecular biology may be used to enhance the efficiency of breeding and selection of microorganisms employed in different areas of food production and processing. By using such methodologies, specific changes can be introduced into microorganisms that would not have been possible through the use of traditional breeding methodologies. The most widely known method of nontraditional genetic modification is recombinant DNA technology, often referred to as genetic engineering (2,5,7). In fact, microorganisms are more amenable to genetic modification than plants or animals for a number of reasons (2), e.g.:

- Gene structure, expression, and regulation are best understood in microorganisms;
- Microorganism cells are not highly differentiated;
- Microorganisms have a relatively simple and small genome; and
- They are easy to handle, grow rapidly on simple media, and allow for easy selection of variants with desired traits.

Recombinant (r) DNA technology will generally be used to introduce one or only a few genes into the genome of a production microorganism strain. The first basic step in genetic engineering is the isolation of the DNA segment that contains the gene(s) of interest, a process known as DNA cloning; this is generally accomplished via its linkage to a vector DNA, usually a plasmid, a circular DNA molecule, that is able to copy and reproduce itself and any gene inserted into the plasmid DNA. The inserted DNA segment is usually limited to its smallest functional size containing the information needed for the production of the desired product. The vector, i.e. the plasmid plus inserted DNA, is then moved into a suitable host organism, where it will continue to reproduce itself (5,10). One difficulty faced during this process is the fact that not every cell exposed to a vector will incorporate the cloned gene. In order to identify cells to which the cloned DNA has been introduced, vectors for genetic modification mostly contain selectable marker genes (e.g. antibiotic resistance) that will provide them with a growth advantage under selective laboratory conditions (e.g. in the presence of antibiotics) (2,5,7). At present methods are being developed that make it possible to remove the selectable marker after it has been used to select the modified organisms.

Apart from rDNA, techniques of cell fusion may be used to overcome barriers in genetic exchange between eukaryotic microorganisms. On the one hand, there are natural barriers to crossing between different strains (e.g. of yeast) (11,12), and, on the other hand, "inbreeding barriers" may develop during extensive selection of highly stable strains. Through fusion of somatic cells (e.g. protoplast fusion), new strains with certain desired traits can be generated.

Recombinant DNA is generally used to move defined genes, whose expression leads to defined gene product, between organisms that are often only distantly related. Cell fusion can only be employed between organisms that are sufficiently related to lead to a viable fusion product, but has the advantage that it facilitates the exchange of traits that are polygenic, or for which the genetic background is unknown.

Applications of techniques of molecular biology to modify microorganisms used for food production cover several aspects. Microorganisms used for the production of fermented foods may be modified to allow for specific improvements. Some typical examples are:

1   Commercial brewing strains may be modified by incorporation of genes for gluco-amylase production, allowing the degradation of remaining unfermentable dextrins and thus the production of low calorie beer (13). Another modification increases the levels of α-acetolactate decarboxylase. This enzyme catalyzes the breakdown of ketonic side products of isoleucine and valine, which are undesirable flavoring substances produced during the process of fermentation (14).

2   In the baking industry, bread baking yeast strains may be modified, for example, to degrade maltose more efficiently. This would result in a reduction of baking time (2). Also, incorporation of DNA coding for lipoxygenase of plant origin into yeast will allow the production of a whiter bread, as lipoxygenase bleaches yellow pigments of wheat flour (15).

3   In the dairy industry, lactic acid-producing bacteria used in the production and preservation of food items like cheese, yogurt, and butter, are modified for the development of improved strains, e.g. with higher phage resistance or with improved capacity for flavor production (16,17).

Besides their use in the reproduction of fermented food items, microorganisms are also a source for a wide range of food additives. The potential of modern techniques based on molecular biology to develop microorganisms strains with a high capacity for the production of food additives is quite considerable (2,7). In particular, the development of microorganism strains with the capability to produce flavours that are currently synthesized chemically or extracted from plants, may have considerable impact. Current applications include the increased yield of existing additives (e.g. L-trytophan, citric acid) and the development of new sources for additives or processing aids, including enzymes (e.g. chymosin) (18).

Protein obtained from microorganisms (single-cell protein) may have applications in both animal and human nutrients. In fact, single-cell protein can be obtained from different microorganisms grown on various substrates, many of which are nonconventional. Examples are bacteria grown on methanol, yeast (e.g. *Saccharomyces*) grown on sugar, starch, whey, and molasses; yeast (e.g. *Candida*) grown on non-conventional substrates such as hydrocarbons; filamentous fungi, capable of converting ligninocellulose into protein, that are grown on starchy substrates; and finally, several photosynthetic algae (4,8,9). The potential of genetic engineering methodology may also be used to develop microorganism strains with the capacity to produce lipids that can be used in food.

Plants

In agriculture, traditional breeding methods have a long history of safe use in the introduction of variability into crop plants. The development of new varieties of food plants has been directed principally toward increasing yields and minimizing losses. In fact, wide, genetically based, differences in the chemical composition of traditionally developed varieties of the same species of food plants are a reflection the genetic diversity attained (5).

One of the most widely used methods in this respect is hybridization, a process during which different cultivars of a species are crossed, yielding improved or unusual types. Genetic variabilities may be even further enhanced by "wide" hybridization, i.e. a cross between cultivars belonging to different species or even genera (19). There are natural limits to this method, i.e. when the genetic differences between the partners in the cross are too large to lead to a viable offspring. Another source of genetic variation is the induction of mutations. It was, however, recognized early that induced mutations occurred at random; therefore, it proved almost impossible to target a particular trait. In addition, many induced mutations proved to be deleterious.

Some of the difficulties and limitations of traditional breeding can be overcome using methodology based on molecular biology. Limitations in genetic exchange due to crossing barriers between unrelated species can be overcome by the use of rDNA methods. Traditional breeding is time-consuming; rDNA methods have the potential to shorten the time needed to incorporate certain traits into a food plant. Difficulties in identifying individual plants in the offspring of a cross that carry the desired genotypic changes can be overcome by the use of molecular methods for the tracing of molecular markers linked to the genotype. Finally, in

207

traditional breeding, a desired trait may be linked to some other, undesired characteristics. The elimination of such traits via the usual backcrossing procedure is often difficult. Again, genetic engineering offers possibilities to overcome such difficulties (2,5).

In fact, genetic engineering appears to be the most powerful technique for the introduction of specific traits into cultivars (18–21). The methodology is basically the same as for rDNA techniques used in microorganisms, but there are some differences. One major difference lies in the fact that there are no plasmid vectors available that can replicate autonomously in plant cells after transfer. The transgenes should therefore always be integrated into the plant genome. The only available natural method to accomplish this takes advantage of a conjugative plasmid of the bacterium *Agrobacterium tumefaciens*, by which specific DNA sequences, located between border sequences on the plasmid, are transferred to plant cells and subsequently integrated at random sites in the plant genome. This technique is limited to plant species susceptible to *A. tumefaciens* (5). Other techniques make use of free DNA delivery treatments (e.g. microinjection, electroporation, biolistics) of plant cells.

Apart from rDNA techniques, a number of tissue culture methods have found application in plant breeding. One such technique is clonal propagation (22), in the course of which plant cell cultures of some genetic variant or hybrid plant are established and scaled up to allow for the generation of "carbon copies" of certain genetic variants. One other technique is the production of somaclonal offspring of a plant (22). This method involves the preparation of tissue explants, usually single cells, from which an entire plant is regenerated. During somaclonal propagation various genome rearrangements may occur; the process is therefore used to obtain genetic variation, especially in plant species where other sources of genetic variation are less easily obtained. It should be kept in mind that somaclonal propagation, which is part of the process of genetic modification in many cases, could be a source of genetic variation in the resulting transgenic plant too. A related technique is that of gametoclonal variation, in which plants are regenerated from cultured pollen cells or from pollen within the anther (22). One further technique is hybridization by protoplast fusion, i.e. fusion of wall-less plant cells. Protoplasts are isolated from cells of parent plants to be hybridized. A multistep chemical treatment is employed to induce fusion, after which the fused cells are identified (by selective markers) and separated. Hybrid cells are then grown in a culture medium permitting the regeneration of the new hybrid plants. Via this methodology, hybrid plants may be obtained from parent plants of different origin, where conventional breeding methodologies would fail (20–23), although in many cases similar barriers appear to exist for crossing and hybridization by cell fusion alike.

Improvements of food plants through methods aimed at the introduction of specific traits using techniques based on molecular biology include modifications of the plant itself but not the product, e.g. the fruit, or of the fruit or other derived products, as well as the development of new products. Among the improvements that might be attained through modification of a single gene, one of the most prominent is the increased resistance to pests and diseases (2). Resistance to insects might be attained, for example, by inserting a gene encoding small polypeptides that inhibit insect trypsin (24), or genes encoding insecticidal substances such as *Bacillus thuringensis* toxin (25). The integration of genes for viral coat proteins into the genome of crop plants results in an improved resistance to viruses (26).

Another improvement of food crops through addition of one (or a few) genes is resistance to herbicides. This result can be obtained in two different ways: the introduction of genes encoding for enzymes capable of detoxifying or degrading the herbicide, or the introduction of a resistant form of the target-protein for the herbicide (27). For example, the resistance of modified soybean (GTS) tolerant to the herbicide glyphosate, is based on an inserted EPSPS gene from bacterial origin, that is resistant to glyphosate. The enzyme EPSPS (5-enolpyruvylshikimate-3-phosphate synthase), involved in the biosynthesis of aromatic amino acids, is normally inhibited. However, the GTS soy contains this EPSPS resistant gene. The herbicide tolerance is usually enhanced by introduction of a *gox* gene, encoding an enzyme that converts glyphosate to non toxic metabolites (3).

Among improvements that affect the quality of the product, the prolongation of shelf-life of tomatoes by induction of delayed softening characteristics is an example (28). Further developments in this field will no doubt have far-reaching consequences for agriculture and the consumer. Varieties of food plants that are well adapted to consumer needs in terms of their nutritional composition, taste and appearance will be used. In addition, crop plants might also be modified to allow for a large-scale production of certain food additives and ingredients (e.g. protein sweeteners).

Animals

Conventional breeding of food producing animals has proven to be useful in modifying their characteristics and increasing their output from both a qualitative and a quantitative point of view. Examples of changes in characteristics of animals used in food production include the breeding of early-maturing chickens, of farm animals producing meat of higher quality due to changes in the protein/fat ratio, and of varieties of cows with higher milk yields, to name but a few (2,5).

In the past decade, novel methods have been developed that permit the introduction of new genetic material into the genome of animals, techniques leading to the development of "transgenic animals." Major advances were made following the initial development of microinjection techniques into mammalian eggs, to introduce genetic modifications into animals (29,30). Among the methods developed, microinjection is the simplest and most widely used (31). In order to apply this method, pronuclei of fertilized eggs have to be injected with copies of the desired gene. In contrast to rodents, the pronuclei of which are easily identifiable, difficulties arise with common farm animals, whose fertilized eggs are opaque, making the pronuclei not clearly visible. Another problem is that few fertilized eggs survive microinjection treatment. Gene transfer into animal genomes may also be accomplished using retroviruses (2). In retroviral vectors used for gene transfer, most of the viral (i.e. vector) genes are replaced by new genetic material. The generation of transgenic animals also may be facilitated by the use of embryonic stem cells (32). With this method, cultures of embryonic stem cells of an animals are established, retaining their karyotype. Genes of interest can be introduce into the genome of these cultured cells using microinjection or retroviral infection. The modified cells can then be incorporated into the embryonic lineage of a developing embryo, resulting in a chimeric animal, in which part of the cells (i.e. those derived from the modified stem cell) is genetically modified. Fully transgenic animals may be produced by crossing the chimeric animal, as part of the germ cells will be transgenic, too.

It must be stressed that the application of novel methods of molecular biology in developing modified varieties of common domestic animals is still at a very early stage in which most efforts are spent on basic research. The major limitations center upon the difficulties in collecting embryonic cells at the desired state of development and in identifying strategies for introduction of advantageous traits into the mammalian genome. Some applications that have been successful so far include the introduction of drug-resistance genes as well as genes responsible for growth enhancement. Other applications include the enhancement of immune response and the insertion of genes whose products block viral infection, alter fat composition and content of meat, or influence the protein production pattern. It should be noted, however, that transgenic farm animals will remain mainly a subject of basic research like laboratory animals in the near future.

SAFETY APPROACHES: THE BASIC ISSUES

In dealing with novel food items and food ingredients produced by the application of novel technologies, a number of safety aspects have to be considered that are similar to the ones considered for "traditional" substances. In part, however, they stem from considerations related to the applied methods as such and therefore require some "novel approaches." In general safety aspects cover nutritional, toxicologic and pathogenic issues (1–7,33–37). The nutritional aspects to be considered encompass factors such as nutrient composition of the food (both macro- and micronutrients) and the presence of antinutritive factors. Toxicologic issues include, apart from generic toxicologic aspects, the possibility of an increased production of natural toxins or of foreign gene products with adverse effects. Pathogenic aspects cover the possible incidence of pathogenicity due to adverse effects of genetically modified microorganisms employed in the production of food or food additives, or due to microbial contamination of nonspecific origin (2,7,34). Although environmental aspects of biotechnology cannot be discussed in the framework of this chapter, it must be stressed that full consideration should be given to the ecological effects resulting from the introduction into the environment, i.e. the agricultural production, of genetically modified organisms intended for use in food production, before assessing their food safety. Data required for the environmental assessment, mainly those related to the characterization of the host and donor organisms and to the method of genetic modification, will also be useful in the course of the food safety assessment process.

Microorganisms

In dealing with safety aspects of food items derived from modified microorganisms, a distinction has to be made between microorganisms that will become constituents of food, such as starter cultures and probiotics, and those used as sources of certain food additives, such as amino acids, organic acids, antioxidants, thickeners, and enzymes, or as sources of food ingredients, such as single cell protein. While microorganisms of the first group are living organisms that are intentionally added to the food, microorganisms of the second group do not become a constituent of the food product and are usually separated from the final product.

Safety assessment of genetically modified foods produced using genetically modified microorganisms needs to be based on an evaluation of single steps in the food production process (1,2,5,7,37,38). These include:

- Knowledge about the identity and characteristics of the source organism and of the steps undergone in modifying its function;
- Knowledge about the chemical processes that occur during the fermentation process (substrates, growth material, growth conditions);
- Where appropriate, knowledge of the purification process and the isolation procedure; and
- Chemical composition of the final product.

With genetically modified microorganisms, the origin and (taxonomic) identity of the production (host) organism should be clearly established, as is the case with non-genetically modified organisms. This is of importance in assessing factors that might be of relevance to safety assessment such as the potential ability to produce toxins or the presence of virulence factors and other impurities. The main additional safety consideration is the nature of the new gene products produced by the modified organism. It is essential to ensure whether the inserted DNA does encode for any harmful substances. Additionally, it should be ascertained that the final construct is stable. In cases in which vectors are used to transfer the genetic information, it is essential to establish their safety for use in food processing and production. They should therefore be, in general, derived from organisms having a history of safe use in food. In introducing genetic material into the vector, it must be ensured that such material is well characterized and represents the smallest fragment necessary to express the desired trait. The vector should be modified in a manner minimizing the possibility of transferring genetic information to other microorganisms.

The use of methodologies of molecular biology may also aim at changing the expression of desired genes. An enhanced gene expression may be achieved through alterations of the regulatory signal sequence, through gene amplification, or through the use of multicopy plasmids, to name a few techniques. In such instances, the increased level of gene products should be taken into consideration (2,7).

Another problem is the fact that the insertion of genetic material into the genome may influence the expression of genes in the region, either by integrative disruption of a gene, or by affecting the expression of neighboring genes. This may result either in down regulation, or in enhanced expression of genes "downstream" of the insertion site due to read through from a promoter on the inserted DNA. These aspects have to be borne in mind when carrying out safety assessments of genetically modified novel foods (2,7).

From the point of view of pathogenicity, it should be recognized that this characteristic is linked to a number of traits appearing in combination, for example, the ability to adhere, to invade the host organism, and to produce toxins. It therefore seems impossible that the insertion of one or a few well-characterized genes in the genome of an organism known to be nonpathogenic will turn it into a pathogen (2,7). A history of safe use and the lack of pathogenic traits of host organisms should therefore be established. One related consideration is the capability of the microorganisms to produce toxins. These include known microbial toxins (exotoxins, endotoxins, mycotoxins) and antimicrobial compounds (2,7,39). As a matter of fact, this aspect is being considered in safety evaluations of conventional foods and food additives and is considered to cause little concern if good manufacturing practices are being followed. The use of methods of molecular biology to modify microorganisms does not raise additional concerns, but attention should be directed toward assessing the potential of bioengineered organisms to produce higher levels of such toxins. This requires consideration of information not only on the identity of the organism, its genetic make-up, and its physiologic characteristics, but also on the growth conditions and on the composition of the final food products (2,5,7).

In summary, there is a need for a careful analysis of microbiologic, molecular, chemical, and toxicologic parameters in order to establish the safety of novel foods produced using (genetically modified) microorganisms. It is a prerequisite that such food items be produced according to current good manufacturing practices and HACCP guidelines (7,40).

Allergenicity

One issue of specific relevance is the question as to whether the ingestion of a new protein (e.g. single-cell protein) or a modified protein would give rise to an increased risk of allergenicity (41). This is more likely to be the case with protein obtained from sources with a known high allergenic potential (e.g. peanuts and

Brazil nuts). The potential allergenicity is difficult to assess as there are no validated predictive tests for determining the allergenicity of proteins from sources that are not commonly recognized as allergens (41). As the major food allergens are known, this could be the basis for structural alerts of proteins formed. On the basis of base pair sequence encoding for proteins known to cause food allergy, similarities of the protein in putative allergenic properties can be assessed for the genetically modified product. In case of similarity specific and immunologic tests for identifying the protein responsible for causing adverse reaction, for example, "Western blotting" or RAST test, may be performed using sera from the protein fraction in question. If these in vitro assays are negative, confirmation of absence of allergenic components can be obtained by an in vivo skin prick test followed by a double blind placebo controlled food challenge under controlled clinical conditions with patients sensitive to the food component in question (35,4,41). No single criterion, or even the complete set of criteria, is sufficient to confirm allergenicity or lack thereof.

Plants

As indicated above, most of the novel techniques used to modify food plants involve transfer of foreign genes, usually a small amount of DNA, into the plant genome. In general, the DNA transferred may be well characterized in terms of nucleotide sequence and the nature of the product(s) encoded for. What is left to chance is the manner in which the external DNA is integrated into the plant genome. Methods available do not (yet) allow for an insertion of DNA at a determined site; DNA will therefore be inserted at random. The DNA inserted may also be rearranged or truncated, or it may lose segments during the insertion procedure. Still, the outcome of the insertion procedure can be assessed via analysis of the modified plants and selection of stable variants carrying the desired traits (2).

Safety aspects specific for novel plant foods are not principally different from those applicable to modified microorganisms (2,7). One issue that has raised concerns is the possible transfer of DNA fragments from plants to bacteria through naturally occurring transfer processes. Although this is improbable, the possibility that DNA released from food of plant origin in the gut will be taken up and integrated by the intestinal microflora cannot be completely excluded. The stable integration and expression of such DNA in the microorganism will, however, occur in multiple steps and will therefore be a very rare event. In this respect, the main perceived concern is the possible transfer of antibiotic resistance genes used as selectable markers.

Other issues of relevance to the safety assessment of novel plant foods are the nature and characteristics of the expression products, the possible occurrence of pleiotropic effects in cases of insertion of one gene that is responsible for two or more distinct and seemingly unrelated phenotypic effects, as well as the induction of elevated or decreased expression rates of other gene products, similar to those encountered in genetically modified microorganisms.

Because of the fact that DNA insertion occurs at random, it is possible that such insertions will occur at regulatory sites of certain genes; this would lead to an alteration in their expression characteristics (2,7). As such sites are quite rare in comparison to the large portion of non-coding genome areas, these events will be rather infrequent. Similar effects would also be expected as a consequence of conventional methodologies of mutation induction, i.e. irradiation- and chemical-induced mutagenesis. An issue of greater relevance is the possibility of activation of silent genes in the harvested part of a crop plant (e.g. the production of inherent plant toxins). Genes encoding for products having adverse effects that are expressed in the non-harvested portion of a food plant but not in the harvested portion might be activated through "insertional mutagenesis" in those parts of the plant also. In such cases, however, the nature of such toxic gene products will generally be known, as they are present in the non-harvested portions of the crop plant and in species that are closely related. The harvested plant product could be easily analyzed for the presence of such toxins.

The application of modern techniques of molecular biology to food plants may yield products with characteristics different from those of traditional cultivars both from a toxicologic and a nutritional point of view. However, it must be noted that similar changes might also occur as consequence of applying traditional breeding methods. Toxicologic changes in modified food plants may include the appearance or the increased levels of natural toxicants, the expression of new toxic compounds as a result of gene insertion, the production of allergenic compounds (see previous paragraph), as well as the increased levels of environmental toxicants as a result of an accumulation process. From a nutritional point of view, modifications of the nature, the levels, or the bioavailability of macro- and micronutrients may become evident, and antinutrient factors may be produced.

Safety assessment procedures of novel foods of plant origin will therefore need to focus on the nature of the host plant, of the modification procedure (including information on any vectors used), and of the modified plants in terms of the stability of gene expression and inheritance and of its general phenotypic characteristics. Detailed information of the modification procedure(s) used should be provided to ensure the safety of the process as such and to assess the possible occurrence of pleiotropic or secondary effects due to gene insertion. The novel gene product(s) should be characterized and carefully assessed for possible toxic effects. Finally, from a nutritional point of view, modified plants should be assessed in terms of modifications in the content of traditional constituents as well as for the possible emergence of any new material (1,2,5–7,36,42).

Animals

Although the basic considerations related to safety assessment of genetically modified animals (transgenic animals) and their products are similar to those applicable to microorganisms and plants (2,7,43,44), specific issues are not as well explored yet because of the fact that, in the case of animals, applications have developed at a much slower pace.

Safety assessment will basically need to consider information on the safety of the product of any newly inserted gene, the safety of the genetic construct, and unintended effects resulting from the DNA insertion. Effects of specific gene products can be expected and tested for. For example, with domesticated animals modified to produce higher levels of growth hormone, the effects of such an elevated content following ingestion of meat or other food products should be investigated. Elevated hormone levels may also disrupt the endocrine system, giving yield, for example, to compensatory increases in the levels of endogenous hormones (2,7).

With respect to the gene construct itself, its consumption *per se* will only cause concern if it is infectious (2,7,43). The major concern will be the possibility of transmitting genetic information to susceptible cells in the gastrointestinal tract, although this is quite improbable, as infectivity requires in general the presence of a specific viral protein on the surface. In cases in which genetic modification has been accomplished using retroviruses, however, there is a possibility that fully functional retroviruses may occur in food material. The application of new technical strategies and the increased application of rDNA methods will limit this possibility, however.

As with plants and microorganisms, one major safety consideration is related to the possible emergence of unintended effects. In contrast to plants, however, the possibility of toxin production (at unexpected sites or levels) is quite unlikely (43,44). Toxin production by food animals is known in fish (e.g. tetrodotoxin in puffer fish), molluscs, and snakes, but not in higher animals used for food production. It is therefore improbable that any silent gene may be activated for an unforeseen toxin production. The occurrence of new food intolerances, in contrast, is possible and should be taken into consideration. The same is true with regard to the nutritional value of food originating from genetically modified animals.

SAFETY APPROACHES: REGULATORY ASPECTS

In general

In evaluating the safety of novel foods, it is imperative to recognize that it is neither possible nor really desirable to set up rigid guidelines or protocols for different parts of the evaluation procedure because of the large diversity of products that will be dealt with within this framework. Safety approaches should not, however, be too open-ended in terms of total reliance on expertise and common sense. More general guidance with respect to the safety requirements for novel food and food additives should be provided, leaving enough freedom for a case-by-case safety assessment (1,2,5–7,45).

215

In general, safety approaches to the assessment of novel foods and food additives produced using novel biotechnologies should be based on the recognition of the fact that the aspect of novelty relates only to a late stage of a continuous development. Thus, genetic modification of organisms using new technologies of molecular biology is not a unique, new branch of science. From this point of view, "traditional" safety approaches are still applicable (5,7). In fact, a strategy aimed at assessing the safety of any new food item, be it modified through traditional breeding or through a novel method, will have several common features. These include knowledge of the biologic and molecular components of the system, and of the potential consequences of any genetic modification introduced, as well as comparison of the final product with traditional equivalent (or a comparable product the safety of which has been established). If they are nearly identical it is called substantial equivalence (4,35). In the concept for safety assessment written by ILSI Europe (46) three categories are distinguished: substantial equivalence, sufficient equivalence and non-equivalence. If no traditional equivalent product exists, the wholesomeness of the novel food item needs to be assessed (7).

In assessing the wholesomeness of novel foods, as is the case with traditional ones, basic information such as on biologic, molecular and chemical characteristics is needed. In this respect, the identity of the organisms employed should be clearly established; in particular, knowledge about their potential to produce toxins (or their relation to organisms known to produce toxins), and of any pathogenic or infective potential should be known. In addition, it is necessary to identify the techniques employed in bringing about genetic information and to characterize any genetic material transferred to the food source organism. Finally, chemical characterization of the final product is a prerequisite for all further steps in the safety evaluation process. In cases in which information is lacking, there will be a need to carry out toxicologic studies in animals, the type and extent of which will depend on the type and extent of the available information (1,2,5–7).

In this context, it must be stressed that major differences exist between the safety evaluation of single chemicals or simple chemical mixtures and that of whole foods (and other "complex mixtures") (34). With simple chemicals and chemical mixtures, in general, purified and chemically characterized ingredients are being dealt with. The type and amount of any impurity can easily be assessed. Based on such considerations, these substances should be subjected to toxicologic evaluation according to established procedures for the safety evaluation of food additives, residues, and contaminants (34,47,48). Generally, no additional toxicologic testing of purified items (which proved to be free of replicable new genetic material) will be needed apart from testing for the possible emergence of antigenicity or allergenicity in the case of proteins (including enzymes).

In contrast, safety evaluation of novel (generally genetically modified) food plants and microorganisms, as well as of complex mixtures of macroingredients isolated from such organisms, is a more complicated issue. Any regulation should be based on some type of decision-tree approach (1,2,4–7) (Figure 1). Such a decision tree should first address questions related to the original host organism, the

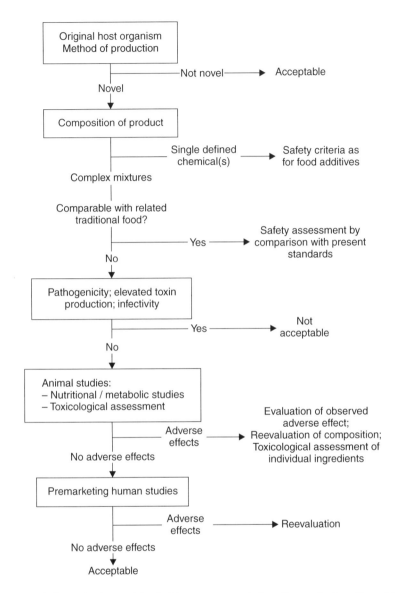

Figure 1 General scheme of the decision tree approach for safety evaluation of novel foods.

source of any introduced genetic material, and the history of human exposure. Obviously, the use of genetic material from a source known to have been part of the human diet will provide more confidence in the safety of a product than would any material for which no dietary experience is available. At this stage, knowledge of the potential of the host and the source organism for toxin production,

217

pathogenicity, or infectivity is essential. The information needed is related to the composition of the novel food (1,2,4–7). In this context, focus should be put on the evaluation of the nutritional value of a novel food in terms of the levels of both essential macro- and micronutrient constituents, as well as on testing for the presence of toxins (also those known to be produced, for example, by the non-edible parts of a food plant) or antinutrient factors. The composition of novel foods should be in the range considered normal in related traditional foods. At this stage, human consumption levels of the food products assessed should be known (1,2,5–7). If data available on the composition and the source of a novel complex food item are not sufficient for safety evaluation, and if available toxicologic data are inadequate, further toxicity testing might be considered. Testing may be conducted on specific constituents of a novel complex food item, if the safety of such a component has not been established (47). In other instances, particularly when it is not possible to isolate such a new food constituent, it might prove necessary to carry out safety testing using the whole complex mixture. However, several problems arise with such approach. Longer-term toxicologic studies on whole foods are generally subject to several confounding factors (2,7,42). The use of complex food mixtures may result in an unbalanced diet and, subsequently, in a distortion of nutritional balance. Besides, dosing at levels high enough to provide a margin sufficient to establish an "acceptable daily intake" is usually not possible. Therefore, toxicity testing of novel foods and complex mixtures will mostly be confined to short-term ingestion studies using doses that approximate the expected human consumption levels (5,34,43). In addition, some specific toxicity issues (allergenicity, mutagenic potential) may be considered.

Following the introduction to the market, one additional aspect contributing to the safety evaluation of novel foods would be the introduction of some mechanism of "postmarketing" surveillance. Reporting on observed adverse effects of novel foods in consumers would allow the identification of possible toxic and/or intolerance reactions not observed in animal studies or in premarketing human studies, and provide the possibility of conducting adequate retrospective epidemiologic studies.

There is a general consensus about the decision-tree type approach (Figure 1) within the scientific community and among regulatory bodies in many countries. In fact, such an approach has been adopted by a number of governmental agencies (1,49,50), and was recommended by national, regional and international expert groups (2,5,7,51). It is generally accepted that there is no need for specific rigid regulations. At present several foods have been evaluated by both national and international bodies among which the Scientific Committee on Food (SCF) (52).

Novel food regulation in the European Union (EU)

In the EU regulation several decision trees are included to assist the safety assessment. However, this regulation makes exceptions for food additives, flavours, and extraction solvents, because they are covered by other Directives. Another definition is that novel foods are those introduced from application of new or modified physical, chemical, and biotechnical treatments performed separately or in combination (53).

Novel foods may be derived from microorganisms, higher plants, or animals. Although there are similarities, it is best to deal separately with these groups since there are a few important differences involved in the safety assessment. As for food use novel foods of animal origin have not yet been the case, the animal products will not be discussed further in this paper. From the risk assessment point of view it is also useful to discriminate between novel foods prepared by modern biotechnology based on molecular biology and products that are novel not involving molecular biology. These aspects are discussed in the context of safety assessment.

Certain background information of the novel food or food ingredient is needed for the safety assessment and to help identify the need, if any, for additional studies to facilitate the assessment of toxicological and nutritional safety. Some of the information involves only GMO derived foods, other are general. A number of important issues in relation to the safety of the novel food will be addressed separately. The safety assessment of novel food is a stepwise approach in which elements, as how well-defined is it and to what extend it is comparable to traditional counterpart, are important for the requirement of further nutritional and toxicological studies (35,46).

Basic information as requested by EU-regulation (4)

Name

This should include, as appropriate, details of the scientific, trivial or chemical name of the novel food or ingredient.

Source

Essential details of the source include whether the novel food is obtained from a plant, animal or microorganism or whether it is a product of chemical synthesis. In the case of novel foods from biological sources the full taxonomic classification should be given.

Origin

The source from which the novel food is obtained, i.e. a plant, animal, or microorganisms should be described, as well as whether it has been developed by traditional breeding and selection techniques or by GMO techniques. In the case of foods, which are, or derived from genetically modified organisms, the data should include characterization of the host organism, the vector, inserted genes and the recombinant organism.

Host organism

Genotypic and phenotypic characteristics should be known (including, in the case of microorganism, any history of pathogenicity); as well as the presence of secondary metabolites or other potentially toxic and/or anti-nutritional components; and any history of use in food production.

Vector/inserted gene(s)

Sequence characterization, size, stability and mobility; the presence of resistance markers; any history of use in food production; and any potential allergenicity of the gene product(s).

Recombinant organism

Genetic stability; specificity of expression of the new gene(s); predicted secondary effects; levels of expression of known toxicants, anti-nutrients and potentially significant nutrients; and phenotype should be compared (agronomic traits, growth characteristics, metabolism, nutritional value, etc.) with the host organism and with other commercially important varieties of the species.

Method of production and/or preparation

For all novel foods it is necessary to describe the method of production and/or processing. It should be sufficiently detailed to enable consideration of potential effects on the composition of the novel food including major nutrients, toxicants and pathogens present in the food as well as contaminants and by-products that might be introduced by the process. Where particular instructions are required at a processing or domestic level to ensure the safe use of the novel food or ingredient, these should be given.

Specification

Full details of the specification of the novel food or ingredient are essential to ensure that the product marketed is the same as that upon which the safety evaluation is based. In all cases the specification should include gross composition. Other aspects that might require analysis should be decided on a case by case basis, taking into account the source of the novel food, it's a processing history and whether substantial equivalence is claimed. They will usually include toxins, nutrients, anti-nutritional factors or other secondary metabolites known to be associated with the source of the novel food or contaminants that might arise from the production process. Examples include: natural toxins and anti-nutritional factors in the case of plant based materials; nucleic acids, D-amino acids or odd carbon chain length fatty acids in microbial derived material; residues of catalysts or solvents in processed products or in the case of novel foods with a modified molecular structure, details of related structures posing potential risks. Where the presence of particular components might give rise to a significant toxicological or nutritional risk, safe limits should be specified. For those novel foods, which are expected to provide a significant dietary source of protein, fat, carbohydrate and/or minerals and vitamins, further detailed analyses will be required.

Purpose

Details of the rationale behind the development should be given, e.g. whether it is for technological reasons, to improve nutritional status of the diet or to reduce existing dietary risks. This information will help establish the role of the novel food in the diet as well as potential intakes and target groups.

Expected use

This should include details of how the product is expected to be processed, prepared and used. It should include the frequency and levels of use by the population as a whole and by particular target groups as well as identifying existing foods that it might be expected to replace in the diet and their nutritional impact in the diet. These intake data are needed for the risk or safety assessment of novel foods.

Some issues deserving special attention

Substantial equivalence

A central issue in the assessment of novel food ingredient safety is the fact that they are, to a certain extent, similar to ingredients already used in food. This idea of using traditional foods accepted as safe in use, as a basis for comparison in the safety assessment of novel foods, was developed into a concept to provide a practical approach to safety assessment of novel foods in 1992 by the Food Safety Working Group of the OECD (54) and a group of national experts for safety in biotechnology and food safety (35). The concept was intended to provide a practical approach to the safety evaluation of foods that are, or are produced from, genetically modified organisms, but it can be applied to other types of novel food. Within ILSI Europe this concept was further developed into three classes: (1) novel foods that are substantially equivalent to a traditional reference food with a history of safe use; (2) novel foods that are sufficiently equivalent; and (3) novel foods that are not equivalent to such traditional reference food. For a single, biochemically defined food or ingredient, substantial equivalence means biochemical identity within the limits of natural diversity of the traditional counterpart of commerce. For a complex food or ingredient, substantial equivalence means identity with traditional food in regard to composition, nutritional value, metabolism, intended use, and the level of undesirable substances contained therein within the limits of known and measured natural diversity of the traditional counter of commerce. In plants, secondary metabolites are important too, and should be compared with limits of natural diversity of the traditional counterpart. In case there is little or no equivalence, nutritional and toxicological studies should be performed. The OECD held a series of meetings to get international consensus on what compounds have to be analysed for food plants (54).

Although this concept appears to be a sensible approach, the limited knowledge that we have about traditional food can be problematic (36). This is the case for plant foods, as well as fungi. For higher plant species there is still little knowledge of the composition of secondary metabolites and whether they have any toxic properties. The limited analytical-chemical and toxicological data on traditional food will greatly restrict evaluation on the basis of substantial equivalence. In other words, if the basis for comparison in not available, it is virtually impossible to evaluate similarities or differences, implying that more nutritional and toxicological studies should be required. To a lesser extent this also holds true for fungi and the possible formation of mycotoxins.

History of safe use

The principle of history of safe use is an important element in the comparison of novel foods with traditional foods and the tiered approach for safety assessment. However, careful consideration should be given before qualifying a food as having a history of safe use. This qualification should be based on retrospective scientific research which addresses safety with respect to possible long-term toxic and nutritional effects. It should not be merely an anecdotal transfer of information. Frequently the knowledge about safe use is limited to acute toxic or nutritional effects, although there are a few exceptions for which some epidemiological studies have been performed (36,52).

Safety consideration of dietary DNA and RNA

As humans have with their food a daily intake of DNA/RNA from variable origin, and rDNA does not differ physically and chemically, the risk of the consumption of rDNA is not considered a safety concern as 95% of it will be digested. ILSI organised a workshop on this topic in 2000, and the outcome of this will be published later. Intake of functional DNA might cause concern about transfer to other cells or organisms. Therefore antibiotic resistance marker genes are mostly a separate safety issue.

Antibiotic resistance marker genes

In the evaluation of the use of marker genes in micro-organisms, particularly antibiotic resistance genes, it is important to understand the host organism, the biological constrainment of the marker gene by the kind of gene construct, the possibility that the genetically modified organism could colonise the gastrointestinal tract, and the connection between the efficacy of the antibiotics and the gained resistance (43). It is also important to know if the antibiotic is used and at what frequency it is used by humans (37,46,55).

Nutritional studies

A number of nutritional parameters are determined as part of the specification of the novel food, and the establishment of whether there is sufficient equivalence. The nutritional consequences of the novel food for the intended consumer should be assessed at normal and maximum probable levels of consumption. Even in traditional staple foods there is often a balance between components with a nutritional benefit and those with adverse effects, and the same is true for novel foods and ingredients. A balanced nutritional risk-benefit evaluation of a novel food or ingredient will need to take into account positive and negative nutritional effects arising to particular groups through novel food-induced nutrient excesses or shortfalls. The nutritional evaluation will have three elements; the composition of the novel food *per se*; the role of the novel food in the diet; and the food(s) in which it will be used. The background information on the novel food will include details of its nature, nutrient composition and purpose, and this will help establish its nutritional role in the diet as well as the foods in which it will be used.

The nutritional consequences for population subgroups such as children, the elderly, people dependent on institutional catering and those particularly susceptible to the novel food require particular consideration. Many of these groups are at risk in respect of certain nutrients.

Some novel foods, particularly some functional foods reduced in fat or intended for blood cholesterol lowering, might impact the bioavailability of certain nutrients. For those nutrients (proteins, lipids, carbohydrates, vitamins and minerals) which are identified as being of particular significance in relation to the introduction of a particular novel food, bioavailability studies may be necessary (35).

Toxicological studies

In the case of novel foods, which on the basis of the basic information are not shown to be substantially equivalent to their traditional counterpart, further toxicological information may be required. The nature and extent of any additional studies should be carefully selected, taking into account the source and composition of the novel food, its potential intake and whether it is intended for a specific application or for more general use in the diet. For example, toxicokinetic studies covering absorption, distribution, metabolism and excretion, genotoxicity studies both in vitro and in vivo, information on potential allergenicity and a sub-chronic (90-day rat study) toxicity study (4) may be appropriate. If the genotoxicity test reveals positive effects, a long-term carcinogenicity studies will be required. The necessity for other toxicity studies including second species toxicity studies, reproduction studies and carcinogenicity studies will depend on the level of concern. The concern level is determined by structure category assignment, the intended levels of human exposure, and, in the case of novel foods obtained by a novel process, the anticipated changes to the chemical structure of known food components.

A critical remark is needed about the value of toxicity studies with whole foods such as plant materials. Due to physical limitations, high dose levels of whole foods to determine toxic effects and to establish a no observed adverse effect level are not feasible. Toxicity studies using achievable dosages of whole foods are instead a kind of wholesomeness testing. Therefore toxicity studies should include as many relevant parameters as possible, including complete histopathology (36,42,56).

CONCLUSION

Novel methods of biotechnology have provided the means to introduce specific traits to microorganisms, plants, and animals used in food production. These methodologies, in fact, simply represent an improvement of methods of traditional breeding that have been used throughout the history of mankind, but are superior to those in terms of speed and adequacy. In the future, traditional breeding methodologies may seem too hazardous, due to the fact that genetic changes introduced were created by chance, and that the phenotypes selected could not be attributed to a well-defined genetic change. Still, the perception of consumers of a risk needs to be adequately considered. In fact, the public will more readily accept risks posed by nature (57) than man-made risks. Therefore, in addition to an assessment of the safety of novel foods, which is not different in principle from the safety assessment of traditional food items, there is a need to consider the needs of the public for authentic and nutritionally adequate food that is subjected only to minimal necessary processing (57). It is within this framework, both of providing adequate food safety practice and of meeting public needs, that novel methodologies can play a vital role in ensuring the food supply of our planet's population and of future generations.

In the safety assessment of novel foods, principles as substantial equivalence and sufficient equivalence in conjunction with a history of safe use may be applied, provided that they are based on facts and not just on anecdotal reports. Although particularly for complex food commodities the principle of substantial equivalence when applicable is the best manner to establish safety, the principle is not precisely defined. As the set of analyses performed can never be complete, they should give a representative fingerprint. For plants, secondary metabolites, irrespective of whether their toxic properties are known or not, should be included in the chemical analyses. In complex foods, it is difficult to establish the safety on the basis of toxicity studies, as the physical-chemical properties will limit the doses to administer. In such cases only confirmatory wholesomeness studies, which should be as complete as possible with respect to health parameters, can be performed. In practice the producer (notifier/sponsor) submits a mixture of data including demonstration of certain equivalence and confirmatory nutritional and toxicological studies. Novel analytical and molecular biologic methodologies might be developed and play an important role in ensuring the safety for now and for future generations.

REFERENCES

1. Advisory Committee on Novel Foods and Processes (ACNFP). *Guidelines on the assessment of novel foods and processes*. London: HMSO Publications Centre; 1991.
2. Nordic Council of Ministers. *Food and new biotechnology*. Copenhagen: Scantyk; 1991.
3. ACNFP. UK Advisory Committee on Novel Foods and Processes (ACNFP) report on herbicide tolerant soy beans. *ACNFP Annual Report*; 1994.
4. EC. EC-regulations novel foods. Recommendations concerning the scientific aspects of information necessary to support application for the placing on the market of novel food and novel ingredients. *Official Journal of the European Communities* Sept 1997; L253/1–36.
5. International Food Biotechnology Council (IFBC). Biotechnologies and food: assuring the safety of foods produced by genetic modification. *Regul Toxicol Pharmacol* 1990; 12:S1–196.
6. Anderson D, Cuthbertson WJF. Safety testing of novel food products generated by biotechnology and genetic manipulation. *Biotechnol Genet Eng Rev* 1987; 5:369–95.
7. World Health Organization. *Strategies for assessing the safety of foods produced by biotechnology*. Geneva: WHO; 1991.
8. Anonymous. Safety checks on single proteins. *Fd Chem Toxicol* 1984; 22:89–90.
9. Solomons GL. Mycoprotein: safety evaluation of a novel food. *Arch Toxicol* 1987; (Suppl 11):191–3.
10. Wu R, Grossman L, Moldave K, eds. *Recombinant DNA methodology*. New York: Academic Press; 1989.
11. Yamamoto M, Fukui S. *Agricult Biol Chem* 1977; 41:1829–30.
12. Panchal CJ, Russell I, Sills AM, Stewart GG. Genetic manipulation of brewing and related yeast strain. *Food Technol* 1984; 38:99–106.
13. Perry C, Meaden P. Properties of genetically-engineered dextrin-fermenting strain of brewer's yeast. *Int J Inst Brew* 1988; 94:64–7.
14. Suihko ML, Blomqvist K, Pentillä M, Gisler R, Knowles J. Recombinant brewer's yeast strains suitable for accelerated brewing. *J Biotechnol* 1990; 14:285–300.
15. Casey R, von Wettstein D, Petersen A. Constitution of baker's yeast secreting legume lipoxygenase during production of bread dough. In: Vassarotti A, Magnien E, eds. *Biotechnology R&D in the EC-Biotechnology action programme (BAP) 1985–1989, vol II*. Paris: Elsevier; 1990; 333–5.
16. Sandine WE. Genetic manipulation to improve food fermentation. *Dairy and food Sanitation* 1986; 6:548–50.
17. Venema G, Kok J. Improving dairy starter cultures. *Trends Biotech* 1987; 5:144–9.

18. Whitaker JR. New and future uses of enzymes in food processing. *Food Biotechnol* 1990; 4:669–97.
19. Rick CM, Deverna JW, Chetelat RT, Stevens MA. Potential contributions of wide crosses to improvement of processing tomatoes. *Acta Horticult* 1987; 200:45–55.
20. Gasser CA, Fraley RT. Genetically engineering plants for crop improvement. *Science* 1987; 244:1293–9.
21. Goodman RM, Hauptli H, Crossway A, Knauf VC. Gene transfer in crop improvement. *Science* 1987; 236:48–54.
22. Sharp WR, Evans DA, Arnmirato PV. Plant genetic engineering: designing crops to meet food industry specifications. *Fd Technol* 1984(Feb); 112–19.
23. Evans DA. protoplast fusion. In: Evans DA, Sharp WR, Arnmirato PV, Yamada Y, eds. *Handbook of plant cell culture*. New York: MacMillan; 1983; 291–321.
24. Hilder VA, Gatehouse AMR, Sheerman SE, Barker RF, Boulter D. A novel mechanism of insect resistance engineered into tobacco. *Nature* 1987; 200:160–3.
25. Fishhoff DA, Bowdish KS, Perlak FJ, *et al.* Insect tolerant transgenic tomatoplants. *Bio/technology* 1987; 5:807–13.
26. Nelson RS, McCormick SM, Delannay X, *et al.* Virus tolerance, plant growth, and field performance of transgenic tomato plants expressing coat protein from tobacco mosaic virus. *Bio/technology* 1988; 6:403–9.
27. De Bolck M, Botterman J, Vandewiele M, *et al.* Engineering herbicide resistance in plants by expression of a detoxifying enzyme. *EMBO J* 1987; 6:2513–18.
28. Shah DM, Horsch RB, Klee HJ, *et al.* Engineering herbicide tolerance in transgenic plants. *Science* 1986; 233:478–81.
29. Gordon JN, Scangos GA, Plotkin DJ, Barbosa JA, Ruddle FH. Genetic transformation of mouse embryos by microinjection of purified DNA. *Proc Natl Acad Sci USA* 1980; 77:3780–4.
30. Van Brunt J. Molecular farming: transgenic animals as bioactors. *Bio/technology* 1988; 6:1149–54.
31. Jaenisch R, Transgenic animals. *Science* 1988; 240:1468–74.
32. Church A, Simons J, Wilmut IP, Lathe R. Pharmaceuticals from transgenic live stock. *TrendsBiotech* 1987; 5:20–4.
33. Knudsen I. Potential food safety problems engineering. *Regul Pharmacol Toxicol* 1985; 5:405–9.
34. Lindemann J. Biotechnologies and food. A summary of major issues regarding safety assurance. *Regul Pharmacol Toxicol* 1990; 12:96–104.
35. Jonas DA, Antignac E, Antoine JM, Classen HG, Huggett A, Knudsen I, Mahler J, Ockhuizen T, Smith M, Teuber M, Walker R, De Vogel. The safety assessment of novel foods. *Food and chemical toxicology*, 1996; 34:931–40.
36. Speijers GJA, Younes M, Van der Heijden CA. Novel Foods. In: Kees van der Heijden, Maged Younes, Larry Fishbein, Sanford Miller, eds. *International Food Safety Handbook*. New York, USA: Marcel Dekker Inc; 1999; 381–96.
37. Jonas D. Safety assessment of viable genetically modified micro-organisms used in food. *Microbial Ecology in Health and Disease* 2000; 1:198–207.
38. Parish WE. Toxicological evaluations of biotechnology food products. *Fat Sci Technol* 1988; 90:93–6.
39. Egmond van HP, Speijers GJA. Natural toxins I. Mycotoxins. In: Kees van der Heijden, Maged Younes, Lawrence Fishbein, Sanford Miller, eds. *International Food Safety Handbook*. New York, USA: Marcel Dekker, Inc; 1999; 341–56.

40. ILSI Europe. A simple guide to understandiing and applying the Hazard Analysis Critical Control Point Concept (HACCP). *ILSI Europe Concise Monograph Series* 1993; 1–12.

41. Fuchs RL, Astwood JD. Allergenicity Assessment of Foods Derived from Genetically Modified Plants. *Food Technology* 1996(Feb); 83–8.

42. Speijers GJA. Toxicological data needed for safety evaluation and regulation on inherent plant toxin. *Natural Toxins* 1995; 3:222–6.

43. Berkowitz DB. The food safety of transgenic animals. *Bio/technology*. 1990; 8:819–25.

44. Acuff GR, Albanese RA, Batt CA, *et al*. Implication of biotechnology, risk assessment and communications for the safety of foods of animal origin. *J Am Vet Med Assoc* 1991; 199:1714–21.

45. FAO/WHO. *Safety aspects of genetically modified foods of plant origin*. Report of a joint FAO/WHO Expert Consultation on Foods Derived from Biotechnology. Geneva, Switzerland: WHO. 29 May–2 June 2000; 1–17.

46. ILSI. Safety Assessment of Viable Genetically Modified Micro-organisms Used in Food. Consensus Guidelines reached at a Workshop held in April 1999. *International Life Science Institute (ILSI) Europe Report Series*. Brussels, Belgium; 1999.

47. International Programme on Chemical Safety (IPCS). Principle for the safety assessment of food additives and contamination in food. *Enviromental Health Criteria* Nr 70. Geneva, Switzerland; 1987.

48. EHC. Assessing human health risk of chemicals; Derivation of guidance values for health-based exposure limits. *Environmental Health Criteria (EHC)* Nr 170. Geneva, Switzerland: WHO; 1996.

49. Moseley BEB. Control of novel, including genetically engineered, foods in the United Kingdom. *Food Control* 1991; 2:199–201.

50. Kessler DA, Taylor MR, Maryanski JH, Flamm EL, Kahl LS. The safety of foods developed by biotechnology. *Science* 1992; 256:1747–832.

51. Barros C, Segura I. Criterios sobre el possible concepto de alimento nuevo. *Alimentaris* 1990; 27:15–20.

52. Kleter GA, Noordam MY, Kok EJ, Kuiper HA. New developments in crop plant biotechnology and their possible implications for food product safety; Literature study under commission of the Foundation "Consument en biotechnologie". *RIKILT report 2000.004*. 2000, State Institute for Quality Control of Agricultural Products (RIKILT), Wageningen, The Netherlands.

53. Knudsen I, Overse L. Assessment of novel foods: a call for a new and broader GRAS concept. *Regulatory Toxicology and Pharmacology* 1995; 21:365–9.

54. OECD-report of the OECD Workshop on the Toxicological and Nutritional Testing of Novel Foods. Aussois, France, 5–8 March 1997. *OECD-report* 1997. Paris, France.

55. Kärenlampi S. Health effects of marker gene in genetically engineered food plants. *Temanord*; Nordic Council of Ministers, Copenhagen, Danmark; 1996; 530.

56. Speijers GJA, Van Egmond HP. Inherent plant toxins. In: Kees van der Heijden, Maged Younes, Larry Fishbein, Sanford Miller, eds. *International Food Safety Handbook*. New York, USA: Marcel Dekker Inc; 1999; 369–80.

57. Busch L. Biotechnology: consumer concerns about risks and values. *Food technol* 1991; 45:96–101.

11

SAFETY, EFFECTIVENESS AND LABELING OF SELF-CARE PRODUCTS

R. William Soller

INTRODUCTION

Self-care is a well developed, yet still evolving, practice in the United States. "Self-care means responsibly taking charge of personal well-being through the application of professional and self-diagnosis, as well as self-recognition of emergent signs and symptoms, the use of conventional and alternative medical modalities (e.g. Rx and OTC drugs and devices, dietary supplements, acupuncture, etc.), and the institution of appropriate life-style changes (e.g. balanced diet, exercise, etc.)." (1,2).

The last 25 years of the Twentieth Century have been called the age of self-care empowerment (3). The seeds of this revolution in health care started with the 1962 amendments to the drug law, requiring pre-marketing proof of effectiveness for drugs.[1] As a result, FDA initiated the OTC Review in 1972, using 17 advisory panels to assess the safety and efficacy, and make recommendations about the labeling, of about 730 active ingredients used in more than 300,000 US drug products, with the net result that consumers have a high level of confidence in OTC medicines. On a parallel track, the consumer became more sophisticated, health-conscious, educated and self-reliant, as consumer empowerment helped break the barriers of medical paternalism. Consumerism strengthened, evidenced by the massive out-pouring from 2 million consumers to their Congressional representatives to support the passage of the Dietary Supplement Health and Education Act in 1994 (4).

Successful marketing of self-care products relies on the confidence of today's consumers. This confidence is born of the scientific evidence supporting the safety, effectiveness and labeling of self-care products. As laws and regulations affecting OTC drugs and dietary supplements have evolved over time, similarities

---

1 Prior to 1962, drugs had to be proven safe, but proof of their effectiveness was not required by law.

and differences in the underlying scientific/regulatory principles and scientific documentation for dietary supplements and OTC drugs have emerged. This chapter examines these similarities and differences.

OTCness

Legal basis of OTC drug availability

OTC availability has been described as "OTCness," the widespread availability of safe and effective nonprescription medicines for responsible self care by the consumer according to label directions, pursuant to the applicable laws, regulations, and voluntary industry codes affecting manufacturing, packaging, labeling, distribution, and sales of quality products and the advertising of those products in all media (2). Although the scientific/regulatory paradigm and regulatory policy for determining the OTCness of an ingredient was developed through the OTC Review, Rx-to-OTC switch has contributed importantly to the enhancement of "OTCness" through FDA/industry interactions in the R&D process and in the public deliberations of switch candidates at FDA advisory committee meetings.

The legal basis for OTCness is 21 USC 353(b)(1) of the Food Drug and Cosmetic Act (FD&C Act), which mandates that a drug is OTC unless it must be dispensed by prescription because it is habit-forming [to which 21 USC 352(d) applies], is not safe to use except under professional supervision due to its toxicity or other potentiality for harmful effect or the collateral measures necessary for its use, or is limited by an approved application under 21 USC 355 to use under supervision of a practitioner licensed by law to administer such a drug. Hence, drugs are prescription by exception, based chiefly on safety considerations.

While drugs are intended for use in the diagnosis, cure, mitigation, treatment, or prevention of disease,[2] some categories of intended use of OTC drugs are natural conditions (i.e. not diseases *per se* but categories of historical use of OTC medicines, including, for example, menstrual pain, hair loss due to aging, noncystic acne, and irregularity). These "natural conditions" (or states) are areas of "claims overlap" for OTC drugs and dietary supplements as a result of FDA's final rule on structure/function claims.

---

2 Section 321 (g)(1) of the FD&C Act defines "drug" as "(A) articles recognized in the official United States Pharmacopoeia, official Homoeopathic Pharmacopoeia of the United States, or official National Formulary, or any supplement to any of them; and (B) articles intended for use in the diagnosis, cure, mitigation, treatment, or prevention of disease in man or other animals; and (C) articles (other than food) intended to affect the structure or any function of the body of man or other animals; and (D) articles intended for use as a component of any article specified in clause (A), (B), or (C). A food for which a claim, subject to sections 343(r)(1)(B) and 343(r)(3) of this title or sections 343(r)(1)(B) and 343(r)(5)(D) of this title, is made in accordance with the requirements of section 343(r) of this title is not a drug under clause (B) solely because the label or labeling contains such a claim."

Scientific documentation as the basis for OTCness

One of the most important regulatory developments stemming from the OTC Review is FDA's longstanding OTC warning policy (5). Under this policy, FDA has broad authority under the FD&C Act to require warnings on OTC labels, or remove a product from OTC availability. Under the Act, an OTC may be found to be misbranded and thus subject to regulatory action (including seizure) if its label does not contain "adequate warnings against use in those pathological conditions or by children where its use may be dangerous to health or against unsafe dosage or methods or duration of administration or application in such manner and form as are necessary for the protection of users." To implement its broad authority, FDA has established as a matter of regulatory policy that: (a) adequate warnings must appear, as needed and deemed appropriate, in OTC labeling; (b) the consumer should be made aware through labeling of the potential risks of OTC drug use; and (c) only essential information should appear on the label.

FDA has further elaborated its warning policy as follows: "warnings should be scientifically documented, clinically significant, and important to the safe and effective use of the product by the consumer (6)." The first criterion, "scientific documentation," means in this context the recording of information or data in a scientifically reasonable and responsible manner, including lack of bias by study investigators, controlling for confounding variables in epidemiological studies, and use of validated research tools (e.g. questionnaires), and other aspects of good scientific practice. The second and third criteria, "clinically significant" and "important to the safe and effective use by the consumer," require that a particular scientific finding has meaning from both a clinical and a use standpoints. By far, the first criterion, "scientific documentation," is the foundation for assessing safety and effectiveness of OTCness, and provides the basis to prevent, for example, broad generalizations from a limited uncontrolled case series of adverse experience reports from being the sole basis for regulatory action.

Importantly, the OTC Review also manifested the important scientific/regulatory paradigm used to assess benefit-risk of OTCness through scientific documentation – i.e. the case-by-case, weight-of-the-evidence, data-driven, dialogue-driven approach (7). This approach was developed through the 17 advisory panel comprising FDA's OTC Review, which were charged by FDA as follows: "*In every instance*, the panel must evaluate whether, balancing the benefits against the risks, the target population will experience a beneficial rather than a detrimental effect" (emphasis supplied) (8). This paradigm has stood the test of time and allows each generation of advisory committees dealing with OTCs to use the best science of the day, as the OTCness of a particular self care category evolves with new findings and understandings, and to assess the uniqueness of each OTC ingredient in its special self-care setting (7). This approach is scientifically sound and should be the core of any self-care ingredient decision. Interestingly for dietary supplements this approach is encompassed in the statute (see p. 233) (21).

Safety, effectiveness and labeling as the cornerstones of OTCness

Although a benefit/risk assessment for assessing the OTCness of a particular ingredient or product is not defined by regulation, it is in practice founded on the regulatory definitions of safety, effectiveness and labeling.

Safety is not defined as "no risk," but rather as "a low incidence of adverse reactions or significant side effects under adequate directions for use and warnings against unsafe use as well as low potential for harm which may result from abuse under conditions of widespread availability ... with proof of safety [consisting] of adequate test by all methods reasonably applicable to show the drug is safe under the recommended conditions of use defined in labeling." (9). Elements of the safety component of the benefit/risk decision include carcinogenic potential; reproductive toxicity potential; side effects relating to the inherent toxicity of the ingredient (e.g. gastric irritation), which should be of low incidence in the general population and of low severity; potential therapeutic hazards, including the potential for misdiagnosis of disease through masking of symptoms or therapeutic suppression of emergent signs or symptoms or signs of toxicity of other drugs, potential treatment failure, and potential incorrect use of the drug, including overdose, misuse, and abuse potential; and, drug–drug (or food) interactions. A finding of either carcinogenicity or reproductive toxicity is not necessarily an impassable hurdle to OTCness, since such toxicity may not be relevant to human use (10). Depending on the nature of side effects, label warnings may be used to caution users, as in the case of different drowsiness warnings for different classes of OTC antihistamines (11). Generally, through the use of indications, directions of use, and warnings, OTC labeling is considered an adequate safeguard against potential therapeutic hazards and incorrect use of the drug.

Effectiveness is not defined in the absolute, but rather as "a reasonable expectation that, in a significant proportion of the target population, the pharmacological effect of the drug, when used under adequate directions for use and warnings against unsafe use, will provide clinically significant relief of the type claimed ... with proof consisting of controlled clinical investigations." (12). Generally, controlled clinical investigations serve as the principal basis of OTCness, with design issues worked out by the sponsor with the Center for Drug Evaluation and Research (CDER) review divisions, in the context of current clinical study guidelines available.

OTC *labeling* is defined as being "clear and truthful in all respects and may not be false or misleading in any particular ... [stating] the intended uses and results of the product; adequate directions for proper use; and, warnings against unsafe use, side effects, and adverse reactions in such terms as to render them likely to be read and understood by the ordinary individual, including individuals of low comprehension, under customary conditions of purchase and use." (13). While the OTC Review defined what elements should be on the label, it was not until 1999 that FDA finalized a rule on the standardized format and language simplification for the OTC label (14), creating requirements for the order of information, specified

231

headings and subheadings, type size, and interchangeable terms. In part, this labeling rule was developed through the experience of labeling requirements for Rx-to-OTC switch products which, because of their increasing complexity and newness in the self-care arena, are the subject of specialized label comprehension studies prior to approval. FDA has no guidelines for label comprehension studies, leaving it to companies to derive knowledge from studies presented by sponsors at FDA advisory committee meetings on a switch candidate.

For Rx-to-OTC switch candidates, FDA often requires an actual use study, which is defined as "a controlled experiment in which a prescription drug or unapproved new drug is used by subjects under OTC-like conditions." (15). Actual-use studies provide information on product self-selection, use of the proposed label, effectiveness and compliance of projected OTC drug use in a simulated OTC setting. Such studies often assess active vs. placebo and/or other currently available Rx or OTC therapies, ranging in size from several to many thousands of patients. In an extreme example, where there was a question about possible rare side effects, the actual-use trial for pediatric ibuprofen included approximately 80,000 children with febrile illnesses to assess the comparative organ-specific side effects of acetaminophen and ibuprofen (16).

In summary, the triad of safety, effectiveness and labeling – as defined by regulation – are generally regarded as the cornerstones of OTCness and thus the basis of Rx-to-OTC switch decisions and post-marketing deliberations on OTC availability and potential labeling changes to currently marketed products. All such deliberations and decisions are made on a case-by-case, weight-of-the-evidence, data-driven (i.e. scientifically documented), dialogue-driven basis – a scientific/ regulatory paradigm that has stood the test of time and helped build consumer confidence in OTC medicines.

DIETARY SUPPLEMENTS

The regulatory environment for the dietary supplement component of the self-care field is still in a maturation phase, with FDA – some five years after the passage of DSHEA – only recently announcing its long-range plan for "a science-based regulatory program that fully implements [DSHEA], thereby providing consumers with a high level of confidence in the safety, composition, and labeling of dietary supplement products." (17). FDA can reasonably predict its long-range goal, since there is an established legal framework for dietary supplements that includes, depending on the newness of the ingredient and the type of claim, pre-marketing notification; post-marketing compliance; and pre-marketing approval.

As the environment for dietary supplements evolves through stakeholder/FDA interactions on general policy and ingredient-specific issues, the nature and extent of the scientific documentation by companies to support product claims and by FDA in the assessment of potential safety issues will increasingly come under scrutiny and will evolve. This was true for the OTC drug category of self-care, as the field transitioned from the pre-OTC Review era through the OTC Review to the current NDA approach for new product development. Similarly, the maturation of the OTC drug sector of self-care included the development of an increasingly data-driven competitive claims arena outside of FDA but within the realm of FTC over-sight and enforcement. With the expansion of any market sector to mainstream consumers and the consumer demands and expectations for quality products mak-ing truthful claims, the pressure on manufacturers to deliver on those demands escalates, with public emphasis on scientific documentation. This can be expected of dietary supplements in the context of current law and developing regulations.

Legal standard for dietary supplement availability

There is an established legal framework for dietary supplements, as there is for OTC drugs, with dietary supplements defined in relation to their statutory defini-tion as foods. Specifically, DSHEA exempts dietary supplement ingredients from the food additive provisions of the Food, Drug and Cosmetic Act and establishes conditions for the marketing of new dietary ingredients not marketed in the United States as dietary supplements prior to October 15, 1994. The provisions of DSHEA define dietary supplements and dietary ingredients; establish a new framework for assuring safety of dietary supplements;[3] outline guidelines for lit-erature displayed at retail; provide for use of claims and nutritional support state-ments; require ingredient and nutrition labeling; and grant FDA the authority to establish good manufacturing practice (GMP) regulations. The law also required formation of an executive level Commission on Dietary Supplement Labels and an Office of Dietary Supplements within the National Institutes of Health (18).

DSHEA defined a dietary supplement as "a product (other than tobacco) that is intended to supplement the diet that bears or contains one or more of the follow-ing dietary ingredients: a vitamin, a mineral, an herb or other botanical, an amino acid, a dietary substance for use by man to supplement the diet by increasing the total daily intake, or a concentrate, metabolite, constituent, extract, or combina-tions of these ingredients; is intended for ingestion in pill, capsule, tablet, or liquid form; is not represented for use as a conventional food or as the sole item of a meal or diet; is labeled as a 'dietary supplement;' includes products such as an approved new drug, certified antibiotic, or licensed biologic that was marketed as

---

3 DSHEA amends the adulteration provisions of the FD&C Act, such that a dietary supplement is adulterated if it or one of its ingredients presents "a significant or unreasonable risk of illness or injury" when used as directed on the label, or under normal conditions of use (if there are no direc-tions). A dietary supplement that contains an ingredient not marketed for dietary supplement use in the US prior to 15 October 1994, is a new dietary supplement and may be adulterated when there is inadequate information to provide reasonable assurance that the ingredient will not present a sig-nificant or unreasonable risk of illness or injury. Manufacturers must notify FDA at least 75 days before marketing products containing new dietary ingredients, providing information supporting a conclusion that the new dietary ingredient "will reasonably be expected to be safe."

a dietary supplement or food before approval, certification, or license (unless the Secretary of Health and Human Services waives this provision)." (19). As such, dietary supplements are foods and may not therefore treat, cure, mitigate, diagnose or prevent a disease However, DSHEA also amends the Food Drug and Cosmetic Act to carve out dietary supplements from the drug definition if such supplements make health (or disease risk reduction) claims, such as folate for the prevention of neural tube defects, solely because the label or labeling contains such a claim [see 21 USC Sections 321 and 343(r)].

Scientific documentation of dietary supplement safety

In reflecting on potential safety issues associated with the passage of DSHEA, Congress concluded that "dietary supplements are safe within a broad range of intake, and safety problems of supplements are relatively rare; and although the Federal government should take swift action against products that are unsafe or adulterated, it should not take any actions to impose unreasonable regulatory barriers limiting or slowing the flow of safe products and accurate information to consumers." (20). Thus, FDA retains substantial enforcement authority post-DSHEA, and may deem a dietary ingredient adulterated if it is or contains an ingredient that: "presents a significant or unreasonable risk of illness or injury under conditions of use recommended or suggested in labeling or, if no conditions of use are suggested or recommended in the labeling, under ordinary conditions of use; is a new dietary ingredient for which there is inadequate information to provide reasonable assurance that such ingredient does not present a significant or unreasonable risk of illness or injury; ... is or contains a dietary ingredient that renders it adulterated ... under the conditions of use recommended or suggested in the labeling of such dietary supplement." (21).

However, while DSHEA gives FDA broad enforcement authority, the law also stipulates that "the United States shall bear the burden of proof on each element to show that a dietary supplement is adulterated." Further, FDA must show affirmatively, in court, that an unreasonable risk is posed by consumption of a dietary supplement. Yet even so, the manufacturer bears the primary responsibility for assuring the safety of dietary supplements, both under the Act and as a matter of good business practice. Further, the case-by-case scientific regulatory paradigm developed through the OTC Review for nonprescription medicines is applied by law for dietary supplements. DSHEA specifically states: "[T]he court shall decide any issue under this paragraph on a de novo basis," with the person against whom such proceeding might be initiated having appropriate notice and the opportunity to present views, orally and in writing, at least 10 days before such notice, with regard to such proceeding (21).

Dietary supplements are grandfathered in terms of requirements for pre-marketing notification of FDA, if they have been marketed prior to 15 October 1994, are present in the food supply as an article used for food in a form in which the food has not been chemically altered, or there is a history of use, or for which some other evidence of safety exists that establishes that there is a reasonable expectation of safety when the product is used according to recommended conditions of use. New uses of an existing dietary supplement, or an increase in the recommended dose, does not make a supplement "new" for purposes of the substantiation requirement. For products containing new dietary supplements introduced after 15 October 1994, FDA must be notified 75 days prior to marketing, including in the notification information supporting a conclusion that the new ingredient "will reasonably be expected to be safe." Any interested party, including a manufacturer of a dietary supplement, may petition FDA to issue an order prescribing the conditions of use under which a new dietary ingredient will reasonably be expected to be safe. FDA may object, on grounds of inadequate information to provide reasonable assurance that the proposed new ingredient does not present a significant or unreasonable risk of illness or injury, with the burden of proof resting with the Federal government (22). There is no guidance to industry as to what constitutes an appropriate "safety package" for the support of new dietary supplement ingredients, so it is prudent for those interested in dietary supplement product development to follow the notification docket for denial letters to 75-day notification letters. As a final note, an alternate route to the market for a dietary supplement, which was available prior to DSHEA, is affirmation of GRAS status of substances as provided for in 21 CFR section 570.30(e). Generally, because dietary supplements are no longer food additives under DSHEA, but foods, and because of the relatively complex route to approval for GRAS affirmation which includes agreement by qualified experts, the 75-day notice appears to be the preferred route for new dietary supplement submissions.

In summary, a system was created under DSHEA that grandfathers dietary supplements marketed pre-DSHEA and notifies FDA of the intended marketing of new post-DSHEA dietary supplements, while maintaining an ability for FDA to protect the public should an unreasonable scientifically-documented hazard arise. If the OTC Drug Review is a foreshadowing of how a self-care sector develops, then with time and through experience with post-marketing safety issues as well as the pre-marketing notification and GRAS affirmation processes for dietary supplements, this dietary supplement system can be expected to be increasingly data-driven, requiring companies to be increasingly scientific in their approach to the agency. Currently, there are no guidelines for the scope and extent of data needed to support the safety of dietary supplements, although one would expect that, as with OTCs, safety of dietary supplements must be considered on a case-by-case, data-driven basis, using similar scientific principles of toxicology and epidemiology as are used for all self-care products.

Application of FDA's OTC warning policy to dietary supplements

As stated earlier, the OTC Review helped solidify a policy that is at the core of safety decisions for OTC drug ingredients. That policy is that label warnings, or as a practical matter the decision about ingredient availability, should be scientifically documented, clinically significant and important to the safe and effective use of the product by the consumer. This policy has also been applied across other FDA regulated product categories, including foods.

For example, as stated by Richard Merrill in relation to an FDA advisory committee deliberation of the safety of an OTC antihistamine (23), FDA has maintained a consistent policy across OTC and food products, such that "if the evidence as a whole does not suggest that an OTC ingredient poses a significant risk, no mention in labeling is required; if the risk is significant, unrestricted availability to consumers is not appropriate." At the core of this determination is the nature, extent and human applicability of the scientific documentation challenging an ingredient's safety. Merrill cites FDA's actions on certain human nutrients that, when isolated and administered to laboratory animals in enormous quantities represented by the maximum tolerated dose, pose significant toxicity, but in the context of an appropriate risk assessment pose no significant risk to humans, therefore requiring no labeling.

In view of the rationale, reasonable, scientifically-sound, and time-proven construct of FDA's warning/availability policy, it should remain at the cornerstone of any decisions about the safety of all self-care products, including both dietary supplements and OTC drugs.

Benefit claims and scientific substantiation

In passing DSHEA, Congress acted to protect "the right of access of consumers to safe dietary supplements" and to remove "unreasonable regulatory barriers limiting or slowing the flow of safe products and accurate information to consumers" that had been imposed by FDA (24). DSHEA established the ability of companies to use the product label and advertising and promotion for direct-to-the-consumer communication about health promotion and maintenance through responsible dietary supplement use. This direct communication was demanded by the consumer, as evidenced by the massive write-in support from constituents to their representatives during the passage of DSHEA. The resulting array of claims that may be made for dietary supplements has thus become quite broad, including product quality claims, nutrient content claims, nutrient deficiency claims, structure/function claims, and health claims.

For all of these claims, there is a post-marketing regulatory enforcement system in place to ensure that the claims are truthful and not misleading. In the case of health claims, FDA approves the claim prior to marketing. Irrespective of the manner whereby the claim is generated for use on product labeling or in advertising, sound scientific documentation is needed to support the claim. However, the scientific documentation used to support claims may vary in relation to the nature of the claim.

For example, for quality claims (e.g. purity, origin, percentage of active constituents or marker compounds, flavor, storage information, etc.) there should be substantial support from validated technical methods and procedures. The analytical methods relating to product quality may derive from United States Pharmacopeia, AOAC International (formerly the Association of Official Analytical Chemists), other compendia, FDA validated methods, or company-validated methods. Although not discussed above for OTC drugs, comparable scientific rigor in developing laboratory-based technical claims on quality would be expected for both OTC drugs as for dietary supplements.

On the other hand, with respect to structure/function claims, it was the intent of Congress that consumers would have more information about the possible health benefits of dietary supplements. As noted by Commissioner Henney: "... [R]egarding labeling, DSHEA seeks to provide consumers with information to help guide personal choice" (25). Since structure/function claims focus on the maintenance and/or promotion of health and general well being, and the scope and extent of the level of support may vary depending on the strength of the association stated in the claim (see also the report of the Commission on Dietary Supplement Labels) (20). DSHEA requires supplement manufacturers to have substantiation of such label claims and to notify FDA within 30 days after first marketing a product with a statement of nutritional support that such a statement is being made. The label must also carry a disclaimer "prominently displayed and in boldface type" that states: "This statement has not been evaluated by the Food and Drug Administration. This product is not intended to diagnose, treat, cure, or prevent any disease."

Further, in the case of health claims, either approved based on a submission from interested party(ies) or sanctioned based on an statement from an authoritative body (e.g. Institute of Medicine), the extent of the scientific documentation needed to make the claim may vary, provided the nature of the link between a substance and a disease is adequately expressed in a disclaimer or qualifier. The example of folate and neural tube defects represents the type of claim that is at one end of this claims spectrum, where the relationship is established to the point that it is likely to not be reversed by evolving science. At the other end of this spectrum are the type of claims that are currently being defined in a regulatory context by FDA, as a result of the *Pearson v. Shalala* (26) court decision that requires FDA to consider permitting health claims with qualifiers, based on First Amendment rights of freedom of commercial speech.

Whatever type of claim is included in the labeling and advertising of a particular dietary supplement product, it is important to be aware of at least three sources of competent information about how to evaluate scientific literature and how to consider whether a claim is truthful and not misleading. The first source is the document prepared by the Foods Advisory Committee (FAC) Working Group at the request of FDA for the purpose of helping to define significant scientific agreement. This document, which was issued by the FAC Working Group and attached to FDA's 22 December 1999 guidance on significant scientific agreement, provides an elementary exposition of principles for evaluating scientific research. The recommendations developed by the FAC Working Group (27) on how to approach the scientific literature to come to a weight-of-the-evidence evaluation provides a valuable perspective, given its apolitical tone and rational and credible approach.

The second source of information derives from the report of the Commission on Dietary Supplement Labels (20). The Commission made recommendations for what information should be in a substantiation file for structure/function claims: Notification Letter; identification of dietary supplement ingredients; evidence to substantiate statements of nutritional support; evidence to substantiate safety; Good Manufacturing Practices; and qualifications of reviewers (22). With respect to scientific documentation, the Commission was quite specific in its recommendations as to "substantiation:"

> "*Evidence to Substantiate Statements of Nutritional Support*: Such evidence should include copies of key references to experimental or clinical data and/or findings of authoritative bodies and other evidence, where appropriate. References should include relevant information, positive or negative. Research or monographs from appropriate foreign sources may be cited, along with evidence that specific uses or claims are approved in other countries. An interpretive synopsis by an individual(s) or group qualified by training and experience to evaluate the evidence should accompany the literature citations and should assess clearly the evidence supporting the statement. Evidence for efficacy should include the dosage at which effects are observed. Where historical use is cited as the evidence for a statement, the composition of the product should correspond with the material for which such claims of historical use may be made. The complexity of a product may affect the substantiation required."

> "*Evidence to Substantiate Safety*: The Commission believes safety is of primary concern in marketing dietary supplements, and the file should indicate the basis of the manufacturer's conclusion that the product can reasonably be expected to be safe at levels of intended use."

The third source of important guidance for those interested in generating claims for dietary supplements is the Federal Trade Commission (FTC). While FDA is principally responsible for claims on the product labeling (i.e. packaging, inserts, and other promotional materials distributed at the point of sale), FTC has primary responsibility for claims in advertising (e.g. print and broadcast ads, infomercials, catalogs, websites, and similar direct marketing materials). FTC recently issued an advertising guidance to the dietary supplement industry in which FTC elaborates its long-standing advertising principles and policies defined in various consumer product categories (28). As noted in the guidance, FTC's truth-in-advertising law can be boiled down to two common-sense propositions: (a) advertising must be truthful and not misleading; and (b) before disseminating an advertisement, advertisers must have adequate substantiation for all objective product claims. The guidance amplifies FTC's view on express and implied claims, when to disclose qualifying information, what constitutes clear and prominent disclosure, and how to interpret the relevance of the evidence to the claim, including considerations of the amount/type, quality, and totality of the evidence, as well as claims based on testimonials, traditional use, third party literature and the FDA disclaimer for structure/function claims.

In summary, consumer confidence is the most important product in the self-care sector – i.e. confidence that the dietary supplement product will deliver on its label and advertising claims, confidence in the quality of the dietary supplement product, and confidence that the dietary supplement product contains the amount and type of ingredients claimed. Hence, basing claims of a product's benefits and/or quality on credible, scientifically-based evidence is a sound approach to building the dietary supplement field.

POST-MARKETING RISK MANAGEMENT OF SELF-CARE PRODUCTS

Post-marketing surveillance for emergent safety and quality issues is an important public health practice and applies to both nonprescription medicines and dietary supplements. In the self-care sector, there is mandatory submission of adverse events reports (AERs) for new OTC drugs covered under a New Drug Application or Supplement, but not for older OTC drugs, which are covered under the OTC Review and have an extensive market history of use. For dietary supplements there is also no mandatory reporting of AERs, like that for OTC drugs covered under the OTC Review.

However, for both OTC drugs and dietary supplements, both FDA and FTC have surveillance systems for consumer and health professional reports on product safety, effectiveness, quality, and labeling. FDA's MedWatch program (29) accepts reports on both OTCs and dietary supplements, with "Safety Information Summaries" on dietary supplement and drug products published periodically on the MedWatch website. The Special Nutritional Adverse Events Monitoring System, maintained by FDA's Center for Food Safety and Applied Nutrition (CFSAN), is an on-line index of AERs from health professionals and consumers on dietary supplements, infant formulas and medical foods (30). CFSAN has placed refinement of its AER systems as a top priority in its Year 2001 program activity.

FTC also actively follows consumer and health professional complaints on self-care products, maintaining an online complaint form, a consumer hotline, and a system for sharing results of its post-marketing surveillance systems with state Attorneys General and other law enforcement agencies (31).

FDA collects information from spontaneous reports of AERs associated with self-care products as well as the published scientific and medical literature in their on-going assessment of product safety and labeling. For example, during 1984 and 1985, seven cases of esophageal obstruction caused by the swelling of tablets containing glucomannan were reported to the Australian Adverse Drug Reactions Advisory Committee (32), with esophagoscopy needed to remove the obstruction in five cases. Between June 1988 and August 1989, FDA's spontaneous reporting system received 16 AERs of esophageal obstruction reportedly associated with a product containing 500 mg guar gum per tablet, and between 1975 and 1989, FDA received 61 AERs of esophageal obstruction from OTC laxative drug products containing a high concentration of psyllium, among other reports. This compilation of AERs led to FDA's proposal and then requirement for a warning and new directions of use for OTC drug products containing water soluble gums pertaining to adequate intake of water during use.

There is also a recent example from the dietary supplement field of post-marketing surveillance as an important monitor of product safety. The recent NIH study on potential drug-dietary supplement interactions between St. John's wort and indinavir, a protease inhibitor used to treat HIV infection, led FDA to issue a Public Health Advisory on 10 February 2000 (33) and then to the adoption of a voluntary labeling program by the Consumer Healthcare Products Association that places the following label statement on St. John's wort-containing supplements: "If you are taking a prescription drug, ask a health professional." (34).

In both the OTC drug and dietary supplement examples cited above, post-marketing surveillance led to important label changes to help ensure that self-care products are safe when used in accordance with label directions. Given the importance of AER sentinel systems as a means to help ensure ongoing product safety, dietary supplement companies and others should be aware of the current AER systems and how to utilize the output of such systems.

CONCLUSIONS

The Congressional findings reported in DSHEA state that "preventive health measures, including education, good nutrition, and appropriate use of safe nutritional supplements will limit the incidence of chronic diseases, and reduce long-term health care expenditures" and that "consumers should be empowered to make choices about preventive health care programs based on data from scientific studies of health benefits related to particular dietary supplements." Successful growth of the dietary supplement market is dependent on consumer confidence in these self-care products, which derives from the scientific documentation supporting their safety and claimed benefits. Thus, for the promise of DSHEA to be fully realized, industry has the responsibility to ensure the appropriate scientific documentation is applied to the safety of, and the claims that are made for, dietary supplements.

ACKNOWLEDGEMENT

I wish to thank Patrice B. Wright, Ph.D., formerly of CHPA, for her review and comments on the manuscript.

REFERENCES

1. Soller RW. Evolution of self-care with over-the-counter medications. *Clinical Therapeutics* 1998; 20:C134–9.
2. Soller RW. OTCness. *Drug Inf J* 1997; 32:555–60.
3. Soller RW. OTCs 2000: Achievements and challenges. *Drug Inf J* 2000; 34:693–701.
4. Clay E. Personal communication. 28 September 1999.
5. Soller RW. When to warn. *RAPS Focus* 1997(Oct); 2(10).
6. Federal Register. Washington; 1988; 53:46213.
7. Soller RW. The OTC scientific/regulatory paradigm. *Drug Inf J* 1999; 33:799–804.
8. Federal Register. Washington; 1972; 37:94649.
9. 21 Code of Federal Regulations 330.10(a)(4)(i).
10. Federal Register. Washington; 28 January 1994; 59:4216–18.
11. Federal Register. Washington; 9 December 1992; 57:58375.
12. 21 Code of Federal Regulations 330.10(a)(4)(ii).
13. 21 Code of Federal Regulations 330.10(a)(4)(v).
14. Federal Register. Washington; 17 March 1999; 64:13254.
15. Bowen D. Points-to-consider for OTC actual-use studies. Draft, 22 July 1994; and FDA Center for Drug Evaluation and Research Manual of policy and procedures (MaPP) 6020.5, 15 January 1997; 3, 15.
16. Lesko SM, Mitchell AA. An assessment of the safety of pediatric ibuprofen. *JAMA* 1995; 273:929–33.
17. Center for Food Safety and Applied Nutrition: Dietary Supplement Strategy (The Ten-Year Plan). Letter from the Director, January 2000.
18. Food and Drug Administration: Dietary Supplement Health and Education Act of 1994 Summary, 1 December 1995.
19. 21 USC DHSEA Section 321 [Section 301] (ff).
20. Commission on Dietary Supplement Labels: Final Report Transmitted to the President, the Congress, and the Secretary of Health and Human Services, US Government Printing Office No. 017-001-00531-2; 24 November 1997.
21. 21 USC DSHEA Section 342 [402](f)(1)(D).
22. 211 USC DSHEA Section 342 [402](f)(1)(B).
23. Merrill RA. Possible labeling of products containing doxylamine. Statement before the Food and Drug Administration Nonprescription Drugs Advisory Committee; 28 June 1993.
24. 21 USC Section 321 note (setting forth Congressional findings).
25. Henney J. Statement by Jane E Henney, MD. Commissioner Food and Drug Administration, Department of Health and Human Services before the Committee on Government Reform, US House of Representatives; 25 March 1999.
26. 164 F.3d 650 (DC Cir 1999)
27. CFSAN Foods Advisory Committee Working Group: Interpretation of Significant Scientific Agreement in the Review of Health Claims; 24–25 June 1999; *http://vm.cfsan.fda.gov/~dms/facssa.html*
28. Federal Trade Commission, Bureau of Consumer Protection: Dietary Supplements: An Advertising Guide for Industry; 1998 (see *www.ftc.gov*).
29. Food and Drug Administration: *www.fda.gov.medwatch/medwatch-online*
30. FDA Center for Food Safety and Applied Nutrition: *http://vm.cfsan.fda.gov/~dms/supplmnt.html*

31. Rusk M. Update on FTC activities relating to dietary supplements. Food Drug Law Institute Conference, Washington, DC, 9 November 1999.
32. Federal Register. Washington; 20 October 1990; 55:45783.
33. Food and Drug Administration: FDA Public Health Advisory Risk of Drug Interactions with St John's Wort and Indinavir and other Drugs. 10 February 2000.
34. Consumer Healthcare Products Association: Voluntary Label Statement Program for St John's Wort. Washington; Adopted 2 April 2000.

12

FUNCTIONAL FOODS: REGULATORY AND SCIENTIFIC CONSIDERATIONS

Maureen Mackey and Frank N. Kotsonis

I. INTRODUCTION

Historically, the relationship between diet and health has primarily been that food provides basic nourishment for growth and maintenance of the body. However, for sometime there has been a growing awareness that food can be important to health beyond this traditional, basic role. Since the 1960s there has been widespread attention given by researchers and the popular media to the recognition that diseases, such as coronary heart disease, are linked to components in the diet, e.g. saturated fat, cholesterol and sodium.

Research on diet and disease accelerated in the 1970s and 1980s, increasing public awareness of this connection. Examples include the epidemiological associations between high-fiber diets and reduced risk for some types of cancer, low calcium diets and osteoporosis, and diets high in fatty fish and reduced risk for coronary heart disease. In the 1990s, anti-oxidants, soy proteins, plant sterols, and polyphenolic compounds are some examples of substances in foods under investigation for potential health benefits.

The increasing awareness among scientists, regulators, health policy experts, and consumers that foods in our diets can have significant impact on health has resulted in the recognition of a new category of foods, "functional foods." While there is no universally accepted definition of a functional food, several definitions have been offered. For example, the Institute of Medicine's Food and Nutrition Board defines such foods as "foods that encompass potentially healthful products which include any modified food or food ingredient that may provide a health benefit beyond the traditional nutrients it contains" (1). The European Commission Concerted Action on Functional Food Science concluded that a functional food is one that has been, "...satisfactorily demonstrated to affect beneficially one or more target functions in the body, beyond adequate nutritional effects, in a way that is relevant to either an improved state of health and well-being and/or reduction of risk of disease." (2). Clearly, the recognition that foods may have roles in disease prevention and/or treatment poses a challenge to the traditional distinction between "food" and "drug."

For our purposes, functional foods are defined as: "*Foods or components of foods that explicitly or implicitly are represented as providing health benefits beyond basic nourishment.*" Examples of functional foods include oat bran, omega-3 fatty acids, carotenoids, and soy isoflavones. As will be discussed in this chapter, functional foods can take many forms, including conventional and processed foods, food ingredients, dietary supplements, and medical foods.

In this chapter we will review the scientific and regulatory considerations in developing and labeling functional foods. We will also suggest several options that could improve the incentives for manufacturers to invest in the research and development of functional foods.

II. REGULATION OF FUNCTIONAL FOODS IN THE US

Foods and drugs

The Food and Drug Administration (FDA) is responsible for regulating foods and drugs by interpreting the relevant laws, by establishing implementing regulations, and by enforcing them. However, the term, functional food, does not appear in any food or drug laws. As a result, functional foods are regulated as either foods or drugs (or both). Since these categories and their respective requirements are critical to the following discussion, it is important to be clear about the legal definition of each, as established in the 1938 Federal Food, Drug, and Cosmetic Act (FD&C Act).

In pertinent part, the FD&C Act provides that the term "food" means:

> "articles used for food or drink for man and other animals."
> 21 USC §321(f)

In pertinent part, the FD&C Act provides two, alternative definitions for drugs:

a. "articles intended for use in the diagnosis, cure, mitigation, treatment, or prevention of disease"; or
b. "articles (other than food) intended to affect the structure or any function of the body."
> 21 USC §321(g)(1)(B) and (C)

Thus, under these definitions, if a functional food makes a claim regarding its effect on the structure or function of the body, but does not make a claim about disease, it probably will be classified as a food. However, if that food is explicitly promoted in labeling to *diagnose, cure, mitigate, treat* or *prevent* disease, the

FDA could assert that the product is a drug (even if it is also a food). For example, if a food, e.g. milk, had a claim on its labeling that "calcium helps build strong bones," this would be acceptable since the claim describes the effect of a nutrient, calcium, on bone structure. If, however, milk were labeled with a claim stating that it was intended to "*prevent* or *treat osteoporosis*," then that milk also would be representing itself as a drug, and could be required by the FDA to comply with all applicable drug requirements, including the submission of a new drug application (NDA) to the FDA. In practice, the NDA-approval process is costly, time-consuming and impractical for most functional foods. Accordingly, if a company is considering the introduction of a new functional food, the more practical and attractive alternative is to develop that product as a food and to avoid claims that would render the product a drug. Other possibilities are discussed below.

Medical foods

In the early 1970s FDA began to accept the premise that a formulation of nutrients could be represented for use under medical supervision to address the special nutritional needs of a person with a disease or a medical condition, without triggering drug status for the product, and, that such a product could be regulated as a food rather than as a drug. One of the first regulations that accepted this view (although without any discussion of the medical food concept), was the 1973 food additive regulation concerning the use of amino acids as "special dietary *foods* that are intended for use solely under medical supervision to meet nutritional requirements in specific medical conditions." (3)

In the Orphan Drug Amendments of 1988 to the FD&C Act, the concept became legally defined:

> The term "medical food" means a food which is formulated to be consumed or administered enterally under the supervision of a physician and which is intended for the specific dietary management of a disease or condition for which distinctive nutritional requirements, based on recognized scientific principles, are established by medical evaluation.
>
> 21 USC §360ee(b)(3)

In regulations [21 CFR §101.9 (j)(8)], FDA has interpreted the meaning of "medical food" to include the following provisions:

- It is a specially formulated and processed product (as opposed to a naturally occurring foodstuff used in its natural state) for the partial or exclusive feeding of a patient by means of oral intake or enteral feeding by tube;
- It is intended for the dietary management of a patient who, because of therapeutic or chronic medical needs, has limited or impaired capacity to ingest, digest, absorb, or metabolize ordinary foodstuffs or certain nutrients,

or who has other special medically determined nutrient requirements, the dietary management of which cannot be achieved by the modification of the diet alone;

- It provides nutritional support specifically modified for the management of unique nutrient needs that result from the specific disease or condition, as determined by medical evaluation;
- It is intended to be used under medical supervision; and
- It is intended only for a patient receiving active and ongoing medical supervision wherein the patient requires medical care on a recurring basis for, among other things, instructions on the use of the medical food.

As an example, Lofenalac® is a product that meets the legal definition of, and FDA's regulatory requirements for, a medical food. Lofenalac® is an infant formula for patients with phenylketonuria, a rare in-born error of metabolism in which phenylalanine metabolism is impaired. The management of this disease requires severe dietary restriction of phenylalanine such that few normal foods can be consumed. Lofenalac® provides a mixture of amino acids with very little phenylalanine and is consumed instead of standard formula.

If a functional food is to be developed under the medical food category, the manufacturer should keep several requirements in mind. First, medical foods are intended for dietary management of *patients* whose diseases cause them to have dietary or nutrient needs that cannot be met by modification of the normal diet alone. Thus, while it could be argued that, e.g. low sodium products help meet the dietary needs of patients with hypertension, it still is possible (although perhaps not so easy), to achieve a low sodium intake through modification of the normal diet. Second, medical foods are intended for use under medical supervision, and labeling and promotional material should clearly reflect this. For example, labels should include a statement, such as, "For use under medical supervision," or "Ask your doctor whether this product is right for you." This is particularly important since medical foods are available to consumers on pharmacy shelves or in a food store without a prescription.

For many years the medical foods industry was largely self-regulated and dominated by a few pharmaceutical companies. In recent years, however, the medical foods category has increased significantly, and there are now many more manufacturers. This recent growth in the medical foods business is one reason why the FDA published an "Advance Notice of Proposed Rulemaking" (ANPR) (61 Fed Reg 60661–71, 1996) to consider re-evaluation of the regulation of medical foods. FDA is concerned about product safety, since, when used as intended, medical food products may be the sole or major source of nutrients for patients. FDA also wants to ensure that the claims made for medical foods are truthful, not misleading, and supported by sound science. A third major concern is that labeling adequately inform consumers about safe and proper use of the products.

As of this writing FDA has not followed up the ANPR with a proposed regulation. However, the ANPR indicated to industry that the regulatory requirements for medical foods and the attractiveness of this route of product development may change in the future.

Dietary supplements

During the early 1990s intense pressure was applied to Congress by the dietary supplement industry and consumers to reduce the regulatory burdens on dietary supplements under the FD&C Act. Thus, the Dietary Supplement Health and Education Act (DSHEA) was passed on 25 October 1994 (4). Insofar as a functional food can be positioned as a dietary supplement, this new legislation offers substantial opportunities for promotion of health-related benefits that might not otherwise be available without compliance with drug standards or approval of a health-claim regulation.

Among the many provisions it established, DSHEA created an expansive definition of "dietary supplement":

> "... a product intended to supplement the diet that bears or contains one or more of the following ingredients: ... a vitamin; ... a mineral; ... an herb or other botanical; ... an amino acid; ... a dietary substance for use by man to supplement the diet by increasing the total dietary intake; or ... a concentrate, metabolite, constituent, extract, or combination [of any ingredient(s) described above]"
>
> 21 USC §321(ff)(1)

However, a dietary supplement *either* must be intended for ingestion in tablet, capsule, powder, softgel, gelcap, or liquid droplet form; *or* if not intended for ingestion in such a form, must not be "represented for use as a conventional food or as a sole item of a meal or the diet."

One of the most significant advantages DSHEA provides to manufacturers of dietary supplements is an exemption from "food additive" status. A "food additive" is a substance added to food that is not "generally recognized as safe" (GRAS) by experts. To place a food additive on the market, a manufacturer often must invest millions of dollars to conduct studies to demonstrate the additive's safety and functionality, then submit a petition to the FDA requesting formal approval of the additive. The review and approval process typically requires several years. Previously, FDA viewed substances in dietary supplements that were not known to be GRAS as unapproved "food additives" and took regulatory action against many such substances (4). Under DSHEA, however, if an ingredient in a dietary supplement is not GRAS, it cannot be treated as an unapproved food additive. The practical result of this provision is that dietary supplements can contain substances that do not meet the "generally recognized as safe" criterion necessary for other food ingredients. As a quid pro quo, DSHEA requires that dietary supplements not present a "significant or unreasonable risk of illness or injury." However, the law places the burden on the FDA to prove that a dietary supplement violates this standard.

The DSHEA also provides another important provision for dietary supplements that is not available for conventional foods: the use of third party literature to tell consumers about the health benefits of products (5). The FDA has on occasion asserted that publications used in connection with the sale of a food product to consumers can be regulated as *labeling* for the product. The DSHEA, however, now provides that publications (including articles, book chapters, or official abstracts of a peer-reviewed scientific publication prepared by the author or the editors of the publication) "shall not be defined as labeling," and may be "used in connection with the sale of a dietary supplement to consumers" if the publication is "reprinted in its entirety," and meets certain criteria. Thus, a retailer of dietary supplements now has the ability under law to guide the attention of a customer to published nutritional, medical or other scientific literature that describes the health benefits of dietary supplements and their ingredients, if certain criteria are met. The publication:

- Must not be false or misleading;
- Must not promote a particular manufacturer or brand of a dietary supplement;
- Must be "displayed and presented ... so as to present a balanced view of the available scientific information";
- If "displayed in an establishment," must be "physically separated from the dietary supplement"; and
- Must not "have appended to it any information by sticker or any other method."

The use of reprints provides an additional opportunity for adjunctive promotional activities. This does not, however, authorize the labeling to contain information about the reprints. The full implication of the use of publications has not been worked out, and it remains to be seen what will be the response of the FDA, on a case-by-case basis.

III. CLAIMS THAT CAN BE MADE FOR FUNCTIONAL FOODS

Two major routes by which claims information can be conveyed to consumers are labeling and advertising. At the outset it is important to distinguish between these claims. Claims made in labeling are regulated by the FDA, and include all claims made on a label or other labeling. A "label" is "a display of written, printed, or graphic matter upon the immediate container" of a product; and "labeling" includes "all labels and other written, printed, or graphic matter ... upon any [product] or any of its containers or wrappers, or ... accompanying such [product]." For simplicity, we will use the term "labeling" to mean both labels and other forms of labeling.

Claims made in advertising are regulated by the Federal Trade Commission (FTC), and include those used in radio, television, newspaper, magazine, or other advertising.

Claims made in labeling

A manufacturer can pursue several options for making labeling claims of a health benefit for a functional food; these include health claims and structure/function claims for both foods and dietary supplements, and medical foods claims.

Health claims

During the past several decades the FDA acted strictly and aggressively on certain claims made in labeling by food manufacturers, based on its authority under the FD&C Act. In many instances claims made regarding the link between food and disease were interpreted by FDA as drug claims, which was sufficient reason, under law, to remove the products from the market.

By the 1980s food manufacturers wanted to develop new products based on the evolving understanding of the relationship between an improved diet and reduced risk for certain diseases. However, new product development activities would be economically justified only if the information could be communicated to the public, and in legal fashion. A new legal climate was needed for industry to be able to use this emerging scientific evidence to represent the health benefits of its products to consumers. The representations of the role of high fiber, low fat diets in reducing risk for some types of cancer, and of the role of oat bran in reducing risk for coronary heart disease on breakfast cereal packages in the mid-1980s forced the issue of such health claims in food labeling onto center stage in the regulatory community. While the FDA at first opposed such claims in labeling, many in the scientific, medical and public arenas, as well as the food industry supported them. In 1990, Congress passed the Nutrition Labeling and Education Act (NLEA), which enabled manufacturers to make certain health claims in labeling to inform consumers of the health benefits of foods, under certain tightly-controlled circumstances (6).

The NLEA defined a health claim as a statement in labeling of a food (including a dietary supplement) that "characterizes the relationship of any nutrient . . . to a disease or a health-related condition." According to FDA regulations, the term, "disease or health-related condition" means "damage to an organ, part, structure, or system of the body such that it does not function properly (e.g. cardiovascular disease), or a state of health leading to such dysfunctioning (e.g. hypertension); except that diseases resulting from essential nutrient deficiencies (e.g. scurvy, pellagra) are not included." Claims of this type are usually worded to state that consumption of a substance in the diet on a regular basis "may reduce the risk of" a named disease.

As implemented by FDA, a health claim will not be a short, bullet-type claim, but rather will be a more lengthy, balanced statement. Examples of such claims include:

"While many factors affect heart disease, diets low in saturated fat and cholesterol may reduce the risk of this disease."

"The development of cancer depends on many factors. A diet low in total fat may reduce the risk of this disease."

Health claims do not encompass general dietary guidance, e.g. claims about good nutrition in general; or about classes of "healthy" foods, without also noting a relationship to some disease, damage or dysfunction. For example, a claim that links a nutrient solely to normal, healthy functioning of the human body, e.g. "protein helps build strong muscles" would not be a health claim.

With the exception of health claims allowed under the FDA Modernization Act of 1997, health claims may be made in labeling only if prior FDA approval has been obtained. Seven claims were approved in 1993 as part of the NLEA implementing regulations. FDA has reviewed petitions for several other claims; for example, in 1999 FDA approved a claim relating consumption of soy protein to reduced risk for heart disease.

A health claim will be approved by the FDA only if it is judged to be adequately substantiated. The requirement for substantiation of a health claim is based on: (1) the totality of publicly available scientific evidence ("including evidence from well-designed studies conducted in a manner which is consistent with generally recognized scientific procedures and principles"); and (2) the existence of "significant scientific agreement" among experts, qualified by scientific training and experience to evaluate such claims. A petitioner for approval of a new health claim must submit evidence meeting these criteria.

Health claim petitions require substantial work on the part of the petitioner, and FDA has published regulations governing their construction and submission. A notable drawback in submitting a health claim petition has been the amount of time required by FDA to review and act on it. In the Food and Drug Administration Modernization Act of 1997, provisions were established for time frames by which FDA is to take certain actions regarding a health claim petition. FDA still has two years to complete its process or explain to Congress why it failed to do so.

The FDA Modernization Act provided a new option for making certain limited health claims without FDA preapproval. These health claims must be the subject of a "published . . . authoritative statement, which is currently in effect" by "a scientific body of the US Government with official responsibility for public health protection or research directly relating to human nutrition, such as the National Institutes of Health or the Centers for Disease Control and Prevention, or the National Academy of Sciences or any of its subdivisions . . ." {§403(r)(3)(C)(i) of the FD&C Act, as amended by the FDA Modernization Act}. At least 120 days before introducing a product with such a claim on the market, a manufacturer must notify FDA that the claim will be made. FDA has the authority to issue a regulation prohibiting or modifying the claim. To date, only one health claim – relating consumption of whole grain foods to reduce risk for heart disease and certain cancers – has successfully gone through the notification process.

By now it should be evident that obtaining approval for use of a new health claim is not rapid, easy or inexpensive. It requires substantial data (publicly available), a well-written petition, and the time requirements associated with FDA rule making. Nonetheless, an approved health claim may be made on any product that meets the qualifications set forth in an approving FDA health claim regulation – use of the claim is not exclusive to the petitioning company.

In the late 1990s the health claims provisions of NLEA and FDA's implementing regulations were challenged in a landmark court case, *Pearson v. Shalala*, 164 F.3d 650 (DC Cir 1999). A lawsuit was filed against the FDA after it had denied approval of four proposed health claims on the grounds that the evidence supporting them did not reach the significant scientific agreement standard. Although the district court ruled for FDA in all respects, the US Court of Appeals for the DC Circuit reversed the lower court's decision, stating that the First Amendment does not permit FDA to reject proposed health claims that it determines to be *potentially* misleading if the potential deception could be remedied by a disclaimer. Thus, proposed health claims that do not meet the substantiation standard of significant scientific agreement may still be approved by FDA if the addition of a qualifier or disclaimer removes any potential deception and makes the claim truthful and not misleading. The court, however, maintained that in cases where the evidence in support of a claim is outweighed by evidence against it, the FDA could ban it outright on the grounds that no disclaimer could remedy its *inherent* misleading nature.

As examples, Table 1 provides two of the four proposed health claims originally denied by FDA and possible disclaimers proposed by the Court of Appeals.

Table 1

Proposed health claims	Disclaimers
"Consumption of antioxidant vitamins may reduce the risk of certain kinds of cancers."	"The evidence is inconclusive because existing studies have been performed with *foods* containing antioxidant vitamins, and the effect of those foods on reducing the risk of cancer may result from other components in those foods." 164 F.3d at 658 (DC Cir 1999)
"0.8 mg of folic acid in a dietary supplement is more effective in reducing the risk of neural tube defects than a lower amount in foods in common form."	"The evidence in support of this claim is inconclusive." 164 F.3d at 659 (DC Cir 1999)

As a result of the Pearson decision, FDA must reconsider approving the four proposed health claims (and presumably all future health claims) with possible qualifiers/disclaimers. While the Court of Appeals noted that under the US Constitution First Amendment principles applicable to potentially misleading advertising, "the preferred remedy is more disclosure, rather than less," and that disclaimers are "constitutionally preferable to outright suppression" of truthful information (164 F.3d at 657), it remains to be seen whether consumers will find qualified health claims credible and will buy products with such claims on their labels.

Structure/function claims for food and dietary supplements

The FD&C Act defines drugs as articles *(other than food)* that are intended to affect the structure or function of the body. Thus, there is no legal reason why a food cannot represent in labeling its ability to affect the structure or function of the body, provided the statement is truthful and non-misleading and is neither a drug claim nor a health claim. Historically food manufacturers were reluctant to make such claims, possibly because they feared the FDA might view them as unapproved drug or health claims. However, a few claims such as, "protein helps build strong muscles" or "calcium builds strong bones" have been made in food labeling and have not been challenged by FDA.

The FDA comments in its regulations regarding dietary supplement labeling are relevant to the use of structure/function claims on conventional foods, as well as supplements (62 Fed Reg 49859–68, 1997; 65 Fed Reg at 1033, 2000). Using cranberry juice as an example of how a food could make a structure/function claim, the agency stated that cranberry juice could include the statement, "cranberry products help to maintain urinary tract health," assuming it is truthful, non-misleading and derives from the *nutritional value* of cranberry juice. A problem here, however, is that FDA did not explain what it meant by "nutritional value." In other contexts, FDA appears to have accepted varied (indeed, inconsistent) meanings for "nutritional value" or "nutritive value," in some instances encompassing a broad array of substances in foods, but in other instances, including only essential nutrients. For example, as part of the health claim regulations, FDA narrowly defined "nutritive value" to mean, "... a value in sustaining human existence by such processes as promoting growth, replacing loss of essential nutrients, or providing energy." (21 CFR §101–14). On the other hand, before passage of NLEA in 1990, Senators Metzenbaum and Symms agreed to a list of a few of the items and foods that would be regarded as "nutritional substances" under the new legislation. The list included black currant seed oil, Coenzyme Q 10, enzymes such as bromelain and quercitin, pollens, garlic, orotates, glandulars, nutritional antioxidants such as superoxide dismutase (SOD) and herbal tinctures [136 Congressional Record S16609 (24 October 1990)]. Since FDA has cited with approval the above legislative history, it appears that the concept of "nutritive" or "nutritional" value may be a broad one that extends well beyond traditional vitamins and minerals.

The meaning of "nutritive value" has continued to raise both scientific and regulatory questions. The central issue is whether nutritive substances have affects only on the structure and function of the body, or whether they also have affects on disease. This issue was discussed at length at a conference at Georgetown University, Washington, DC, in 1999 (7). A case in point are phytostanol esters, substances present in small quantities in vegetable foods (and trees) that, when consumed in much larger amounts, reduce serum cholesterol. In the mid-1990s the Raisio Company in Finland developed a margarine product, Benecol®, containing phytostanol esters and marketed it for its cholesterol-lowering effects. Later, the Johnson & Johnson Company introduced Benecol® margarine in the US. Among the regulatory issues the Company addressed with the FDA was the "nutritive value" of phytostanol esters. Phytostanol esters are added to margarine to help lower blood cholesterol levels, an effect the FDA has maintained is a drug effect. However, the FDA did not disagree with the company's position that phytostanol esters have nutritive value. This example illustrates that as more is learned about the effects of substances in foods on disease, it will become increasingly difficult to differentiate between nutritive effects and drug effects.

While the freedom to use a structure/function claim in labeling on a food is inferred from the definitions of "food" and "drug" in the FD&C Act, the DSHEA makes specific provisions for four types of "statements of nutritional support" on labeling of dietary supplements. They are:

- A statement that "claims a benefit related to a classical nutrient deficiency disease and discloses the prevalence of such disease in the United States";
- A statement that "describes the role of a nutrient or dietary ingredient intended to affect the structure or function in humans";
- A statement that "characterizes the documented mechanism by which a nutrient or dietary ingredient acts to maintain such structure or function"; and
- A statement that "describes general well-being from consumption of a nutrient or dietary ingredient."

Statements 2 and 3 are structure/function claims, similar in concept to those for foods. Thus, structure/function claims can be a type of statement of nutritional support.

DSHEA imposes further qualifications for these statements:

- The manufacturer must have substantiation that the claim is truthful and not misleading;
- The labeling must contain, prominently displayed, the following text:

  "This statement has not been evaluated by the Food and Drug Administration. This product is not intended to diagnose, treat, cure, or prevent any disease";

- The manufacturer must notify FDA no later than 30 days after the first marketing of the dietary supplement that such a statement is being made.

Although the law states that a manufacturer must have "substantiation" that a statement of nutritional support is truthful and non misleading, the law does not explain what evidence constitutes adequate substantiation. In its implementing regulations (62 Fed Reg at 49884, 1997), FDA provided no guidance as to what constitutes adequate substantiation, stating instead that DSHEA is clear on this point: "the manufacturer must have substantiation that the statement is truthful and not misleading."

In its final rule on structure/function claims in dietary supplement labeling (65 Fed Reg 1000, 6 January 2000) the FDA made several comments about the scope of proper structure/function claims that it derived from the definition of "disease." The FDA concluded it would retain (instead of broaden, as it had proposed) the definition of "disease" as established in its 1993 implementing regulations for health claims under NLEA. "Disease" would continue to mean "damage to an organ, part, structure, or system of the body such that it does not function properly (e.g. cardiovascular disease), or a state of health leading to such dysfunctioning (e.g. hypertension); . . . " [new 21 CFR §101.93(g)].

Based on this definition, FDA concluded that "disease claims" would have the following characteristics:

- Explicitly or implicitly claim to diagnose, mitigate, treat, cure or prevent a specific disease of class of diseases;
- Claim to effect the characteristic signs or symptoms of a specific disease or class of diseases, using scientific or lay terminology;
- Claim to have an effect on an abnormal condition associated with a natural state or process if the abnormal condition is uncommon or can cause significant or permanent harm;
- Claim to have an effect on a disease or diseases through one or more of the following factors:

  - The name of the product;
  - A statement about the formulation of the product, including a claim that the product contains an ingredient that has been regulated by the FDA as a drug;
  - Citation of a publication or reference that implies treatment or prevention of disease; and
  - Use of the term "disease" or "diseased," except in general statements about disease prevention that do not refer explicitly or implicitly to a specific disease or class of diseases or to a specific product or ingredient;

- Claim that the product belongs to a class of products that is intended to diagnose, mitigate, treat, cure or prevent disease;
- Claim to be a substitute for a product that is a therapy for a disease;
- Claim to augment a particular therapy or drug action that is intended to diagnose, mitigate, treat, cure, or prevent a disease or class of diseases;
- Claim to have a role in the body's response to a disease or to a vector or disease;

- Claim to treat, prevent, or mitigate adverse events associated with a therapy for a disease, if the adverse events constitute diseases; and
- Otherwise suggests an effect on a disease or diseases.

Since structure/function claims cannot be disease claims, FDA ruled that claims with any of the above-described attributes could not be acceptable structure/function claims. Examples of acceptable structure/function claims and similar, but impermissible disease claims provided by FDA in the final rule are provided in Table 2.

From this final rule, FDA made clear that it would not allow even implied references to diseases in structure/function claims. It did, however, acknowledge that certain claims allowed for OTC drugs also could be made for dietary supplements if they describe an effect on a nonspecific group of conditions that have a variety of causes, but may not be disease-related. Thus, "relief of occasional heartburn," or "relief of occasional sleeplessness," could be allowable structure/function claims. In addition, FDA recognized that there are a number of conditions associated with natural states that are not disease states. Structure function claims, such as those referring to leg edema associated with pregnancy, wrinkles and hair loss associated with aging, and cramps associated with the menstrual cycle could be allowable structure/function claims.

Medical food claims

Claims on medical foods can be similar to health claims on foods and dietary supplements, but there are distinct features. For example, the labeling for a medical food should be clear that the product is to be administered under a physician's supervision, and that it is helpful solely in the *dietary* management of any disease or medical condition to which reference is made. A further distinction is that a medical food claim requires no approval by, or notification to, the FDA, because medical foods are exempt from NLEA.

This does not mean that medical food labeling is immune from FDA regulation (see Section II). The company responsible for a medical food must possess substantiating data to show that the claim is not "false or misleading in any particular." If any claim on a medical food label is "false or misleading in any particular,"

Table 2

Acceptable structure/function claim	Impermissible disease claim
Helps to maintain cholesterol levels that are already within the normal range	Lowers cholesterol
Helps support cartilage and joint function	Improves joint mobility and reduces joint inflammation and pain
Use as part of your diet to help maintain a healthy blood sugar level	Use as part of your diet when taking insulin to help maintain a healthy blood sugar level
Supports the immune system	Supports the body's antiviral capabilities

the product can be deemed to be misbranded and can be removed from the market. Furthermore, the labeling claim must not be a drug claim. Rather, the claims may only address a patient's special *dietary* needs that exist because of a disease.

Medical foods are currently marketed that carry claims describing how they meet the distinctive nutritional requirements associated with kidney or liver disease, lung disease, hypermetabolism (generally, critically ill patients), malabsorption syndromes, and altered amino acid metabolism. A generic example of a medical food claim is: "For use under medical supervision. Product 'X' can be helpful in the dietary management of nutrient malabsorption by providing sources of nutrients that are pre-digested, and that are more easily absorbed than are conventional foods."

Claims made in advertising

Advertising is the second major route for communicating health benefits to consumers. Claims made in advertising are regulated by FTC (except for claims made about prescription drugs, which are regulated by the FDA). This agency enforces the Federal Trade Commission Act, which declares it unlawful to use false advertising or "unfair or deceptive acts or practices" in commerce. When an advertisement is found to be unlawful, the FTC can seek a court order either to prohibit the claims, or to impose substantial fines.

The FDA and the FTC have overlapping jurisdiction to regulate advertising and labeling of foods and other articles. The two agencies have established an agreement to the effect that FDA regulates labels and labeling for foods and drugs, while the FTC focuses on non-prescription drug claims made in advertising media, including radio, television, newspapers, and magazines (8). Where the inevitable overlaps occur, FTC tries to coordinate with FDA, and to be consistent with FDA rulings.

In general the FTC is not concerned about whether advertising claims might technically create "drug" status under existing laws and regulations. Rather, the FTC applies two major principles in its enforcement of advertising claims:

- Advertising must be truthful and not misleading.
- Advertisers must have substantiation for all objective claims, before they are disseminated.

The FTC requires that both the explicit claims *and* the implied claims in an advertisement be truthful and not misleading. Advertisements can be misleading by virtue of what is omitted from them, as well as by what is stated. Moreover, an advertiser is liable for a misleading claim whether or not there was *intent* to mislead.

The FTC applies a high standard for substantiation; it must consist of "competent and reliable scientific evidence," further characterized to mean "tests, analyses, research, studies, or other evidence based on the expertise of professionals . . . ." Specific studies will be evaluated in the context of all relevant scientific information, not just on the study in isolation. The specific type and amount of claim substantiation required is decided on a case-by-case basis. FTC has rejected many studies presented to it, including university research offered by companies in support of advertising claims. A company must also possess the required substantiation at the time of dissemination of the advertising; i.e. no "claim now, verify later." Finally, substantiation must support the advertising claims. While this last point may seem unnecessary to state, it is the most frequently cited basis for rejection of substantiation by FTC (8).

In 1998, the FTC issued it advertising guidance to the dietary supplements industry (10). It reiterated the same principles applied to all advertising, and stated that it would apply to dietary supplements the substantiation standard of "competent and reliable scientific evidence" that it applies to foods and drugs. The FTC also explained the appropriate use of clear and prominent disqualifiers in connection with claims that would otherwise be misleading. For example, a claim on a dietary supplement about its dramatic effects on body weight loss, as demonstrated in clinical studies, might need a disclaimer indicating that the user must follow a diet and exercise regimen to achieve the same weight loss.

Recently, some dietary supplement companies have included statements in advertisements that their health-benefit claims have not been reviewed by the FDA, and that their products are not intended to cure/treat/mitigate/prevent disease. That is, these companies have used in *FTC-regulated advertisements*, text that would be required in *FDA-regulated labeling* for certain statements of nutritional support. In its guidance to the dietary supplement industry, the FTC stated that there are instances where this disclosure could be necessary to make the advertisement non-misleading. For example, if an advertisement for a dietary supplement could lead consumers to believe that the product has undergone the same or similar safety and efficacy review the FDA conducts on new drugs, the disclaimer could be added to mitigate this potential deception. On the other hand, as the FTC has pointed out, that the inclusion of the DSHEA disclaimer will not cure an otherwise deceptive advertisement.

IV. ESTABLISHING THE SAFETY OF FUNCTIONAL FOODS

Functional foods, as described above, can take many forms, including traditional foods, dietary supplements and medical foods. In the future, we may see functional foods with unique attributes, such as an increased concentration of components with health enhancing properties or entirely new components added to, or bioengineered into, foods that traditionally did not have these components. Examples could include canola oil with long chain omega-3 fatty acids for enhancing cardiovascular health, tomatoes with enhanced lycopene concentration that may reduce the risk for cancer, and a dietary supplement bar containing soy isoflavones for helping to maintain bone health.

A critical element in the research and development of such products is the assurance that they will be safe at the intended level of intake. In this section we briefly review the concepts and processes that are in place to establish the safety of foods, food additives, and GRAS substances, and how they can be applied to establish the safety of functional foods. A more in depth review of these concepts and processes can be found in Kotsonis *et al.* (11).

Under the FD&C Act, traditional foods, such as vegetables, meat, milk, and grains with a long history of use, may be marketed (and in essence are presumed to be safe) unless they contain poisonous or deleterious substances in amounts which make them *ordinarily injurious to health*. This presumption of safety is necessary on a practical basis because foods are complex mixtures of thousands of substances, and it would be impossible to test the vast array of these substances for safety. However, when a substance not naturally present in food is "added" to food, the FD&C Act requires a higher safety standard. Such a substance can be banned by the FDA if it *may render* food to which it is added *injurious to health*. The chapter by Flamm *et al.* in this volume also includes a discussion of the safety standard of traditional foods.

A "food additive" is defined in the FD&C Act as, "any substance the intended use of which results ... in its becoming a component ... of any food ... if such substance is *not generally recognized* ... to be safe under the conditions of its intended use." In defining a "food additive," the law, in fact, created *two* categories: (1) food additives, those substances which are not generally recognized as safe (not GRAS); and (2) GRAS substances, those substances which are generally recognized as safe.

The safety assessment of a food additive is a complicated, timely and expensive undertaking, especially when the additive is widely used in a variety of foods. Each food additive can pose unique safety questions depending on its chemistry, stability in use, metabolism, and toxicity study results. Nonetheless, it should be recognized that technically, the safety standard for food additives and GRAS substances is the same. The distinction between the two lies in whether or not the safety of the substance is "generally recognized" by experts as having been adequately shown through scientific procedures or, in the case of a substance in food prior to 1 January 1958, through either scientific procedures or experience based on common use in food. The scientific procedures to establish the safety of a food additive can take many years and tens of millions of dollars. In addition, food additives may not be used until the FDA has published a regulation specifically approving their use in foods; in contrast, GRAS substances can be used without prior notice to, or approval by the FDA.

The determination of GRAS status is made by scientists with expertise in food safety, nutrition, and other relevant areas, based on knowledge and experience of prior safe use of the substance, generally available data published in scientific literature, and other corroborative data. It is possible to conclude that a food ingredient is GRAS based entirely on published literature. In other cases, extensive additional studies, including animal toxicology studies, may be necessary. A more in depth discussion of the GRAS concept is in the chapter by Burdock.

Another important point regarding the safety of food additives and GRAS substances is that their safety is established at particular levels of usage and for certain food applications. This means, for example, that a food additive may be approved by FDA for use in only certain food products at specified concentrations, or similarly, that a substance can be concluded to be GRAS only for certain uses. These conditions of use stem from the fact that the safety of GRAS substances and food additives is determined *at a particular level of intake*.

In many instances, functional foods will consist of substances naturally present in the diet, but their intended intake will be higher than what is typically obtained from the diet. For example, suppose a concentrated fraction of flax seed oil was shown in animal studies to help maintain healthy blood vessels. A manufacturer might wish to develop a new functional food from concentrated flax seed oil and to test whether this benefit occurs in humans. In other instances, a functional food could be developed from new, non-traditional food sources, for example, a dietary supplement of algae rich in micronutrients.

Whatever the particular circumstances, the safety evaluation of a functional food must include an evaluation of its composition and manufacturing process. The evaluation also should include a thorough literature review to determine what is already known about the safe use of the functional food and its components. Data from dietary surveys, such as those conducted by the USDA, should be consulted to determine the current intake of the functional food or its components. Using the flax seed oil example above, one could determine the current per capita intake of flax seed and the per capita intake of components in flax seed oil, such as fatty acids and sterols, from other dietary sources, such as vegetable oils, to arrive at the total per capita intake of flax seed oil and its components. These data are then compared against estimates of likely intake once the functional food is introduced into the market. If the increase in intake of flax seed oil and its components is well outside of the range of intakes existing in the population, then additional studies are likely necessary.

The type of studies needed will be dictated by the composition and intended intake of the product and could include in vitro studies of genotoxicity, acute exposure studies in rodents, subchronic exposure studies in rodents, and reproduction and teratology studies in two species. Other studies, such as a two generation rodent study, also could be required. The amount of test material fed to the animals should be large enough to allow for a generous margin of safety compared to the intended level of intake by humans. Because the data from these studies provide evidence of the safety of the product and could form the basis of a regulatory submission, the studies should meet current Good Laboratory Practices standards.

Thus, based on information from the literature, estimates of likely intake vs. current intake of the product or its components, and the results of appropriate safety studies, a conclusion can be made about the safety of the functional food at the intended level of intake.

V. DEVELOPMENT OF FUNCTIONAL FOODS

The success of functional foods requires that consumers have confidence in the health value of these products and that the products be accepted by the health care and regulatory communities. At the same time, manufacturers must have an incentive to invest in the necessary research to substantiate the health value of such products, and must be able to communicate this value to consumers via labeling and advertising. In this section we evaluate the current regulatory challenges to development of functional foods and suggest approaches that will encourage greater innovation in this area.

Harmonization of standards applied to claims

Since the passage of NLEA in 1990 and publication of implementing regulations in 1993, there has been continued debate about the meaning of the statutory standard of "significant scientific agreement" to substantiate a health claim. Such a standard has been difficult to interpret, particularly since the nature of the body of evidence to support health claims can be widely variable in extent and composition. As part of its plan to implement the Pearson decision, (see p. 251) FDA published a guidance document for industry, describing its interpretation of the significant scientific agreement standard for approval of a health claim (12). The attributes of a claim that meets this standard include:

- "...qualified experts likely would agree that the scientific evidence supports the substance–disease relationship..."
- "...the validity of the claim is not likely to be reversed by new and evolving science, although the exact nature of the relationship may need to be refined."
- "...does not require a consensus of agreement based on unanimous and incontrovertible scientific opinion, however, on the continuum of scientific discovery that extends from emerging evidence to consensus, it represents an area on the continuum that lies closer to the latter than to the former."

FDA's guidance is a positive step in providing clarity to the appropriate substantiation of health claims. A further step would be for the FTC and FDA to harmonize the standards applied to health claims with those applied to advertising claims, and to apply this standard equally to structure/function claims. There is no reason to have different substantiation standards in labeling than those required in advertising; consumers rely on both in making purchasing decisions. Furthermore, from a scientific perspective it is reasonable to expect that all claims describing the relationship between a food or dietary supplement and health or disease be based on consistent and reliable scientific information. While the type of information needed to support a health claim likely will be different from that to support a structure/function claim, the evidence for both should reach the same standard of reasonable certainty.

Exclusivity in the use of health claims

Given the current requirement that data to support a health claim must be publicly available and the fact that under NLEA, once a health claim is approved, any company can use it, it would seem reasonable for approval of a health claim to be granted for a limited time solely to the petitioner who provides the substantiating research. This would create real incentive and reward for the conduct of expensive, quality research on the health-related benefits of non-proprietary food substances. It would also seem reasonable to allow a successful petitioner to sublicense to others the right to make the same claim.

FDA notifications for health claims

As described earlier in this chapter, the health claim approval process has required, and likely will continue to require, at least 2 years from the time a petition is submitted to the FDA. Such a long and unpredictable time frame undoubtedly is another factor discouraging manufacturers from investing in the research to identify and establish the health benefits of functional foods. There are alternate approaches that could greatly expedite the appearance of credible claims in labeling.

First, instead of being required to await FDA approval of a health claim regulation, manufacturers could be permitted simply to submit a notification to the FDA regarding their intentions to make a health claim in labeling, based on evidence presented in the notification. The notification could be submitted at the time, or even shortly after, the product is placed on the market, similar to what is required under DSHEA for statements of nutritional support on dietary supplements labels. Alternatively, the notification could be required *prior to* marketing, giving the FDA a short but reasonable period to object. In the absence of such an objection, the company could proceed with using the claim.

Another approach could be to allow independent expert panels to evaluate the substantiation for a claim and provide a statement indicating their rationale for concluding that the claim is supported by reliable scientific information. This statement could then be included in the (premarket or otherwise) notification to the FDA.

VI. CONCLUSIONS

Functional foods are exciting new opportunities at the interface between foods and drugs that are intended to provide health benefits beyond basic nourishment. There is no special regulatory category for functional foods; however, they can fit into one or more of several existing categories, including traditional foods, dietary supplements and medical foods. It should be emphasized that functional foods are still *foods, not drugs*, and are intended to be part of an overall healthful diet.

The safety of functional foods must be established to a reasonable certainty. In addition, claims for these products in labeling and advertising must be based on adequate and appropriate scientific research such that a reasonable certainty of benefit to health is established. Finally, development of functional foods would be enhanced if it could occur in a regulatory environment that rewards investment in research and facilitates responsible communications about benefits.

REFERENCES

1. Committee on Opportunities in the Nutrition and Food Sciences. Food and Nutrition Board, Institute of Medicine. Thomas PR, Earl R, eds. *Opportunities in the Nutrition and Food Sciences: Research Challenges and the Next Generation of Investigators.* Washington, DC: National Academy Press; 1994.
2. Diplock AT, Aggett PJ, Ashwell M, Bornet F, Fern EB, Roberfroid MB. Scientific Concepts of Functional Foods in Europe: Consensus Document. *Br J Nutr* 1999; 81(Suppl 1):S1–27.
3. McNamara SH. FDA regulation of "Medical Foods" – an industry perspective. *Clin Res Reg Affairs* 1997; 14:15–31.
4. McNamara SH. FDA regulation of ingredients in dietary supplements after passage of the Dietary Supplement Health and Education Act of 1994: An update. *Food Drug Law J* 1996; 51:313–18.
5. McNamara SH. Dietary supplements of botanicals and other substances: A new era of regulation. *Food Drug Law J* 1995; 50:341–8.
6. Geiger CJ. Health claims: History, current regulatory status, and consumer research. *J Am Diet Assoc* 1998; 98:1312–22.
7. Georgetown University. "What is a Nutrient? Defining the Food–Drug Continuum". *Proceedings Georgetown University Center for Food and Nutrition Policy.* Washington, DC; 1999.
8. Peeler CL, Cohn S. The Federal Trade Commission's regulation of advertising claims for dietary supplements. *Food Drug Law J* 1995; 50:349–55.
9. Federal Trade Commission. *Enforcement Policy Statement of Food Advertising*; 1994; http://www.ftc.gov/bcp/policystmt/ad-food.htm
10. Federal Trade Commission. *Dietary Supplements: An Advertising Guide for Industry*; 1998; http://www.ftc.gov/bcp/conline/pubs/buspubs/dietsupp.htm
11. Kotsonis FN, Burdock GA, Flamm WG. Food Toxicology. In: Klaassen CD, ed. *Casarett and Doull's Toxicology: The Basic Science of Poisons*, 6th edn. McGraw-Hill New York; 2000.
12. Food and Drug Administration Center for Food Safety and Applied Nutrition. Office of Special Nutritionals. *Guidance for Industry. Significant Scientific Agreement in the Review of Health Claims for Conventional Foods and Dietary Supplements*; 2000; http://vm.cfsan.fda.gov/~ dms/

13

ESTIMATION OF FOOD CHEMICAL INTAKE*

David Tennant

Governments worldwide have a responsibility to ensure that a safe and nutritious food supply is available to enable the public to choose a healthy and varied diet. As part of that responsibility there is a need for governments to conduct risk assessments for food chemicals such as additives, contaminants, and pesticides in order that appropriate risk management strategies can be applied.

A fundamental part of the risk assessment process is the estimation of the intakes of food chemicals. The accuracy of intake estimates must be sufficient to allow an optimum balance between the potential risks from chemicals and their potential benefits to be achieved. In recent years greater emphasis has been placed on improving the accuracy of dietary intake assessments, either by estimating "actual" rather than "potential" or "worst-case" intakes (1) or by starting with over-estimates and then using available information to make more realistic estimates (2,3).

It is also becoming clear that risk assessments must be based on the intakes of those groups of individuals who are at greatest risk rather than on the intakes of the general population. More emphasis has therefore been placed on protecting those "critical" or "nonaverage" groups of individuals who may be more suscep-tible to the specific toxic effects of a food chemical, those who consume greater quantities, or those who regularly consume foods with higher concentrations of the food chemical.

Different systems of varying degrees of complexity have evolved throughout Europe and North America to estimate the dietary intake of food chemicals and to assess risk to the "average," "nonaverage," or "critical" groups of consumers. This chapter aims to review the available methodology and to suggest a structured or "hierarchical" approach to the estimation of food chemical intakes.

---

* The opinions presented in this chapter are those of the authors and do not necessarily represent the policies of the MAFF Food Safety Directorate.

FOOD CHEMICAL RISK ASSESSMENT

Intakes of food chemicals are usually estimated in order to conduct a risk assessment. The three essential elements of the risk assessment process are the hazard evaluation, where all the available toxicologic or epidemiologic data are considered and a relationship between dose and effect established, the occurrence assessment, which seeks to establish the concentration or incidence of the hazard of concern in foods, and the consumption estimate, which seeks to establish how much of the foods of concern consumers actually eat (Figure 1). The occurrence assessment and consumption estimate are brought together to estimate the intake of the food chemical by the consumers of the foods in question.

The intake estimation can be expressed in different ways depending on the population examined. For example, intakes can be averaged for all the members of the study [those who are consumers (eaters) and those who are not (noneaters)] or averaged for the consumers (eaters) only. The intake estimation can differ depending on whether the intake is averaged over the study period (usually 1–14 days), or is expressed by the actual number of consuming days (eater-days). Finally the intake can be expressed as any percentile of the range of intakes (usually the 90th, 95th, 97.5th, or 99th percentile) or by body weight (usually by dividing through by each person's body weight or an average figure of 60 kg).

The intake assessment and the hazard evaluation are considered together to estimate risk by comparing the likely intake with an acceptable level of exposure derived from the hazard evaluation. If there is a possible risk associated with the use of a food chemical then risk management is required.

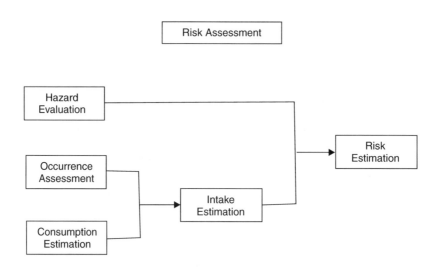

Figure 1 General scheme for food chemical risk assessment.

In order to protect all consumers, risk assessments for additives, contaminants, or pesticides are usually based on intake estimates at the upper end of the range of normal consumption patterns. This is usually achieved by estimating measures of intake at the 90th, 95th, 97.5th, or 99th percentiles of the range of consumptions. These percentiles are used to define the upper limit of the range of normal consumption patterns. The underlying assumption behind their use, rather than the maximum, is that dietary habits that give rise to intakes higher than such levels are unlikely to be maintained over a significant part of any individual's lifetime. This assumption is reasonable, for when a population is free to choose a varied diet, it is likely that the average consumption will stay fairly similar, but the individuals at the tail of the distribution probably vary.

It is important, however, that consumers who have intakes above the "top centile" are considered closely. If investigations confirm that these consumers are exhibiting dietary behavior that they are unlikely to continue on a regular basis, the "top centile" intake is probably sufficient for a risk assessment. However, if the individuals at the tail of the distribution form a distinct group, exhibiting a regular pattern of dietary behavior, then risk management based on this group may be required. Such atypical groups could be defined by age or sex, ethnicity, need for special diets like diabetics and vegetarians, special needs like pregnant or lactating women, or any combination of the above.

While the "top centile" approach is satisfactory for food chemicals that have chronic effects for which concern is about exposure over a lifetime, for chemicals that have acute toxic effects the amount consumed in 1 day or in one meal may be important. Under these circumstances the use of percentile estimate is not appropriate and the maximum amount likely to be consumed during a short time period must be considered. These maximum intakes would be compared to hazard evaluations based on acute exposure.

RISK ESTIMATION

For noncarcinogenic food chemicals and nongenotoxic carcinogens, the estimated dietary intake is often compared to an acceptable daily intake (ADI). The ADI is defined as "the amount of additive, expressed on a body weight basis, that can be ingested daily over a lifetime without appreciable health risk" (4). ADIs are based on the highest intake (expressed as mg/kg body weight/day) that gives rise to no observable adverse effect (NOAEL), usually in laboratory animals. Appropriate safety factors are then applied to allow for interspecies extrapolation and interindividual variation. The tolerable daily intake (TDI) is set in a similar fashion, for chemical contaminants.

The Joint FAO/WHO Expert Committee on Food Additives (JECFA) was the first to express its assessment of a compound in terms of the ADI, in 1961 (5), and in Europe, the European Community Scientific Committee for Food (SCF) has adopted the same approach. In 1989, the UK Committee on Toxicity of

Chemicals in Food, Consumer Products and the Environment (COT) started providing its advice in terms of ADIs (6). However, several authors have criticized the use of ADIs. For example, the NOAEL may be sensitive to sample size and can vary from experiment to experiment (7), and safety factors may not sufficiently protect vulnerable groups (8).

The regulation of carcinogens is considered a unique issue in risk assessment because of the lack of evidence of a threshold dose. A major influence on US policy in the regulation of carcinogens was the 1958 Delaney Amendment and the Color Additive Amendments of 1960 to the US Food and Drug Act of 1954. These legislative acts required that no chemical determined to be carcinogenic in either humans or animals be allowed as a food additive or color. However, because of the potential benefits of some food chemicals, the Environmental Protection Agency (EPA) has subsequently introduced the concept of "socially acceptable risk" (9). In these cases, the EPA uses quantitative risk assessment for carcinogenic compounds (including those believed to be nongenotoxic). This involves fitting a dose–response curve and extrapolating to predict an acceptable dose for humans (10). Intake studies are required to translate the acceptable dose into an acceptable concentration in food.

In other countries, including the United Kingdom, known genotoxic carcinogens are not permitted for use in foods. Thresholded, nongenotoxic carcinogens are allocated an ADI as for other food chemicals.

METHODS USED TO ESTIMATE FOOD CHEMICAL INTAKES

The estimation of intakes of food chemicals is a complex activity and no single approach is suited to all circumstances. Information on dietary patterns and on the levels of chemicals in food is often difficult to obtain and many uncertainties can be introduced. Intake assessments have been conducted differently in different countries because there are different risk management policies and legislation within each country. More importantly the approach chosen will depend on the type of food chemical and, in particular, the information resources available.

Some general strategies for estimating food chemical intakes that adopt a progressive or "hierarchical" approach have been described before (2,3,11). This hierarchical approach is particularly useful as it can be used to highlight the characteristics of each method, as we progress from the relatively simple to the very sophisticated (Figure 2). The collection of detailed intake data is time-consuming and expensive. However, it may not be necessary to undertake detailed and costly estimates of intake for each risk assessment. The hierarchical approach offers a cost-effective stepwise and flexible framework within which to estimate the intake of food chemicals. Resources can therefore be diverted to problems for which more detailed analysis is required.

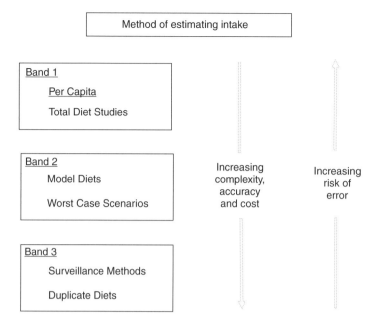

Figure 2 A hierarchical approach to the estimation of dietary intakes of additives and contaminants.

Each approach in Figure 2 will be reviewed in turn and where possible examples will be given of the ways these methods have been used in various countries. Methodologic difficulties in assessing food consumption have been reviewed extensively elsewhere (2,3,11–13), and these possible sources of error are discussed only briefly.

Per capita method for intake assessments

The per capita approach is the first method described in Figure 2 and the simplest. The estimate gives the average intake of every individual in the population. A per capita estimate can be used to explore likely intakes and is particularly useful for highlighting cases in which dietary intakes are high and more detailed assessment is required or to prioritize food chemicals for more rigorous investigation. Although per capita intake estimates do not permit analysis of variations because of different dietary habits, they do allow comparison between countries (14).

Most countries have sufficient information to make per capita intake estimates. These estimates can be made either by multiplying per capita food consumption by estimated concentrations and then summing the intakes or by dividing the total available food chemical by the number of individuals in the population. Modifications may include the use of more refined food consumption estimates or analytical data for the major source foods, and these approaches can allow a range of likely intakes to be estimated (Table 1).

Per capita intakes can be estimated by collecting manufacturer's information on the quantities available for use and then used to prioritize large numbers of additives for further work. Such a prioritization exercise may consist of four main steps. First, the per capita intake is expressed as a proportion of the ADI and the additives listed in descending order. Second, some assessment of additive use is made. For example, those additives found in a small range of foods likely to be consumed by a small group of the population would be assigned a high priority. Third, the results of any previous national and international studies on specific additives would be assessed. Finally, additives for further consideration would be selected within the constraints of available resources.

Per capita estimates have been used in several European countries to estimate the intakes of additives and contaminants. In Belgium, production, import, and export data were used with analytical data to estimate per capita intake of lead, cadmium, and mercury and were found to be in good agreement with intake estimates based on average household consumption (15). In Finland, intakes of 30 additives were estimated (16) using analytical and per capita consumption data. The results indicated that most dietary intakes were well below the ADI, although intakes of some additives, including nitrates, nitrites, saccharin, cyclamates, carrageenan, and benzoic and sorbic acid, were above or close to the ADI. These were prioritized for further work. Analytical data and an industrial survey were also used to estimate the per capita intake of several food colors (16). In Italy, per capita estimates were used to assess the population risk from intake of polycyclic aromatic hydrocarbon (PAH) derived from olive oil (17). In Spain, pesticide intakes were estimated (18) using analytical and per capita consumption data.

Although per capita intake assessments are valuable as a prioritization tool they do not take into consideration variations in consumption patterns of food or the distribution of intakes of food chemicals within the population. In particular,

Table 1 The per capita approach to the estimation of food chemical intake

1.	Multiply the average food consumption of the whole population by anticipated or actual levels of the food chemical. This method can estimate the intakes of additives and contaminants.
2.	Divide the total available food chemical by the number of individuals in the population. This method is more suited to estimate the intakes of additives.

Advantages	–	prioritizing food chemicals for more rigorous investigation.
	–	allows comparison across countries.
	–	cost effective exploratory tool.
Disadvantages	–	cannot be used to express "nonaverage" intake.
	–	relies on good manufacturing and demographic data.
Possible modifications	–	can be adjusted to take into account the consuming population, if these data are available.

they assume that all members of the population are consumers even though some individuals may not consume the source foods at all. They therefore cannot be used to estimate intakes of the "nonaverage" consumer or possible "at-risk" groups.

Total diet study method for intake assessments

In total diet (market basket) studies, data on food purchases obtained from national surveys are used to determine the average consumption of various representative groups of foods. Samples of each food are then purchased from retail outlets throughout the country and prepared (as if) ready for eating. They are combined into groups of like foods and analyzed to determine the food chemical concentrations. The intake from each food group is calculated by multiplying the concentration of the chemical in the food group by the amount of that food consumed, as determined from the food purchase survey. The intake from each food group can then be added to give the overall "average" intake. However, while it may be possible using this method to identify those food groups that contribute most to the intake, it is not possible to identify specific source foods within the groups (Table 2).

Total diet studies (TDSs) have been carried out in many countries since the early 1960s. Initially the purpose was to estimate background exposures of the population to pesticide residues (19) and radioactive contamination (20). The range of chemicals has since broadened to include toxic heavy metals and more recently has included a variety of trace elements and organic contaminants (21).

Table 3 gives a summary of the main characteristics of TDSs carried out in Europe and the United States. The dietary information used to formulate the "market basket" has been derived from different types of studies and may or may

Table 2 The Total Diet Study (market basket) approach to the estimation of food chemical intake

A "market basket" of food reflecting the total diet of consumers, as determined from expenditure surveys, is purchased, prepared as if for consumption, and analyzed in groups.	
Advantages	– if repeated can provide useful trend information on dietary intakes of additives, pesticides, or contaminants.
	– can determine the food group contributing the major source of food chemical.
	– provides "background" data in the event of a local contamination.
Disadvantages	– intake usually relates to average consumer.
	– cannot pinpoint the individual foods that are the major sources of the food chemical.
	– may not mirror preparation habits of consumers.
Possible modifications	– the major sources of food chemical in diet may be analyzed (e.g. fish).
	– the consumers may be restricted to at-risk groups.

Table 3 Total diet (market basket) studies in Europe and the United States

Country	Group	No. of food items (groups)	Reference
United Kingdom			
1966–1975	Average person within a	78 (11)	(19)
1976–1981	household	68 (9)	(22)
1981–present		115 (20)	(23)
United States			
1961–1971	Males 16–19 yrs	82 (11)	(20)
1971–1975	Males 15–20 yrs	120 (12)	(25)
1975–1982	3 age/sex groups	50–120 (11–12)	(26)
1982–present	8 age/sex groups	201–230 (234)	(21,27)
The Netherlands			
1976–1978	Males 16–18 yrs	126 (12)	(28,29)
1984–1986	Males 18 yrs	221 (23)	(30)
Spain			
1988–present	25–60 yrs	91 (16)	(31)
Italy			
1980–1984	Average person within a household	130 (11)	(32)

not have included water, alcoholic drinks, or foods eaten outside the home. Grouping foods prior to analysis is required for the TDS to be cost effective, but too few groups may lead to the dilution of the food chemical below the detection limit of the analytical method. The TDSs reported here either use 9–23 food groups or analyze individual foods. The UK TDS (19,22,23) is based on the National Food Survey (24) and the US TDS (20,21,25–27) is based mainly on the Household Food Consumption Survey, both of which are household food disappearance studies. The first Netherlands TDS (28,29) was based on several pieces of information, including food balance sheets; the second (30) was based on dietary information collected from a diet history study over 14 days. The Spanish TDS was based on a 24-hour recall (31). An Italian reference diet (32) has been constructed to monitor food chemical intake. There is only limited published information on the TDSs carried out in Finland (33), Denmark (34), and Germany (35).

It is possible to modify the TDS approach in several different ways in order to estimate the intake of the "nonaverage" consumer:

1   Determine the "at-risk" group and conduct a food purchase study on this group.
2   Calculate intakes from the maximum chemical concentration levels and assume that the "top centile" consumer always has an intake at this level.
3   Assume that the 97.5th percentile intake is approximately three times the mean TDS intake. This relationship has been found to hold fairly well for individual foods (36), but it is less likely to be reliable when the chemical is present in several foods.

None of these approaches can be recommended if an accurate estimate of intake is needed, particularly when the chemical occurs at different concentrations in various foods. Safety factors or some modeling may be used to improve the estimate.

Model diet method for intake assessments

In the absence of detailed dietary data several modeling approaches have been developed to estimate food chemical intakes. A model diet may simply consist of typical serving sizes of the foods of interest. If the concentration of the food chemical is known then the intake from one serving of a food product can be expressed as a percentage of the ADI or TDI. This method is particularly useful when few consumption data exist or when the major source of the food chemical is derived from one food (Table 4). Modeled intakes should not normally be used as the sole basis for a risk assessment. However, provided that the modeled intake is based on sound information that provides a conservative estimate and there are no known "at-risk" groups, it may be adequate as a technique for investigating chemicals whose intakes may approach the ADI. For example, in the United Kingdom, as part of the Colour Usage Survey (1980) based on per capita data the potential intake of erythrosine was estimated to be higher than the COT ADI of 0.1 mg/kg body weight/day. Erythrosine is especially important for its use in glacé cherries. By considering the likely portion size of glacé cherries, the Food Advisory Committee recommended that erythrosine should only be used in this food and limited to a maximum level of 200 mg/kg if the ADI was not to be exceeded.

The Danish Budget method (37) is a modeling approach that has been widely applied to food additives. It is based on the assumption that the intake of additives is limited by the amount of energy required and the quantity of food and beverage one individual can consume each day. By making assumptions about average energy requirements, food and beverage consumption and the proportion of processed to fresh items in the diet, a maximum permitted level can be estimated.

Table 4 The "model diet" approach to the estimation of food chemical intake

Available information is used to construct a hypothetical or "model" diet. The veterinary medicine diet (38) and the Dutch Budget Method (37) could be considered under this approach.

Advantages	–	can be a cost-effective way of estimating dietary intakes.
Disadvantages	–	subject to error when many foods are involved.
Possible modifications	–	different proposed level of use or additional information can be incorporated into the model.
	–	different age groups can be modeled.

Sources: Hansen SC (37).
        Joint Expert Committee on Food Additives (38).

A hypothetical adult diet of meat and animal products was proposed by JECFA and confirmed at its 36th meeting in 1990 (38). In order to protect all segments of the population, exaggerated consumption values were used. This diet has been used to estimate maximum residue levels (MRLs) for veterinary drug residues and comprises 300 g lean muscle, 100 g liver, 50 g kidney, 50 g fat, 1.5 l milk or milk products, 100 g eggs, and 20 g honey.

If this hypothetical diet is compared to the consumption data collected from the UK Dietary and Nutritional Survey of British Adults (39) it can be shown that it contains at least twice the amount of animal products eaten by the "top centile" (97.5th percentile) adult consumer in the United Kingdom. A few adults consume quantities of individual foods that exceed the amount of that food in the hypothetical diet but no adult consumes *all* food items in the quantities suggested by the hypothetical diet. Furthermore the hypothetical diet supplies more calories from these animal products than expected for the entire adult diet. The model therefore probably provides an overestimate of the true intakes of veterinary drug residues in the United Kingdom.

Scenarios used for intake assessments

In some cases, it may be possible to construct a range of "scenarios" that reflect the least to the worst possible case in terms of food consumption, anticipated concentrations, or both. This is a slightly more refined method than modeling as it assumes that more information is available. The approach attempts to extend the usefulness of the data by predicting trends or intakes of "at-risk" groups (Table 5).

There is scope to incorporate a large number of variables in order to make the estimates more realistic and to generate a range of possible intakes. These factors may include, for example, brand loyalty if considering additives, differing levels of migration if considering the potential intake of packaging material, or varying preparation habits if considering pesticide intake. The output from this kind of

Table 5 The "scenarios" approach to the estimation of food chemical intake

A range of "scenarios" is constructed that reflects average to the worst possible case in terms of food consumption, likely concentration, or both. This is a useful starting point and can be used to screen potential problems for further work.

Advantages	–	useful when there are limited data or the food chemical is relatively new.
	–	powerful tool to address difficult or new problems that may require creative solutions.
Disadvantages	–	by definition these estimates are imprecise.
	–	relies on reliable information on the food chemical use as well as consuming patterns of the population of interest.
Possible modifications	–	these include methods for making the estimates more realistic, and will vary.

Occurrence

(or concentration of food chemical)

		"Typical"	"Worst case"
Food Consumption	"Typical"	likely	possible
	"Worst case"	possible	unlikely

Figure 3 Matrix of the intake estimates using the "scenario" approach.

analysis usually comprises a matrix (Figure 3) that reflects the range of values from the typical (or likely) case to the worst possible case (and least likely to occur). The probable intake of the "top centile" consumer will probably lie between these limits.

Surveillance methods for intake assessments

If the methods described above produce inconclusive results, it may prove necessary to look at particular food chemicals in greater detail. Reliable food consumption data are required for the population of interest and this is combined with data on anticipated concentrations in order to estimate intakes (Table 6). If the food chemical is present in a limited range of foods then it may be possible to identify a particular subpopulation who are consumers of these foods and to conduct a targeted survey of their dietary habits. For example, it is likely that diabetics or individuals on weight-loss programs will have higher intakes of sweeteners than the average consumer.

Some food additives, however, have several functions and some contaminants may be found in a wide range of foods. It would not be feasible in such circumstances to conduct a specific diary study for each additive or contaminant. In these circumstances national dietary surveys of individuals may be a more cost-effective means of obtaining information on food consumption patterns.

Several countries have collected or are in the process of collecting consumption and analytical data suitable for this method of intake assessment. In Finland, in connection with the Multicenter Study on Atherosclerosis Precursors in Finnish

273

Table 6 The surveillance approach to the estimation of food chemical intake

Consumption data of individuals are combined with residue data. Consumption data can be collected by various methods (e.g. food frequency or weighed food diary). Exposure data may be maximum residue (permitted) levels, manufacturing levels, or analyzed values	
Advantages	– if the dietary survey covers a large representative sample of the population, and records of all foods eaten are taken, the resulting consumption data can be used to estimate the intake of a wide range of additives and contaminants.
	– the intake of "average" and "nonaverage" consumers can be estimated.
Disadvantages	– critical group may not be adequately represented.
	– insufficient data will be collected on the rarely consumed foods.
	– expensive.
Possible modifications	– the definition of "nonaverage" consumer may differ.
	– a range of intakes can be estimated by considering different combinations of concentrations and consumption values.
	– specific surveys can also be conducted on the range of foods or subgroup of the population.

Children, the dietary intakes of mercury, lead, cadmium, and arsenic were assessed (40). In Germany, mercury, lead and cadmium intakes in 2–14-year-old German children were estimated using food consumption data and data on the heavy metal content of German foods (41).

Several dietary studies have been conducted in the United Kingdom (42) and further dietary studies designed to assess intakes by at-risk groups are proposed. A major source of consumption data at present is the Dietary and Nutritional Survey of British Adults (39). This study was carried out between 1986 and 1987 on behalf of Ministry of Agriculture Fisheries and Food and the Department of Health (39). A nationally representative sample of over 2,000 adults aged 16–64 completed a 7-day weighed diary of foods consumed inside and outside the home. Over 4,500 foods have been coded and a recipe database can be used to express consumption data on the basis of complete foods or individual food components. Estimates can be made of the mean, median, and any other percentile of food consumption for single foods or groups of foods. Similar statistics may be derived for food chemical intake given their known or estimated concentrations in each food. Intakes can then be computed for each individual on a body weight basis. The dietary intakes are sorted so that any percentile intake can be estimated. In future, it will be possible to use dietary data from other age groups as the National Diet and Nutrition Survey (NDNS) comprises a continuous program of diary-based surveys of infants, school-children, adults, and the elderly.

There are two major sources of consumption data used in the United States, the Market Research Corporation of America (MRCA) food consumption data and the United States Department of Agriculture (USDA) Nationwide Food Consumption Survey (NFCS) data. MRCA provides consumption data to the Food and Drug Administration (FDA) under contract. The 1977–8 MRCA survey collected frequency data over 14 days from approximately 12,000 individuals. This is combined with portion size data from the 1977–8 USDA survey to estimate consumption of the individual. Technical Assessment Systems Inc., under contract from the EPA, has developed a menu-driven computer system to estimate intake. The food consumption data are derived from the USDA 1987–8 NFCS. The NFCS was designed as a stratified survey using 3-day dietary records to collect consumption data. There are 22 subpopulations defined in terms of demographic variables such as age, gender, and race.

Two main types of intake assessments are made in the US system, depending on whether the food chemical exhibits chronic or acute toxic response. Chronic intake assessments use average consumption of the sample population. Acute intake assessments use average consumption of the consumers on the days they consume. In the past the FDA commonly used the "estimated daily intake for the 90th percentile consumer of a food additive as a measure of high chronic exposure because that level of consumption is significantly above the intake of the average consumer and provides an added conservatism with respect to the ADI for those individuals who regularly consume food containing the additive" (43).

Duplicate diet method for intake assessment

In the most refined studies of intakes, individuals most "at-risk" of high dietary intakes may be requested to supply a duplicate of their meals to be analyzed (Table 7). Analyses of duplicate diets are valuable in determining the dietary intake from foods prepared as they would be eaten and provides an almost ideal means of assessing intakes; however, the participants may not correctly divide the meals between themselves and the "duplicate plate" and it is also difficult to

Table 7 The duplicate diet study approach to the estimation of food chemical intake

Duplicate diet studies provide a means of looking at diets of individuals identified as potentially most at risk because they are thought to be (1) more susceptible to the food chemical, (2) more likely to consume foods with a higher concentration of the food chemical, or (3) consume greater quantities of the source foods.

Advantages	–	directly measures the intakes of "at-risk" groups.
	–	probably the best way to measure the intake of contaminants, especially if they are widespread at low concentrations.
Disadvantages	–	"at-risk" groups may be very different depending on the food chemical and can change over time.
	–	expensive.
Possible modifications	–	a cross-section of the population may be studied or the duplicate diet study could be carried out in specific sites with the highest contamination levels.
	–	specific surveys can also be conducted on a narrower range of foods or subgroup of the population.

Table 8 Duplicate diet studies in Europe

Country	Group	Food type	Reference
United Kingdom			
1979	Fishing communities	Fish	(44)
1979	Community	Local fruit & vegetables	(45)
1982	Children	Whole diet	(46)
1984	Preschool children	Whole diet	(47)
1988	Community	Local produce	(48)
The Netherlands			
1976–1978	Volunteers	Whole diet	(49)
1984–1985	Adults	Whole diet	(50)
Belgium			
1983	30–65-year-olds	Whole diet	(51)
Finland			
1987	Middle-aged men	Whole diet	(52)
Denmark			
1986–1987	Community	Water & vegetables	(53)

identify the individual food sources contributing to any contamination. Table 8 shows a summary of the main characteristics of some duplicate diet studies carried out in Europe (44–53). Previous work is often used to indicate which groups are "at-risk" of high intakes and the foods selected for investigation will depend on the food chemical of concern; however, it may not be possible to identify the group or groups most "at-risk" and the identity of any such group may change over time. Duplicate diet studies are probably the best way to estimate contaminant intake of "at-risk" individuals but they are a very expensive approach.

ESTIMATING LEVELS AND OCCURRENCES OF FOOD CHEMICALS

The methods described so far have shown how food consumption data of increasing levels of detail can be used to produce more accurate estimates of dietary intakes when these are required. Of course, it is also possible to use different data on the levels and occurrence of food chemicals. The simplest methods will use levels set in legislation to investigate likely intakes; however, not all food will contain chemicals at the maximum levels set in legislation. More accurate estimates can be based on the levels that are likely to be actually present. This information can be obtained from manufacturers or could relate to the level required in a food to achieve a given technological function. The most accurate method is to obtain data on actual levels and occurrences through food surveillance programs (54,55). Food surveillance is very expensive, however, and has only proved practicable for contaminants for which there is no other way of obtaining the data. Here again the level of detail required will depend on the purpose to which the risk assessment is to be put.

POSTMARKETING SURVEILLANCE

Dietary intakes of food chemicals have to be regularly updated and monitored in order to ensure that the risk assessment remains valid. Trade, market, and surveillance data become increasingly important for monitoring any changes in the pattern of use of food chemicals, and can be used to indicate the need to revise estimates of intake. Any additional information on dietary patterns of "at-risk" groups and data suggesting uneven distribution of the food chemicals in food need to be fed into this postmarketing evaluation process.

Postmarketing surveillance exercises must be carefully planned because the consumption patterns of food in which the chemical is presently used can vary along with the range of foods in which the chemical is used. Consumption patterns can be influenced by changing dietary habits, social circumstances, or even the presence of a certain additive in food. For example, the introduction of a fat replacer into certain fatty foods could result in a significant increase in the consumption of those foods if they are now perceived to be "low fat." Patterns of additive use can also be interrelated. For example, a move away from synthetic colors by the food industry could result in an increase in the use of natural colors and indicate a need to reevaluate intakes.

Over a number of years the intake of a food chemical can be expected to reach a stable level; however, this should not necessarily be taken as a sign that further postmarketing surveillance is unnecessary. The introduction of one new use might not affect the average intake but if that product is consumed by a critical group it could result in a significant change to the risk estimation.

Postmarketing surveillance is particularly important for novel products for which market predictions are very unreliable. The lack of historical data on either the likely consumers or their patterns of consumption may indicate the need for a comprehensive postmarketing surveillance exercise.

INTERNATIONAL GUIDELINES

International guidelines for additives, pesticides, and contaminants have adopted progressive or "hierarchical" approaches (2,3), and have provided a useful framework with which to compare dietary intake estimates across the EC (56).

Additives

The Codex Committee on Food Additives and Contaminants (CCFAC) has developed guidelines for a simple evaluation of additive intake (3). The guidelines suggest that the first step should consider per capita consumption and maximum permitted levels. This has been called the theoretical maximum daily intake (TMDI). A more refined estimate would then consider per capita consumption and analytical data; this has been called the estimated daily intake (EDI).

Pesticides

The Codex Committee on Pesticide Residues (CCPR) has also developed guidelines for estimating pesticide intake (2) using per capita consumption values derived from the most recent FAO food balance sheets. Nine "cultural" diets have been developed; African, cereal-based; African, root and tuber-based; North African; Central American; South American; Chinese; Far Eastern Mediterranean; and European. A "global" model diet has been constructed by using the highest average food consumption value for individual foods from each of the "cultural" diets and then adjusted to give a total of 1.5 kg/day of solid food (57). The guidelines (2) suggest the first estimate uses the "global" or national diet and assumes all source foods contain the highest residue level (TMDI), then a "cultural" or national diet with any possible processing losses (called estimated maximum daily intake, EMDI), and ends by estimating intakes using national consumption data and monitoring data (EDI).

Contaminants

The JECFA has evaluated a large number of food contaminants and established "provisional tolerable weekly intakes" (PTWIs). The Joint UNEP/FAO/WHO Food Contamination Monitoring Programme or GEMS/Food provides information used in these safety considerations (56). There are 39 countries participating in GEMS and the information collected (57–59) on contaminant intake has been mainly derived from total diet "market basket" studies. The composition of the diet and preparation for analysis vary from country to country but the method allows the monitoring of trends within each country.

DIETARY INTAKE OF FOOD ADDITIVES

Estimating the intake of additives presents unique problems, as the consumption pattern and market use can change relatively quickly. This is particularly so for new additives as new products are developed and marketed. When a submission for a new additive enters the regulatory process for approval, an assessment of the safety-in-use of the additive is considered. However, at this early stage it may not be possible to accurately estimate the potential intake of the additive, and interim methods, which are sufficiently conservative, are desirable. Aspartame is a good example with which to illustrate these issues because its intake has been estimated in several countries in different ways preceding approval and during the first years of use. After approval, various more specifically designed dietary surveys were conducted to estimate the intake of aspartame in potentially "at-risk" groups.

Table 9 shows the estimated intakes of aspartame within the context of the hierarchical approach to food chemical intake. The hierarchical approach is a stepwise progression from relatively simple methods to the very sophisticated, from "average" estimates to "at-risk" consumer estimates, from "pen and paper" exercises to specifically designed studies, and from using maximum permitted levels to analyzed values.

The per capita estimates derived in two different ways in the United Kingdom and Canada (60) suggest intakes in the range 2.5–4 mg/kg body weight/day. The JECFA ADI for aspartame is 40 mg/kg body weight/day (the FDA ADI is 50 mg/kg body weight/day). The SCF and the COT ADI for aspartame is 40 mg/kg body weight/day. Considering the per capita estimate is 6–10%, the JECFA/SCF/COT ADI and the intake of aspartame is unlikely to be evenly distributed across all consumers, therefore further work was required.

Table 9 Intake of aspartame in various countries

Approach	Intake of aspartame (mg/kg body weight/day)[a]				
	Canada (60,62)	*United Kingdom*[b]	*United States (61,63)*	*Germany (64)*	*Finland (65)*
Per capita					
Food balance × MPL[c]	4				
Total available/population		2.5			
Model					
Aspartame replaces sugar	11		8.3		
Scenario					
One serving of each food		21.1			
Aspartame replaces all carbohydrates			25.0		
Surveillance					
24-hour recall × man. use[d]	3.25				
7–14-day diary × man. use					
General population	6.8–7.7 (95[e])	1.8 (97.5[e])	1.6–2.3 (90[e])	1.2 (mean)	
Diabetics	6.2–14.4 (95[e])	9.9 (97.5[e])	2.1–3.3 (90[e])		1.15 (mean)
On weight loss diets	7.5–9.4 (95[e])		1.6–2.6 (90[e])		
On sugar-avoiding diets	8.5–10.4 (95[e])				
Pregnant females			1.3–2.7 (90[e])		

Notes
a JECFA ADI for aspartame is 40 mg/kg body weight/day, FDA ADI is 50 mg/kg body weight/day.
b UK unpublished data from 1987 sweetener study (42).
c Maximum permitted level.
d Intake of population (eaters and noneaters).
e Percentile intake of consumers only (eaters only).

Some models have been developed in North America (61) and Canada (60). They assume that aspartame would replace sugar consumption (assuming 1 g sugar is equivalent to 5 mg aspartame). These suggest an average intake in the range of 8.3–11 mg/kg body weight/day.

"Scenario" estimates by definition are usually high and several have been conducted in North America (61) and the United Kingdom as "premarketing" estimates of potential intake. Assuming the consumption of one serving (portion) of each source food or that aspartame replaces all dietary carbohydrate, "scenario" estimates suggested a maximum intake in the range 21.1–25 mg/kg body weight/day.

All the "pen and paper" estimates mentioned above served a useful purpose in the interim period between the time aspartame was first approved and sufficient information became available to conduct more detailed intake assessments. Some projected "average" intakes were sufficiently close to the ADI to raise concerns about consumers with potentially high intakes such as individuals on weight-loss programs or diabetics. These concerns stimulated the regulatory authorities in several countries to request dietary surveys of the general population and the potential "at-risk" groups to monitor intakes of aspartame.

Dietary intakes of sweeteners have received much detailed attention recently, and several targeted dietary surveys have been conducted (42,62–65). These have provided some evidence to suggest that diabetics, high consumers of soft drinks or table-top sweeteners, individuals on weight-loss programs, and children may have higher intakes of sweeteners than typical adults. These more detailed intake assessments were based on consumption data and manufacturers' use of aspartame.

Table 10 gives a summary of the main characteristics of the dietary studies conducted in five different countries (42,62–65) designed to estimate the intake of aspartame. The method of data collected differed across studies; for example, some studies were based on 24-hour recalls while others collected consumption data over 14 days. Some of the surveys relied on serving information (then matched to portion size data) while others encouraged the participants to weigh all food items. Sample selection and the size of the studies differed. Most of the studies collected brand data and obtained aspartame use information from labels or manufacturers.

The second half of Table 9 shows the range of estimated aspartame intakes from five different countries (42,62–65). In three of these countries the aspartame intakes for "top-centile" groups were estimated. For comparative purposes the intakes of the consumers only (eaters only) have been listed in Table 9. Various percentiles of intake (90th, 95th, and 97.5th percentiles) are quoted to indicate the range at the upper end of the intake distribution. A range of values is given in the Canadian study (62) that coincides with Wave I (Feb/April 1987) and Wave II (July/Sept 1987). The range in the US study (63) coincides with data from 1984 to 1988.

Taken together these three studies have investigated intakes of aspartame over a number of years and in different seasons, over populations in three countries, three dietary habit groups (diabetics, individuals on weight-loss diets, and

Table 10 Specific surveillance studies designed to estimate the intake of sweeteners

Country	Source of information	Sample size	Output variables	Reference
Canada	24-hour dietary recall. Level of use from food labels.	13,000	Mean intake (eaters & noneaters) for 1–4 years old (M&F), 12–19 (M) and all age groups (M&F).	(60)
Canada	7-day diary survey in Feb/April and Jul/Sept 1987. Brand data collected. Level of use from manufacturers for 145 products. Body weight recorded.	5,544 4,872	Intake (eaters & noneaters), Intake by consumers (eaters), Intake by consumers on consuming days (eater-days) for eight age-sex groups & three dietary habit groups.	(62)
United Kingdom	7-day weighed diary in Sept 1987 in general population. 4-day weighed diary in July/Aug 1988 in diabetic population. Brand data collected. Most level of use from man. Body weight recorded.	681 89	Intake by consumers (eaters only)[a] for six to eight age/sex in both populations. Diabetic group not representative.	
United States	14-day frequency survey. Manufacturers data on level of use. Data on intake collected from 1984–1988.	2,000/yr	Intake by consumers (eaters), Intake by consumers on consuming days (eater-days) for six age/sex groups, two dietary habit groups, and pregnant females.	(63)
Finland	Two 48-hour dietary recalls on adolescents aged 11.5–17.5 years. Level of use from labels or manufacturers.	152 diabetics; 74 nondiabetics	Mean intake (eaters & noneaters), Intake of consumers (eaters only) for both groups.	(64)
Germany	1-day survey of representative sample in Sept 1988. 7-day survey on sample with high intake (75% ADI) in May 1989. Data on level of use from manufacturers.	2,291 40	Mean intake of all sweetener users and specific sweeteners (eaters only).	(65)

Note
a UK unpublished data from 1987 sweetener study (42).

individuals on sugar-avoiding diets), and pregnant females, and in total involved over 36,000 individuals (households). Despite the differences in study methodology the estimated intakes of aspartame by "average" and "top-centile" consumers all fell below the ADI, even within subpopulations considered "at-risk" for high intakes.

Additive use can change quite rapidly, however, and these changes may influence the intake assessment. Some of the surveys mentioned above collected dietary information on a small range of foods, while others were more extensive. These smaller studies will have limited usefulness if the additive is used in an increasing number of food products. With greater additive use, it is likely that a greater proportion of a consumer's diet will contain the additive. Some consumers may regularly choose branded foods with the highest permitted concentrations of the additive. They are therefore more likely to have high intakes.

Postmarketing surveillance is necessary to ensure that the risk assessment made prior to placing a new sweetener on the market is valid. In the early stages this might need to be highly intensive. However, if there is evidence that the dietary intakes have stabilized, less comprehensive surveillance may be acceptable. It would need to be shown, from dietary intake estimates over several years and for different age groups, that there are no "new" pathways for intake, that the intake of the "at-risk" groups had not increased, that no "new" "at-risk" groups were identified, and that trade and market data on additive use showed the same degree of leveling-off.

CONCLUSIONS

There is a need for simple approaches to estimate dietary intake of food chemicals in several situations:

1 To prioritize food chemicals for further investigation;
2 To monitor trends and changes in dietary intake;
3 To provide estimates when detailed information is not available;
4 To facilitate negotiations on acceptable levels between countries; and
5 To make preliminary risk assessments when the food chemical is found in a limited number of foods.

An increasingly complicated approach is needed when the additive, contaminant, or pesticide is found in a wide range of foods or enters the food chain by different pathways such as via veterinary drugs that are also used as pesticides. Natural background levels are usually considered when estimating intakes of contaminants. Greater sophistication is also required when considering food chemicals with chronic effects or when lifetime exposure to local contamination is thought possible. Risk assessments for food chemicals must also consider potential "critical" groups, and this usually requires detailed information.

For risk assessment purposes it is not sufficient to consider average levels of intake only. The likely levels of intake of those who consume large amounts of the source foods, particularly when these consumers form distinct population subgroups, must also be considered. Information on dietary patterns and on the levels of chemicals in food is often difficult to obtain and many uncertainties can be introduced. The chosen approach should therefore take into account the quality of the data and the purpose to which the intake assessment will be put, so that an appropriate level of accuracy can be achieved.

In summary, the estimation of intakes of food chemicals is a complex activity and no single approach is suited to all circumstances. The various approaches employed must be designed to set consumer safety as the highest priority. While simple methods can be used for prioritization, the use of crude and inaccurate intake estimates in food chemical risk assessments can result in suboptimal risk management solutions being adopted. The cost of such errors is borne by the consumer either in an increased health risk from food or in restrictions in the supply and variety of food. It is therefore the duty of regulators and of the food industry to ensure that accurate estimates of intake are available whenever they are required.

REFERENCES

1. Fisher CE, Norman JA. Do actual intakes ever equal potential intakes? In: MacDonald I, ed. *Monitoring dietary intakes*. New York: Springer-Verlag; 1991; 213–20 (ILSI Monographs).
2. WHO Guidelines for predicting the dietary intakes of pesticide residues. Prepared by the Joint UNEP/FAO/WHO Food Contamination Monitoring Programme in collaboration with the Codex Committee on Pesticide Residues. Geneva: World Health Organisation; 1989.
3. WHO Guidelines for simple evaluation of food additive intake. Prepared by the Joint FAO/WHO Food Standards Programme Codex Alimentarius Committee on Food Additives. Geneva: World Health Organisation; 1989.
4. WHO Principles for the safety assessment of food additives and contaminants in Food. Environmental Health Criteria document no 70. Geneva: World Health Organisation; 1987.
5. Lu FC. Acceptable daily intake: inception, evolution, and application. *Regul Toxicol Pharmacol* 1988; 8:45–60.
6. Rubery ED, Barlow SM, Steadman JH. Criteria for setting quantitative estimates of acceptable intakes of chemicals in food in the UK. *Food Addit Contam* 1990; 7(3): 287–302.
7. Liesenring W, Ryan L. Statistical properties of NOAEL. *Regul Toxicol Pharmacol* 1992; 15:161–71.
8. Babich H, Davis DL. Food tolerances and action levels: Do they adequately protect children? *Bioscience* 1981; 31(6):429–38.
9. Whysner J, Williams GM. International cancer risk assessment: the impact of biologic mechanisms. *Regul Toxicol Pharmacol* 1992; 15:41–50.

10. Krewski D, Wargo J, Rizek R. Risk of dietary exposure to pesticides in infants and children. In: MacDonald I, ed. *Monitoring dietary intakes*. New York: Springer-Verlag; 1991; 75–89 (ILSI Monographs).

11. WHO Guidelines for the study of dietary intakes of chemical contaminants. Prepared by the Joint UNEP/FAO/WHO Global Environmental Monitoring Programme. Offset Publication no 87, Geneva: World Health Organisation; 1985.

12. Krantzler NJ, Mullen BJ, Comstock EM, Holden CA, Schutz HG, Grivetti LE, Meiselman HL. Methods of food intake assessment – an annotated bibliography. *J Nutr Educ* 1982; 14(3):108–18.

13. Stockley L. Changes in habitual food intake during weighed inventory surveys and duplicate diet collections. A short review. *Ecol Food Nutr* 1985; 12:263–9.

14. Louekari K, Salminen S. Intake of heavy metals from foods in Finland, West Germany and Japan. *Food Addit Contam* 1986; 3:355–62.

15. Fouassin A, Fondu M. Estimation of the average intake of lead and cadmium from the Belgium diet. *Arch Belg Med Soc Hyg Med Trav Meg Leg* 1980; 38:453–67.

16. Penttilä PL, Salminen S, Niemi E. Estimates on the intake of food additives in Finland. *Z Lebensm Unters Forsch* 1988; 186:11–5.

17. Menichini E, Bocca A, Merli F, Ianni D, Monfredini F. Polycyclic aromatic hydrocarbons in olive oil on the Italian market. *Food Addit Contam* 1991; 8(3):363–9.

18. Carrasco JM, Cunat P, Martinez M, Primo E. Pesticide residues in Total Diet Samples, Spain – 1971–72. *Pestic Monit J* 1976; 10(1):18–23.

19. Harris JM, Jones CM, Tatton J O'G. Pesticide residues in the Total Diet in England and Wales 1966–1967. Organisation of a Total Diet Study. *J Sci Fd Agric* 1969; 20: 242–5.

20. A-Laug EP, Mikalis A, Bollinger HM, Dimitroff JM; B-Deutsch MJ, Duffy D, Pillsbury HC, Loy HW; C-Mills PA. Total Diet Study: A. Strontium-90 and Caesium-137 content. B. Nutrient content. C. Pesticide content. *J Assoc Off Agric Chem* 1963; 46: 749–67.

21. Pennington JA, Gunderson EL. History of the Food and Drug Administration's Total Diet Study – 1961 to 1987. *J Assoc Off Anal Chem* 1987; 70(5):772–82.

22. Buss DH, Lindsay DG. Reorganisation of the UK Total Diet Study for minor constituents of food. *Fd Cosmet Toxicol* 1978; 16:597–600.

23. Peattie ME, Buss DH, Lindsay DG, Smart GA. Reorganisation of the British Total Diet study for monitoring food constituents from 1981. *Fd Chem Toxic* 1983; 21(4): 503–7.

24. Household Food Consumption and Expenditure. *Annual report of the national food survey Committee*, UK: HMSO; 1989.

25. Duggan RE, Cook HR. National food and feed monitoring program. *Pestic Monit J* 1971; 5:73–212.

26. Johnson RD, Manske DD, New DH, Podrebarac DS. Food and feed: pesticides and other chemical residues in infant and toddler Total Diet samples-(I)-August 1974–July 1975. *Pestic Monit J* 1979; 13:87–98.

27. Gunderson EL. FDA Total Diet Study, April 1982–April 1984, dietary intakes of pesticides, selected elements, and other chemicals. *J Assoc Off Anal Chem* 1988; 71(6): 1200–9.

28. de Vos RH, Van Dokkum W, Olthof PDA, Quirijns JK, Mugs T, Van Der Poll JM. Pesticide and other chemical residues in Dutch Total Diet samples (June 1976–July 1978). *Fd Chem Toxic* 1984; 22,1:11–21.

29. Van Dokkum W, de Vos RH, Cloughley FA, Hulshof KFAM, Dukel F, Wijsman JA. Food additives and food components in total diets in The Netherlands. *Br J Nutr* 1982; 48:223–31.

30. Van Dokkum W, de Vos RH, Muys TH, Wesstra JA. Minerals and trace elements in total diets in The Netherlands. *Br J Nutr* 1989; 61:7–15.

31. Urieta I, Jalon M, Garcia J, de Galdearo LG. Food Surveillance in the Basque country (Spain). I. The design of a Total Diet Study. *Food Addit Contam* 1991; 8(3):371–80.

32. Turrini A, Saba A, Lintas C. Study of the Italian reference diet for monitoring food constituents and contaminants. *Nutr Res* 1991; 11:861–73.

33. Varo P, Koivistoinen P. Mineral element Composition of Finnish foods. X11. General discussion and nutritional evaluation. *Acta Agricult Scand* 1980; 22:165–71.

34. Solgaard P, Aarkrog A, Fenger J, Flyger H, Graabæk AM. Lead in Danish food stuffs, evidence of decreasing concentrations. *Danish Med Bull* 1979; 26:179–82.

35. Diehl JF, Frinclik O, Müller H. Radioactivity in total diet before and after the Chernobyl reactor accident. *Z Lebensm Unters Forsch* 1989; 189:36–8.

36. Coomes TJ, Sherlock JC, Walters B. Studies in dietary intake and extreme food consumption. *R Soc Health* 1982; 102:119–23.

37. Hansen SC. Conditions for use of food additives based on a budget for an Acceptable Daily Intake. *J Food Protect* 1979; 42(5):429–34.

38. WHO Evaluation of certain veterinary drug residues in food. Thirty-fourth report of the Joint FAO/WHO Expert Committee on Food Additives. WHO Technical Report Series no 788 Geneva; 1989.

39. Gregory J, Foster K, Tyler H, Wiseman M. *The dietary and nutritional survey of British adults*. UK: HMSO; 1990.

40. Mykkänen H, Räsänen L, Ahola M, Kimppa S. Dietary intakes of mercury, lead, cadmium and arsenic by Finnish children. *Hum Nut Appl Nutr* 1986; 40A:32–9.

41. Stolley H, Kersting M, Droese W. Spurenelement- und Schwermetallaufnahme mit der Nahrung von 2–14 Jahre alten Kindern. *Monatsschr Kinderheilkd* 1981; 129: 233–8.

42. Ministry of Agriculture, Fisheries and Food. Intakes of Intense and Bulk Sweeteners – the UK 1987–1988. *Food Surveillance paper no 29*, UK: HMSO; 1990.

43. Food and Drug Administration: Food additives permitted for direct addition to food for human consumption, aspartame (final rule). *Fed Regist* 1986; 51:42999–43002.

44. Haxton J, Lindsay DG, Hislop J *et al*. Duplicate diet study in fishing communities in the United Kingdom: Mercury exposure in a critical group. *Environ Res* 1979; 10: 351–8.

45. Sherlock JC, Smart GA, Walters B, Evans WH, McWeeny DJ, Cassidy W. Dietary surveys on a population at Shipham, Somerset, United Kingdom. *Sci Total Environ* 1983; 29:121–42.

46. Sherlock JC, Barltrop D, Evans WH, Quinn MJ, Smart GA, Strehlow C. Blood lead concentration and lead intake of children of different ethnic origin. *Human Toxicol* 1985; 4:513–9.

47. Smart GA, Sherlock JC, Norman JA. Dietary intakes of lead and other metals: a study of young children from an urban population in the UK. *Food Addit Contam* 1987; 5(1):85–93.

48. Walker MI, Walters B, Mondon KJ. The assessment of radiocaesium intake from food using duplicate diet and whole-body monitoring techniques. *Food Addit Contam* 1991; 8(1):85–95.

49. Stephany RW, Schuller PL. Daily intakes of nitrate, nitrite and volatile N-nitrosamines in The Netherlands using the duplicate portion sampling techniques. *Oncology* 1980; 37:203–10.

50. Ellen G. Dietary studies in The Netherlands: duplicate diet approach. In: Van Dokkum W, de Vos RH, eds. *Dietary studies in Europe*. Euronut report no 10, Wageningen; 1988; 22–36.

51. Buchet JP, Lauwerys R. Oral daily intake of cadmium, lead, manganese, copper, chromium, mercury, calcium, zinc and arsenic in Belgium: A duplicate meal study. *Fd Chem Toxic* 1983; 21(1):19–24.

52. Louekari K, Jolkkonen L, Varo P. Exposure to cadmium from foods, estimated by analysis and calculation – comparison of the methods. *Food Addit Contam* 1987; 5(1): 111–7.

53. Møller H, Landt J, Pedersen E, Jensen P, Autrup H, Jensen OM. Endogenous nitrosation in relation to nitrate exposure from drinking water and diet in a Danish rural population. *Cancer Res* 1989; 49:3117–21.

54. Ministry of Agriculture, Fisheries and Food. Food Surveillance 1985 to 1988. *Food Surveillance paper no. 24*. UK: HMSO; 1988.

55. Knowles ME, Bell JR, Norman JA, Watson DH. Surveillance of potentially hazardous chemicals in food in the United Kingdom. *Food Addit Contam* 1991; 8(5):551–64.

56. Herman JL. Use of Intake data in Risk assessments by JECFA/JMPR and in Codex decisions. In: Macdonald I, ed. *Monitoring dietary intakes*. New York: Springer-Verlag; 1991; 90–8. (ILSI Monographs).

57. Galal-Gorchev HG. Dietary intake of pesticide residues; cadmium, mercury and lead. *Food Addit Contam* 1991; 8,6:793–806.

58. Galal-Gorchev HG, Jelinek CF. A review of the dietary intakes of chemical contaminants. *Bull WHO* 1985; 63,5:945–62.

59. Galal-Gorchev HG. WHO International Co-operation in Exposure Studies. In: Macdonald I, ed. *Monitoring dietary intakes*. New York: Springer-Verlag; 1991; 231–9 (ILSI Monographs).

60. Lauer BH, Kirkpatrick DC. Food additive intake: estimated versus actual. In: Macdonald I, ed. *Monitoring dietary intakes*. New York: Springer-Verlag; 1991; 213–20 (ILSI Monographs).

61. Food and Drug Administration: Aspartame: commissioner's final decision. *Fed Regist* 1981; 46:38285–308.

62. Heybach JP, Ross C. Aspartame consumption in a representative sample of Canadians. *Rev Assoc Canad diétét* 1989; 50(3):166–70.

63. Butchko HH, Kotsonsis FN. Acceptable Daily Intake vs actual intake: The aspartame example. *J Am Coll Nutr* 1991; 10:258–66.

64. International Sweeteners Association. Sweetener intake in Germany is within recommended limits. Sweetener update. *Q J Int Sweetener Assoc* 1990; 1–5.

65. Virtanen SM, Räsänen L, Paganus A, Varo P, Åkerblom HK. Intake of sugars and artificial sweeteners by adolescent diabetics. *Nutr Rep Int* 1988; 38(6):1211–18.

14

POSTMARKETING SURVEILLANCE IN THE FOOD INDUSTRY: THE ASPARTAME CASE STUDY

Harriett H. Butchko and W. Wayne Stargel

The Federal Food, Drug, and Cosmetic Act of 1938 gave the US Food and Drug Administration (FDA) its regulatory authority over foods and food additives. Twenty years later, this act was amended to require that the FDA be provided with proof of reasonable certainty that any substance added to food was safe for its proposed uses (1). Often such proof requires extensive toxicologic testing (2,3). Regulatory agencies require prospective demonstration that the additive is safe for its proposed use and that its probable consumption will be less than the acceptable daily intake (ADI). Another useful tool in further ensuring the safety of a food additive is monitoring it following approval, or postmarketing surveillance. With widespread use of an additive, postmarketing surveillance of consumption levels can determine whether actual consumption exceeds the ADI established from safety studies. Furthermore, postmarketing surveillance of anecdotal reports of health effects may identify issues for further evaluation that had not been identified in earlier animal and human studies.

The value of using postmarketing surveillance to assess the safety of approved food additives has been recognized by regulatory authorities throughout the world. For example, the Canadian Health Protection Branch (HPB) has indicated that postmarketing surveillance, in particular for sweeteners with low ADIs, will be a general policy in Canada (4). In addition, the FDA initiated an Adverse Reaction Monitoring System (ARMS) in 1985 to collect and evaluate anecdotal information regarding adverse health reports allegedly due to foods or food additives (1,5,6).

Aspartame (L-aspartyl-L-phenylalanine methyl ester) is an example of a widely used food additive that has undergone extensive evaluation during the postmarketing period through both monitoring of consumption levels as well as evaluations of anecdotal reports of health effects. As a result, it will be referenced throughout this chapter.

Aspartame is metabolized by digestive esterases and peptidases to its three constituents: aspartic acid, phenylalanine, and methanol. Scientific investigations have demonstrated that the body uses these naturally occurring dietary substances in the same way as when they are derived in larger amounts from other food sources (7). Extensive toxicologic and pharmacologic research was done in animals using much greater doses of aspartame than people could ingest (8–14). From the results of the toxicology studies, a no-observed-effect level (NOEL) of greater than 2,000–4,000 mg/kg body weight was established for aspartame. The animal toxicology data were used by the Scientific Committee for Foods of the European Economic Communities (SCF) (15), the Joint FAO/WHO Expert Committee on Food Additives (JECFA) (16) and the Canadian HPB (17) to establish an ADI of 40 mg/kg body weight/day for aspartame. When aspartame was first approved in the United States in 1974, the FDA authorized an ADI of 20 mg/kg body weight/day for aspartame (18).

The concept of the ADI was developed almost 40 years ago by JECFA (19). The ADI is based on lifetime studies in animals and indicates the amount of an additive (in mg/kg body weight) that may be safely consumed every day for a lifetime. It is not the maximum amount that is safe to consume during any 24-hour period. The ADI for a food additive is generally 1% of the NOEL. It is reasoned that the 100-fold safety factor takes into consideration the species differences between humans and animals and the inter-individual variation among humans (20,21).

The safety of aspartame and its metabolic constituents was assessed further in several human subgroups: infants, children, adolescents and healthy adults, obese individuals, individuals with diabetes, postpartum lactating women, and individuals heterozygous for the genetic disease, phenylketonuria (PKU), who have a decreased ability to metabolize phenylalanine. These and longer-term studies showed no untoward health consequences from aspartame (22–37). The results of the human studies, along with the animal research, provided convincing evidence that aspartame was safe for use by the general population, including pregnant women and children. The FDA responded to these additional data by increasing the ADI for aspartame to 50 mg/kg in 1983 (38). The importance of human studies in establishing an ADI, as was done with aspartame, has been emphasized by several authors (20,39).

POSTMARKETING SURVEILLANCE: CONSUMPTION

As part of the safety evaluation for a food additive, regulators evaluate projected use levels relative to the ADI. If projected intake levels approach or exceed the ADI, restrictions may be imposed by regulatory agencies, such as limiting approvals for some categories of use to decrease potential exposure in the general population. Various methods for estimating food additive intake have been detailed by Rees and Tennant (40).

Postmarketing surveillance is utilized to ensure that the intake of an approved additive does not consistently exceed the ADI. The FDA required such follow up as a condition of approval for aspartame in 1981 (41) as did the HPB (17). The

United Kingdom adopted a similar stance in 1983 by recommending that, once markets for sweeteners had stabilized, there should be postmarketing evaluation of actual intake levels (42).

At the time of approval of aspartame, the FDA considered the 99th percentile estimated intake as representative of high-level consumers. Since that time, the FDA has determined that the 99th percentile is unduly conservative and probably unrealistic (43). The very small number of consumers included in the 99th percentile may have large and variable intakes, which may skew the data markedly (44). Thus, the FDA now uses projections at the 90th percentile as the benchmark of high-level consumers (43). The more conservative 97.5th percentile is used in the United Kingdom (42).

Aspartame consumption in the United States

Before approval, projected intake levels of aspartame in the United States ranged from 8.3 mg/kg/day if all sucrose in an average-sized (60 kg) person's diet was replaced by aspartame to 25 mg/kg/day if all dietary carbohydrate could be replaced by aspartame. Upon evaluation of 2-week dietary records of close to 12,000 individuals, it was estimated that, if all possible foods were replaced with aspartame-containing foods, the 99th percentile daily consumption of aspartame would be 34 mg/kg (41).

Actual aspartame consumption was tracked in the United States by MRCA Information Services (Northbrook, IL) (45–50) from 1984 to 1992 through detailed collection of data by means of annual menu census surveys from over 2,000 households a year. The company's procedures ensured that the approximately 5,000 people surveyed are a valid, representative sample of the general US population. During the 14-day survey, all foods eaten both at home and away from home were recorded.

Aspartame intake levels were calculated from: (1) the frequency with which a person ingests an aspartame-containing product; (2) the average size of a portion based on the person's age and gender; and (3) the amount of aspartame in the product. Ingestion is calculated in two ways: overall mg of aspartame/day and mg of aspartame/kg body weight per day. The mean amount consumed and various percentiles (50th, 90th, 95th, and 99th) were tabulated as averages over 14 days, both for "eaters," who consumed aspartame at least once during the survey, and the population at large.

Because of their smaller body weights, children may consume more of an additive on a mg/kg basis than adults. To evaluate intake by children specifically, data also were recorded by age group: 0–23 months, 2–5 years, 6–12 years, 13–17 years, 18 years and over, as well as all age groups together. In addition, intake by special population subgroups such as diabetics and people on weight-reduction programs, who might be enthusiastic users of aspartame with potentially higher intakes, and women of childbearing potential and pregnant women was also monitored (45–50).

The continuing survey in the United States (MRCA Information Services) demonstrated that, from 1984 to 1992, the average intake over the 14-day period for

Table 1 Aspartame intake in the United States: 90th percentile, 14-day average, "eaters" only

Population	Survey dates							
	7/84– 6/85	7/85– 6/86	7/86– 6/87	7/87– 6/88	7/88– 6/89	7/89– 6/90	7/90– 6/91	7/91– 6/92
				(mg/kg/day)				
All ages	1.6	2.1	2.1	2.3	2.2	2.5	2.8	3.0
2–5 years	3.1	4.8	3.7	2.6	4.0	3.1	3.5	5.2
Diabetics	2.1	2.2	3.0	3.3	2.6	2.7	3.4	3.3
Reducing diet	1.6	2.2	2.3	2.6	2.5	2.7	2.8	3.3
Childbearing age	2.0	2.2	2.5	2.8	2.6	3.2	3.7	4.2

aspartame "eaters" (at the 90th percentile) has ranged from 1.6 to 3.0 mg/kg/day. As shown on Table 1, intake of aspartame, even by children, diabetics, people on weight-reduction diets, and females of childbearing age, was only approximately 5–10% of the ADI in the United States at the 90th percentile.

Data from other types of consumption evaluations in the United States corroborate these results. Upon analysis of 1-day diary data from the US Department of Agriculture (USDA) Continuing Survey of Food Intakes by Individuals (CSFII) from over 1,500 women, aspartame intake ranged from 0 to 16.6 mg/kg/day; over 90% of the women who reported aspartame consumption had intakes less than 5 mg/kg/day (51). Although per capita disappearance data may underestimate consumption since both "eaters" and "noneaters" are included, aspartame consumption for the total population (based on a 50-kg person) can be estimated to be about 1.6 mg/kg/day based on USDA per capita disappearance data (52).

Aspartame consumption in other countries

Canada

Aspartame consumption in Canada was evaluated using a 7-day diary survey, similar in design to the US survey discussed above, completed by about 5,000 individuals during both cold- and warm-weather months in 1987 (53). The general population of aspartame "eaters" in Canada consumed 5.5 mg/kg/day during cold weather months and 5.9 mg/kg/day during warm weather months (7-day average, 90th percentile). Children and special populations, who might consume higher amounts of aspartame, consumed from 5.5 to 11.4 mg/kg/day ("eaters" only, 90th percentile, 7-day average).

Germany

In 1988–9, consumption of the sweeteners, aspartame, cyclamate, and saccharin, was evaluated in Germany by means of a 24-hour dietary recall questionnaire completed by 2,800 individuals. The average daily intake (90th percentile) for aspartame was 2.75 mg/kg/day (54).

United Kingdom

Intakes of both high intensity sweeteners and bulk sweeteners were evaluated in the United Kingdom in 1987–8 in the general population (7-day diaries) and diabetics (4-day diaries). Aspartame intake ("eaters" only, 7-day average) was 16, 109, and 372 mg/day at the median, 97.5th percentile, and maximum, respectively. Intake by children aged 2–5 years was 1.0 mg/kg/day at the median and 1.6 mg/kg/day at the maximum. Diabetics had median and maximum aspartame consumption of 128 mg/day and 643 mg/day, respectively. At the 97.5th percentile, diabetics consumed 596 mg/day of aspartame (42). In another 7-day sweetener consumption survey done in the United Kingdom in 1988, aspartame consumption (90th percentile) was 4% of the ADI, or about 1.6 mg/kg/day. Children and diabetics ingested only 7% and 6%, respectively, of the ADI at the 90th percentile (44).

In 1994, MAFF undertook to survey the intake of intense and bulk sweeteners among diabetics (55). Diabetics studies ranged from 2 to 65 years of age. Data were collected by a Food Frequency Questionnaire which included interviews on meal patterns, recall of food intake, and a checklist of usual consumption of a food. Information on content of sweeteners in various products was obtained from the manufacturers. The 97.5th percentile was assumed to be representative of the upper limit of usual dietary patterns. For aspartame, the 97.5th percentile was 10.1 mg/kg/day, only about 25% of the ADI, even among diabetics who would likely be frequent consumers of aspartame.

Finland

Sweetener intake by 152 diabetics was evaluated in Finland using two 48-hour (2 school days and 2 weekend days) dietary recall questionnaires. Nearly three-quarters (73%) of the diabetic children consumed aspartame-containing products, and their mean intake was 1.15 mg/kg/day, less than 3% of the ADI (56).

Italy

Aspartame intake among Italian teenagers who were known to be users of diet products ($n = 212$) aged 13–19 years was estimated using a 14-day food intake diary. Average aspartame intake was only 0.03 mg/kg/day, and the maximum aspartame intake was 0.39 mg/kg/day (57).

France

In 1992, 6,914 households (including 19,158 individuals) were surveyed for food eaten at home. Over 56% of the households reported consumption of aspartame.

Aspartame intake was 0.6 and 1.0 mg/kg/day at the 90th and 95th percentiles, respectively, and maximum intake was 10.4 mg/kg/day (58). That no data were collected for food eaten away from home was a limitation of this study. More recently, aspartame intake was evaluated in insulin-dependent diabetic children (ages 2–20 years). Based on a 5-day diary questionnaire, mean, 97th percentile, and maximum aspartame intakes were 2.4, 7.8, and 15.6 mg/kg/day, respectively (58a).

Norway

Based on food intake surveys of the amounts of various products containing aspartame and the amounts of aspartame in those products, aspartame consumption was estimated. The average estimated intake of aspartame was 0.9 and 1.0 mg/kg/day for men and women, respectively, and 3.4 and 2.9 mg/kg/day for boys and girls, respectively (59). The estimated maximum intake was below the ADI for both adults and children.

Netherlands

The Dutch National Food Consumption Survey in 1992 included a 2-day food intake record for 6,218 individuals and an evaluation of food frequency questionnaires from 6,060 individuals. Aspartame content of foods and beverages was assumed to be the maximum use levels in the EU directive. Aspartame was the most frequently used sweetener. Based on the food intake records, estimated mean and 90th percentile aspartame intakes were 1.9 mg/kg/day and 5.2 mg/kg/day, respectively, for the total population. Based on food frequency questions, the results for the total population were very similar with mean and 90th percentile intakes of 2.4 and 5.2 mg/kg/day, respectively (60).

Brazil

Intake of intense sweeteners was evaluated during both winter (1990) and summer (1991) months in three cities. The 673 participants were randomly chosen from users of intense sweeteners. Median aspartame intake by the general population was 2.9% of the ADI; median intakes by diabetics and individuals on weight control regimens were 1.02 (2.6% of the ADI) and 1.28 mg/kg/day (3.2% of the ADI), respectively (61).

Australia

A 7-day diary survey of 128 individuals who consumed sweeteners in Australia was conducted in 1994. Consumption levels of aspartame were 6% and 7% of the ADI for all respondents and total consumers, respectively. The 90th percentile consumption was 23% of the ADI, however, the small sample made a precise estimate of 90th percentile intake difficult (62).

Conclusions: postmarketing surveillance of aspartame consumption

The results of the above evaluations indicate that, in the case of a high-intensity sweetener with widespread use, such as aspartame, a number of different methods yield comparable results. Thus, in the case of aspartame, the different surveys from the United States, Canada, Brazil, Australia, and seven European countries resulted in remarkably consistent intake levels of aspartame that are well below the ADI (Figure 1).

POSTMARKETING SURVEILLANCE: ANECDOTAL REPORTS OF HEALTH EFFECTS

The 1940s and 1950s were a time when many new drugs were developed and marketed. The rapidly expanding therapeutic horizons led to the observation that the full spectrum of adverse reactions was not always apparent until a pharmaceutical product had been used by many patients over time (63). Even extensive research in animals and humans could not always predict infrequent or uncommon effects, especially if they took a long time to develop. Thus, it was concluded that, along with pre-approval studies, a passive postmarketing surveillance system was required to document and evaluate spontaneous reports of adverse reactions associated with marketed drugs. This system has now become an integral part of the complete safety evaluation of marketed drugs. While data from such a system may be useful for identifying trends or areas for further study, anecdotal data

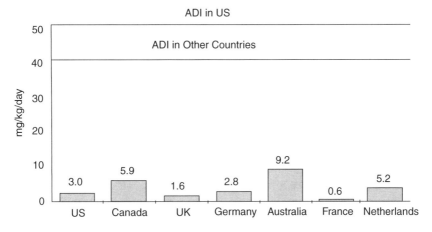

Figure 1 Surveys from the United States, Canada, several European countries, and Australia consistently demonstrate that aspartame consumption by the general population is well below the ADI.

cannot establish a definite cause-and-effect relationship. Confounding variables, such as underlying medical conditions or other medications, may distort the picture. This is especially true when the product is used by many people and the symptoms reported are common in the general populace, circumstances that make it difficult to differentiate between adverse effects of the compound and expected "background" occurrences in the general population. In addition, various biases, such as the suggestibility of negative media coverage, can affect the numbers and types of reports (63).

Although postmarketing surveillance for potential health effects is not required after approval of foods or food additives, regulatory agencies, such as the FDA (1,5,6) and the HPB (4), have recognized its usefulness to expand the safety database, assess patterns of associations, and identify areas for further study. In 1984, the FDA asked the Centers for Disease Control (CDC) to evaluate reports of adverse health effects anecdotally associated with aspartame consumption. In 1985, the FDA's Center for Food Safety and Applied Nutrition (CFSAN) started its own process, the Adverse Reaction Monitoring System (ARMS), to monitor accounts of health problems anecdotally associated with consumption of foods, food and color additives, and vitamin/mineral supplements (1,5). The Nutra-Sweet Company also developed and maintained a postmarketing surveillance system for aspartame.

Centers for Disease Control (CDC) evaluation

Anecdotal reports from some consumers that aspartame caused health symptoms increased following approval of aspartame for use in carbonated beverages in 1983. In response, the FDA asked the CDC epidemiologists to investigate these reports to determine if there was a specific pattern of symptoms suggesting a causal relationship with aspartame that would point to a need for further research (64,65).

More than 500 reports were analyzed by the CDC, and almost half underwent detailed follow-up and evaluation. Most complainants were white women aged 21–60 years, and they were randomly distributed from all over the United States with one exception. Aspartame had been subjected to a negative media barrage in Arizona, so, not surprisingly, proportionately more reports came from that state. While reports were received about a variety of different symptoms, two-thirds fell into the neurologic/behavioral category. These consisted mostly of headache, mood alterations, insomnia, dizziness, and fatigue. About a quarter of the reports were gastrointestinal, including abdominal pain, nausea, diarrhea, and vomiting (64,65). The CDC reported "Despite great variety overall, the majority of frequently reported symptoms were mild and are symptoms that are common in the general populace" (64). No specific clinical syndromes that suggest a causal relationship with aspartame were observed. The CDC concluded that focused clinical studies would be the only way to address thoroughly the issues raised by the anecdotal reports.

FDA Adverse Reaction Monitoring System (ARMS) evaluation

In 1985 CFSAN started a passive surveillance system, ARMS, to collect and evaluate anecdotal accounts of health problems associated with consumption of foods, food and color additives, and vitamin/mineral supplements (1,5). Unlike in the case of pharmaceuticals, where most information is received from physicians, such information regarding food additives is largely obtained directly from consumers. In the case of aspartame, the information in ARMS was largely obtained from The NutraSweet Company (see following discussion).

Depending on the severity of symptoms, reports are categorized as Type I (serious) and Type II (moderate). Type I includes significant breathing disturbance, anaphylaxis, chest pain or heart irregularity, continuous vomiting or diarrhea with dehydration, seizures, and symptoms requiring emergency medical treatment. Deaths, cancer, and fetal malformations are always Type I. Type I symptoms are investigated by FDA field inspectors by means of interviews and medical record review. Type II symptoms include headache, nausea, hives, and fatigue (1,5).

Reports are classified on the basis of the consistency and frequency with which they occur (1,5). This classification code (A–D) assists in weighing the likelihood of whether a given symptom is related to a specific food or additive. Table 2 depicts the classification system used by the FDA, which is similar to that developed by the CDC for its 1984 evaluation of aspartame (64,65). All reports are evaluated for demographic variables and symptoms. Reports are entered by a descriptive term, e.g. nervousness, headache, or rash. In addition, Type I reports are coded using a body organ system derived from COSTART (Coding Symbols for Thesaurus of Adverse Reaction Terms), the same system used by the FDA for coding adverse drug or biologic reactions.

On a periodic basis, CFSAN publishes its evaluations of reports about specific additives. The data also are evaluated by the Health Hazards Evaluation Board (1,5), composed of FDA scientists from a variety of backgrounds. In a recent report on aspartame (66), a total of 7,232 consumer reports had been received since marketing, and only 11% were classified as Type I. Headache topped the list of symptoms reported, followed by dizziness, mood changes, and nausea/vomiting. The report noted the decline, since the peak in 1985, of reports from consumers regarding aspartame and further stated, "In summary, the number of adverse reaction complaints received by the FDA and the nature of these reports in terms of demographic distribution, severity, strength of association with the

Table 2 ARMS classification to evaluate relative likelihood of causality

Group A	Multiple occurrences with multiple products containing the additive
Group B	Multiple occurrences with a single product containing the additive
Group C	Single occurrence but no re-challenge data available
Group D	Symptoms did not occur with every exposure to additive or physician stated opinion that reported symptoms were unlikely to be related to the additive

295

product, and symptoms remain comparable to those from previous analyses." (66). It is also of interest that only about 300 of the 7,000 food-associated reports received by the FDA per year are anecdotally associated with aspartame (L. Tollefson, personal communication).

FDA also separately analyzed the 251 reports of seizure anecdotally associated with aspartame consumption received through ARMS from 1986 to 1990 and concluded that almost half were highly unlikely to be related to aspartame (67). Furthermore, the FDA could not exclude the possibility that the remaining reports had not simply occurred by chance. FDA concluded that the anecdotal reports "did not support the claim that the occurrences of the seizures were linked to consumption of aspartame." (67). It was further concluded that the data did not suggest the need for a controlled clinical study to evaluate this issue.

Company system for health report evaluation

Despite the extensive pre-approval safety testing demonstrating that aspartame was not associated with adverse health effects, some anecdotal reports of health effects were received early on by The NutraSweet Company. As a result, the company developed a postmarketing system to collect and evaluate these reports (49, 50,68).

The NutraSweet Company postmarketing surveillance system was a collaborative effort between the Consumer Center, where the staff was responsible for data collection, documentation, and follow-up, and the Clinical Research Department, where physicians provided medical expertise for evaluation of the reports. Individuals who contacted the Consumer Center and believed that they had symptoms associated with aspartame were entered as reports. This initiated a process for documenting, following-up, and evaluating the reports. All reports were evaluated by a physician who assigned a symptom code(s) reflective of the specific symptom and body system category, as determined from a modified version of the World Health Organization's adverse reaction coding system.

Through follow-up correspondence, attempts were made to collect details about the complaint. The information requested included: (1) symptom description; (2) timing of symptom after ingestion of aspartame; (3) products associated with symptom; (4) whether symptoms occurred every time the product was consumed, disappeared upon cessation of consumption, and/or recurred on re-challenge; (5) whether the consumer consulted a physician; (6) whether there were concurrent medical conditions or medications; and (7) whether the consumer was on a weight-reduction regimen. Information was also obtained from the consumer's physician whenever possible. Once the follow-up process was completed, the file was closed with an outcome designation.

The anecdotal reports were tabulated monthly and reported to the FDA. In fact, more than 70% of the aspartame reports in the ARMS have come from The NutraSweet Company's postmarketing surveillance system. In-house reviews of the data were completed quarterly and yearly. Reviews included an analysis of the numbers and types of symptoms, demographics, and an analysis of any trends. A

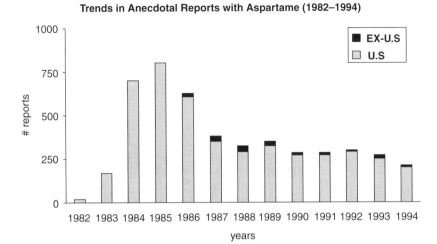

Figure 2 Trends in reports of health effects anecdotally attributed to aspartame 1982–1994.

committee, consisting of a physician, a biostatistician, an attorney, a regulatory affairs professional, and the manager of the Consumer Center, also reviewed the data.

There was only a small number of reports during 1982 and 1983 when aspartame was first marketed (Figure 2). However, when aspartame was approved for use in carbonated soft drinks, the numbers of reports increased, prompting the CDC evaluation discussed above. Also during this same time, there were several negative media stories that markedly affected the reporting of anecdotal reports. For example, a three-part series on the CBS Evening News in January, 1984 reported allegations of adverse health effects from aspartame. After this series, more than 200 reports were received in February and March of 1984 compared with about a third that number in each of the other months (Figure 3). Since the mid-1980s, although there was a marked increase in the number of aspartame-containing products on the market, the number of anecdotal reports actually declined, remained stable at about 300 per year for several years, and dropped to 201 reports in 1994, the last year of the program. (Figure 2). In addition, as noted on Figure 2, the reports are almost exclusively from the United States. Reports from countries in the rest of the world combined represent only about 3% of the total.

As noted in the CDC and FDA reports discussed above, symptoms allegedly associated with aspartame tend to be mild and common in the general population. The data collected by The NutraSweet Company are consistent with this observation. For example, headache was the most frequent symptom reported by consumers, accounting for about one-fourth of total symptoms reported; it is also one of the most common symptoms reported by people in their everyday lives.

There are more than 100 million aspartame users in the United States, thus, it is inevitable that some of them will experience medical ailments temporally associated

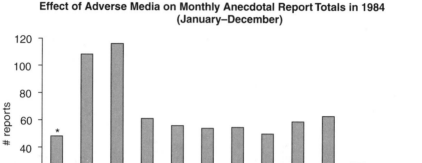

**Effect of Adverse Media on Monthly Anecdotal Report Totals in 1984
(January–December)**

* Three part CBS Evening News report alleging adverse side effects of aspartame aired January 16–18, 1984.

Figure 3 Biases introduced by negative national media stories have a major impact on the numbers of anecdotal reports reported.

with consumption of an aspartame-containing product simply by chance. This temporal association does not mean that the symptom is causally related to aspartame. Analysis of the reported symptoms and trend evaluation of demographics, symptoms, and types of products associated with these reports over the 12 years of the program did not suggest a causal relationship between the symptoms and aspartame.

Conclusions: health report evaluations

Passive surveillance systems of health reports in the postmarketing period are useful to identify or exclude possible safety issues. Needless to say, as the number of consumers increases with no significant findings observed over time, the value of continuing such a system diminishes.

BEYOND POSTMARKETING SURVEILLANCE

Research to evaluate allegations of health effects from aspartame

A number of studies, including focused clinical studies in humans, have been done to expand the knowledge base about aspartame and address scientific issues, including the anecdotal reports of alleged health effects associated with aspartame consumption. Long-term clinical studies with high doses of aspartame (75 mg/kg/day for 24 weeks, or about 25 times current consumption levels at the 90th percentile) resulted in no changes in clinical or biochemical parameters or adverse experiences compared with a placebo (69). The focused clinical studies evaluated whether aspartame causes headache, seizures, or allergic-type reactions and have included individuals identified through the company's medical postmarketing surveillance system who were convinced that aspartame caused their symptoms.

Headaches

Koehler and Glaros (70) did an outpatient study to evaluate the occurrence of migrane headache in migraneurs and concluded that aspartame caused a significant increase in the frequency of headaches but not in the intensity or duration of headaches. However, there were some statistical difficulties with this study; data from only 11 of the 25 subjects was reported, and the effects on frequency of headaches can be attributed largely to data from only 2–3 subjects (71). In contrast, when individuals who were convinced that aspartame had caused their headaches were evaluated in a randomized, double-blind, placebo-controlled study in the controlled environment of a Clinical Research Unit at Duke University, aspartame was no more likely than a placebo to elicit headache (72).

Allergenicity

Kulczycki had reported a single case report of an individual he felt was allergic to aspartame (73). However, from the results of a multicenter, randomized, double-blind, placebo-controlled, crossover study with individuals who were convinced they were allergic to aspartame, the investigators concluded that aspartame and its conversion products are no more likely than a placebo to cause allergic-type reactions (74). Another study also demonstrated that alleged allergic-type reactions to aspartame were not reproducible under blinded conditions (75).

Brain function: neurotransmission, cognition, behavior, mood, and seizures

A number of the allegations about aspartame centered on brain function. The underlying hypothesis was that aspartame, as a source of phenylalanine without other large neutral amino acids which compete for transport across the blood–brain barrier, would increase the serum ratio of phenylalanine to the other large amino acids (Phe/LNAA). It was further hypothesized that this increase in Phe/LNAA would selectively increase brain phenylalanine concentrations, thus, resulting in disturbances in monoaminergic neurotransmission (76). However, the results of animal studies demonstrated that increases in brain phenylalanine concentrations after enormous doses of aspartame do not affect brain monoaminergic neurotransmission (77–80). Furthermore, any effect that aspartame may have on the selective entry of phenylalanine into the brain is not unique to aspartame. For example, consumption of equisweet amounts of sugar has similar effects on the Phe/LNAA, through insulin-mediated changes in the serum concentrations of these amino acids (81–84). Studies done in humans evaluated and discounted allegations that aspartame may affect brain function, i.e. that it may cause or exacerbate seizures or affect memory, mood, or behavior. These studies demonstrated that even massive doses of aspartame have no effect on cognitive performance, mood, or behavior compared to a placebo (85–100) and that aspartame is no more likely than a placebo to cause or exacerbate seizures (101,102).

Evaluation of allegations that aspartame is associated with brain tumors

Olney *et al.* (103) claimed that the rate of brain tumors increased in the US concurrent with aspartame approval, that brain tumors increased in rats fed aspartame, and that nitrosation of aspartame is a putative mechanism for the increase in brain tumor rates. Specifically, Olney and colleagues describe a "surge in brain tumors in the mid-1980s" based on their analysis of selected data from the US Surveillance, Epidemiology and End Results (SEER) tumor data base.

Epidemiological trends in brain tumors

The arguments of Olney *et al.* (103) implicitly require two biologically indefensible assumptions: first, that a certain factor (aspartame) could cause an observed increase in the incidence of brain cancer in less than 4 years and second, that even more widespread exposure to this factor would cause no further increase in the incidence of that cancer in subsequent years. As it were, the trend of increased brain tumor rates started well before aspartame was approved, and overall brain tumor rates have actually been decelerating (104). More recently SEER data (105) show that the overall percent change in incidence of brain tumors from 1975 to 1979, before aspartame was marketed, was 1.6% whereas from 1992 to 1996, after aspartame was marketed, the percent change was –6.6%. Further, the estimated annual percent change was 0.6% for 1975–79 vs. –2.2% for 1992–96, a statistically significant decrease. Thus, this deceleration of brain tumor rates has continued (Figure 4).

Further, the pattern of increased brain tumor rates primarily in the elderly as referenced by Olney *et al.* (103) has been earlier described by several investigators (106–109). For example, Greig *et al.* (107) reported a 500% increase in people 85 years and older. There has been considerable debate regarding whether the reported increases in brain tumor rates in the elderly actually reflect a real increase (106–112). Modan *et al.* (113) critically reviewed brain tumor mortality data and concluded that the increases observed were not genuine but related to

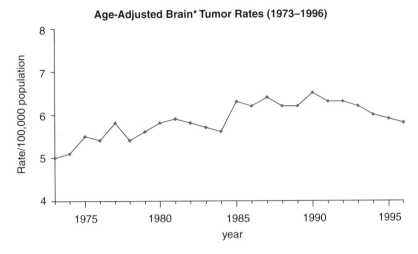

Figure 4 Yearly age-adjusted all brain tumor rates (*includes all CNS) from the National Cancer Institute SEER registry (1973–1996). Aspartame was approved for dry use in 1981 and for carbonated soft drinks in 1983. Rates of brain tumors have continued to decline since that time.

enhanced detection resulting from "the availability of more sophisticated non-invasive diagnostic technology; change in the attitude toward care of the elderly; and introduction of support programs such as Medicare that facilitate diagnostic procedures in the elderly." More recently, Legler *et al.* (114) evaluated brain and other CNS tumor incidence from the SEER results from 1975 to 1995. These authors stated that, in contrast to earlier reports, their evaluation found a level or declining trend for all but the most elderly people. A recent SEER report (105) further confirms that the elderly largely account for brain tumor increases. The overall percent change in brain tumor incidence from 1973 to 1996 was 15.8%; however, for ages under 65 years, the percent change was only 6.0%, whereas it was 46.5% for ages 65 and over during this time period (Figure 5).

Since the report by Olney *et al.* (103), Smith *et al.* (115) have reported a 35% increase in brain tumors in children from 1973 to 1994. In children, these authors noted a lower rate in brain tumors before 1984–5 with a subsequent step increase. Increases in low grade gliomas was an important factor in the overall increase in malignant childhood brain tumors. However, such tumors are detected more readily by MRI (very restricted availability prior to 1985) than by CT; there was also a change in the childhood brain tumor classification system during the mid-1980s whereby some tumors (e.g. low grade gliomas) previously classified as benign were then classified as malignant and therefore subsequently included in the SEER database; and there were changes in neurosurgical practices (e.g. stereo-tactic biopsies) in the mid-1980s that could have led to increased diagnosis and reporting of brain tumors in children. Thus, these authors attributed this finding

Percent Change in Brain Tumor Rates by Age Group (1973–1996)

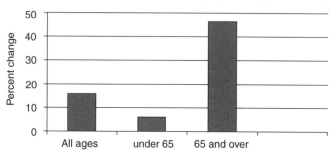

Figure 5 Brain tumor rates in all ages combined, under 65 years, and 65 years and older for the years 1973–1996 from the National Cancer Institute SEER report.

to changes in the mid-1980s in the detection and/or reporting of brain tumors in children. A subsequent evaluation of SEER data for childhood brain tumors by Linet *et al.* (116) supported this conclusion.

Olney *et al.* (103) also concluded that there was an increase in the malignancy of brain tumors after aspartame marketing. As with brain tumors in general, increases in glioblastomas were primarily in the elderly in the SEER data base (104). The increased rate of glioblastomas in the elderly has also been reported by Werner *et al.* (108). Thus, increases in glioblastomas are largely accounted for by increases in the elderly and may be the result of the factors described by Modan *et al.* (113). The increase in anaplastic astrocytomas referenced by Olney *et al.* (103) is likely an artifact of recent shifts in classification terminology. Werner *et al.* (108) cited wider acceptance of a three-tier classification system (astrocytoma, anaplastic astrocytoma, and glioblastoma) during the mid-1980s. Thus, some tumors classified as astrocytomas in the early 1980s would have been classified as anaplastic astrocytomas by the midlate 1980s.

Further, if there was truly an increase in the malignancy of brain tumors, one would have expected that mortality from brain tumors would have risen. On the contrary, the percent increase in mortality for all ages from 1975 to 1979 was 2.5% compared to -2.1% from 1992 to 1996, and 5 year relative survival rates are significantly higher from 1989 to 1995 compared to 1974–76 (105) (Figure 6). Thus, declining mortality is not consistent with a true increase in malignancy of brain tumors but rather can be explained largely by changes in classification system of brain tumors and the ability to diagnose tumors earlier due to the availability of advanced neuroimaging techniques and changes in neurosurgical practices, such as the availability of stereotactic biposies, as described above.

In addition to the above studies, a case-control study specifically evaluating aspartame consumption and the risk of childhood brain tumors was published by Gurney *et al.* (117). In this study, case patients were 19 years of age or older and were diagnosed with a primary brain tumor between 1984 and 1991. The results

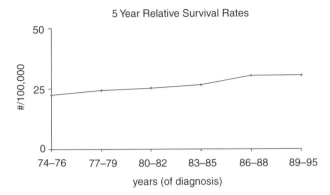

5 Year Relative Survival Rates

Figure 6 Five year relative survival rates for brain and other central nervous system tumors diagnosed in the years 1974–1995 (all races, males and females) from the National Cancer Institute SEER registry. There is a statistically significant increase in 5-year survival after aspartame marketing (1989–95) compared to before marketing (1974–76).

of the study showed that children with brain tumors were no more likely to have consumed aspartame than control children nor was there any elevated risk from maternal consumption of aspartame during pregnancy. The authors further found no evidence of an association of specific types of brain tumors with aspartame. The authors also found no evidence to support a hypothesis of an association between aspartame consumption and brain tumors in children. Gurney and coworkers concluded, "...it appears unlikely that any carcinogenic effect of aspartame ingestion could have accounted for the recent brain tumor trends as Olney *et al.* contend" (117).

Animal carcinogenicity studies

Olney *et al.* (103) claimed that aspartame may be a carcinogen based on results from earlier animal studies. The three rat and one mouse studies evaluating the carcinogenicity of aspartame demonstrate that aspartame is not a carcinogen, even at dosages hundreds to thousands of times higher than the 90th percentile of human consumption (41,118–121).

A number of characteristic features are recognized for neurocarcinogens, and aspartame has been evaluated based on these criteria (118,122). These criteria include the ability to consistently increase tumor incidence, the demonstration of a dose–effect relationship, the appearance of tumors at a younger age, a greater effect on embryonal and fetal cells than on adult neuroectodermal cells, and a shift to more anaplastic tumor types. After thorough analysis of aspartame studies, aspartame did not meet any of these criteria. Further, it was noted that there was no evidence that aspartame increased the incidence of higher malignancy-type brain tumors.

Nonetheless, Olney *et al.* argued that aspartame-fed rats had a higher rate of brain tumors than control rats in one study and that a later lifetime rat study including in utero exposure to aspartame was unreliable because the brain tumor incidence was too high in the control group. The underlying basis for such claims is the incorrect assertion that the background incidence of brain tumors in Sprague-Dawley (SD) rats is 0.1%; the actual background incidence is at least 20–30 times higher (122,123).

Using accepted statistical methods, FDA concluded that there was no dose-dependent increase in brain tumors or any expected characteristic of carcinogens in the first 2-year rat study (41). This conclusion was corroborated by the results of a second lifetime rat study, incorporating in utero exposure to aspartame, done in the same rat strain (122). The overall brain tumor rates of 2.7% and 2.9% represent the expected brain tumor incidence in 2-year old Sprague-Dawley rats without exposure to a carcinogenic substance. The results of both rat studies were considered valid by the FDA in supporting its conclusion that aspartame is not carcinogenic in rats. An additional study done in mice further demonstrated that aspartame was not carcinogenic (41) and a subsequent third, 2-year Wistar rat study also confirmed that aspartame was not carcinogenic (120).

Although the Public Board of Inquiry (PBOI) appointed by FDA to review Olney's concerns of neurotoxicity and brain tumors prior to the approval of aspartame initially could not reach a decision regarding aspartame and brain tumors, the additional considerations and findings in animals summarized above prompted a letter (124) dated 6 August 1981, to FDA Commissioner Hayes from Dr. Nauta, Chairman of the PBOI, who stated in regard to aspartame's approval by FDA: "...had we known earlier about the reassuring outcome of the recent Japanese oncogenicity studies, our recommendation would doubtless have been for unqualified approval... we wish to express our endorsement of your final decision in this matter."

Putative mechanism

Olney *et al.* (103) claim that nitrosation studies by Shephard *et al.* (125) may provide a mechanistic explanation for how aspartame could act as a carcinogen. However, the study cited and Shephard's earlier work show that any nitrosation of aspartame is infinitesimally small and insignificant under conditions relevant to human consumption of aspartame. Meier *et al.* (126) made the point in their first paper on nitrosation of aspartame that any formation of nitrosoaspartame under standard conditions of use as a sweetener will be "negligible in comparison with the intake of pre-formed nitroso compounds." Furthermore, Shephard *et al.* (125) pointed out that dietary tryptophan would contribute more to the amount of nitrosated products in the body than would aspartame. As a practical matter, the studies show that nitrosation of aspartame in the human stomach is negligible and will have no effect on the total body burden of nitrosamines whether they be pre-formed or formed in the stomach (127).

304

Reviews by epidemiologists

Epidemiologists (128,129) have also questioned the conclusions of the report by Olney *et al.* (103). Davies *et al.* concluded, "This report is most disturbing for those who practice epidemiology (this study was not done by an epidemiologist). The take-home message is completely speculative (as well as volatile), but to undo the damage that has already been done in the media is likely impossible" (128). Ross concluded, "From an epidemiologic perspective, the conclusion of the report may well represent a classic example of 'ecologic fallacy' [reference 6 in the paper by Ross], because the Olney *et al.* study was a correlative analysis (i.e. ecologic analysis) that demonstrated that two events occurred during roughly the same time period. There is no information available regarding whether the individuals who developed brain tumors consumed aspartame. For example, one might also invoke: (a) cellular phone, home computer, and VCR usage; (b) depletion of the ozone layer; or (c) increased use of stereo headphones as potentially causative agents to argue trends in brain tumors and the changing environment. All such events could potentially be positively correlated with brain tumor incidence, and some or all of these possibilities may or may not have any biological plausibility to the observed associations." (129)

Government and Regulatory evaluations

The allegations regarding aspartame and brain tumors have also been evaluated by scientists at regulatory agencies in the US, the United Kingdom, the European Union, Australia/New Zealand, and Brazil. The US National Cancer Institute (NCI) evaluated the statistics on cancer incidence and concluded that the analysis "does not support an association between the use of aspartame and an increased incidence of brain tumors" (129a). After evaluation of the National Cancer Institute's data in the SEER program, the FDA concluded that the analysis "does not support an association between the use of aspartame and increased incidence of brain tumors" (130). The Committee on Carcinogenicity at the Department of Health in the UK (131,132) arrived at the same conclusion as FDA stating, "The Committee concluded that the data published by Olney *et al.* did not raise any concerns with regard to the use of aspartame in the United Kingdom." (131). The Scientific Committee for Food of the European Union agreed with the FDA regarding Olney's findings and concluded, "... the data do not support the proposed biphasic increase in the incidence of brain tumors in the USA during the 1980s" (133). From their analysis, the Australia/New Zealand Food Authority (ANZFA) concluded, "From the extensive scientific data available at this stage, the evidence does not support that aspartame is carcinogenic in either animals or humans. There appears to be no foundation to recent USA reports of increased brain tumors in humans" (134). Finally, after an expert analysis, the Brazilian Ministry of Health concluded that aspartame does not cause brain tumors (135).

Conclusions: Aspartame does not cause brain tumors

Each of the claims used by Olney *et al.* to support their hypothesis is wrong. First, there is no association between aspartame consumption and an increase in human brain tumor rates. Second, aspartame is not an animal carcinogen. And third, the speculated biochemical mechanism is not relevant to aspartame consumption. Thus, the evidence is compelling that aspartame is not a carcinogen.

The Internet and misinformation

Recently, there has been a marked increase on the Internet of articles and personal testimonials alleging that aspartame is responsible for a wide variety of symptoms and diseases. Although the Internet can be a wonderful source of information, there is no way to assure that the information is factual – any individual or group can say just about anything without being held accountable. Unfortunately, the Internet has given a forum for several people to widely disseminate false and misleading information about the safety of aspartame.

For example, one letter circulated by "Nancy Markle" alleges that aspartame is associated with multiple sclerosis, among a number of other diseases. The medical advisor for the Multiple Sclerosis Foundation (136,137) has evaluated the data on aspartame and concluded that there is no association of aspartame with multiple sclerosis. He went further to state, "This campaign by the "aspartame activists" is not innocent drum banging" as they have created a danger for some individuals who need appropriate medical treatment for their problems rather than blaming aspartame.

Allegations regarding aspartame safety on the Internet have sometimes been based on a book by Roberts (138). Roberts' allegations, which are based on his personal opinions and anecdotes, are not consistent with the peer-reviewed scientific literature about aspartame. Roberts' allegations were best summed up by a book review published in the New England Journal of Medicine (139) where Rolla stated, "There is no place for a publication such as this one. It only adds to public misinformation, confusion, and mistrust."

Another study, that has recently received attention on the Internet, by Walton *et al.* (140) concluded that individuals with mood disorders are especially sensitive to aspartame. Their conclusion was based on results of adverse experience information from an incomplete study with data from only 13 of the 80 individuals who were to be enrolled. The authors combined individual symptoms in order to find statistical differences between aspartame and placebo. However, even if the data were appropriately combined to show a statistical significance, no statistically valid conclusions can be drawn from such an incomplete study (141,141a).

Walton's "review" of the aspartame scientific literature sent to the producers of the US television program, "60 Minutes," in 1996 has also been widely discussed on the Internet. Walton alleged that studies "not funded by industry" suggest that aspartame is "not safe" or causes "problems," with the implication that studies funded by industry were done in a manner to obtain favorable results. However, Walton neglected to mention that: (1) he had omitted, for no apparent reason, at least 50 published peer-reviewed safety-related aspartame studies; and (2) most of the "non-industry sponsored research" were letters to the editor, single patient case reports, review articles, and book chapters while the vast majority of the industry sponsored studies were published in peer-reviewed scientific journals. In order to accept Walton's allegations, one would also have to believe that scores of scientists who have worked on aspartame at leading institutions around the world would knowingly put consumers at risk and that the numerous regulatory scientists around the world, who have studied the data and concluded that aspartame is safe, are somehow in collusion with these scientists. As it were, the results of numerous clinical studies (22–37,69,72,74,75,85–102) have clearly demonstrated aspartame's safety.

The safety of methanol derived from aspartame has also recently received new and increased attention due to the proliferation of misinformation on the Internet. Trocho *et al.* (142) published the results of a study in rats from which they concluded that aspartame may be hazardous because formaldehyde adducts in tissue proteins and nucleic acids from aspartame may accumulate. However, according to Tephly (143), the doses of aspartame used in the study do not even yield blood methanol concentrations outside control values. In addition, the amounts of aspartame equal to that in about 75 cans of beverage as a single bolus result in no detectable increase in blood formate concentrations in humans, whereas increased urinary formate excretion shows that the body is well able to handle even excessive amounts of aspartame. Further, there is no accumulation of blood or urinary methanol or formate with long term exposure to aspartame. Tephly concluded that "the normal flux of one-carbon moieties whether derived from pectin, aspartame, or fruit juices is a physiologic phenomenon and not a toxic event."

Regulatory agencies in several countries, e.g. the United Kingdom (144), the United States (145) and Brazil (135) have recently evaluated various allegations about aspartame on the Internet, including allegations of a number of serious diseases such as multiple sclerosis, lupus erythromatosis, Gulf War syndrome, and brain tumors. The conclusions are that the information is anecdotal and that there is no reliable scientific evidence that shows that aspartame might be responsible for any of these conditions, thus re-affirming the safety of aspartame.

Conclusions: beyond postmarketing surveillance

Postmarketing surveillance and proactively identifying and addressing scientific issues is important in providing added assurance for the safety of new products. With the assistance of information from postmarketing surveillance, the scientific issues raised regarding aspartame were evaluated using controlled scientific studies. The results of these studies did not show a causal relationship between aspartame and any alleged adverse effects (146–157).

REFERENCES

1. Tollefson L. Monitoring adverse reactions to food additives in the US Food and Drug Administration. *Regul Toxicol Pharmacol* 1988; 8:438–46.

2. US Food and Drug Administration Bureau of Foods. *Toxicological principles for the safety assessment of direct food additives and color additives used in food*; 1982.

3. Miller SA. Food additives and contaminants. In: Amdur MO, Doull J, Klaassen CD, eds. *Casarett and Doull's toxicology*. New York: Pergamon Press; 1991; 819–53.

4. Gunner SW. Novel sweeteners – regulatory issues and implications. In: Walters DE, Orthoefer FT, DuBois GE, eds. *Sweeteners: discovery, molecular design, and chemoreception*. Washington, DC: American Chemical Society; 1991; 302–12.

5. Tollefson L, Barnard RJ, Glinsmann WH. Monitoring of adverse reactions to aspartame reported to the US Food and Drug Administration. In: Wurtman RJ, Ritter-Walker E, eds. *Dietary phenylalanine and brain function*. Boston: Birkhauser; 1988; 317–37.

6. Food and Drug Administration. A plan for action, phase II. *FDA Drug Bulletin*; 1987 (May); 1–51.

7. Ranney RE, Oppermann JA, Muldoon E. Comparative metabolism of aspartame in experimental animals and humans. *J Toxicol Environ Health* 1976; 2:441–51.

8. Molinary SV. Preclinical studies of aspartame in nonprimate animals. In: Stegink LD, Filer LJ Jr, eds. *Aspartame physiology and biochemistry*. New York: Marcel Dekker; 1984; 289–306.

9. Kotsonis, FN, Hjelle, JJ. The safety assessment of aspartame: scientific and regulatory considerations. In: Tschanz C, Butchko HH, Stargel WW, Kotsonis FN, eds. *The clinical evaluation of a food additive: assessment of aspartame*. Boca Raton, FL: CRC Press; 1996; 23–41.

10. Aspinall RL, Saunders RN, Pautsch WF, Nutting EF. The biological properties of aspartame V. Effects on a variety of physiological parameters related to inflammation and metabolism. *J Environ Pathol Toxicol* 1980; 3:387–95.

11. Bianchi RG, Muir ET, Cook DL, Nutting EF. The biological properties of aspartame II. Actions involving the gastrointestinal system. *J Environ Pathol Toxicol* 1980; 3:355–62.

12. Potts JW, Bloss JL, Nutting EF. Biological properties of aspartame I. Evaluation of central nervous system effects. *J Environ Pathol Toxicol* 1980; 3:341–53.

13. Saunders FJ, Pautsch WF, Nutting EF. The biological properties of aspartame III. Examination for endocrine-like activities. *J Environ Pathol Toxicol* 1980; 3:363–73.

14. Lennon HD, Metcalf LE, Mares SE, Smith JH, Nutting EF, Saunders FJ. The biological properties of aspartame IV. Effects on reproduction and lactation. *J Environ Pathol Toxicol* 1980; 3:375–86.

15. Scientific Committee for Food. Food-science and techniques. Reports of the Scientific Committee for Food (16th series). *Comm Eur Communities*; 1985.

16. Joint Food and Agriculture Organization/World Health Organization Expert Committee on Food Additives. *Evaluation of certain food additives*. (WHO Tech Rep Ser No. 653). Geneva: WHO; 1980; 20–1.

17. Health Protection Branch Canada. Proposal on aspartame. Information Letter No. 564. 12 September 1979.

18. Food and Drug Administration. Food additives permitted in food for human consumption: aspartame. *Fed Regist* 1974; 39:27317–9.

19. Lu FC. Acceptable daily intake: inception, evolution, and application. *Regul Toxicol Pharmacol* 1988; 8:45–60.

20. Renwick AG. Acceptable daily intake and the regulation of intense sweeteners. *Food Addit Contam* 1990; 7:463–75.

21. Renwick AG. Safety factors and establishment of acceptable daily intakes. *Food Addit Contam* 1991; 8:135–49.

22. Stegink LD, Filer LJ Jr, Baker GL. Effect of aspartame and aspartate loading upon plasma and erythrocyte free amino acid levels in normal adult volunteers. *J Nutr* 1977; 107:1837–45.

23. Stegink LD, Filer LJ Jr, Baker GL. Plasma, erythrocyte and human milk levels of free amino acids in lactating women administered aspartame or lactose. *J Nutr* 1979; 109: 2173–81.

24. Stegink LD, Filer LJ Jr. Baker GL. Plasma and erythrocyte concentrations of free amino acids in adult humans administered abuse doses of aspartame. *J Toxicol Environ Health* 1981; 7:291–305.

25. Stegink LD, Filer LJ Jr, Baker GL, McDonnell JE. Effect of aspartame loading upon plasma and erythrocyte amino acid levels in phenylketonuric heterozygotes and normal adult subjects. *J Nutr* 1979; 109:708–17.

26. Stegink LD, Filer LJ Jr, Baker GL, McDonnell JE. Effect of an abuse dose of aspartame upon plasma and erythrocyte levels of amino acids in phenylketonuric heterozygous and normal adults. *J Nutr* 1980; 110:2216–24.

27. Stegink LD, Brummel MC, McMartin K, *et al*. Blood methanol concentrations in normal adult subjects administered abuse doses of aspartame. *J Toxicol Environ Health* 1981; 7:281–90.

28. Stegink LD, Brummel MC, Filer LJ Jr, Baker GL. Blood methanol concentrations in one-year-old infants administered graded doses of aspartame. *J Nutr* 1983; 113:1600–6.

29. Filer LJ Jr, Baker GL, Stegink LD. Effect of aspartame loading on plasma and erythrocyte free amino acid concentrations in one-year-old infants. *J Nutr* 1983; 113:1591–9.

30. Koch R, Shaw KNF, Williamson M, Haber M. Use of aspartame in phenylketonuric heterozygous adults. *J Toxicol Environ Health* 1976; 2:453–7.

31. Frey GH. Use of aspartame by apparently healthy children and adolescents. *J Toxicol Environ Health* 1976; 2:401–15.

32. Knopp RH, Brandt K, Arky RA. Effects of aspartame in young persons during weight reduction. *J Toxicol Environ Health* 1976; 2:417–28.

33. Stern SB, Bleicher SJ, Flores A, Gombos G, Recitas D, Shu J. Administration of aspartame in non-insulin-dependent diabetics. *J Toxicol Environ Health* 1976; 2:429–39.

34. Langlois K. *Short term tolerance of aspartame by normal adults*. GD. Searle & Co. Research Report (E-23), Submitted to the FDA 30 November 1972.

35. Hoffman R. *Long term tolerance of aspartame by obese adults*. GD. Searle & Co. Research Report (E-64), Submitted to FDA 14 June 1973.

36. Hoffman R. *Short term tolerance of aspartame by obese adults*. GD. Searle & Co. Research Report (E-24), Submitted to FDA 30 November 1972.

37. Frey GH. *Long term tolerance of aspartame by normal adults*. GD. Searle & Co. Research Report (E-60), Submitted to FDA 9 February 1973.

38. Food and Drug Administration. Food additives permitted for direct addition to food for human consumption; aspartame. *Fed Regist* 1984; 49:6672–82.

39. Rubery ED, Barlow SM, Steadman JH. Criteria for setting quantitative estimates of acceptable intakes of chemicals in food in the UK. *Food Addit Contam* 1990; 7:287–302.

40. Rees N, Tennant D. Estimation of food chemical intake. In: Kotsonis FN, Mackey M, Hjelle JJ, eds. *Nutritional toxicology*. New York: Raven Press; 1994; 199–221.
41. Food and Drug Administration. Aspartame: commissioner's final decision. *Fed Regist* 1981; 46:38285–308.
42. Ministry of Agriculture, Fisheries and Food. Intakes of intense and bulk sweeteners in the UK 1987–1988. *Food Surveillance Paper No. 29*; 1990.
43. Food and Drug Administration. Food additives permitted for human consumption; aspartame. *Fed Regist* 1986; 51:42999–3000.
44. Hinson AL, Nicol WM. Monitoring sweetener consumption in Great Britain. *Food Addit Contam* 1992; 9:669–81.
45. Abrams IJ. Using the menu census survey to estimate dietary intake. In: Finley JW, Robinson SF, Armstrong D, eds. *Food safety assessment. (ACS Symposium Series 484)*. Washington, DC: American Chemical Society; 1992; 201–13.
46. Abrams IJ. Using the menu census survey to estimate dietary intake – post market surveillance of aspartame. In: *Proceedings of the international aspartame workshop*, Marbella, Spain; 1986.
47. Butchko HH, Kotsonis FN. Acceptable daily intake vs. actual intake: the aspartame example. *J Am Coll Nutr* 1991; 10:258–66.
48. Butchko HH, Kotsonis FN. Acceptable daily intake and estimation of consumption. In: Tschanz C, Butchko HH, Stargel WW, Kotsonis, FN, eds. *The clinical evaluation of a food additive: assessment of aspartame*. Boca Raton, FL: CRC Press; 1996; 43–53.
49. Butchko HH, Kotsonis FN. Postmarketing surveillance in the food industry: the aspartame case study. In: Kotsonis FN, Mackey M, Hjelle JJ, eds. *Nutritional toxicology*. New York: Raven Press; 1994; 235–50.
50. Butchko HH, Tschanz C, Kotsonis FN. Postmarketing surveillance of food additives. *Reg Toxicol Pharmacol* 1994; 20:105–18.
51. Heybach JP, Smith JL. Intake of aspartame in 19–50 year old women from the USDA continuing survey of food intakes by individuals (CSFII 85). *FASEB J* 1988; 2: A1615.
52. Heybach JP, Allen SS. Resources for inferential estimates of aspartame intake in the United States. In: Wurtman RJ, Ritter-Walker E, eds. *Dietary phenylalanine and brain function*. Boston: Birkhauser; 1988; 365–72.
53. Heybach JP, Ross C. Aspartame consumption in a representative sample of Canadians. *J Can Diet Assoc* 1989; 50:166–70.
54. Bar A, Biermann C. Intake of intense sweeteners in Germany. *Z Ernahrungswiss* 1992; 31:25–39.
55. Ministry of Agriculture, Fisheries and Food (UK). Survey of the intake of sweeteners by diabetics. *Food Surveillance Info Sheet*. London; 1995; 76.
56. Virtanen SM, Rasanen L, Paganus A, Varo P, Akerblom HK. Intake of sugars and artificial sweeteners by adolescent diabetics. *Nutr Rep Int* 1988; 38:1211–18.
57. Leclercq C, Berardi D, Sorbillo MR, Lambe J. Intake of saccharin, aspartame, acesulfame K and cyclamate in Italian teenagers: present levels and projections. *Food Addit and Contam* 1999; 16:99–109.
58. Collerie de Borely A, Renault C. Intake of Intense Sweeteners in France (1989–1992). Paris: CREDOC, Departement Prospective de la Consommation; 1994. (a) Garnier-Sagne I, Leblanc JC, Verger PH. Calculation of the intake of three intense sweeteners in young insulin-dependent diabetics. *Food Chem Toxicol* 2001; 39:745–9.
59. Bergsten C. Intake of Acesulfame-K, aspartame, cyclamate and saccharin in Norway. SNT Norwegian Food Control Authority; 1993; Report 3:1–56.

60. Hulshof KFAM, Bouman M. Use of various types of sweeteners in different population groups: 1992 Dutch National Food Consumption Survey. TNO Nutrition and Food Research Institute, Netherlands; 1995; 1–49.

61. Toldeo MCF, Ioshi SH. Potential intake of intense sweeteners in Brazil. *Food Addit and Contam* 1995; 12:799–808.

62. National Food Authority, Australia. Survey of intense sweetener consumption in Australia: final report; 1995; 1–91.

63. Faich GA. Adverse-drug-reaction monitoring. *N Engl J Med* 1986; 314:1589–92.

64. Centers for Disease Control. Evaluation of consumer complaints related to aspartame use. *MMWR* 1984; 33:605–7.

65. Bradstock MK, Serdula MK, Marks JS, *et al.* Evaluation of reactions to food additives: the aspartame experience. *Am J Clin Nutr* 1986; 43:464–9.

66. Food and Drug Administration. *Summary of adverse reactions attributed to aspartame.* 20 April 1995.

67. Tollefson L, Barnard RJ. An analysis of FDA passive surveillance reports of seizures associated with consumption of aspartame. *J Am Diet Assoc* 1992; 92: 598–601.

68. Butchko HH, Tschanz C, Kotsonis FN. Postmarketing surveillance of anecdotal medical complaints. In: Tschanz C, Butchko HH, Stargel WW, Kotsonis FN, eds. *The clinical evaluation of a food additive: assessment of aspartame.* Boca Raton, FL: CRC Press; 1996:183–193.

69. Leon AS, Hunninghake DB, Bell C, Rassin DK, Tephly TR. Safety of long-term large doses of aspartame. *Arch Intern Med* 1989; 149:2318–24.

70. Koehler SM, Glaros A. The effect of aspartame on migraine headache. *Headache* 1988; 28:10–13.

71. Schiffman SS. Aspartame and Headache. *Headache* 1988; 28:370.

72. Schiffman SS, Buckley CE, Sampson HA, *et al.* Aspartame and susceptibility to headache. *N Engl J Med* 1987; 317:1181–5.

73. Kulczycki A. Aspartame-induced urticaria. *Ann Int Med* 1986; 104:207–8.

74. Geha R, Buckley CE, Greenberger P, *et al.* Aspartame is no more likely than placebo to cause urticaria/angioedema: results of a multicenter, randomized, double-blind, placebo-controlled, crossover study. *J Allergy Clin Immunol* 1993; 92: 513–20.

75. Garriga MM, Berkebile C, Metcalfe DD. A combined single-blind, double-blind, placebo-controlled study to determine the reproducibility of hypersensitivity reactions to aspartame. *J Allergy Clin lmmunol* 1991; 87:821–7.

76. Wurtman RJ. Neurochemical changes following high-dose aspartame with dietary carbohydrates. *N Engl J Med* 1983; 309:429–30.

77. Garattini S, Caccia S, Romano M, *et al.* Studies on the susceptibility to convulsions in animals receiving abuse doses of aspartame. In: Wurtman RJ, Ritter-Walker E, eds. *Dietary phenylalanine and brain function.* Boston: Birkhauser; 1988; 131–43.

78. Reilly MA, Debler EA, Fleischer A, Lajtha A. Lack of effect of chronic aspartame ingestion on aminergic receptors in rat brain. *Biochem Pharmacol* 1989; 38: 4339–41.

79. Reilly MA, Debler EA, Lajtha A. Perinatal exposure to aspartame does not alter aminergic neurotransmitter systems in weanling rat brain. *Res Commun Psychol Psychiatry Behav* 1990; 15:141–59.

80. Perego C, DeSimoni MG, Fodritto F, *et al.* Aspartame and the rat brain monoaminergic system. *Toxicol Lett* 1988; 44:331–9.

81. Martin-Du Pan R, Mauron C, Glaeser B, Wurtman RJ. Effect of various oral glucose doses on plasma neutral amino acid levels. *Metabolism* 1982; 31:937–43.
82. Burns TS, Stargel WW, Tschanz C, Kotsonis FN, Hurwitz A. Aspartame and sucrose produce a similar increase in the plasma phenylalanine to large neutral amino acid ratio in healthy subjects. *Pharmacology* 1991; 43:210–9.
83. Stegink LD, Wolf-Novak LC, Filer LJ Jr, *et al.* Aspartame-sweetened beverage: effect on plasma amino acid concentrations in normal adults and adults heterozygous for phenylketonuria. *J Nutr* 1987; 117:1989–95.
84. Wolf-Novak LC, Stegink LD, Brummel MC, *et al.* Aspartame ingestion with and without carbohydrate in phenylketonuric and normal subjects: effect on plasma concentrations of amino acids, glucose and insulin. *Metabolism* 1990; 39:391–6.
85. Lapierre KA, Greenblatt DJ, Goddard JE, Harmatz JS, Shader RI. The neuropsychiatric effects of aspartame in normal volunteers. *J Clin Pharmacol* 1990; 30: 454–60.
86. Trefz F, de Sonneville L, Matthis P, Benninger C, Lanz-Englert B, Bickel H. Neuropsychological and biochemical investigations in heterozygotes for phenylketonuria during ingestion of high dose aspartame (a sweetener containing phenylalanine). *Human Genetics* 1994; 93:369–74.
87. Wolraich M, Milich R, Stumbo P, Schultz F. Effects of sucrose ingestion on the behavior of hyperactive boys. *J Pediatr* 1985; 106:675–82.
88. Wolraich ML, Lindgren SD, Stumbo PJ, Stegink LD, Appelbaum MI, Kiritsy MC. Effects of diets in sucrose or aspartame on the behavior and cognitive performance of children. *N Engl J Med* 1994; 330:301–7.
89. Ryan-Harshman M, Leiter LA, Anderson GH. Phenylalanine and aspartame fail to alter feeding behavior, mood and arousal in men. *Physiol Behav* 1987; 39:247–53.
90. Saravis S, Schachar R, Zlotkin S, Leiter LA, Anderson GH. Aspartame: effects on learning, behavior, and mood. *Pediatrics* 1990; 86:75–83.
91. Stokes AF, Belger A, Banich MT, Taylor H. Effects of acute aspartame and acute alcohol ingestion upon the cognitive performance of pilots. *Aviat Space Environ Med* 1991; 62:648–53.
92. Stokes AF, Belger A, Banich MT, Bernadine E. Effects of alcohol and chronic aspartame ingestion upon performance in aviation relevant cognitive tasks. *Aviat Space Environ Med* 1994; 65:7–15.
93. Dodge RE, Warner D, Sangal S, O'Donnell RD. The effect of single and multiple doses of aspartame on higher cognitive performance in humans. *Aerospace Medical Association 61st Annual Scientific Meeting*; 1990; A30.
94. Lieberman HR. Caballero B, Emde GG, Bernstein JG. The effects of aspartame on human mood, performance, and plasma amino acid levels. In: Wurtman RJ, Ritter-Walker E, eds. *Dietary phenylalanine and brain function*. Boston: Birkhauser; 1988; 196–200.
95. Kruesi MJP, Rapoport JL, Cummings EM, *et al.* Effects of sugar and aspartame on aggression and activity in children. *Am J Psychiatry* 1987; 144:1487–90.
96. Ferguson HB, Stoddart C, Simeon JG. Double-blind challenge studies of behavioral and cognitive effects of sucrose-aspartame ingestion in normal children. *Nutr Rev* 1986; 44:144–50.
97. Milich R, Pelham WE. Effects of sugar ingestion on the classroom and playgroup behavior of attention deficit disordered boys. *J Consult Clin Psychol* 1986; 54: 714–18.

98. Spiers PA, Sabounjian LA, Reiner A, Myers DK, Wurtman J, Schomer DL. Aspartame: neuropsychologic and neurophysiologic evaluation of acute and chronic effects. *Am J Clin Nutr* 1998; 68:531–7.

99. Goldman JA, Lerman RH, Contois JH, Udall JN. Behavioral effects of sucrose on preschool children. *J Abnorm Child Psychol* 1986; 14:565–77.

100. Shaywitz BA, Sullivan CM, Anderson GM, Gillespie SM, Sullivan B, Shaywitz SE. Aspartame, behavior, and cognitive function in children with attention deficit disorder. *Pediatrics* 1994; 93:70–5.

101. Rowan JA, Shaywitz BA, Tuchman L, French JA, Luciano D, Sullivan CM. Aspartame and seizure susceptibility: results of a clinical study in reportedly sensitive individuals. *Epilepsia* 1995; 36(3):270–5.

102. Shaywitz BA, Anderson GM, Novotny EJ, Ebersole JS, Sullivan CM, Gillespie SM. Aspartame has no effect on seizures or epileptiform discharges in epileptic children. *Ann Neurol* 1994; 35:98–103.

103. Olney JW, Farber NB, Spitznagel E, Robins LN. Increasing brain tumor rates: is there a link to aspartame? *J Neuropathol Exp Neurol* 1996; 55:1115–23.

104. Levy PS, Hedeker D. Letter to the Editor. *J Neuropathol Exp Neurol* 1996; 55:1280.

105. National Cancer Institute. *SEER Cancer Statistics Review 1973–1996*.

106. Muir CS, Storm HH, Polednak A. Brain and other nervous system tumours. *Cancer Surv* 1994; 19/20:369–92.

107. Greig NH, Ries LG, Yancik R, Rapoport SI. Increasing annual incidence of primary malignant brain tumors in elderly. *J Natl Cancer Inst* 1990; 82:1621–4.

108. Werner MH, Phuphanich S, Lyman GH. The increasing incidence of malignant gliomas and primary central nervous system lymphoma in the elderly. *Cancer* 1995; 76:1634–42.

109. Davis DL, Ahlbom A, Hoel D, Percy C. Is brain cancer mortality increasing in industrial countries? *Am J Ind Med* 1991; 19:421–31.

110. La Vecchia C, Lucchini F, Negri E, Boyle P, Maisonneuve P, Levi F. Trends of cancer mortality in Europe, 1955–1989:IV, urinary tract, eye, brain and nerves, and thyroid. *Eur J Cancer* 1992; 28A:1210–81.

111. Boyle P, Maisonneuve P, Saracci R, Muir CS. Is the increased incidence of primary malignant brain tumors in the elderly real? *J Natl Cancer Inst* 1990; 82:1594–6.

112. Marshall E. Experts clash over cancer data. *Science* 1990; 250:900–2.

113. Modan B, Wagener DK, Feldman JJ, Rosenberg HM, Feinleib M. Increased mortality from brain tumors: A combined outcome of diagnostic technology and change of attitude toward the elderly. *Am J Epidemiol* 1992; 135:1349–57.

114. Legler JM, Gloeckler Ries LA, Smith MA, *et al.* Brain and other central nervous system cancers: recent trends in incidence and mortality. *J Natl Cancer Inst* 1999; 91:1382–90.

115. Smith MA, Freidlin B, Gloeckler Ries, LA, Simon R. Trends in reported incidence of primary malignant brain tumors in children in the United States. *J Nat Cancer Inst* 1998; 90:1269–77.

116. Linet MS, Ries LAG, Smith MA, Tarone RE, Devesa SS. Cancer surveilance series: recent trends in childhood cancer incidence and mortality in the United States. *J Natl Cancer Inst* 1999; 91:1051–8.

117. Gurney JG, Pogoda JM, Holly EA, Hecht SS, Preston-Martin S. Aspartame consumption in relation to childhood brain tumor risk: results from a case – control study. *J Natl Cancer Inst* 1997; 89:1072–4.

118. Koestner A. Aspartame and brain tumors: pathology issues. In: Stegink LD, Filer LJ Jr, eds. *Aspartame: physiology and biochemistry*. New York: Marcel Dekker; 1984; 447–57.

119. Cornell RG, Wolfe RA, Sanders PG. Aspartame and brain tumors: statistical issues. In: Stegink LD, Filer LJ Jr, eds. *Aspartame: physiology and biochemistry*. New York: Marcel Dekker; 1984; 459–79.

120. Ishii H. Incidence of brain tumors in rats fed aspartame. *Toxicol Lett* 1981; 7:433–7.

121. Ishii H. Chronic feeding studies with aspartame and its diketopiperazine. In: Stegink LD, Filer LJ Jr, eds. *Aspartame: physiology and biochemistry*. New York: Marcel Dekker; 1984; 307–19.

122. Koestner, A. Letter to the Editor. *J Neuropathol Exp Neurol* 1997; 56:107–9.

123. Borzelleca JF, Hogan GK, Koestner AA. Chronic toxicity/carcinogenicity study of FD&C Blue No. 2 in rats. *Food Chem Toxicol* 1995; 23:551–8.

124. Nauta, WJH. Letter to FDA Commissioner Hayes. 6 August 1981.

125. Shephard SE, Wakabayashi K, Nagao M. Mutagenic activity of peptides and the artificial sweetener aspartame after nitrosation. *Food Chem Toxicol* 1993; 31:323–9.

126. Meier I, Shephard SE, Lutz WK. Nitrosation of aspartic acid, aspartame, and glycine ethylester. Alkylation of 4-(p-nitrobenzyl)pyridine (NBP) in vitro and binding to DNA in the rat. *Mutat Res* 1990; 238:193–201.

127. Flamm WG. Letter to the Editor. *J Neuropathol Exp Neurol* 1997; 56:105–6.

128. Davies SM, Ross JA, Woods WG. eds. Aspartame and brain tumors: junk food science. *Causes of childhood cancer newsletter*. University of Minnesota; 1996.

129. Ross JA. Editorial: Brain tumors and artifical sweeteners? A lesson on not getting soured on epidemiology. *Med Pediatr Oncol* 1998; 30:7–8. (a) National Cancer Institute. Cancer Facts. Artificial Sweeteners, April 9, 1997 (http://cis.nci.nih.gov/fact/3_19.htm).

130. Food and Drug Administration. FDA statement on aspartame. *FDA Talk Paper*; 18 November 1996.

131. Department of Health. *1996 Annual Report of the Committees on toxicity, mutagenicity, and carcinogenicity of chemicals in food, consumer products and the environment*. United Kingdom; 1998.

132. Food Standards Agency, Additives and Novel Foods Division. Aspartame. *JFSSG*; November, 1999. http://www.foodstandards.gov.uk/maff/archive/food/asp9911.htm

133. European Commission. Scientific Committee for Food: Minutes of the 107th meeting of the Scientific Committee for Food held on 12–13 June 1997. Brussels; 1997.

134. Australia New Zealand Food Authority (ANZFA). Information paper. *Aspartame: information for consumers*; 1997.

135. Agencia Saude. Sao Paulo, Brazil. Forum of scientific discussion: aspartame; 29 October 1999. http://anvs1.saude.gov.br/Alimento/aspartame.html

136. Squillacote D. Aspartame (Nutrasweet): no danger. The Multiple Sclerosis Foundation; 12 January 1999. www.msfacts.org/aspartame.htm

137. Squillacote D. Aspartame: come in from the ledge. The Multiple Sclerosis Foundation; 26 January 1999. www.msfacts.org/aspart2.htm

138. Roberts HJ. Aspartame(NutraSweet): is it safe? Philadelphia: Charles Press; 1990.

139. Rolla AR. Aspartame (NutraSweet): is it safe? (book review) *New Engl J Med* 1990; 323:1495–6.

140. Walton RG, Hudak R, Green-Waite RJ. Adverse reactions to aspartame: double-blind challenge in patients from a vulnerable population. *Biol Psychiatry* 1993; 34: 13–17.

141. Butchko HH. "Adverse reactions to aspartame: double-blind challenge in patients from a vulnerable population" by Walton *et al. Biol Psychiatry* 1994; 36:206–7. (a) Schomer DL, Spiers P, Sabounjian LA. Evaluation of behavior, cognition, mood, and electroencephalograms in normal adults and potentially vulnerable populations. In: Tschanz C, Butchko HH, Stargel WW, Kotsonis FN, eds. *The Clinical Evaluation of a Food Additive: Assessment of Aspartame*, New York: CRC Press; 1996; 217–33.

142. Trocho C, Pardo R, Rafecas I, *et al.* Formaldehyde derived from dietary aspartame binds to tissue components in vivo. *Life Sci* 1998; 63:337–49.

143. Tephly TR. Comments on the purported generation of formaldehyde and adduct formation from the sweetener aspartame. *Life Sci* 1999; 65:157–60.

144. Food Standards Agency. Food safety information bulletin No 117, February 2000. http://www.foodstandards.gov.uk/maff/archive/food/bulletin/2000/no117/asparte.htm

145. Henkel, J. Sugar substitutes: Americans opt for sweetness and lite. *FDA Consumer* 1999; 12–16.

146. JAMA. American Medical Association. Council on Scientific Affairs. Aspartame: review of safety issues. *JAMA* 1985; 254:400–2.

147. Steginik LD, Filer LJ Jr, eds. *Aspartame: physiology and biochemistry*. New York: Marcel Dekker; 1984.

148. Steginik LD. Aspartame: review of the safety issues. *Food Technol* 1987; 41:119–21.

149. Steginik LD. The aspartame story: a model for the clinical testing of a food additive. *Am J Clin Nutr* 1987; 46:204–15.

150. Tschanz C, Butchko HH, Stargel WW, Kotsonis FN, eds. *The clinical evaluation of a food additive: assessment of aspartame*, Boca Raton, FL: CRC Press; 1996.

151. Butchko HH, Kotsonis FN. Aspartame: review of recent research. *Comments Toxicol* 1989; 3:253–78.

152. Sze PY. Pharmacological effects of phenylalanine on seizure susceptibility: an overview. *Neurochem Res* 1989; 14:103–11.

153. Fisher RS. Aspartame, neurotoxicity, and seizures: a review. *J Epilepsy* 1989; 2:55–64.

154. Jobe PC, Dailey JW. Aspartame and seizures. *Amino Acids* 1993; 4:197–235.

155. Janssen PJCM, van der Heijden CA. Aspartame: review of recent experimental and observational data. *Toxicology* 1988; 50:1–26.

156. Fernstrom JD. Central nervous system effects of aspartame. In: Kretchmer ND, Hallenbeck CB, eds. *Sugars and sweeteners*. Boca Raton: CRC Press; 1991; 151–73.

157. Ohnhaus EE. Aspartame: the profile of a unique food additive. In: McLean A, Wahlqvist ML, eds. *Current problems in nutrition pharmacology & toxicology*. London: John Libbey; 1988; 116–25.

15

REGULATION OF FLAVOR
INGREDIENTS

George A. Burdock

INTRODUCTION

An understanding of food flavor ingredients reveals some interesting paradoxes. First, of the approximately 3,000 ingredients added to food in the US, the majority (2,000) are flavor ingredients. Second, although flavor ingredients are quantitatively minor components of food (75% present at less than 100 ppm (1)), they are likely the single most important determinant in food choice, because the expectation of a specific and pleasing flavor plays a major role in the ultimate decision to consume a product. Indeed, an evolutionary mechanism for survival is based on flavor, with animals rejecting foul-tasting substances as being potentially poisonous (2). It is clear that although the flavor of food has a profound impact on caloric and nutritional intake, little thought is given by the consumer to what constitutes approximately two-thirds of the substances available for addition to food. Further, the composition of flavors is a relatively arcane subject – although some flavors are single-substance, naturally derived entities, flavors are more often made of a combination of substances, as many as 20 or more, and produced through a variety of physical, chemical, or biotechnological means and each requiring regulatory oversight.

Despite the complex nature of flavors, the great number of flavor substances and the importance of their impact on food and nutrition, the regulation of these substances remains a model of government and industry cooperation.

DEFINITION OF THE TERM "FLAVOR"

The answer to the question, "what is a flavor?," is not as straightforward as might be expected and varies according to the organization queried (Table 1). The US Food and Drug Administration (FDA) defines flavoring agents and adjuvants as "substances added to impart or help impart a taste or aroma in food." FDA

316

identifies flavor enhancers as "substances added to supplement, enhance, or modify the original taste and/or aroma of a food, without imparting a characteristic taste or aroma of its own."[1]

The question then arises as to what is a natural and what is an artificial flavor? According to the FDA, a chemical is natural as described in 21 CFR §101.22(a)(3)

> The term natural flavor or natural flavoring means the essential oil, oleoresin, essence or extractive, protein hydrolysate, distillate, or any product of roasting, heating or enzymolysis, which contains the flavoring constituents derived from a spice, fruit or fruit juice, vegetable or vegetable juice, edible yeast, herb, bark, bud, root, leaf or similar plant material, meat, seafood, poultry, eggs, dairy products, or fermentation products thereof, whose significant function in food is flavoring rather than nutritional. Natural flavors include the natural essence or extractives obtained from plants listed in §§182.10, 182.20, 182.40, and 182.50 and part 184 of this chapter, and the substances listed in §172.510 of this chapter.

Such a broad definition also embraces thermally processed flavors and smoke flavors. Artificial flavors are as described in 21 CFR §101.22 (a)(1)

> The term artificial flavor or artificial flavoring means any substance, the function of which is to impart flavor, which is not derived from a spice, fruit or fruit juice, vegetable or vegetable juice, edible yeast, herb, bark, bud, root, leaf or similar plant material, meat, fish, poultry, eggs, dairy products, or fermentation products thereof. Artificial flavor includes the substances listed in §§172.515(b) and 182.60 of this chapter except where these are derived from natural sources.

This distinction between artificial and natural serves as a basis for labeling, and in a market where the concept of "natural" is preferred, manufacturers will go to great lengths to avoid disclosure of "artificial flavors" on the label.

Until recently, Europeans embraced the nature-identical concept, a term used to identify a substance, which although may have been synthesized, was also found in nature. For example, a substance identified as natural benzaldehyde may only have come from natural sources, such as amygladin from peach pits, but benzaldehyde produced from petroleum sources is nature-identical; although produced synthetically, it has a counter-part in nature. Substances not found in nature, such as

---

1 Flavor enhancers include a range of substances such as inositol with no specific taste of their own, but also commonly used flavors or sweeteners (such as cinnamon or aspartame), that, when used below the thresholds of their own activity, may enhance the intensity of other flavors.

Table 1 Definition of a flavor

US Food and Drug Administration 21 CFR §170.3(o)(12)	Flavoring agents and adjuvants: Substances added to impart or help impart a taste or aroma in food.
US Food and Drug Administration 21 CFR §170.3(o)(11)	Flavor enhancers: Substances added to supplement, enhance, or modify the original taste and/or aroma of a characteristic taste or aroma of its own.
Council of Europe (Committee of Experts on Flavouring Substances)	A flavouring substance is a chemically-defined compound which has flavouring properties. It is obtained either by isolation from a natural source or by synthesis. Flavouring properties are those which are predominantly odour-producing and which may also affect the taste.
European Communities 88/388/EEC	1. This Directive shall apply to 'flavourings' used or intended for use in or on foodstuffs to impart odour and or taste, and to source materials used for the production of flavorings
	2. For the purposes of this Directive:
	(a) 'flavouring' means flavouring substances, flavouring preparations, process flavourings, smoke flavourings or mixtures thereof;
	(b) 'flavouring substance' means a defined chemical substance with flavouring properties which is obtained:
	(i) by appropriate physical processes (including distillation and solvent extraction) or enzymatic or microbiolgical processes from material of vegetable or animal origin either in the raw state or after processing for human consumption by traditional food-preparation processes (including drying, torrefaction and fermentation),

(ii) by chemical synthesis or isolated by chemical processes and which is chemically identical to a substance naturally present in material of vegetable or animal origin as described in (i),

(iii) by chemical synthesis but which is not chemically identical to a substance naturally present in material of vegetable or animal origin as described in (i)

(c) 'flavouring preparation' means a product, other than the substances defined in (b)(i), whether concentrated or not, with flavouring properties, which is obtained by appropriate physical processes (including distillation and solvent extraction) or by enzymatic extraction) or by enzymatic or microbiological processes from material of vegetable or animal origin, either in the raw state or after processing for human consumption by traditional food-prepartion processes (including drying torrefaction and fermentation);

(d) 'process flavouring' means a product which is obtained according to good manufacturing practices by heating to a temperature not exceeding 180 °C for a period not exceeding 15 minutes a mixture of ingredients, not necessarily themselves having flavoring properties, of which at least one contains nitrogen (amino) and another is a reducing sugar;

(e) 'smoke flavouring' means a smoke extract used in traditional foodstuffs smoking processes.

3. Flavourings may contain foodstuffs as well as other substances as described [elsewhere]

ethyl vanillin, obviously could not be identified as nature-identical. This distinction was never recognized by FDA, but held great importance in pre-European Community days when many European countries did not require regulatory notification of nature-identical ingredients. Lately, the term "defined chemical substance" has generally superceded the term, nature-identical.

Another point of divergence is the use of the dual or single listing system. For example, FDA has a "positive" and "negative" list system, wherein approved substances are listed in 21 CFR parts §172, §182 and §184, but substances forbidden for use (e.g. safrole, coumarin) are listed in 21 CFR §189. Several European countries used this system as well. Industry groups use a positive list system, wherein only approved (i.e. GRAS) substances are listed and those no longer approved are simply dropped from the industry list.

The EC has had the benefit of 40 years of regulatory experience in the US and longer in most European countries. As a result, EC regulations (developed by the Scientific Committee on Food) are more specific, including distinctions in their definitions that have been provided for in the US by fine-tuning with notices in the Federal Register. As can be seen in Table 1, Directive 88/388[2] is simultaneously broad and very specific, addressing "flavouring," "flavouring substance" and even "flavouring preparation" which has often been identified as the "gray area" between naturally-derived and modified or process flavors, such as enzymatically developed flavors in cheese. In all, the directive extends over several pages and includes definitions, specifications and provisions for further action on items, including limits on production methods and non-flavor ingredients in flavors (including preservatives, solvents and processing aids). This directive is binding on member states, with the exception that if a member state determines that the regulation or regulated substance may constitute a danger to the health of the citizens or environment of the Member State, it may suspend or restrict the use of that chemical within the borders of the Member State. The Council of Europe (CE) uses a fairly broad definition of flavor and this reflects the earlier start the CE had in flavor review (Table 1).

Non-regulatory organizations have slightly different views of the term "flavor." The International Organization of Flavor Industries (located in Switzerland) has a very utilitarian definition: a flavor is a "concentrated preparation, with or without solvents or carriers, used to impart flavor, with the exception of only salty, sweet, or acid tastes. It is not intended to be consumed as such." In other words, a flavor is never eaten in its natural state, it is always incorporated in a matrix or delivery system (at the very least as a flavor in mineral water or as a sweetened flavor poured over shaved ice). Not surprisingly, the Society of Flavor Chemists uses a more organoleptic definition and considers flavor to be "a substance that may be

---

2 A Directive sets out minimum standards and when adopted by the member states may be made more stringent. A Regulation is directly applicable to the member states without any need to transpose into national law.

a single chemical entity or a blend of chemicals of natural or synthetic origin whose primary purpose is to provide all or part of the particular effect to any food or other product taken into the mouth."

One of the best definitions offered to date came from Richard Hall: "Flavor is the sum of those characteristics of any material taken in the mouth, perceived principally by the senses of taste and smell, and also the general pain and tactile receptors in the mouth, as received and interpreted by the brain."(3). This definition is a reminder there is more to flavor than taste and smell. Consider, for example, the pain, bite, and heat associated with capsicum (red pepper), piperine (black pepper), and allylisothiocyanate (horseradish). Hall's definition also relates to the involvement of learned behavior and associations. For example, it is possible to make banana-flavored potato chips, but the crunchy texture would be incongruous with the taste. Conversely, we have long enjoyed pineapple-flavored hard candy, which in reality, tastes nothing like the fruit but has long been accepted as how a pineapple candy should taste.

CREATION OF A FLAVOR

As noted, a flavor is the sum of one or as many as 20 ingredients that may be used to create it. These ingredients can include spices (whole or ground), extracts (botanical or animal products), or synthetic products/nature-identical substances. A finished flavor can be created in the laboratory by using organoleptic building blocks, including those flavors with a characteristic key or only providing supplemental effects. Organoleptic building blocks are single substances or combinations of substances that produce an organoleptic sensation associated with a generalized taste sensation, such as "herbaceous" or "creamy." When these various and often subtle, sensations or building blocks are combined in the right proportions, a flavor evolves. For example, the right combination of green, waxy, creamy, and nutty, may produce a "butter" flavor (Table 2). Because these subtler building blocks differ from substances with a characteristic chemical key in that they do not have a "stand-alone" taste, their use often requires considerable fine-tuning to achieve the desired result. Flavors with a characteristic key include methyl eugenol, responsible for the characteristic flavor of cloves, and cinnamaldehyde, which is the basis of any cinnamon taste (Table 3).

Some flavors require secondary influences of thermal (Table 4) or biologic (Table 5) origin for full development. For example, if not roasted, coffee and cocoa beans are extremely bitter; ammonia and sugar must be heated together to produce a caramel taste and; the combination of carbohydrate and protein heated together results in a pleasing cooked meat flavor. Biosynthetic production of flavor is nearly as straightforward. For example, microorganisms can be used to produce a ripe cheese flavor or to eliminate the undesirable taste of substances.

Table 2 Organoleptic sensations and representative flavor substances

Sensation(s)	Substance(s)
Burning, biting	Allylisothiocyanate
	Piperine
Heat	*Capsicum* oleoresin
Cool	L-Menthol
	3-L-Menthoxypropane-1,2-diol
Cheese	Butyric acid
	4-Pentenoic acid
	Caprylic acid
	Capric acid
	5,6,7,8-Tetrahydroquinoxaline
Green, herbaceous	Diphenyl ether
	Ethyl levulinate
	Ethyl 2-methylbutyrate
	1-Furfurylpyrrole
Waxy	Octyl-2-furoate
	10-Undecenal
Creamy	*delta*-Undecalactone
	Tributyrin
Nutty	2-Hexenal
	3-Octanol
	Pyrazines

Table 3 Flavors with a characteristic chemical key

Flavor(s)	Chemical
Anise, fennel, licorice	*trans*-Anethole
Almond, cherry, fruital	Benzaldehyde
Caraway, dill	D-Carvone
Spearmint	L-Carvone
Cinnamon	Cinnamaldehyde
Peppermint	L-Menthol
Wintergreen	Methyl salicylate

FLAVOR REGULATION

Flavor regulation in the United States

There are two basic mechanisms by which an appropriate substance may be legitimately added to food i.e. as a food additive or as a generally recognized as safe

Table 4 Flavors produced by heating

Flavor	Component(s)
Coffee	Coffee beans
Cocoa	Cocoa beans
Caramel	Ammonia and sugar
Roasted or cooked meat	Carbohydrate and protein

Table 5 Flavors produced biosynthetically

Flavor	Microorganism	Substrate	Product generated
Pleasant acidic taste	*Aspergillus niger*	Simple sugars	Citric acid
Nonbitter taste	*Aspergillus niger*	Naringin	Naringenin
Coconut-like	*Trichoderma reesei*		6-Pentyl-*alpha*-pyrone
Peach-like	*Sporobolomyces odorus*		4-Decanolide *cis*-6-Dodecen-4-olide
Pineapple, ethereal	*Candida utilis*	Glucose	Ethyl acetate
Cheese	*Aspergillus oryzae*	Fatty acids	Hydrolysis products

(GRAS) food ingredient. The food additive petition route is that taken by most food ingredients and the requirements are described in part §171 of section 21 of the Code of Federal Regulations. Many flavor ingredients are, in fact, food additives and therefore, are not GRAS. A list of flavors is provided in 21 CFR §172 (subpart F). However, the majority of flavor ingredients are added to food on the basis of their status as generally recognized as safe (GRAS). Although as provided by statute (§201(s) of the Food Drug and Cosmetic Act, as amended), there may be many origins of GRAS status, but the best known are the Flavor and Extract Manufacturers' Association's (FEMA) Expert Panel and the Federation of Associated Societies of Experimental Biology (FASEB). Also, a large number of flavor ingredients have been approved GRAS by the FDA, in 21 CFR §182 as Generally Recognized As Safe (GRAS) and in 21 CFR §184 as affirmed GRAS. How then, did all this come about? What is GRAS and how does it work?

The original Pure Food and Drug Act of 1906 underwent its first major overhaul in 1938, but by the mid-1950s, food ingredient regulation had again become problematic. The House Select Committee on the use of chemicals in foods discovered that approximately half of the additives examined by the Committee were not considered safe. The Committee discovered that however well-intentioned the existing law was, the regulatory process was a burden to both FDA and to industry. For example, on one hand, a food processor could add ingredients without regard for technical function or even margin of safety; conversely, FDA was placed in the position of "policeman" and was required to demonstrate the lack of safety of an ingredient before a producer could be forced to remove it. Further, once the FDA declared an ingredient injurious to health, the producer was forced to prove that it was "harmless *per se*" (i.e. harmless at all doses) or forego the use of this substance in any amount. The Committee was faced with a difficult task, how could they ensure consumer safety for these 2,000+ food ingredients without placing an unmanageable safety testing requirement on industry, and yet not "grandfather" previously used ingredients for which the safety data was often non-existent?

As irreconcilable as these concerns may have seemed, the Food Additives Amendment of 1958 successfully addressed the problem and the result was born from pragmatism: first, recognition of the fact that not all substances added to food were harmful, many had a long history of safe use and did not require extensive testing; second, recognition by Congress that not all expertise in food safety was confined to government experts. This philosophy is reflected in the FFDCA (Section 201(s) (Definitions)):

> The term "food additive" means any substance the intended use of which results or may reasonably be expected to result, directly or indirectly, in its becoming a component or otherwise affecting the characteristics of any food... if such substance is not generally recognized among experts qualified by scientific training and experience to evaluate its safety, as having been adequately shown through scientific procedures (or, in the case of a substance used in food prior to January 1, 1958, through either scientific procedures or experience based on common use in food) to be safe under the conditions of its intended use....

This definition accomplished several ends: (1) it changed the meaning of the term "food additive" from a food technology adjective to one of legal status by defining this category as those substances *not* generally recognized as safe; (2) it created an entirely new class of substances, those that *are* generally recognized as safe (later to be known by the acronym, GRAS); (3) further, it defined *who* can determine what is GRAS (i.e. experts) and; the *criteria* by which these experts would determine a substance is GRAS (i.e. (a) through scientific procedures or (b) for those substances used prior to 1 January 1958 the option of scientific procedures or previous use in food; and finally (4) this section abolishes the safety *per se* concept with the statement "...safe under the conditions of intended use..."

Thus the GRAS mechanism accomplished three major goals and, in so doing, responded to the concerns of the public, industry and FDA: (1) the public was assured that there would be no "blanket grandfathering" of previously used substances, that all substances would be subject to an expert review;

(2) industry would be assured that a history of safe use would be taken into account and that the potentially ruinous burden of extensive testing of all substances would not be necessary; and (3) the regulatory agencies were assured that they would not be overwhelmed with hundreds of petitions for food additive status.

The Generally Recognized As Safe (GRAS) concept

Eligibility standards

By definition, a food additive is a substance not included in the category of substances regarded as GRAS. However, many substances used as food additives for one purpose (restricted for use in particular foods and/or for specific technical effect) may well be GRAS for use in another food or for another effect. For example, an artificial sweetener, used at a concentration below the sweetness threshold, may have functionality as a flavor enhancer and may be GRAS for this purpose.

The safety standard

The statutory mandate for safety is laid out in §402 of the FFDCA and refers to substances that *may render it* [the food] *injurious to health*. Because this is unclear, FDA regulations (21 CFR §170.3(i)) state the following:

> *Safe* or *safety* means that there is a reasonable certainty in the minds of competent scientists that the substance is not harmful under the intended conditions of use. It is impossible in the present state of scientific knowledge to establish with complete certainty the absolute harmlessness of the use of any substance ...

Thus, the basis for the standard is a *reasonable certainty* of safety. This section (21 CFR §170.3(i)) also lays out what considerations must be observed in determining safety:

(1) The probable consumption of the substance and of any substance formed in or on food because of its use.
(2) The cumulative effect of the substance in the diet, taking into account any chemically or pharmacologically related substance or substances in such diet.
(3) Safety factors which, in the opinion of experts qualified by scientific training and experience to evaluate the safety of food and food ingredients, are generally recognized as appropriate.

Therefore, following consideration of these three factors, a substance then may be conferred GRAS status by experts on the basis of two criteria:

> ...[1] through scientific procedures (or, [2] in the case of a substance used in food prior to January 1, 1958, [2a] through either scientific procedures or [2b] experience based on common use in food) to be safe under the conditions of its intended use....

On the basis of the above a substance must have demonstrated safety through scientific procedures (i.e. testing) or through a history of use or both. Above all, GRAS status may only be conferred by experts following a review, there is no "grandfathering."[3]

The experts and general recognition

No substance may be conferred GRAS status unless first recognized as such by *experts qualified by scientific training and experience to evaluate its safety.* Importantly, GRAS status does not require unanimous recognition among experts (as not all experts might agree) and does not require that the substance be safe *per se*, only safe for its intended use. Substances are not GRAS when the case for recognition is weak or conflicting, or when the experts are not generally aware of the substance (4).

Although the statute does not further define the experts, the regulations (21 CFR §170.3(i)), refer to *competent scientists* and in a later section (21 CFR §170.3(i)(3)), makes reference to...*experts qualified by scientific training and experience to evaluate the safety of food and food ingredients...*, but only in the context of determining appropriate safety factors. This regulation is insightful because it allows input from experts from various fields, only requiring that the experts be *qualified* or *competent* – not an unreasonable expectation. However, when it comes to the application of safety factors, expertise in the safety of food is required. The upshot allows a toxicologist, not familiar with food toxicology, to render an opinion on the general safety of the substance, based on his qualifications and competency. Similarly, a veterinary pathologist could render an opinion on animal histological data, a physician on the possible effects in humans and a chemist on possible contaminants involved in the synthesis or extraction of the product (5). However, the Agency retains ultimate authority and may overturn the decision of an expert body, as 21 CFR §170.38 permits the Commissioner to determine that a substance is not GRAS, but is a food additive and may place restrictions thereupon.

---

3 With the exception of "prior sanctioned" substances as permitted by Section 201(a)(4) of the Act and as listed in 21 CFR 181, although no flavor ingredients are prior sanctioned.

Common use in food

The 1958 Amendment rightfully recognized the validity of a prior (i.e. prior to 1 January 1958) history of safe use of a substance. The regulation (21 CFR §170.3(f)) defined the concept as follows:

> Common use in food means a substantial history of consumption of a substance for food use by a significant number of consumers.

Therefore, "commonness" of use depends on two factors: a meaningful and sustained (i.e. substantial) history of consumption and that a significant number of consumers be exposed (thus accounting for the presence of a range of responses, especially among susceptible subpopulations). One factor cannot be emphasized to the exclusion of the other. For example, one manufacturer was unable to claim GRAS status for a substance, which although it had been in use for 40 years, it was held that the population base exposed to this chemical was inadequate to define it as having been in common use (4).

At one time, "common use" was considered to pertain only to the American or Western European diets. A case in point is Fmali Herb, Inc. v. Heckler in which FDA maintained the honey-like product (renshenfengwangjiang) which normally contains schizandra seed, could not be sold in the US if it contained schizandra seed, because the seed had not been tested nor did it have a history of safe use anywhere but in China (6). Therefore, while a foreign use of a substance may be corroborative of safety, it is not dispositive. However, now that Americans are consuming more ethnic food, the FDA has expanded the "commonness" of use concept to include non-Western countries. Logically, more rigorous proof of long and widespread use is required of such a substance, and the substance cannot have been used as a tonic, folk remedy, or drug(4). In addition, US experts must provide the general recognition of safety.

Documentation of scientific procedures and the burden of proof

If a product does not meet the "commonness" of use standard, it can become GRAS through *scientific procedures* (including testing).[4] The Code of Federal Regulations (21 CFR §170.30(b)) describes the requirements as being the "same quality and quantity of scientific evidence" for a food additive regulation which includes, but are not limited to the identity, method of synthesis and detection of the substance, manufacturing process, date of first use, reports on other past uses in food and, information to establish the safety and functionality of the substance in food (21 CFR §170.35(b)). Further, information supporting safety data

---

4 Although as noted, 21 CFR §170.30(i) does not provide for history of use alone to determine GRAS status.

must be published in the scientific literature (21 CFR §170.30(b)) and while unpublished studies may be used to corroborate certain points, thus preserving proprietary information, otherwise critical information may not be concealed under the guise of propriety. This requirement for GRAS by *scientific procedures* clearly changed the role of FDA from that of "policeman" to that of "gatekeeper" (7), by shifting the burden of proof of safety to the petitioner (or GRAS experts).

Restrictions ensuring safe conditions of use

An important concept in general recognition of safety is the enunciation of the intended conditions of use. In the case of flavor ingredient review, the experts evaluating the substance already know the technical effect is as a flavor, but they must also know at what level and in which of the 43 food categories (21 CFR §170.3(n)) the flavoring substance will be used. In turn, the rate at which these categories are consumed by the average and the 90th percentile users is established and substance consumption may be determined. These consumption levels are then compared to the appropriate safety factors described earlier.

It is clearly stated in §409 of the Act that the Secretary may impose restrictions on the use of GRAS substances, i.e. the Secretary may prescribe . . . *the conditions under which such additive may be safely used* . . . including, but not limited to, particular technical effects, food categories, specifications and restrictions on packaging and labeling. In this case, 21 CFR §170.38 permits the Commissioner to determine that a substance is not GRAS, but is a food additive and subject to §409 of the statute. The position of the Agency was reasserted in 1974 with the following statement:

> It has been too often assumed that the GRAS substance may be used in any food, at any level for any purpose. As a result, the uses of some GRAS food ingredients have proliferated to the point where the GRAS status was brought into serious question.
>
> (39 FR 34194 (1974))

Notification of a claim for GRAS exemption and FDA response

In April 1997, FDA published the proposed rule for the GRAS claim notification process (62 FR 18938 (1997)). This proposal keeps the Agency informed, allows public scrutiny of the supporting data and allows the manufacturer to provide assurance to his customer that the Agency is aware of the use of the substance. The notice requires that the same type of information be presented that would otherwise be found in a GRAS affirmation petition or, for that matter, a petition for a food additive regulation (see above, *Documentation of scientific procedures and the burden of proof*).

Receipt of the notice for claim of GRAS exemption must be acknowledged by the Agency within 30 days and within 90 days of receipt of the notice FDA must respond to the notifier, in writing, of any objection. Explicit in the proposed regulation is that the notification is placed on public display upon receipt (except any proprietary information) and that the Commissioner may, in accordance with 21 CFR §170.35(b)(4), determine that a substance is not GRAS, but is a food additive subject to §409 of the FAA.

The role of the independent expert panels

As noted earlier, the Flavor and Extract Manufacturers Association (FEMA) has the longest continuous history producing industry lists of GRAS substances. The FEMA program began in 1959 with an industry survey to identify flavor ingredients and quantities in use. The FEMA Expert Panel, consisting of scientists with expertise in relevant fields, was first appointed in 1960 to evaluate existing flavoring ingredients for GRAS status. Substances submitted by manufacturers without sufficient data were taken off the list, as were ingredients not in general use. The FEMA panel presented its conclusions to the FDA prior to publication (*Recent Progress in the Consideration of Flavoring Ingredients Under the Food Additives Amendment. III GRAS Substances*) (8). Following a decade of review to clear out the backlog of flavor ingredients in use, the FEMA panel published its first report to include new flavor ingredients in 1970 (GRAS IV). The FEMA panel has published 15 subsequent lists, the total of which includes over 1,900 substances. Of this extensive list of substances, only nine have had their GRAS status revoked (alkanet root, brominated vegetable oil, calamus and calamus oil, 2-hexyl-4-acetoxytetrahydrofuran, musk ambrette, 3-nonanon-1-yl acetate, 2-methyl-5-vinylpyrazine and *o*-vinylanisole).

The criteria used by the FEMA panel in determining whether a flavor ingredient is GRAS include the following: *exposure* to the substance in specific foods, total amount in the diet, and total poundage; *natural occurrence* in food; *chemical identity* (including purity and method of preparation) and specific chemical structure; *metabolic and pharmacokinetic characteristics* and; animal *toxicity* (5). All criteria must be mutually supportive and unanimously accepted by the FEMA Panel for it to conclude that a substance is GRAS under the conditions of proposed use.

The concept of GRAS was not intended to relieve anyone from the production of data and as mentioned earlier, not much safety information was available for the tens of hundreds of substances whose safe use FEMA must support. This did not present much of a problem for the first FEMA GRAS list (GRAS III), because the majority of substances were well-known herbs and spices and their extracts, with a safe history of use of hundreds, if not thousands of years. However, with subsequent FEMA GRAS lists, the proportion of natural to artificial substances rapidly changed and FEMA was confronted with a dearth of safety data and the requirement for public disclosure of scientific data upon which the

FEMA panel had based its decisions. The response by FEMA was the production of the Scientific Literature Reviews (or SLRs) in the 1970s. This series of reviews divided flavor ingredients into categories by chemical structure (e.g. aliphatic ketones, secondary alcohols and related esters; aliphatic acetals; aliphatic lactones; aromatic ethers; furfuryl alcohol and related substances, etc.). These reviews treated similar chemical entities as groups with (logically) presumed similar biological activities. With each new GRAS candidate, the structure was compared to similar substances and any new data incorporated into the "database" upon which a decision could be made. Occasionally, a new GRAS candidate would have to provide supporting data that it indeed "behaved" biologically as did other members of the group. Also, the FEMA panel has requested FEMA perform additional testing in support of the safety of a particular group, nominating substances representative of the group. Although not all data were published in the open scientific literature, summaries of the data were incorporated into the SLRs and provided to FDA. FDA was also provided with updates of individual reviews at the time of subsequent GRAS lists. Because these reviews have become dated and public availability limited, FEMA is attempting a new format, with a published review in the literature on alicyclic flavor ingredients in 1996 (9) followed by publications on furfural (10), lactones (11) and *trans*-anethole (12).

As implied by the title of this section, FEMA is but one industry-convened group of experts, most of which have been formed on an *ad hoc* basis for the review of a specific food or flavor ingredient. However, with the promulgation of the GRAS notification procedure, new independent groups have been formed for review of candidate GRAS substances. Among these new groups is Recognized Experts on GRAS Substances, Inc. (REGS, Inc.), which provides various types of expertise from a pool of recognized experts and an infrastructure and archiving capabilities hitherto not generally available with *ad hoc* groups. Unlike FEMA, which has restricted itself to flavor ingredients, these newer groups review other food ingredients as well and, given the burgeoning dietary supplement industry, may play a significant role in the review of the safety of dietary supplements although GRAS determination of a dietary supplement is not necessary.

The role of other agencies in flavor regulation

Although the FDA Center for Food Safety and Applied Nutrition plays the lead role in oversight of flavor ingredients, other parts of FDA and other branches of the government are involved as well. The Drugs division of FDA is involved, because many flavors are used as excipients in drugs, e.g. flavors in cough syrup. The Department of Agriculture is involved when flavoring substances are used in such foods as processed meats or the flavors contain significant quantities of meat and/or poultry. The Bureau of Alcohol, Tobacco, and Firearms plays a role in regulating flavored liqueurs, liquors (e.g. vodka), wine coolers and other types of alcoholic drinks. Finally, in addition to the uses of flavors provided for by the

FDA in 21 CFR §500 and overseen by the FDA Center for Veterinary Medicine, the American Association of Feed Control Officials (AAFCO) publishes a manual listing those flavoring ingredients which may be used in feed.

Flavor regulation throughout the world

Joint (FAO/WHO) committee on food additives and contaminants

At the apex of international efforts to regulate flavors is the Codex Alimentarius (Latin for *food code*), a joint venture of the Food and Agriculture Organization and the World Health Organization. The purpose of Codex Alimentarius is "to guide and promote the elaboration and establishment of definitions and requirements for foods, to assist in their harmonization and, in doing so, to facilitate their international trade." One of Codex's general subject committees is the Committee on Food Additives, which in turn oversees the Joint Expert Committee on Food Additives and Contaminants (known by the acronym, JECFA). This Committee examines food additives, flavorings, and related substances nominated by member governments and provides recommendations on their average daily intake (ADI). Some substances have a specific ADI (mg/kg body weight/day) assigned, while others have no defined restrictions (ADI not specified). Related substances may be assigned a group ADI (e.g. benzoic acid and its calcium, potassium and sodium salts, and benzyl acetate, alcohol and benzoate; with the ADI expressed as benzoic acid).

For some time, JECFA has debated as to how best to approach evaluation of flavor substances. The obstacles are described in *Principles for the Safety Assessment of Food Additives and Contaminants in Food* (13). Principal among the difficulties was a recognition of the large number of substances (approximately 2,500 chemically-defined substances in use in Europe and the US), often with little data to support their safety, but compelling, mitigating arguments that many flavors have a long history of use, they are used in very small amounts, are self-limiting and are largely volatile substances whose concentrations are reduced during processing and storage. Because no workable solution for review of flavor substances had been devised, flavor substance review still necessitated comparatively extensive data submission (e.g. *trans*-anethole), at a cost that could only rarely be afforded. Finally, in 1995, at the 44th meeting of JECFA, a method by which flavor safety assessment might be approached was proposed (14) but has since been revised (15,16). The solution consists of an algorithm (Figure 1) with strong emphasis on structural class and the predictability of toxicological effects of the class (including a threshold of concern for structure classes, I–III (17)), possible metabolites (and a determination if the metabolites could be considered innocuous and/or endogenous) and the amount consumed (both US and European data). This method allowed division of substances into three chemical classes:

Safety Evaluation Procedure for Flavoring Agents

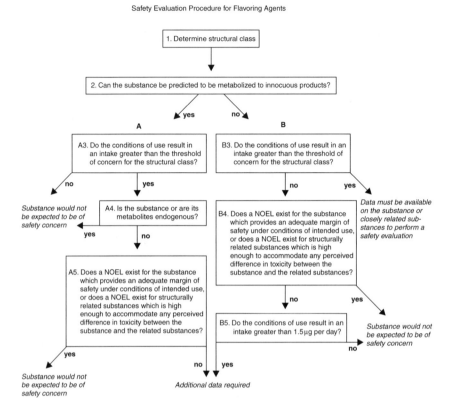

Figure 1

- Class I. Substances with simple chemical structures and efficient modes of metabolism which would suggest a low order of toxicity by the oral route. Exposure threshold for this class is 1800 μg/person/day.
- Class II. Substances with structural features that are less innocuous than those of substances in Class I but are not suggestive of toxicity. Substances in this class may contain reactive functional groups. Exposure threshold for this class is 540 μg/person/day.
- Class III. Substances with structural features that permit no strong initial presumption of safety, or may even suggest significant toxicity. Exposure threshold for this class is 90 μg/person/day.

Exposure thresholds were derived from a large database of subchronic and chronic animal studies, representing 613 substances. The thresholds represent a 100-fold safety factor from the 5th centile of either (a) the no observed effect levels (NOELs) from chronic studies or (b) the NOELs from the subchronic studies divided by a factor of three (i.e. to approximate the most likely NOEL that would be derived from a chronic study) (14).

At this point, JECFA has evaluated 626 substances, expressing no concern for safety at current levels of intake for 620 and deferring decision on six substances pending additional data. Thus, precedent has been set for the review of flavor substances with a determination on the basis of existing consumption, rather than granting a specific ADI using safety factors based on substance-specific animal study data.

EC scientific committee on food

The European Union has formed the Scientific Committee for Food (SCF), which was key in advising on the definitions published as Directive 88/388. In 1991, the SCF adopted guidelines for the evaluation of flavors and by 1995 had reviewed 426 substances, reporting on 148. Of these 148 chemically defined substances, 141 were included in a positive list: 122 determined as safe in use (Category 1), 19 as temporarily safe in use (Category 2), two substances were considered as having insufficient data for a determination (Category 3), one substance not acceptable due to evidence of toxicity (Category 4) and four substances classified as either mixtures or not flavors (Category N). The process was stopped however, with the EC Regulation No. 2232/96 (28 October 1996) which (1) introduced a concept of evaluation for all flavoring substances (i.e. encouraging SCF to take into account approaches for evaluation and evaluations already performed by JECFA and the Council of Europe) and (2) a "register" was established for all flavor substances. The register was published in the Commission Decision 1999/217/EC of 23 February 1999 and contained a list of about 2,800 substances. Within 10 months of the adoption of the register, a program for toxicological evaluation will be adopted and within 5 years of its adoption, a positive list of flavor substances (*to the exclusion of all other substances*) will be adopted (18).

If all goes according to plan, the SCF will have completed its review of the register by March of 2004, a task that could be made easier by adopting determinations made by JECFA and the Council of Europe as encouraged by the Commission Decision 1999/217/EC, a suggestion with which the SCF largely concurs (18). However, the SCF has identified four areas of concern that may impede the hope for rapid progress. First, the SCF is concerned that the methods for estimating intake, the TAMDI[5] and the MSDI[6] (excluding the DINFO[7]) are only gross

---

5 TAMDI (Theoretical Added Maximum Daily Intake) which is calculated on the basis of upper use levels and the estimated daily intakes of flavored beverages, foods and those particular foods with upper uses in variance to normal use levels.

6 MSDI (Maximum Survey-derived Daily Intake) which is based on manufacturers' data, which assumes that only 60% of the actual amount is consumed and that 10% of the population consumes 100% of the calculated amount.

7 DINFO (Daily Intake via Natural Food Occurrence) is intake due to the presence of substance as an intrinsic part of food.

approximations and the MSDI especially. The concern is that these methods underestimate the consumption of population subgroups and that inadequate consideration is given to localized cultural preferences for certain flavors. All of these approximations are especially critical when the level of consumption is at or near one of the exposure threshold decision points. A second area of concern for the SCF is the fact that the thresholds of the JECFA model are predicated upon commonly encountered systemic toxicity and discount the importance of low doses in neurotoxicity, immunotoxicity, developmental toxicity and endocrine effects. This concern is mitigated by the initial screening for chemical structure and a consideration of metabolic products. As relates to these two concerns, the SCF appears content with the JECFA procedure as "a reasonable and pragmatic approach." The third area of concern, however, takes on considerably more weight and resides in the fact that the JECFA procedure does not explicitly address genotoxicity. The SCF asks if it might be worthwhile to set two separate threshold values – one for substances known to be non-genotoxic and a second for substances whose genotoxic potential is unknown. Lastly, the SCF has opined that the JECFA practice of acceptance of substances on the basis of consumption being less than 1.5 µg/person/day, is a criterion best left for indirect food contact items.

Further, the SCF sees a review of substances by both SCF and JECFA as a needless duplication of effort and suggests that the burden be shared by dividing groups of flavorings between the two. The SCF could then concentrate its efforts on flavorings which would not be considered by JECFA and on flavor substances which have a high priority for review. The SCF has emphasized that it may only proceed with its work on the register when all existing chemical, toxicological, production volume, use levels and other information that may be relevant are provided by industry.

The Council of Europe

The Council of Europe, consisting of many, but not all members of the European Union, charged its Committee of Experts on Flavouring Substances to review scientific considerations relating to flavor ingredients. In 1992, for the fourth edition of the Council of Europe "Blue Book," the Committee took a different approach than had been taken in previous years. The Committee determined first to review chemically-defined substances and at a later time to review the natural occurring substances. The chemically defined substances were divided into 16 groups according to structure and individual determinations of safety made for each substance. Considerations by the Committee included the following:

- Toxicity in animals and/or man, including data on mutagenicity, carcinogenicity and teratogenicity where available;
- Metabolic fate and biotransformation to compounds of known toxicological properties;

- Levels of use in various dietary components;
- Natural occurrence in normal food;
- Comparison of the chemical structure with that of compounds of known toxicological and biochemical properties.

These chemically-defined flavoring substances were classified into the following categories based on the information available:

List A – flavoring substances which may be used in foodstuffs;
List B – flavoring substances for which further information is required before the Committee of Experts is able to offer a firm opinion on their safety-in-use. These substances can be used provisionally in foodstuffs.

The Committee reviewed approximately 900 substances and determined which were approved for use (List A, 390 substances and included all JECFA reviewed substances for which ADIs had been published) and which were provisionally approved (List B, 504 substances). Further, the Committee set upper limits on use in food and beverage categories. Certain exceptions (higher or lower than for food or beverage categories) were placed on some substances for the categories of candy and confectionery, condiments and seasonings, alcoholic beverages and, soups and savouries. The results of the Committee's updated findings have also been published (19). The Committee also published a list of studies required to support continued use of List B substances and warned that if data were not submitted, the Committee reserved the right to delete the substance in question from any approved list, provisional or otherwise.

The Council of Europe Committee of Experts is now examining enzymatic flavors, process flavors, natural sources of flavors (e.g. species and parts of plants that produce flavor ingredients), and tolerance levels for active substances. They have also begun examining issues associated with biotechnology and plant tissue cultures as a source of flavors, but because regulation threatened to get ahead of the technology, work on the latter is being held in abeyance. A volume on natural sources of flavorings has recently been published (20).

Other nations

Lists of approved flavor substances include that of MERCOSUR, a free-trade zone in South America consisting of the member states Argentina, Brazil, Paraguay and Uruguay. These countries have established the "Lista de Base" composed of "Flavor and Fragrance Materials 1991" (Allured Publishing Co.) and the lists of Codex Alimentarius, Council of Europe, US FDA and FEMA. The Asian states of China, Hong Kong and Chinese Taipei also recognize the FEMA list and natural flavor ingredients approved by any one of the three member countries (21).

Industry organizations

Industry organizations cover international issues as well. The International Organization of Flavor Industries, of which FEMA is a member, incorporates the European Flavor and Fragrance Association (EFFA), which advises European agencies on scientific matters concerning flavor ingredients. The European Flavor and Fragrance Association, while similar to FEMA, has been less active than it's American counterpart.

Remaining issues

In the last 10 years, quantitative steps forward have been taken to "globalize" approval of flavor substances. The first significant movement toward globalization is the FDA nomination of US industry lists to JECFA. With the new procedure for flavor substance review in place, the capacity for JECFA to review substances has increased markedly. If the SCF takes into consideration the JECFA approvals, the list of approved flavor substances for EC use will increase at a rate not seen in the past and will make the SCF goal of the year 2004 for review of the register feasible. A significant potential problem exists in the possible financial burden of additional substance testing requested by industry groups, JECFA or SCF. It is likely that those substances nominated so far are those with the clearest case for approval and it is conceivable that requests for additional test data may pose a financial burden on the industry that it may not want to undertake. Another significant potential problem is the concern of SCF that more emphasis should be placed on genotoxicity data and the possible dual-system of categorizing flavor substances into "non-genotoxic" and "other." A third problem exists within the SCF itself, that member states may create barriers to commerce with objections to specific substances on grounds of safety or environmental concerns. Also within the SCF is the potential that the register may be closed and no new substances admitted until approval of the current register substances, now scheduled for 2004.

Substances derived through biotechnology are still generally considered natural – *to a degree* – that is, unless the process produces a substantive change in the ingredient during its liberation from its source. Enzymatic conversion of substances is allowable on a case-by-case basis, such as the conversion of constituents of cassia to benzaldehyde. However, often the production of a natural substance through these means is much costlier than its synthesis in the laboratory. As a responsible industry association, FEMA established the Isotopic Studies Committee (ISC) in 1985. To safeguard the integrity of the flavor industry, the ISC was charged with developing analytical methods to verify the origin of flavor materials and in so doing, the ISC established a program at the Center for Applied Isotopic Studies at the University of Georgia. Among other methods established (e.g. optical rotation), sophisticated isotopic ratio methods were developed in which the ratios of carbon, oxygen, nitrogen (when present), and hydrogen can be determined to identify whether a product is natural or synthesized. From this work, the "Naturalness Decision Tree" has been developed to help guide the flavor industry in determining what can legitimately be identified as "natural" (22).

The area of biotechnology and the status of food and food ingredients produced by genetically modified organisms remains an area of significant disagreement. There has not been much public activity in this area with flavor ingredients, save for the determination of the sweetener Recombinant Thaumatin B as GRAS as a flavor by the FEMA panel in 1996 (23). Clearly, any such substance produced by a genetically modified organism must be scrutinized for contaminants including toxins, allergenic or pharmacologically active proteins. The FDA has already determined the DNA introduced into organisms (plants) to produce new substances is generally recognized as safe (24). The basis of some arguments against the use of genetically modified organisms for the production of food ingredients rests on the use of antibiotic resistance genes. This potential difficulty has been addressed by FDA which determined there was no realistic threat to public health or the environment, although researchers were advised to avoid using marker genes to manufacture clinically important antibiotics. Although genetically modified plants and organisms for the production of food ingredients are generally well received in the US, considerable resistance has been voiced in other parts of the world.

CONCLUSION

The flavor of food is the result of a total organoleptic experience, which is the product of flavor ingredients, ingredients that serve ancillary functions (e.g. carriers, preservatives, emulsifiers) and those subsequent processing treatments both by the manufacturer (e.g. storage, aging of cheese or wines) and in the home. Despite the wide range of ingredients and methods for producing or modifying flavor ingredients, the safety of the consuming public has never been at demonstrable risk. The Food and Drug Administration, FEMA and other food and flavor industry groups have worked for more than 40 years to ensure the safety of flavoring substances using a framework for safety assessment that has been a model for other countries and other industries. This well-tested framework has proved adaptable to new technologies, including substances produced by genetically modified organisms, and has provided a stable and reliable basis to ensure the safety of the consumer and access to flavorful foods.

REFERENCES

1. Hall R, Oser B. The safety of flavoring substances. *Residue Reviews* 1968; 24:1.
2. Bernstein IL. Taste aversion learning: a contemporary perspective. *Nutrition* 1999; 15:229–34.

3. Hall RL. Food flavors: benefits and problems. *Food Technology* 1968; 22:1388.
4. Hallagan JB, Hall RL. FEMA GRAS – A GRAS assessment program for flavor ingredients. *Regulatory Toxicology and Pharmacology* 1995; 21:422.
5. Woods LA, Doull J. GRAS evaluation of flavoring substances by the Expert Panel of FEMA. *Regulatory Toxicology and Pharmacology* 1991; 14:48.
6. Hutt PR, Merrill RA. *Food and Drug Law, Cases and Materials.* 2nd edn. Westbury, NY. Foundation Press, University Casebook Series 1991; 340–1.
7. Raiten DJ, ed. *The LSRO report on alternative and traditional models for safety evaluation of food ingredients.* Life Sciences Research Office. American Society for Nutritional Sciences. Bethesda, MD; 1999; 112p.
8. Hall RL, Oser BL. Recent progress in the consideration of flavoring ingredients under the food additive amendment. III. GRAS substances. *Food Technology* 1965; 253:151.
9. Adams TB, Hallagan JB, Putnam JM, Gierke TL, Doull J, Munro IC, Newberne P, Portoghese PS, Smith RL, Wagner BM, Weil CS, Woods LA, Ford RA. The FEMA GRAS assessment of alicyclic substances used as flavour ingredients. *Food and Chemical Toxicology* 1996; 34:763.
10. Adams TB, Doull J, Goodman JI, Munro IC, Newberne P, Portoghese PS, Smith RL, Wagner BM, Weil CS, Woods LA, Ford RA. The FEMA GRAS assessment of furfural used as a flavour ingredient. *Food Chemical Toxicology* 1997; 35:739.
11. Adams TB, Greer DB, Doull J, Munro IC, Newberne P, Portoghese PS, Smith RL, Wagner BM, Weil CS, Woods LA, Ford RA. The FEMA GRAS assessment of lactones used as flavour ingredients. *Food Chemical Toxicology* 1998; 36:249.
12. Newberne P, Smith RL, Doull J, Goodman JI, Munro IC, Portoghese PS, Wagner BM, Weil CS, Woods LA, Adams TB, Lucas CD, Ford RA. The FEMA GRAS assessment of trans-anethole used as a flavouring substance. *Food Chemical Toxicology* 1999; 37:789.
13. International Programme on Chemical Safety. *Environmental Health Criteria 70. Principles for the Safety Assessment of Food Additives and Contaminants in Food.* Geneva: World Health Organization. 1987; 88–92.
14. JECFA. Evaluation of certain food additives and contaminants. Forty-sixth report of the Joint FAO/WHO Expert Committee on Food Additives. *WHO Technical Report Series,* No. 868, 1997.
15. JECFA. Safety evaluation of certain food additives and contaminants. Forty-ninth Meeting of the Joint FAO/WHO Expert Committee on Food Additives. *WHO Food Additives Series,* No. 40; 1998.
16. Munro IC, Kennepohl E, Kroes R. A procedure for the safety evaluation of flavouring substances. *Food and Chemical Toxicology* 1999; 37:207.
17. Cramer GM, Ford RA, Hall RL. Estimation of toxic hazard-A decision tree approach. *Food and Cosmetics Toxicology* 1978; 16:255.
18. Scientific Committee on Food. Opinion on a programme for the evaluation of flavouring substances. *SCF/CS/FLAV/TASKF/11 Final. Annex I to the minutes of the 119th Plenary meeting,* 2 December 1999.
19. Council of Europe. Chemically-defined Flavouring Substances. *Strasbourg* 2000; 606p.
20. Council of Europe. Natural Sources of Flavourings. *Strasbourg* 2000; 276p.
21. Schrankel KR. Flavor safety and international harmonization of flavor regulations. In: Kong LK, Seng LY, eds. *Proceedings of the 6th ASEAN Food Conference.* Singapore Institute of Food Science and Technology; 1997; 193–8.

22. Mussinan CJ, Hoffman PG. Naturalness decision tree. *Food Technology* 1999; 53:54–8.
23. Smith RL, Newberne P, Adams TB, Ford RA, Hallagan JB and the FEMA Expert Panel. GRAS flavoring substances 17. *Food Technology* 1996; 50:72.
24. FDA. Statement of policy: Foods derived from new plant varieties. *Federal Register* 1992; 57:104; 22984.

16

FOOD IRRADIATION

Peter S. Elias

Various physical, chemical, and biologic methods have been used to prevent food spoilage by undesirable tissue changes or microbial action and thus permit food storage for extended periods. These efforts would have been self-defeating, however, if the longer shelf-life had resulted in a deterioration of the quality and wholesomeness of the preserved food. The traditional methods employed in the first half of the Twentieth Century included heat treatment (e.g. cooking, pasteurization, or canning), cold storage, deep-freezing, chemical treatment with preservatives, and biologic procedures such as fermentation.

The discovery of the bactericidal action of X-rays in 1896 (1) opened the way for a novel approach to food preservation, but its exploitation was hampered by the low output of radiation energy from the radiation sources then available. In 1905, a British patent was granted for the use of ionizing radiation from radium or other radioactive substances to improve the condition and keeping qualities of foodstuffs (2). The radium sources then proposed for this purpose, however, were not available in quantities sufficient to permit the commercial application of this patent. Although a practical use for food irradiation, the killing of *Trichinella spiralis* in pork, was established in the United States in 1921, the power output of the available machine sources was inadequate for the commercial exploitation of this process (3). In 1930, a French patent for preserving food sealed in metal containers by irradiation with X-rays was granted to O. Wüst but never applied in practice (4). The first report on the practical application of the biologic effects of ionizing radiation to a foodstuff appeared in 1943 and dealt with the successful preservation of hamburgers by irradiation. Subsequently other reports were published on the use for food preservation of the energy of ionizing radiation emanating as γ-rays from certain radioactive isotopes, or as X-rays or accelerated electrons from machine sources (5).

In the mid-1950s research into food irradiation was sponsored by the US Army Quartermasters Corps as part of the "Atoms for Peace" policy. From 1970 to 1982 the International Project in the Field of Food irradiation, located at Karlsruhe, Germany, financed and coordinated worldwide efforts to establish the wholesomeness of irradiated foodstuffs. The task has been continued since 1984 by the International Consultative Group for Food Irradiation, which is sponsored by the Food and Agriculture Organization (FAO) of the United Nations. Rome, the International Atomic Energy Agency (IAEA), Vienna, and the World Health Organization (WHO), Geneva.

Few safety assessments for foodstuffs, either prepared ready-for-use or chemically preserved, decontaminated, or disinfected, have ever been carried out with the help of a data base that was as extensively and as thoroughly researched as the evaluation of foods treated with ionizing radiation. Food irradiation is still a subject of worldwide, often heated, discussion in the government ministries, parliaments, and consumer organizations directly involved. Consumer organizations in particular are generally opposed to this technology, while other groups consider food irradiation to constitute an important and progressive step in the conservation of foodstuffs and to be an effective public health measure. Unfortunately many of the publicly debated views are characterized by deliberate exaggerations, misconceptions, willful ignoring of the results of extensive scientific research, and, last but not least, inadequate knowledge of the established scientific and technological bases for this process. Despite many years of extensive scientific research, public debate still fastens on a few apparently unclarified aspects and reported adverse experimental findings which are not easily interpretable. Yet in the context of the total knowledge available about the wholesomeness of irradiated foodstuffs these debatable facts can be regarded as worthy of neglect.

PHYSICAL ASPECTS

The basic physical principle underlying the technology of food irradiation is the absorption of energy quanta of electromagnetic radiation by the food exposed to it. In practice, only three types of radiation are employed: continuously emitted γ-radiation arising from the radioactive decay of cobalt 60 (^{60}Co) or cesium 137 (^{137}Cs), discontinuously emitted radiation of X-ray sources, and discontinuous irradiation with beams of accelerated electrons.

Only electromagnetic radiation with wavelengths between 10^3 and 10^{-1} nm is suitable for treating foodstuffs. This corresponds to radiation energies between 10^2 and 10^6 eV[1] possessed by photons arising either within or without the atomic nucleus. The radiation dose, which is the amount of radiation energy absorbed by the treated matter, is measured in Grays or rads.[2] All irradiation effects are in fact due to fast electrons, either produced when γ-rays or X-rays enter any matter, or issuing as such from electron accelerators. The absorbed radiation energy causes virtually no increase in the temperature of the treated product.

---

1 1 eV is the energy absorbed by an electron when accelerated through an electric field potential of
   1 volt.
2 1 Gy = 100 rad = 1 J absorbed by 1 kg irradiated matter.

The quantum energy of the γ- or X-ray photons or of the accelerated electrons must be sufficiently high to induce ionization of atoms or molecules in the irradiated foodstuff. It must, however, not be so high as to create radioactive isotopes in measurable amounts by induced nuclear reactions. For almost all atoms likely to occur in food this upper limit lies between 13 and 16 MeV (1 MeV $= 10^6$ eV). For this reason the energy output of electron accelerator beams is restricted to 10 MeV and that of γ- and X-ray sources to 5 MeV.

The question of the radioactivity of irradiated foodstuffs cannot be answered by a simple yes/no statement, because all foodstuffs naturally contain the radioactive isotopes of various constituent elements (e.g. potassium 40 (^{40}K), carbon 14 (^{14}C), hydrogen 3 (^{3}H)) as well as traces of radium, polonium, and other radioactive elements present in the environment.

The normal adult ingests via food about 150–200 Bq³ of natural radioactivity. Food irradiation does not add any measurable amounts of radioactive isotopes if the radiation energies employed are kept to the prescribed 5 or 10 MeV. The γ-rays of ^{60}Co and ^{137}Cs have radiation energies insufficient to induce artificial radioactivity in foodstuffs. Any discussion of this aspect is therefore superfluous (6).

TECHNOLOGICAL ASPECTS

Insect and parasite infestation, microbial spoilage, and premature maturation of foodstuffs have been estimated to cause postharvest losses of between 25% and 40% of the total world food production, particularly in zones with tropical climatic conditions. Preventing or reducing these losses would contribute enormously to an adequate availability and a wider distribution of food supplies, particularly in developing countries. It would also impact positively on exports from third-world countries and provide a more varied and nutritious diet for indigenous populations.

Hitherto the use of high-dose irradiation for sterilization of meat, vegetables and fish products was of little commercial interest in developed countries because of the high processing costs compared to heat treatment alone or in combination with deep-freezing. However, a recent review by a Study Group, convened by WHO, IAEA and FAO in 1997 (111), indicated that high-dose irradiation could be used increasingly to ensure an adequate hygienic quality of food of animal origin and to overcome quarantine trade barriers for fresh fruits and vegetables. Outbreaks of infections with enterohaemorrhagic *Escherichia coli* O157:H7, *Listeria monocytogenes* or *Yersinia enterocolitica* have led to recalls of many thousands of tons of ground beef and dairy products, in which contamination with these pathogens could not be excluded. The development of high quality shelf-stable convenience foods for general use and for specific target groups such as immunosuppressed individuals, astronauts, military personnel and outdoor

---

3 becquerel (Bq) = 1 radioactive disintegration per second = 1 unit of radioactivity.

Table 1 Effects and suggested applications of food irradiation

Suggested application	Dose range (kGy)	Foods involved
Inhibition of germination	0.05–0.15	Potatoes, onions, garlic, ginger
Insect control by sterilization and destruction; parasite control	0.15–0.50	Cereals, legumes, fresh fruits, dried fruits, dried meat, fresh pork
Delayed ripening	0.10–1.0	Fresh fruit, fresh vegetables
Improved keeping on storage; destruction of microbial commensals	1.0–3.0	Fresh fish, strawberries
Pasteurization, except spores	1.0–8.0	Shellfish, crustaceans, raw and frozen poultry and meat
Improved processing	2.0–10	Increased fruit juice yield lower cooking time of dried vegetables
Decontamination, including spores	10–50	Spices, enzyme preparations, natural gums
Sterilization under special conditions	30–50	Meat, poultry, shellfish, special hospital food
Inactivation of toxins and enzymes	70–100+	Special preparations

Source: Modified from Helle *et al*. (18).

enthusiasts provides new applications for irradiation sterilized foods, provided they can be considered as safe, nutritionally adequate and wholesome.

The high radiation doses and special techniques required for such sterilizing processes necessitate a thorough appraisal of the nutritional and toxicological attributes of foods so treated. These fields of application for irradiation have now been recommended by WHO for niche markets of various high-dose irradiated products, including meat items and whole meats, in view of the lack of any health problems resulting from their consumption. Whatever the processing conditions, only fresh food of good quality is suitable for irradiation treatment. It is usually packaged and frozen or chilled when transported to the irradiation installation. Table 1 lists some applications and the corresponding recommended irradiation doses.

Applications of low-dose irradiation (≤1 kGy)

Low-dose food irradiation generally employs radiation doses between 0.03 and 1 kGy. It is used predominantly (1) to prevent the sprouting of potatoes, onions, and garlic, thereby extending their storage life; (2) to delay ripening of fruits; and (3) to kill insect pests in grains, spices, and some fruits to reduce losses of these foodstuffs.

The biologic basis for these effects is the irreversible inhibition of cell division by radiation doses between 0.03 and 0.2 kGy. Treatment is applied at any time post-harvest and before storage. Most of these agricultural products are subsequently processed further. Usually radiation doses between 0.05 and 0.15 kGy are adequate for inhibiting the sprouting of potatoes. Damaged potatoes can only be treated after healing. Excessive doses would increase spoilage by microbial invasion at the site of damage. Hence only best-quality produce combined with a minimum of handling is suitable for irradiation. Only γ- and X-rays are suitable for treating potatoes, because these radiations penetrate sufficiently deeply to prevent internal sprouting. Irradiation has the advantages over chemical antisprouting treatment of not leaving any residues and of being irreversible.

Onions also require the more penetrating radiation of γ- or X-rays to inhibit the quiescent germinal centers. They can be stored after treatment for up to 7 months. Cold storage will prevent internal browning.

Cereals and cereal products can be disinfested easily by low-dose irradiation, particularly in tropical and subtropical climates. These products can be treated by accelerated electrons because of their small size. The average doses required to kill larvae and insect eggs lie between 0.03 and 0.05 kGy. Rice and pulses require larger doses up to 1 kGy, but do not suffer any significant losses in nutritional value or product quality. As irradiation does not protect against reinfestation, care has to be taken with the storage of these irradiated products.

Doses of 0.2–0.5 kGy can be used to delay the ripening of fruits and vegetables. Subtropical fruits may carry important insect pests and are therefore subject to quarantine regulations when imported into many countries. Citrus fruits, papayas, and mangoes are therefore irradiated for hygiene purposes with doses of 0.3–0.5 kGy, which suffice to kill the eggs and implanted pests. To reduce the microbial load requires higher doses up to 1 kGy. Citrus fruit can be protected against attack by the Mediterranean fruitfly (*Ceratitis capitata*) by irradiation as a replacement for the carcinogenic fumigant ethylene dibromide. Dried dates and other dried fruits can be disinfested with doses up to 1 kGy.

Parasites, such as the larvae of *Trichinae* and *Toxoplasma*, are sensitive to irradiation doses of 0.5–1 kGy but can also be eliminated by deep-freezing. Disinfestation of dried fish in tropical countries requires doses up to 1 kGy.

Applications of intermediate-dose irradiation (1–10 kGy)

Irradiation with doses in this range eliminates nonspore-forming pathogens (e.g. *Salmonella*, *Yersinia*, *Campylobacter*, *Staphylococcus aureus*, *Listeria monocytogenes*, and *Clostridium perfringens*). It also reduces the microbial load of foodstuffs by several magnitudes. It may be combined with heat treatment to achieve synergistic effects.

Salmonellosis is a common cause of food poisoning that may even be fatal in elderly individuals. This pathogen can be eliminated completely from packaged food, such as deep-frozen poultry or froglegs by irradiation with doses ranging from 3.5 to 7.5 kGy, while simultaneously reducing the bacterial load by a factor of 10^3. This treatment may be the only effective public health measure capable of reducing food-borne salmonellosis, because up to 90% of deep-frozen poultry can be contaminated with this pathogen (7). However salmonellae in eggs cannot be removed, as very low doses cause a breakdown of egg white mucoproteins, causing a loss of viscosity and inducing an off flavor in the egg yolk (8).

Cured meat and ham products frequently carry *C. botulinum* spores, which are being prevented from germinating by the nitrite added during the curing process. Comparatively large doses of nitrite (e.g. 120 mg/kg for bacon; 156 mg/kg for ham, corned beef, and certain sausages) are needed to be effective against the production of botulinum toxin in cured foods by germinating clostridial spores. Irradiation with doses ranging from 7.5 to 15 kGy eliminates viable spores of *C. botulinum*, thus permitting a reduction in added nitrite to about one-third of the normally required amounts. This has the advantage of also reducing the possible formation of carcinogenic nitrosamines in these cured products without causing any increased hazard from botulism.

The shelf-life of most fresh fruits and vegetables cannot be prolonged significantly by irradiation without at the same time lowering the quality and inducing poor taste. Strawberries, bananas, fresh figs, papayas and mangoes are exceptional insofar as irradiation with doses of 2.5–4 kGy can extend their shelf-life by 3–7 days depending on the adequacy of the postirradiation storage. The judicious application of 1–3 kGy doses can delay the postharvest maturation of apples, pears, and tomatoes. A combination of a 5 min dip in 55 °C water with an irradiation dose of 7.5 kGy will enable mangoes to be stored subsequently for up to 4 weeks at 10 °C. Irradiation of cultivated mushrooms with doses of 1–2.5 kGy prevents the opening of the head and the development of lamellae and spores during storage for 6–9 days.

Fish and other seafood are liable to spoilage by contaminating microbial flora as soon as they are caught. Prepacked fish and fish fillets can be preserved by storage on ice in combination with irradiation doses of 1.5–2.5 kGy because of the reduction in the microbial load. Deep-frozen shrimp are particularly suitable for decontamination by irradiation because of their frequent contamination with food-borne pathogens.

Irradiation with doses of 7.5–10 kGy is extremely useful for the decontamination of dry products, such as spices, herbs, dried vegetables, and enzymes. In particular, the numbers of common spore-formers and thermophilic bacteria can be reduced as well as pathogens eliminated, thus allowing these products to be used safely in food processing. Bacterial counts of the order of 10^8/g can be reduced to 10^4/g. Irradiation is preferable to heat treatment because of the latter's deleterious effects on the flavoring components in these foodstuffs.

Irradiation with 10 kGy has been used experimentally but not commercially to alter the permeability of plant cells and thus improve the extractability of cane sugar and beet sugar (9) or to reduce the cooking time of dried vegetable soups (10).

Applications of high-dose irradiation (>10 kGy)

High-dose irradiation can be used for the complete sterilization of foods and feeds, killing bacteria and most viruses, but requires special techniques to obtain products with acceptable physical and organoleptic properties. The technology has been applied to high-protein foodstuffs, such as meat, poultry, and fish, particularly if precooked and prepacked in individual portions for indefinite storage at ambient temperatures. With these foods the most important safety parameter is the reduction of *C. botulinum* by 10^{12} which requires radiation doses of 25–45 kGy. To avoid adverse organoleptic changes it is necessary to carry out the irradiation at temperatures of $-40\,°C$ and to inactivate proteolytic enzymes by a preheating phase at a temperature of 70–75 °C. Treated foods must be kept at 20 °C and remain sealed in vacuum packs to prevent oxidative changes from aerial oxygen and recontamination. Under these post-irradiation conditions a storage life of up to 2 years can be achieved.

Early in 1974 radiation-sterilized food was offered in the US to patients with compromised immune systems to maintain their nutrition and to prevent the ingestion of food-borne pathogens (72). Deep-frozen meals for hospital patients, irradiated to an average dose of 75 kGy have been produced in the Netherlands (111). Astronauts and personnel on shuttle flights in the US have consumed irradiated foods while in space since 1972. Examples are: irradiated grilled beef-steaks, smoked turkey slices, corned beef, chicken, pork chops, pizza, breaded chicken breast (112). In South Africa novel convenience foods that cannot be prepared by alternative methods have been developed. Examples are grilled chicken, curried chicken, bacon and curried beef. These foods were produced for expeditions and the Armed Forces in quantities of several hundred thousands of tons. These rations provided the entire intake of protein for long periods to special forces personnel (113). Since 1989, shelf-stable meat items were sold at civilian outlets to yachtsmen and outdoor enthusiasts. There are also many foods which are technologically unsuitable for heat sterilization to destroy microbial contaminants. Examples are breads, pancakes, crackers, stuffing, pastries, cereal, dry beverages, snacks, candies, nutritional supplements, meats and condiments. For these irradiation is the only practical alternative (111).

Irradiation facilities

The industrial use of ionizing radiation requires specially constructed facilities and strict observance of safety and radiologic protection measures. Codes of Practice have been developed (12) to ensure international acceptance of irradiated foods processed in facilities listed as internationally recognized for carrying out the irradiation of foodstuffs. The development of modern high-energy electron accelerators, of X-ray machines with adequate power output, and of radioactive sources with an adequate capacity for large-scale commercial irradiation has paved the way for the commercial application of food irradiation (13).

Food irradiation facilities are basically of three types (11,14):

1. *Gamma cells.* Such an irradiation facility has a relatively simple layout. The isotope sources are contained in zirconium/stainless steel rods in the case of ^{60}Co,

or in plain stainless steel tubes in the case of ^{137}Cs. These rods or tubes are mounted in a frame to produce a uniform radiation field and the frame is kept under about 33 feet of water in a pit within the irradiation cell when not in use. On activation the source is raised into the center of an automatic transport system, which surrounds the source at a constant predetermined distance. This arrangement avoids any contact between the food and the radioactive material.

γ-rays, being very penetrating, can be used for irradiating products already packaged ready for shipment after treatment. The penetration power of the radiation determines the size and thickness of the containers stacked with the food to be treated. The dose is controlled by appropriately distributed dosimeters. The pathway of the transport system is designed so that all goods are uniformly irradiated from all sides. Goods are usually transported by pneumatic systems. The duration of irradiation is controlled to deliver the precise dose required for the food being treated.

^{60}Co is formed by neutron bombardment from ^{59}Co, has a half-life of 5.27 years, and produces γ-radiation with energies of 1.17 MeV and 1.33 MeV. Every year about 25% of the source must be replaced by new isotopic material. ^{137}Cs is a fission product in the fuel elements of nuclear reactors. It has a half-life of 30 years and produces γ-radiation with an energy of 0.66 MeV. The radiation energy of 1 kW suffices in theory to irradiate 3.6 kGy tons/hour, but in practice only about 30% of this energy is utilized.

Sterilization by irradiation may take up to 15 hours in a high-capacity facility, assuming an energy utilization efficiency of 30%, for each throughput of 3.2 tons in a pallet-type plant. The Japanese facility for irradiating potatoes, in operation since 1973, handles about 350 tons/day during a 3-month harvesting period (a total of 30,000 tons/year) with an energy utilization efficiency of 20%. Because irradiated potatoes tend to heal poorly, if wounds have been sustained during harvesting, the Japanese plant uses special containers, each holding 1.5 tons, which can be stored without further handling of the potatoes. The process is expensive and the utility of the irradiation facility therefore depends on the market situation.

2. *Linear electron accelerators.* Modern installations have a radiation output ranging from 20 to 200 kW, the maximum accelerating potential being limited to 5 MeV. Acceleration of electrons is achieved indirectly by microwaves, which permits the generation of pulsed electron beams up to 10 MeV with a beam power spread of 25–75 kW. Electrons having such energies can penetrate matter to a depth of 3.8 cm. A 50-kW beam can deliver a radiation dose of 10 kGy to 18 tons/hour, but again only about 30% of this energy output is utilized.

Accelerated electrons can be used alternatively to generate X-rays from a metal target, such as bremsstrahlung, with a broad energy spectrum and a penetrating power of 30 cm. However only a small percentage of the electron beam energy is converted to X-rays, making the energy utilization of this process very inefficient.

The advantage of linear electron accelerators lies in the very high dose rate delivered by electron beams with large radiation power in fractions of a second. This offers the advantage of short irradiation times, which allows such installations to be part of a continuous production line. The disadvantage is a possible temperature rise in the irradiated product that cannot be easily dissipated. Thus a radiation dose of 25 kGy can raise the temperature of water by 6 K or of steel by 54 K, while a dose of 10 kGy could raise the temperature of water by 2.4 °C.

3. *X-ray generators.* Machine sources generating X-rays with energies of 5–10 MeV have only recently been developed and are not yet in commercial use.

Dosimetry

The process of food irradiation requires reasonably accurate methods for determining the applied radiation dose in order to avoid under- and overtreatment with consequent deleterious alterations in the quality of the treated food. The absorbed radiation dose is a fraction of the total radiation energy delivered into a unit volume of mass and is measured in J/kg. There are many procedures for measuring the absorbed radiation dose (12). They employ essentially dosimeters, placed strategically within the bulk of the food being irradiated, and permit the statistical calculation of the average minimum, average maximum, and overall average radiation dose applied. Dosimeters vary in their composition depending on the dose range to be measured. Because the dose distribution is necessarily nonuniform because of the geometry of the product and the source arrangement, some small fraction (about 2.5%) of the irradiated food will always absorb a dose that is about 50% either above or below the overall average dose (12).

Identification of irradiated food

No common radiation-specific products appear in irradiated foods which would by themselves uniquely identify foods as having been treated with ionising radiation. The yield of such radiation-induced compounds is low, because the amount of radiation energy absorbed by the foodstuff is small, and there is also competition for the absorbed energy among the various food components. Since labeling requirements for irradiated foods have been introduced by the Codex Alimentarius Commission a number of methods have been developed which enable the identification of individual irradiated foods with almost absolute certainty but no single method of detection covers the whole range of foodstuffs likely to be irradiated. This situation creates some difficulties for inspecting authorities wishing to control the marketing of irradiated products or of irradiated foods in transit. Individual irradiated and non-irradiated foods can be distinguished easily by physical or chemical methods, if both types of products are available for simultaneous comparison. Some of the available methods are sufficiently reliable for proper legal enforcement. However, a quantitative estimate of the applied radiation dose is in many cases not achievable.

The availability of validated detection methods would strengthen national regulations, would help the enforcement of labeling requirements, and would enhance consumer confidence in already existing regulatory mechanisms. In the

European Community only five standardized methods are officially recognized since 1994. These are:

1   The detection of irradiated bone-containing foods using ESR spectroscopy.
2   The detection of irradiated cellulose-containing foods using ESR spectroscopy.
3   The detection of irradiated foods containing mineral contaminants using thermoluminescence
4   The detection of irradiated fat-containing foods using gaschromatographic analysis for volatile hydrocarbons.
5   The detection of irradiated fat-containing foods using gaschromatographic analysis for 2-alkyl-cyclobutanones:

Foods containing irradiated ingredients cannot yet be identified with any degree of certainty. Detection methods may be grouped for discussion purposes into five large groups as follows:

1. Chemical methods of detection

These methods of detection have to rely mainly on methodologies involving multicomponent analysis combined with multivariate statistical analysis of the relative concentrations of key components of the foodstuffs under examination, e.g. the gaschromatographic analysis of volatile hydrocarbons, the determination of organic peroxides. They also include the specific determination of 2-dodecyl-cyclobutanone, 2-tetradecylcyclobutanone, and in meat products of o-tyrosine. At one time o-tyrosine was considered to be a unique radiolytic product of the OH$^\bullet$ Radical attack on phenylalanine and therefore useful as a biomarker. Subsequently o-tyrosine was also detected in unirradiated meat and meat treated by UV irradiation, making this approach legally unsuitable.

Lipids in animal tissues yield a typical spectrum of radiolytic products on irradiation. The determination by gaschromatography/mass spectroscopy (GC–MS) of relatively volatile hydrocarbons having one C-atom less than the natural fatty acid present, of aldehydes, of 2-alkylcyclobutanones, and of other radiation-induced changes in the lipid components may be promising approaches for foods with high levels of lipids (17). These radiolytic products can be extracted with non-polar solvents, cleaned up by high performance liquid chromatography (HPLC), and identified through single-ion monitoring. On the other hand, malon-aldehyde, a radiolysis product of carbohydrates, was found to be an unreliable indicator of irradiation (16).

Irradiated proteins might be chemically analysed for non-volatile carbonyl compounds, ninhydrin-reactive substances, crosslinked products, or specific changes in amino acids. None of these methods are as yet sufficiently validated for use as control procedures.

Chemical changes in nucleic acids could theoretically provide useful clues to irradiation treatment of food, because the main radiation-induced DNA lesions are base damage, single and double strand breaks and crosslinking reactions between DNA bases, and between these bases and other proteins. GC–MS analysis of base damage, fluorimetric assays of decomposition products produced by OH• attack on DNA bases (e.g. thymolic glycol), and postlabeling assays for modified bases may be promising techniques.

Changes in vitamins and flavouring components due to irradiation are difficult to distinguish from natural variations and changes due to other technical treatments. They are too non-specific for control purposes (16).

2. Physical methods of detection

Measurements of electrical conductivity

Damage to cell membranes by radiation causes secondary alterations in physical properties by interference with ion transport mechanisms. Impedance and conductivity measurements have been used to estimate the radiation dose absorbed. This method has been employed to identify irradiated potatoes. Two steel electrodes are inserted and an alternating current (50 Hz–100 kHz), amplitude (mA), is applied at a temperature of 22–25 °C. In this manner changes in permeability or the electrolyte content resulting from irradiation can be determined. Comparative measurements of a fall in conductance over 3 minutes were found to be useful and practical in estimating the radiation dosage applied. The magnitude of the impedance depends strongly on the radiation doses employed at frequencies below 20 kHz. To eliminate the influence of the nature of the electrodes a normalize parameter Z_{5K}/Z_{50K} should be used thus obtaining a greater specificity (114). Using the phase angle ratio ϕ_{15k}/ϕ_{80} yields more exact measurements of the irradiation dose (115).

Viscosity measurements

These have been found useful in some circumstances for identifying the application of irradiation. Starch suspensions containing irradiated spices were found to have a lowered viscosity due to entry of the solvent into the cells following permeability changes in the cell wall as a result of irradiation (116,117). Viscosity changes also depend on associated irradiation changes in proteins, starches, pectins and cellulose. However these changes were found to be unreliable indicators. Viscosity is measured in a rotation viscometer in aqueous, alkaline and heat gelatinised solutions. Both increases and decreases in viscosity can be observed and correct identification requires the availability of an unirradiated reference sample. These methods are used essentially for screening of white and black pepper. Using as normalized parameter the ratio of viscosity to starch content increases the specificity of the measurement and can thereby detect irradiation with doses of 5 kGy or higher. The method is applicable to many other spices, e.g. cinnamon, cardamom, curcuma, mustard, ginger and nutmeg (118).

Differential scanning calorimetry

The use of this method for detecting irradiated meat products had only limited success (16).

Determination of supercooling temperature

Cooling of water-containing foods at a constant rate can delay actual freezing until a very low temperature, the supercooling temperature, is reached at which water then crystallises. This difference between the normal freezing point of pure water (0 °C) and the supercooling temperature is increased in irradiated products (119). The supercooling temperature depends on several factors, e.g. cooling rate, contact surface, potential places for ice nuclei formation such as solids and gas bubbles in the liquid phase. Irradiation inactivates ice nuclei centres and destroys gas bubbles thus lowering the supercooling temperature by about 5 K independent of the radiation dose absorbed (116). The method is applicable to chicken meat, certain fish and mushrooms.

Electron spin resonance spectroscopy

Free radicals formed in dry foodstuffs by irradiation can be detected provided they persist long enough to be measured by various methods. They disappear rapidly due to their fast reaction with any water normally present in most foods. Free radicals are, however, trapped in the bones in meat or fish, in the shell of crustaceans, and in the seeds or stones of fruits and are then detectable by electron spin resonance (ESR) signals. This methodology is based on the absorption of radiation by spin changes of unpaired electrons found in free radicals or other paramagnetic species, induced in samples irradiated with microwaves in an external magnetic field. As doses as low as 0.2–0.3 kGy produce measurable signals, this methodology shows promise as a semiquantitative indicator of the applied radiation dose, being sensitive to free radical concentrations as low as 10^{-9} mol/L. The intensity of the microwave energy absorption is proportional to the radical concentration in the sample examined.

Signals from a radiation-specific radical have been observed in several fruits, particularly in the seeds or the shell of nuts, but the results are controversial. In practice a compact spectrometer is used which is relatively easily handled and requires only a 100 mg sample and takes a few minutes for a determination (120).

The best field of application is meat containing bones, in which the signals increase with the radiation dose absorbed and with bone age, and reduce only little with cooking or processing. Irradiation of bones forms two paramagnetic species, an unstable symmetrical doublet originating from collagen radicals, and a stable radiation-specific, asymmetrical singlet originating from the mineral components, e.g. hydroxyapatite (121). This signal is characterized by its position

351

and structure (an asymmetric 2-component peak), while unirradiated samples only show a very weak symmetrical signal. Direct heating does not induce a similar signal. The intensity of the signals increases with increasing radiation dose but an exact determination of the dose is not possible, only a differentiation between mean doses (1–3 kGy) and high doses (7–10 kGy). Quantitation of the radiation dose in bone-containing meat has been attempted by plotting signal amplitude vs. additional irradiation dose applied (e.g. 1, 2, 4, 8 kGy) and extrapolating the graph back to 0 amplitude on the abscissa (18). Pork and chicken bones show linear graphs, while fish bone graphs are nonlinear at the higher doses. The method does not require the availability of control samples nor knowledge of the sample parameters. Exponential back extrapolation yields relatively accurate values up to 7 kGy (122). Recently a simpler method for the dose determination, applicable to chicken bones only, has been developed. Because the mass-production of chickens worldwide has homogenized the chicken bone composition, it is possible to produce a calibration curve by using shape and size standardized bone samples, examined always in the same position within the instrument, which is linear between 1–10 kGy (123). The method is still applicable to processed products, if they contain bone fragments that can be isolated by alcoholic KOH digestion and then be examined by ESR spectroscopy. The detection limit for bone splinters from duck, frog, goose, chicken, rabbit, lamb, turkey, beef and pork is about 0.5 kGy (116).

Similar signals, though of lower amplitude, have been detected in fish bones and are identical with that of irradiated fish meat. The intensity varies with the different fish species. The lower intensity is probably related to the lower density of the crystalline matrix of fish bones. The exact identification of the radiation dose requires measuring the crystalline structure by X-ray diffraction. The detection limit for eel, trout, sardine, halibut, herring, salmon, cod, mackerels is about 1 kGy.

Additional peaks have been identified in irradiated scampi and shrimp shells. The signals seen are very complex because of superposition of the spectra of several radicals. They originate mainly from the chitin and Mn^{2+} contained in the shells. Irradiated mussels produce a very strong signal approximately 1,000 fold that of shrimp shells and about 300 fold that of fish bones. The signals tend to decay slowly with storage but remain sufficiently distinct for identification of irradiation (124). In crustaceans identification of irradiation is less clear, because Mn^{2+} causes complex signals even in unirradiated samples (116). For certain identification it is necessary to have the irradiated shell available. The ESR spectrum contains an additional peak due to chitin which increases nonlinearly with irradiation dose. ESR spectra differ widely, particularly if meat is still attached to the shell, hence only dry clean shells should be tested. In practice no correlations with radiation dose are possible. ESR signals persist, however, for several months. The ESR signals in the case of lobsters vary with the body site and age, the older males showing higher intensity of signals. Cooking before irradiation increases signal intensity.

ESR spectroscopy of irradiated cellulose-containing foods is only useful in case of hard components (seeds, kernel). In dried irradiated fruits signals derive from low molecular crystalline sugars (116). Irradiated nuts can now be correctly identified up to a radiation dose of 1 kGy by using a microwave power output of 0.25 mW. The signal shows a radiation-specific peak pair on either side of the main peak due to the cellulose radical. The main peak may be a quinone radical but it is also found after heating. Different types of nuts show different signal intensities.

The irradiation of fruits and vegetables leaves signals in seeds and kernels similar to those seen in irradiated nuts and are probably due to the cellulose radical (116). Doses down to 0.5 kGy can be identified. Melanin-type pigments, formed by the action of sunlight in seeds and cereal grains, can affect the ESR signal (116). ESR signals from the dried outer skin of irradiated onions, garlic and shallots can be used for identification of radiation treatment (125). The ESR signal of the exocarp of irradiated citrus fruits can also be used for the identification of treatment (126). Strawberries are frequently irradiated and yield a complex ESR signal partially derived from Mn^{2+} in the seeds on the fruit. Unirradiated strawberries also have an ESR signal which increases in intensity on irradiation while the Mn^{2+} signal remains unchanged. These signals persist on storage but cannot be used to quantify the radiation treatment received. Similar signals derive from irradiated raspberries and blackberries. Unirradiated sultanas and papayas have no ESR signal but irradiated fruits show a complex signal.

Many but not all spices show the characteristic peak of the cellulose radical in their ESR spectra, which loses intensity when the spices become moist. The ESR signals of Cu^{2+} and Fe^{3+} remain unaltered by re-irradiation for the purpose of testing but if they change, then the spices have not been irradiated before testing. ESR signals show a decay varying from 10 days to 3 months (16).

Low molecular sugars show a characteristic ESR spectrum after irradiation and this test is useful to identify radiation treatment especially of dried fruits usually so treated for desinfestation purposes. No signals occur in untreated dried fruits but ESR signals appear immediately even after low dose treatment (say 0.5 kGy) It is due to radicals formed from mono- and disaccharides (116). Dried pears and plums do not show any ESR signals on irradiation.

Irradiated eggs, treated to reduce *Salmonella* contamination, show the typical Mn^{2+} signal when their shells are examined. This is usually accompanied by an asymmetric second signal of low intensity probably derived from $CO_3^{\bullet}$, CO_3^{3-}, and $CO_2^{\bullet}$ radicals (127). These signals are greatly intensified at doses of 3 kGy and persist so that irradiation can be identified after 10 days storage even after some early loss of intensity. Dosimetry can be carried out using the re-irradiation methodology.

Irradiation is not the only process generating free radicals, however. Energy may be absorbed during grinding, heat treatment, photolysis, catalysis by metal ions, ultrasound application, oxidation and peroxide action during storage yielding free radicals, thus creating difficulties in using ESR as unique evidence for irradiation (16).

Chemiluminescence

Stable positive ions and free electrons in solid parts of food recombine, if irradiated, with emission of light of wavelength 400–700 nm. If energy is supplied by dissolution in a liquid lyoluminescence results. Being rather weak it can be stimulated by the addition of chemical enhancers such as luminol or lucigenin resulting in chemiluminescence. The chemical reaction between matter activated by absorbed energy and a selective light-emitting compound yields brief emission of light impulses on contact with water. The solution of crystal lattices in samples releases the free electrons produced by irradiation and subsequently entrapped in the crystal lattices forming hydrogen peroxide. This latter then reacts with the enhancer which is oxidized in alkaline medium in the presence of a transition metal as catalyst to emit light. This amplification by enhancers then enables measurement by sensitive light detectors (128). Integrating the glowcurve over 180 seconds results in a measure of signal intensity which is then plotted against the radiation dose. For oysters and turkey bones this dose–response is linear in the lower dose ranges, hence dose estimations are possible in these two instances. The main application has been the detection of some irradiated spices, herbs, dried vegetables, dried milk and cacao powder. Serious drawbacks of the method are the large batch variations encountered, the absence of good correlation with the irradiation dose, the poor reproducibility and interference by UV irradiation (16). If the difference in chemiluminescence intensities of untreated and treated samples is small, other confirmatory methods are needed. When applied for the detection of irradiation of biominerals in animal samples, it is necessary to use EDTA or NTA to form alkali-soluble complexes from calcium carbonate in mussel and crab shells or from calcium hydroxyapatite in bones before recordable light signals are formed. Radiation increases chemiluminescence except of the shells of oysters which do not show any light emission. Turkey bones also show weak chemiluminescence.

Photostimulated luminescence

In this phenomenon light energy is used to release the crystalline matrix energy. The method is non-invasive and can be used several times on the same sample. Irradiated samples emit light of higher wave frequency than the stimulating light frequency due to the released energy which is not present in unirradiated samples. The method is highly specific and applicable to the whole sample or to its mineral fraction. The use of a dye laser instead of a mercury vapour lamp as light source increases the sensitivity (129).

Thermoluminescence

This method uses the emission of visible light when trapped charge carriers, representing matter activated by absorbed thermal energy, are released from impurities or defect sites in crystalline lattices by heating to a temperature of 200–400 °C. It is simple to apply and economic. It has been best applied to identify irradiated herbs, spices, dried vegetables but also fresh fruits (strawberries, mangoes, papayas), fresh vegetables and mushrooms and can identify irradiation even after 6 months. The background signal found in unirradiated samples can be quenched by nitrogen or removal of oxygen during the measurement, as the signal arises from chemical changes during the heating process such as oxidation. Thermoluminescence measurements are more reliable than chemiluminescence determinations because of less interference. UV treatment and grinding can affect the results. There are large variations in the response of various spices, but better discrimination is achieved, if the contaminating mineral impurities are separated from the irradiated food and subsequently analyzed separately. The dry material is heated at a rate of 5–10 °C/sec to 3–400 °C and the emitted light is registered in a detector. The results are plotted as a graph (glowcurve). The light originates in the mineral contaminants (quartz) which contain charge carriers in the crystalline matrix. Hence the same plant material treated with the same radiation dose can give variable results. It is therefore best to use this method for screening of a large number of samples. An approximately 100-fold increase in specificity is obtained when the inorganic matter is separated by centrifugation in sodium polytungstate solution because absorption by the organic matter is avoided and no non-specific luminescence interferes (130). It is best to integrate the glow-curve over a section of the temperature scale used. The thermoluminescence intensities are normalized by re-irradiation with 1 kGy and determining the ratio between the original and re-irradiation measurements. Good results are obtained with herbs, spices and spice mixtures. If the original radiation dose is unknown, it is best to re-irradiate with assumed lower and higher doses to find treated samples for more accurate determinations. Fruits and vegetables with smooth skin are treated with surface active materials to separate off any mineral contaminants. In case of mushrooms it is necessary to remove carbonates by acid treatment before making measurements. Kiwis, fresh dates and strawberries are relatively easy to investigate. Relatively good results are also obtainable with crustaceans, using the sand in their intestines and the calcium carbonate of the shell. Calcium carbonate in chicken, lamb, beef bones and egg shells is also a good material for analysis. Spice mixtures and salted pistacio nuts on the basis of salt also yield good glow curves. Powdered milk protein concentrate shows glow curves with two peaks. Incontrovertible evidence of irradiation can only be established if contemporary unirradiated control samples are available (16,19). Thermolumnescence is presently being employed by some food inspection authorities in Germany.

3. Biological and microbiological methods

The combined use of the direct epifluorescent filter technique and the aerobic plate count has been found suitable for irradiated herbs and spices, chicken meat

(116) and liquid egg albumin (131). Irradiation reduces the microbial load considerably requiring 4–5 kGy. APC estimates the viable organisms in the sample while DEFT determines the total number of microorganisms. If the difference between the log numbers for APC and DEFT exceeds 3.5 log units, this implies irradiation with doses of 5–10 kGy, for chicken meat a difference of 2 log units implies irradiation with doses of 3–7 kGy. The method is useful for screening and for assessing the hygienic status of the sample but is not absolutely definitive, if too many or too few micro-organisms are present (132).

The Limulus Amoebocyte Lysate (LAL) test measures the lipopolysaccharides of the cell walls of living and dead G –ve bacteria in terms of endotoxin units/ml or g. As the endotoxin is unaffected by heat treatment or irradiation, it is a useful estimate of the G –ve bacterial burden before irradiation. If this is combined with a G –ve bacterial count, it permits the identification of irradiated chicken meat, because high endotoxin content plus few G –ve bacteria points to irradiation. Similarly, if the difference between log endotoxin content and log G –ve bacterial count exceeds 0 value, this indicates irradiation treatment (116). Storage at 15 °C can lead to problems, if the initial count of G –ve bacteria is high because of the rapid multiplication of survivors and the small difference from untreated samples. The method is therefore only valid for screening.

Shifts in the composition of the microflora, particularly the absence of radiation-sensitive commensals, have been proposed for identification of irradiation. The occurrence of similar changes by some culturing methods make this parameter unreliable. Thus irradiated strawberries might be found to have low counts of *Enterobacteriaceae* and *Pseudomonas* but high counts of yeasts. Irradiated fish would show predominantly *Moraxella* types in their bacterial flora. Similar considerations have been applied to irradiated chicken (16).

4. Tissue histology

Inhibition of germination, rootlet and hyphae formation, and the detection of morphological changes in tissue culture would be reliable indicators of irradiation of potatoes and onions, but these methods take rather a long time. Pesticides or chemical inhibitors of sprouting can interfere with the interpretation of the results. Embryo development testing using germinated half embryos of fruits have been suggested as a comparatively simple technique, applicable to citrus fruits (16). In this test the seeds of irradiated citrus fruits germinate much more slowly. The seeds are removed from the fruits flesh and their shells stripped. The isolated embryo is allowed to germinate at 35 °C for 3–4 days or for 2 days after addition of gibberillinic acid. A minimum of 10 embryos are needed. Irradiation doses >0.15 kGy may show complete inhibition of germination. Doses <0.15 kGy show retardation (130). Investigations on rice, peanuts, maize, soya beans, red beans, and mungo beans treated with 1 kGy show only 50% germination capability. The drawback are the early recovery which occurs when the appropriate low radiation doses are used. The method can be applied to wheat but here irradiation reduces growth of rootlets and shoots rather than inhibiting germination. Doses of 0.5–1 kGy reduce rootlet growth to a maximum of 20 mm by the fourth day. To inhibit germination needs 5–10 kGy. Fumigated wheat shows strongly reduced germination capability without any effects on root growth. Proof of irradiation is possible even after years as the radiation damage is irreparable.

5. Biochemical methods

Protein electrophoresis

This methodology has been used for detecting irradiated chicken meat. The sample is first defatted, dried and pulverized and the proteins separated by SDS/PAGE (sodium dodecylsulphate)/polyacrylamide gel electrophoresis yielding protein bands of different molecular weight. Comparison with unirradiated meat shows absence of or reduction in intensity of some bands. Irradiation removes the α-macroglobulin and the lysozyme bands due to cleavage of the polypeptide chains. Plasminogen, pyruvate kinase, haemoglobin and α-lactalbumin are also radiation sensitive. Transferrin needs about 20 kGy doses for disappearance. Lactoperoxidase disappears after a radiation dose of 10 kGy but reappears at 20 kGy due to recombination. Myoglobin disappears at 6 kGy and reappears at 10 kGy while avidin disappears at 6 kGy and reappears at 20 kGy. Between irradiation doses of 6 kGy and 20 kGy some 10 protein bands disappear and some 25 new protein fractions become recognisable. However, prolonged storage of untreated samples also causes some breakdown of proteins without disappearance of the bands. Irradiation delays this breakdown and at very high doses proteins reform from peptide radicals with a reduction in the number of bands. This method would be useful for screening of stored chicken.

Immunochemical methods

For detecting irradiated eggs use is made of immunoblotting analysis of peptide fragments using rabbit antibodies against ovalbumin, ovotransferrin and ovomucoid after these proteins have been isolated by SDS/PAGE electrophoresis. The hybridized antibodies are visualized by peroxidase-antirabbit IgG and reacting of the peroxidase with 4-chloro-naphthol (133). The use of the ovalbumin/anti-ovalbumin system gives the clearest difference between irradiated and untreated samples and this difference does not disappear on heating. It is also useful for detecting irradiated egg and egg proteins in processed foods.

DNA investigations

These methods detect the effect of irradiation on DNA bases, DNA sugars and strand breaks. Modified bases resulting from OH$^{\bullet}$ attack are 8-hydroxyadenine, cis- and trans-thymidinglycol, detectable by GC–MS, HPLC or electrochemically. The simultaneous presence of dihydrothymidin and thymidinglycol in a sample makes irradiation treatment highly probable, especially if ELISA methods are used. DNA single or double strand breaks can be demonstrated by separating the smaller fragments by filtration from intact DNA, if the DNA is denatured at alkaline pH. The amount of intact DNA retained on the filter is reduced after irradiation. Thus untreated Norwegian lobster shows 50% DNA retention but after treatment with 3 kGy only 5% DNA is kept on the filter. Strand breaks can also be demonstrated by pulsed gelelectrophoresis. In this the cell nuclei are separated from the cells, the nuclear membrane is broken down and the nuclear DNA separated by pulsed gelelectrophoresis showing a large peak due to intact DNA and a tail due to small fragments. To exclude the interference by enzymic degradation it is necessary to check also the smaller mitochondrial DNA which is more stable and not enzyme-degraded. Normal mitochondrial DNA is in a superhelix conformation. Irradiation transforms it into a circular strand and linear double strand breaks so that after irradiation with doses >3 kGy no superhelical DNA remains. This is good evidence for irradiation treatment.

Microelectrophoresis of single cells is based on uniformity of tissue damage to all cells at radiation doses >100 Gy. Single cells are isolated from a cell suspension of the tissue examined, are then suspended in low melting agarose gel on a microscope slide, stained with fluorochrome and then form a specific pattern visible in a fluorescent microscope (comet assay). The length and shape of the comet tail depend on the irradiation dose. Untreated samples show round or elliptical shape but no comet tail. The detection limit is 2–3 kGy (116).

Other methods

To detect irradiated pollen, live pollen is treated with triphenyltetrazolium chloride. This dye is converted to a red colour by the dehydrogen-coenzyme of unirradiated pollen. Irradiation inactivates this enzyme system so that no colour develops. A pink colour indicates partial inactivation. About 100 pollen grains must be examined (134).

Living insects in food irradiated for quarantine purposes can be shown to have become nonproliferative by an increase in size of the supraoesophageal ganglion. However, in eggs and larvae, the test needs careful validation (16).

The development of an acceptable, practical, and reliable system of monitoring and control for detecting food irradiation, possibly involving a combination of several test procedures, still remains a precondition for the successful introduction of this technology (18,20). A coordinated effort is ongoing between the IAEA and the EU Bureau of Reference to develop identification techniques. Table 2 summarizes the most promising methods for detecting irradiated foods.

RADIOCHEMICAL ASPECTS

When the energy quanta of radiation photons are absorbed by the atoms and molecules of exposed matter, changes are induced in their chemical structure.

Table 2 Promising methods for detecting irradiated foods

Food products	Methodology
Spices, herbs, dried vegetables	Thermoluminescence, chemiluminescence, ESR
Meat, fish, shellfish containing bone or shell	ESR, chemical analysis of lipids
Tubers, onions	Inhibited germination, thermoluminescence of contaminating mineral matter, electrical impedance measurements for potatoes
Fresh vegetables, fresh fruits	Thermoluminescence of contaminating mineral matter, ESR of fruit stones, inhibited germination of fruit seeds

Source: Modified from Delincée (20).

Primarily electrons will be liberated, leading to the formation of ions and excited molecules. These subsequently dissociate or interact independent of temperature to produce chemically reactive but short-lived free radicals. The latter undergo temperature-dependently secondary reactions, resulting ultimately in the formation of stable radiolytic compounds. Radiochemical changes therefore often involve decomposition reactions. The overall radiation damage to a molecule is approximately proportional to its molecular weight. Thus, a radiation dose sufficient to destroy bacterial DNA and cause a D_{12}[4] reduction in clostridial spores would only cause a 0.14% change in proteins, a 0.3% change in carbohydrates, and a 0.4% change in lipids (21).

There is commonality in the chemical and biochemical consequences between high-dose and low-dose irradiation. Low dose primarily involves pasteurization, improved hygiene and enhanced shelf life, high doses are applied either to dry foods at room temperature or to enzyme-inactivated, high moisture muscle foods at subfreezing temperature to sterilize these foods and render them shelf-stable. Because the radiolytic mechanisms transforming constituents, the dependence of radiolytic products on absorbed dose, and the effect of processing conditions on product yields are known, it is possible to extrapolate the results from one particular food to a class of foods, from one dose regime to another, and from one particular set of conditions to another set (135). Apart from the calculability of the extent of chemical change and the predictability of the nature of this change there is a significant reduction in the overall change in constituents associated primarily with the aqueous phase, when irradiation involves frozen food. The direction and extent of the reactions by which primary and secondary chemical entities form stable products depends on the composition of the atmosphere in contact with the food, the temperature and phase of the food, the dose rate and the total absorbed dose (111).

---

4 D_{12} reduction: reduction of the original bacterial spore population by 88%, allowing 12% survival.

The chemical effects of irradiation can be quantified by determining the G-values. A G-value is defined as the number of molecules changed by every 100 eV of absorbed radiation energy. For the usual radiation doses employed in food irradiation, the G-values lie between 1 and 3. Assuming a G-value of 2 and an absorbed radiation energy of 10 Gy $(=1 \text{ krad} = 6.25 \times 10^{17} \times (100 \text{ eV}))$ some $6.25 \times 10^{17} \times 2 \times 6.02^{-1} \times 10^{-23}$ mol/kg $(=2.08 \times 10^{-6}$ mol/kg) substance will be changed. A dose of D krad would therefore generate $D \times G \times 10^{-3}$ mmol/kg substance of radiochemical products.

Individual G-values can be derived from the analysis for radiolytic breakdown products of irradiated solutions of single compounds or simple mixtures. They are then summed to obtain the G-value of the foodstuffs. A dose of 10 kGy would therefore produce 300 mg/kg radiochemical products from a substance with an average molecular weight of 300 and a G-value of 1. As most of the hitherto identified radiolytic breakdown products also occur in unirradiated processed foods or are formed during digestion, only about 10% of the radiolytic products in irradiated foods are likely to be unique (22). The chemical changes caused by low and medium radiation doses are therefore almost negligible. Generally heat treatment causes damage to food constituents similar to that induced by appropriate irradiation doses. The chemical reactions resulting from the absorption of either type of energy have been found to be similar, though there are possibly fewer reactions following irradiation.

The free radicals, most frequently formed by irradiation of food, derive from the water present in it, although other energy sources, such as mechanical friction of dry foods or exposure of fats to light and air, can also generate free radicals. The most common free radicals detected are $OH^{\bullet}$, $H^{\bullet}$, $O_2^{\bullet}$, e_{aq}, $H_2^{\bullet}$, $H_2O_2^{\bullet}$, and H_3O^+, while the nature of the stable radiochemical products depends on whether the isolated food component or the complex food is irradiated (23,24).

The triglycerides undergo bond disruption between the fatty acid and the glycerol moieties, thus forming a dominant tryglyceride radical but leaving an unpaired electron on the α-C atom next to the carbonyl group. Evidence for this comes from sequential ESR investigations (111). Free fatty acid radicals are also formed as well as C_{n-1}-alkanes and -alkenes, alcohols and carbonyl compounds, the latter preferably in the presence of oxygen (25). Alkylcyclobutanones appear to be specific to irradiation and do not occur in unirradiated foods. At 5 kGy some 0.5 μg/g lipid are formed (111). Any palmitic acid present yields 2-dodecylcyclobutanone. Lipids heated to high temperatures may also yield these cyclic compounds but these subsequently decompose and can be no longer identified. Very similar reactions occur in the triglycerides of complex muscle food. Fatty acids yield CO_2, H_2, CO, C_{n-1} alkanes and C_n-aldehydes or undergo C–C chain scission next to the carbonyl bond, while unsaturated fatty acids form hydroperoxides at the sites of unsaturation. Triglycerides tend to form free fatty acid radicals, C_{n-1}-alkanes and -alkenes, alcohols, and carbonyl compounds, the latter preferably in the presence of oxygen (25).

Single amino acids and peptides release NH_3, keto acids, and α,α-diamino acids when irradiated, while peptide chains are deaminated and split to form backbone and side-chain radicals. Aromatic and thioamino acids form radicals. Alanine releases ethylamine and propionic acid, while glycine decomposes to methylamine and acetic acid (24).

Proteins undergo reductive deamination, decarboxylation, and scission of disulfide bridges, as well as unfolding of the conformational structure. Fibrous proteins tend to degrade, while globular proteins aggregate. Protein solutions are protected against radiation damage by carbohydrates and lipids. Deep-frozen and dry proteins are radiation resistant, which explains the persistence of enzyme activity despite irradiation. The irradiation of meat products usually liberates S-containing breakdown products, while the meat fat behaves similar to other lipids by releasing alkanes, alkenes, and carbonyl derivatives, irrespective of the nature of the meat. Most of the volatiles are also liberated when meat is heated. High-dose irradiation may release benzene in ppb amounts (26). Proteins in high-dose irradiated frozen meats would be only slightly altered by some aggregation and fragmentation because of the low G-values for primary radical formation. There would be only slight discrimination among the amino acid moieties affected as shown by electrophoresis, digestibility and amino acid analysis. There is no dose-related change in amino acid composition over the does range used in sterilization (111). The presence within a protein molecule of a metal ion provides opportunities for reaction with primary and secondary entities especially in the case of small globular proteins such as myoglobin. The metal ions influence reactions only if exposed and accessible and are therefore limited to short-range reactions (111).

Irradiation of carbohydrate molecules like starch and the reaction of primary entities with soluble mono- or polysaccharides results in breaking of C–H bonds and of ether linkages. Solids, such as starch and cellulose, show bond breakage at glucosidic linkages resulting in depolymerization and formation of C-1 and C-6 radicals of simpler structure. The radicals are identical irrespective of the originating starch as shown by similar ESR spectra and independence of the presence of oxygen or the temperature. The simpler compounds produced are glucose, maltose, dextrins, various simple aldehydes, formic acid and hydrogen peroxide (27,28). The quantities are proportional to the absorbed dose, e.g. up to 15 kGy for formic acid, up to 40 kGy for malonaldehyde, up to 50 kGy for dehydroxy-acetone and glyceraldehydes, up to 80 kGy for soluble dextrins (111). Individual monosaccharides are oxidized and broken down during irradiation with the release of carbonyl compounds. Pure carbohydrate solutions are more susceptible to radiolysis than solutions also containing proteins or amino acids. In solution all the carbon sites could be radical sites. Glucose yields gluconic acid, 2-deoxyglu-conic acid and 5-deoxygluconic acid (111). On irradiation fruits and fruit juices form essentially free primary radicals from the water component, which then produce secondary reaction products with the fruit sugars (27,29,30). However the chemical consequences of carbohydrate radiolysis for high-dose irradiation of muscle food are minor. Because these are present only at levels of about 0.5%, carbohydrates do not compete for primary radicals. Glucose radicals can react with cysteine moieties in albumin to regenerate glucose (111).

The vitamins in food display differential stability when irradiated. The most radiation-sensitive vitamins are E, B_1, C, and K. Radiation doses of 10 kGy and above destroy vitamins C and B_1 to the same extent seen after cooking. Vitamins B_2, B_6, B_{12}, A, and D are relatively radiation resistant (31,32).

Most of the common salt added to foods does not react with primary radicals from water. However, nitrates react with e_s^- to form nitrite but this reaction hardly occurs in frozen solutions. Hence radiation sterilization of bacon and ham in a frozen system is unlikely to produce nitrite radicals because of the competition of the large number and levels of other constituents reactive towards free electrons (111).

Although nucleic acids are only a very small fraction of the food mass, they are of interest because of their relevance to microbial destruction. Reactions affect the purine and pyrimidine DNA bases and the sugar–phosphate backbone of the DNA. The main reaction occurs with $OH^•$ damaging the bases and causing single and double strand breaks. The $OH^•$ derives either from water bound to DNA or from the surrounding bulk water. The preferred addition site is the 5,6 double bond and the radical formed leads to scission of the sugar–phosphate link. The irradiation of aqueous solutions of DNA liberates mostly damaged purine and pyrimidine bases (33). Altered bases in food are unlikely to be incorporated into human DNA because DNA synthesis uses enzymes acting on base precursors and not on the bases themselves. DNA polymerase would excise any incorrectly matched base (111).

Analytical data obtained in irradiated model systems generally agree with theoretically predicted concentrations of reaction products if radiation dose, temperature, and food composition are taken into consideration (34). For high-dose irradiated precooked moist muscle foods commonality of intermediates and similarity of chemical responses among foods so processed has been demonstrated. ESR examination to detect radicals and chromatographic analysis to quantify product yields in diverse foods, irradiated over a wide dose range, has confirmed the usefulness of chemiclearance as a concept for the safety of irradiated food. Irrespective of the nature and condition of muscle foods, the radicals derived from the protein and the lipids in them show the same behavior and are of the same type. The evidence comes from consideration of the ESR spectra of samples irradiated with 50 kGy at $-40\,°C$. The gelelectrophoretic pattern of the extracted proteins are also similar. The same holds for the ESR spectra of the fats. Similar findings were seen with cooked and uncooked foods, irradiated with low dose at different temperatures, although radicals decay more rapidly in chicken than other meats (111).

Lipid-derived volatile products also show commonality. The yields of C_5 to C_8 hydrocarbons relate to the total fat as they derive from the same C–C scission. The yields of C_{n-1} and C_{n-2} hydrocarbons with an additional double bond depend on the level of precursor fatty acids as shown by analysis. Thus heptadecadiene ($C_{17:2}$) derives from linoleic acid and its yield varies linearly with the linoleic acid content even if uncooked products are irradiated when chilled to $-30\,°C$. The propanedioldiester yields also increase linearly with dose over the range 30–90 kGy which confirms the commonality of the chemistry for diverse triglycerides (111).

The studies reported in the literature on diverse foods irradiated with low and high doses show consistent commonality of the radiation chemistry for proteins, lipids, starches and glucose oligomeres. The ESR spectra have similar characteristics and decay properties on storage. Variations in the food matrix did not alter the reactions occurring and thus would not affect safety. If irradiation is carried out on frozen samples in the absence of oxygen, only 20% of the normal yield of primary entities in non-frozen materials occurs. Absence of oxygen reduces the formation of lipid oxidation products and prevents flavor loss in spices. Food constituents of raw or processed food, irradiated chilled or at ambient temperature in the presence of oxygen up to 10 kGy shows the same effects as precooked vacuum-packaged foods irradiated frozen to 50 kGy. This allows extrapolation of results from radiolysis of model compounds and meals to absorbed doses above 10 kGy. Although the level of radiolytically generated products will increase, the spectrum remains unchanged. Therefore irradiation of other foods than meat to high doses alone or as part of a frozen meal or as ingredient of meat will not produce chemical entities previously not identified and would thus not require separate testing for wholesomeness (111).

TOXICOLOGIC AND SAFETY ASPECTS

The toxicologic assessment of the safety of any substance likely to come into contact with humans, is based on the examination of laboratory animals, such as mice, rats, dogs, and other mammals, exposed to the substance and acting as models for humans. The effects being observed concern appearance, behavior, growth, food and water intake, and survival, and are supplemented by full investigations of all relevant hematologic, clinicochemical, and functional urinary parameters. These studies are further complemented by investigations of the metabolic fate of the test substance and its effects on reproductive function, and the macroscopic and histopathologic appearance of all organs of the test animals. Usually laboratory animals are exposed to several different dose levels of the test substance over various time intervals by the same route by which humans are likely to be exposed.

The evaluation of all results, after statistical analysis, is expressed for ingested substances in terms of an "acceptable daily intake" (ADI) for humans. This value is calculated from the no-adverse-effect level, determined in life-span feeding studies by using an appropriate safety factor, usually 100. The ADI thus represents the amount of a substance, expressed in mg/kg body weight, that may be ingested by humans over their whole lifetime without the occurrence of any observable adverse effects. The ADI is usually 1% of the no-adverse-effect dose level. On the basis of the ADI it is also possible to set legally enforceable maximum limits and guideline levels for substances in food so as to prevent consumers inadvertently exceeding the ADI (35).

Animal feeding studies with irradiated foods present, in practice, methodologic problems not encountered when these same feeding studies are carried out on individual, well-defined, chemical substances. A safety factor of 100 necessitates a study design, in which, for example, a 1% use level in the human diet demands a 100% dose level in the laboratory animal feed. Such a safety factor can never be achieved if the use level in the human diet exceeds 1%, which would almost always occur with irradiated food ingredients. It is also technically impossible to incorporate 100% of an irradiated foodstuff into the diet of laboratory animals without already causing adverse effects by the ensuing nutritional imbalance of that animal diet. The establishment of an ADI for irradiated foodstuffs with a safety factor of 100 is therefore illusory.

Most toxicologic studies have to use standard laboratory rodents, which are essentially herbivores requiring specially balanced diets to maintain good health over an adequate life span. For example, irradiated meats could be incorporated into animal test diets only to a limited extent, to avoid the deleterious effects of excessive protein intake on the renal system of the test animals. The extent of incorporation of irradiated foods into laboratory feeds thus had to be tailored so as to avoid any gross disturbances in the nutritional balance of the test diets. In some cases it was also difficult to provide in long-term feeding studies an appropriate control diet consisting of unpreserved yet easily spoiled foodstuffs.

For years it was not recognized that this type of toxicologic study with irradiated foodstuffs would fail to detect the toxicologically interesting organs or clinical parameters likely to be affected by the minute amounts of radiolytic products present in the foodstuffs. This difficulty could not be overcome by the use of excessive doses of irradiation, because this would result in the production of totally different radiolysis products as well as excessive destruction of vitamins and other essential nutrients. On the other hand, the toxicologic considerations relating to irradiated foodstuffs are identical whether irradiation treatment is carried out by γ-rays, X-rays or accelerated electrons, because the chemical effects of irradiation arise in all three cases fundamentally through the action of fast electrons.

The complications experienced with animal feeding studies were eventually partially circumvented by replacing feeding studies with genotoxicity studies, which are based on the detection of mutagenic activity in irradiated foodstuffs. These studies were conducted on bacterial cultures, mammalian cell cultures, on *Drosophila melanogaster*, and, in vivo, on the bone marrow of treated rats, mice, and hamsters using a methodology developed by the International Project in the Field of Food Irradiation (36).

For the above-mentioned reasons, the results of animal studies could never provide a no-adverse-effect level for the radiolytic substances of real toxicologic importance in irradiated foods and thereby enable a safe exposure level to be estimated (37).

Animal feeding studies carried out in the beginning of the Twentieth Century are only of historical interest. Interestingly, the adverse reactions then observed in mice and guinea pigs fed irradiated vegetable products were correctly explained as consequences of reduced vitamin intake due to the irradiation of the food and were not ascribed to the production of radiotoxins (38). Over the past 40 years innumerable feeding studies have been carried out, mainly in the United States, in the United Kingdom, and between 1970 and 1981, under the auspices and coordination of the International Project in the Field of Food Irradiation, in Germany. These safety studies were later complemented by similar investigations on identified radiolytic products (39).

About 20 individual foods out of some 60 different foodstuffs have been investigated in depth in feeding studies on laboratory animals, including studies extending over the animals' whole life span and involving reproductive function. These studies showed almost without exception a noticeable absence of any significant adverse toxicologic effects. Indeed, in 1965, a statement by the Surgeon General of the US Army was published, which declared as wholesome, safe, and nutritionally adequate any food irradiated up to absorbed radiation doses of 5.6 kGy from either a ^{60}Co source or from accelerated electrons with energies up to 10 MeV. This opinion was based on the results of both extensive feeding studies in animals and clinical studies done on human volunteers, conducted in the United States between 1948 and 1965 (40).

In 1980, the Joint FAO/IAEA/WHO Expert Committee on Wholesomeness of Irradiated Food, an internationally recognized authority on food safety, concluded that the irradiation of any food commodity up to an overall average radiation dose of 10 kGy presented no toxicologic hazard nor did it introduce any special nutritional or microbiologic problems; hence toxicologic testing of foods so treated was no longer required. This opinion was based on a comprehensive review of the extensive data base collected up to that time (41).

After a review of the studies using high-dose irradiation in 1994 (136) WHO concluded in 1999 (111) that animal models could be regarded as suitable models and that predictions made from them were supported by human studies. Many of the studies used higher doses than 10 kGy and larger amounts of irradiated food in order to increase the sensitivity of the responses. Most of these studies, including carcinogenicity bioassays and multigeneration reproduction studies, did not demonstrate any short term or long-term toxicity as a result of high-dose treatment. Except for a few easily explained positive results the overwhelming majority of mutagenicity studies, covering all the usual genetic endpoints, were negative. Therefore foods, appropriately prepared, packaged and irradiated to high-doses to sterilize them, were deemed to be safe (111).

Toxicology of radiolytic products

A few selected radiolytic products have been investigated toxicologically. A mixture consisting of some 26 radiolysis products, derived from beef fat irradiated at

y and representing an estimated human daily intake of 0.77 mg/kg body
t, was fed to female mice in a three-generation reproduction study at 0.55–
_._% of the diet, while additional groups were given other combinations of these
radiolysis products at concentrations varying from 0.76% to 2.1% of the diet. The
adverse effects noted were reduced survival and body weight gain of the F_3 pups
of both sexes at weaning. There was a dose-related increase in small necrotic
hepatic foci in the test groups compared to controls. The body weight of male F_3
pups was reduced in the middle dose group. Hematocrit values were decreased
but not consistently. No data were provided on urinalysis or clinical chemistry.
The only treatment-related histopathologic finding in the nine major organs
examined were the hepatic lesions. Combinations of C_{13}, C_{14}, and C_{17}-1-alkenes
adversely affected reproductive function when fed at 3.82% of the diet, the F_2
generation showing infertility, increased pup mortality, and absent litters (42).

An aqueous mixture of nine identified radiochemical compounds, extracted
from starch irradiated with 3 kGy, was administered to rats in their drinking
water in acute and subacute studies. The 3-week subacute study used dose levels
from 0.015 to 0.63 g/kg body weight, the 6-month subchronic study used dose
levels of 0.072–0.3 g/kg body weight. A reduced fluid consumption at the top
dose levels but no hematologic or clinicochemical abnormalities were observed.
The highest dose level of the 3-week study produced epithelial hyperplasia of the
forestomach only (43).

Toxicology of irradiated carbohydrates

Aqueous solutions of simple sugars, irradiated at doses up to 20 kGy, have been
shown to be cytotoxic and mutagenic when tested immediately after irradiation in
prokaryotic systems or in cultures of mammalian cells. The mutagenic activity
appeared to depend on the amount of carbonyl compounds produced and could be
reduced or removed by heating. The absence of in vivo genotoxic activity may be
due to rapid biotransformation of the mutagenic radicals into nongenotoxic substances (44). Irradiated sucrose solutions were mutagenic in bacterial assays but not
genotoxic in a host-mediated assay in mice (45). Solutions of fructose, glucose,
sucrose, maltose, and ribose, irradiated with doses of 10–25 kGy, were mutagenic
for *Salmonella typhimurium* TA 100 when tested by preincubation in the presence
of oxygen but were hardly active in the absence of oxygen (44). When anhydrous
glucose was irradiated with doses up to 50 kGy, no mutagenic activity was detected
in assays using drosophila or in tests for dominant lethal effects in mice (46,47).

Similarly, fruit juices or the supernatant of whole fruit pulp, irradiated up to
20 kGy, had no mutagenic activity (48). The addition of supernatant also reduced
the mutagenic activity of arabinohexo-2-ulose. Pineapple, citrus, and apple juice,
irradiated with high doses, caused chromosomal aberrations in onion root cells.
Apple juice, irradiated with doses of 10 kGy, contained very little radiolysis products of glucose. Glyoxal, D-erythrohexo-2,3-diulose and D-arabinohexo-2-ulose
were identified as the mutagenic agents in irradiated sugars.

Irradiated solutions of 2-deoxy-D-ribose and ribose were genotoxic for *S. typhimurium* TA 100 and TA 98, but not solutions of nucleic acid bases irradiated with doses of 10 kGy. However, irradiated nucleosides were mutagenic for *S. typhimurium* TA 100 (49).

Toxicology of irradiated fruits and vegetables

Irradiated mangoes, dates, strawberries, and papayas have been extensively investigated. No adverse effects were found in chronic feeding studies with mangoes, dates, and papayas irradiated up to 1 kGy, and with strawberries irradiated up to 3 kGy (50,51). Other less comprehensive feeding studies with oranges, apples, bananas, apricots, and peaches irradiated with doses from 0.3 to 3 kGy also showed no adverse toxicologic effects (52).

Life span, reproduction, and genotoxicity studies with onions irradiated up to 0.15 kGy showed no adverse effects at feeding levels up to 2% in the diet of laboratory animals. Higher levels produced hemolysis and anemia due to naturally occurring toxic constituents (50,51). Mushrooms irradiated at 3 kGy were well tolerated when fed to rats in reproduction and teratology studies. Other studies were inadequate for evaluation and those using high dietary levels were not well tolerated by rats and dogs (50). In vitro studies in bacterial systems with irradiated lettuce, celery, carrots, and cauliflower produced no evidence of genotoxicity (53).

Toxicology of irradiated cereals, pulses, and other plant foods

Many extensive short-term, long-term, teratogenicity, and mutagenicity studies in laboratory animals, fed wheat, rice, and maize irradiated at doses up to 1 kGy, have shown no adverse effects if the cereals were stored after irradiation (50,51,54). However, some contradictory results were obtained when these cereals were subjected to different processing conditions. Feeding tests and mutagenicity tests with cooked irradiated potatoes did not give rise to any toxic effects (48). Rice irradiated up to 1 kGy has been administered to rats, mice, and dogs in chronic feeding studies as well as in multigeneration reproduction studies. Genotoxicity was also investigated. No adverse reactions were noted (50,51,54). A three-generation reproduction study in mice fed maize irradiated up to 1 kGy also showed no adverse effects. Short-term feeding studies with irradiated pulses showed no toxic effects, except a reduction in growth rate, when high doses of irradiated and unirradiated beans were fed. No genotoxicity was detected in beans irradiated at 1 kGy (51).

Toxicology of irradiated spices and condiments

No significant toxic effects were observed in life span feeding studies nor in reproduction, teratology, and mutagenicity studies with onion powder and spice mixtures, containing mainly paprika and pepper, irradiated up to 15 kGy. Dietary levels above 10% caused reduced food intakes, reduced body weight gain, and increased liver weights (51,55–57). Irradiated cocoa beans were not genotoxic (51).

Toxicology of irradiated fish, fish products, and shellfish

These foodstuffs contain polyunsaturated fatty acids, which are liable to form hydroperoxides and carbonyl compounds in the presence of oxygen. Irradiation causes similar effects, particularly in those polyunsaturated fatty acids that are constituents of mitochondrial, lysosomal, and endoplasmic reticulum membranes. Although irradiated herring oils have been shown in model systems to reduce mixed-function oxidase activity, to increase P_{450} and P_{448} activity, and to change the spectrum of fatty acids in the endoplasmic reticulum, particularly of the rat liver, these changes are too small to interfere with xenobiotic metabolism (58).

Feeding studies were carried out with cod, haddock, and mackerel, irradiated up to 2 kGy and then boiled, and with fish paste, irradiated at 4.5 kGy, when incorporated at a 45% dose level in the diet of mice and rats. None of the subchronic, longterm, multigeneration, reproduction, and teratogenicity studies disclosed any adverse effects. In some feeding tests in rats, a rise occurred inconsistently in the serum alkaline phosphatase level, which was not noted in mice and dogs. Cod irradiated up to 6 kGy was found not to be genotoxic (50,51). Subchronic, long-term, multi-generation, and mutagenicity studies with mackerel, irradiated up to 2 kGy, showed no adverse effects (59).

Shrimp irradiated up to 3 kGy were fed subacutely at 28% of the diet to rats without causing any toxic effects (60). Dehydrated shrimp irradiated at 2.5 kGy were fed to rats over four generations (61) and to dogs for 2 years without any adverse effects (62).

Toxicology of irradiated meat and poultry

Because of the gross contamination of freshly slaughtered chicken with salmonellae, it would be a useful public health measure to irradiate these products in order to eliminate human pathogens. Chicken meat irradiated at 7 kGy was found to cause no adverse effects when fed to rats and mice for their life span, to rats in a multigeneration reproduction study, to dogs and rodents in several subchronic studies, and when tested in mutagenicity studies (50,51). An elaborate feeding study in mice with chicken meat, irradiated at $-25\,°C$ with a dose of 58 kGy by either γ-rays or accelerated electrons, used enzyme-inactivated frozen meat or heat-sterilized meat as controls. Survival and reproductive function were investigated. An apparent increase in unilateral testicular interstitial (Leydig) cell tumors could not be confirmed statistically nor was there an associated interstitial cell hyperplasia or progression to neoplasia. A decreased fertility was observed only in the group fed heat-sterilized meat (63,64). The study was complemented by several mutagenicity studies, all of which showed no evidence of genotoxicity.

(fruit flies)

However, tests with drosophila showed reduced progeny in all test groups, particularly where γ-irradiated food was used. This effect could not be explained satisfactorily (65).

Feeding studies, ranging from a few weeks to 2 years and including multi-generation reproduction, dominant lethality, drosophila, and cytogenetic studies, were also carried out with beef and beef products. pork, ham, bacon, and mixed offal. Meat was treated with 6–8 kGy or with 28 and 56 kGy, the control product being raw or cooked. Cured meats received various doses of 20, 50, or 74 kGy at −10 °C to −30 °C. Dietary levels were 35% of the rodent or dog feed or 200 g/day for dogs. No significant adverse effects were noted in any study (66–68).

Toxicology of irradiated animal feeds

Breeding colonies of laboratory animals and animal husbandry of livestock under farm conditions have used irradiation-sterilized feedstock, treated with radiation doses of 15–25 kGy. Many generations of rats and mice have been reared without any reproduction problems on these diets compared to controls raised on auto-claved diets. An unexplained effect, noticed at times, was a 15–20% reduction in lymphocytes in male rats (69–71).

Toxicology of irradiated human diets

Patients suffering from immunologic deficiencies through cytotoxic drug treatment or after organ transplantation have been fed over many years heat-sterilized diets and diets sterilized by irradiation with doses of 25 kGy or more. Similarly treated diets have been employed in space travel. No overt adverse effects have been reported up to now (72). Studies on volunteers are difficult to conduct for ethical reasons, not easy to control, and expensive. They can involve only comparatively few individuals and extend only over weeks to a few months. The experimental population must exclude for obvious reasons babies, pregnant women, and the elderly. Nevertheless, some eight experiments with irradiated foodstuffs were carried out recently on several hundred volunteers over 7–15 weeks, without any significant clinical or hematologic adverse effects having been observed (73).

Special investigations into irradiated foods

Apart from the traditional feeding studies already discussed there have been a number of special investigations to support the safety of irradiated foodstuffs. One study looked at the fatty component of irradiated smoked bacon for the possible presence of toxic radiolytic products, another examined a mixture of nine identified radiolysis products of starch. Other aspects investigated were possible combination effects of irradiation and environmental contaminants, possible increases in allergenicity of irradiated milk proteins, and a nine-generation reproduction study in rats for carcinogenic and genotoxic effects of a dry laboratory animal diet with a high free-radical content following irradiation with doses of 45 kGy. None of these special studies gave any indication of toxic adverse effects (37).

Studies showing apparent toxic effects

Although the vast majority of feeding studies with irradiated foodstuffs have shown no evidence of any adverse toxic effects, it would be unscientific not to discuss those publications that report toxic effects, and not to attempt to explain the observed discrepancies.

An early study reported the appearance of cardiac muscle lesions in mice fed irradiated laboratory diet. A repeat experiment with 5,000 animals of the same strain, including very thorough histopathology, failed to confirm this finding (74). Rats fed irradiated beef suffered internal hemorrhages explainable by an inadequate dietary content of vitamin K, due to destruction by this treatment. These effects could be prevented by supplying additional adequate amounts of vitamin K (75). Pigs fed irradiated fish meal showed an apparent increase in the rate of mitosis of mucosal cells in the jejunal crypts. This finding was not statistically significant in view of the large normal variability of this parameter in pigs (76). Even the recent extensive study on irradiated chicken meat presented initially an apparently increased rate of testicular tumors and increased renal damage in mice fed this material in their diet. The finding could not be supported by a statistical analysis of the data (65). Some studies on rats fed diets containing meat or fish treated with doses of 8 and 6 kGy claimed a variety of adverse effects, such as altered tributyrinase activity and disturbances in growth and reproductive function. A careful review of the designs and execution of these studies revealed serious shortcomings in the composition and handling of the basal and test diets; in addition, the conclusions were not supported by the data (77,78). Repeat studies confirmed the safety of the irradiated foodstuffs tested earlier (79,80).

When goods baked from freshly irradiated unstored wheat were fed to a small number of malnourished children, monkeys, and rats, an increased incidence, about 2%, of polyploidy in the circulating lymphocytes was reported. No such effect was observed after feeding baked goods made from wheat stored for 3 months after irradiation (81–83). Enormous publicity was accorded to the claims of adverse toxicity being caused by food irradiation based on this finding. Yet usually no mention is made of the fact that most of the published data in this study were considered to be inadequate, that the statistical analysis was questionable because of the poor design, and that there was the curious finding of a zero incidence of polypioldy after feeding goods made from stored irradiated wheat, when the normal occurrence of polyploidy in the peripheral leucocytes of healthy people ranges from 1% to 4%. Such a study cannot be repeated for obvious ethical reasons nor is any specific danger to health or any biologic role attributable to polyploidy (84). Other studies in laboratory animals could not confirm these findings (85–87). Some reduction but no gross impairment in the responsiveness of the immune system to antigen challenge was reported in rats fed 70% freshly irradiated wheat in their diet, suggesting that the functional integrity of the immune system had not been significantly affected (88).

Among the numerous mutagenicity studies with irradiated foods or food components there have been a number of published positive findings. Close scrutiny of these studies has shown that these early findings were usually made using aqueous solutions of individual food components. If these studies were repeated with complex foods, no genotoxic effects could be demonstrated. Alcoholic extracts of freshly irradiated raw potatoes were found to be genotoxic, causing chromosomal aberrations and dominant lethality in some feeding tests (89). Subsequent repetition of these tests failed to reproduce the reported effects (50). As in the case of freshly irradiated wheat, the positive genotoxic findings with freshly irradiated foodstuffs using higher than technologically useful radiation doses probably result from free radicals not normally formed except under these more extreme conditions or occurring in biologically insignificant amounts or being inactivated by reaction with food components (90).

Growth depression, reduced food intake, and adverse effects on fetal development and survival were noted when untreated cocoa beans and cocoa beans treated with 5 kGy were fed to rats at high dietary levels. These effects were found to be due to the excessive intake of theobromine present naturally in cocoa beans (51).

Studies on irradiated mixtures of polyunsaturated fatty acids, starch and high doses of benzo[a]pyrene showed an increased rate of peroxidation and the formation of benzo[a]pyrene oxidation products. The absence of oxygen or the presence of antioxidants prevented these reactions (91). Observations made in such a model system cannot be used to support conclusions regarding the situation existing in irradiated foods, because they fail to take into account the modifying or protective effects of the other food components, such as proteins or natural antioxidants. Moreover benzo[a]pyrene occurs naturally in foods, though only in minute amounts.

Only very few studies exist which claim that irradiated foods are carcinogenic. Apart from those concerned with the detection of mutagenic radiolytic products in irradiated aqueous solutions of simple carbohydrates or with the identification of peroxidation products in irradiated polyunsaturated fatty acids, there have been reports of the detection of ppb quantities of benzene, a human carcinogen, in irradiated meat (92). However, benzene occurs naturally in many untreated foodstuffs (e.g. fish, vegetables, nuts, and milk products) in much larger quantities and has also been identified in heat-treated food (93).

NUTRITIONAL ASPECTS

Investigations into the effect of irradiation on the nutritional quality of the treated food have not shown a degree of deterioration of the nutritional value that does not also occur on cooking and on other heat treatments. The observed degree of change depended on the radiation dose, the composition of the treated foodstuff, and such other factors as temperature and oxygen content of the ambient air during the treatment. Evidence has come from large-scale feeding trials, in which laboratory rodents have been raised for many generations on laboratory diets irradiated up to 50 kGy that have been controlled with respect to adequacy of their vitamin content. Only when radiation doses of 15 kGy or more were applied for sterilization purposes was it necessary to supplement the vitamin content to meet the laboratory animal requirements. At doses of 1 kGy no significant losses were noted in nutrient content.

Retention of the sensory quality of food to be irradiated at doses above 10 kGy requires irradiation in the absence of oxygen and at cryogenic temperature except for dry products. Nutritionally these high-dose-treated foods are substantially equivalent to thermally sterilized foods. Animal feeding studies have shown no adverse effects on biological availability of macronutrients following irradiation at 56 kGy and in human studies the consumption of foods irradiated with 28 kGy revealed no effects on metabolizable energy, nitrogen balance or digestibility (111).

Proteins as sources of amino acids show little change at doses up to 10 kGy nor are there significant losses in essential amino acids, digestibility, or biological value (94). No effect on amino acid pattern was noted in chicken meat radiation-sterilized with 59 kGy. Similarly no adverse effects on net protein utilization and amino acid pattern was seen, when mackerel, irradiated up to 45 kGy was fed to rats. The biological value of corn protein or wheat gluten, irradiated to 28 kGy, was unaffected nor was the digestibility and biological value of animal feed proteins adversely affected by radiation treatment up to 70 kGy (111).

The nutritionally most important change in irradiated foods is the loss of vitamins, an event that also occurs when food is thermally processed. The results of experimental irradiation of isolated vitamins differ from those found when vitamins are irradiated in food because of the protection by the food matrix and other constituents. The published data are to some extent contradictory for these reasons.

Although vitamin losses generally increase with increasing radiation dose, the use of cryogenic temperatures and the absence of oxygen reduce vitamin losses. With the exception of thiamine none of the fat-soluble and water-soluble vitamins in chicken meat irradiated up to 59 kGy were significantly diminished. If conditions for irradiation are not kept ideal, some 60–70% of thiamine in beef can be lost by doses up to 30 kGy.

Of the water-soluble vitamins, ascorbic acid is the one most sensitive to irradiation. There is an immediate reduction in the ascorbic acid content of most products due to the conversion of some of the ascorbic acid to dehydroascorbic acid. The losses due to low-dose treatment are comparable to those found after cold storage, while higher doses have the same effect as cooking. The apparent losses disappear again in tubers and fruits when these are stored after irradiation, because of their continued metabolic activity.

Among the fat-soluble vitamins, the tocopherols (vitamin E) and vitamin K are the most easily affected (31,32). Vitamin E appears to be the most sensitive to irradiation. When dissolved in unsaturated solvents destruction is increased. A similar increase in loss was noted when irradiation was carried out in the presence of air or at elevated ambient temperatures, and when the irradiated food was further processed after irradiation. Vitamin K is also destroyed by irradiation, the natural forms being more stable than vitamin K_3. Clinical evidence for the loss of vitamin K in irradiated beef came from feeding tests, in which hypoprothrombinemia and hemorrhages were observed.

The irradiation of meat and fats reduces the content of vitamin B_1 and destroys some of the polyunsaturated fatty acids. These losses can be reduced by irradiating these foods at low temperatures of -30 to $-40\,°C$ and by vacuum packaging under nitrogen. These effects are temperature and dose dependent. Some vitamin B_1 is also lost during cooking and storage.

Riboflavin is the vitamin most stable to irradiation in food substrates, although a somewhat increased extractability can be demonstrated after irradiation, probably due to an alteration in protein binding. It was unaffected by doses of $40\,kGy$ at $-78\,°C$ in a nitrogen atmosphere. Niacin also appears to be reasonably stable due to the protective effect of other tissue components. No loss was seen in radiation-sterilized beef at $28\,kGy$ or chicken meat irradiated at $59\,kGy$. No significant losses in biotin, pyridoxin, and pantothenic acid have been observed in those irradiated foods specifically examined for destruction of these vitamins. No loss of folic acid was seen in radiation-sterilized beef at $28\,kGy$ or chicken meat irradiated at $59\,kGy$ nor any other irradiated foods specifically examined for effects on this vitamin. Cyanocobalamin (B_{12}) is quite insensitive to irradiation even at doses up to $40\,kGy$. Losses of vitamin A and of carotenes in dairy and meat products can be significant and depend on the irradiation dose, storage temperature, and the oxygen concentration in the ambient air. Little change occurs in fruits and vegetables, unless these are canned. Irradiation of Vitamin D did not produce any toxic steroid radiolytic products (95).

MICROBIOLOGIC ASPECTS

The deleterious effects of irradiation on living matter occur essentially at the level of the genome through alterations of the macromolecular components, particularly the DNA and RNA: Radiation chemistry has shown that molecular damage due to photonic energy is approximately proportional to the molecular weight of the irradiated substance. It follows, therefore, that DNA with a molecular weight about 106 times larger than that of the biologic building blocks, like amino acids, fatty acids, or monosaccharides, would be about 106 times more sensitive to radiation damage than these basic molecular species present in food. Similar considerations apply to vegetative forms of bacteria, which consist largely of DNA and are thus radio-sensitive, while bacterial spores are more radiation-resistant.

The common organisms responsible for human food poisoning (e.g. salmonellae, staphylococci, *Shigella*, *E. coli*, *Brucella*, vibrios, *Campylobacter*, *Yersinia*, *Listeria monocytogenes*) are radiation sensitive, while spore formers (e.g. clostridia, *Bacillus cereus*) are mostly radiation resistant. The reason for using irradiation to process food is its power to reduce the microflora responsible for food spoilage and to eliminate human pathogens and toxin producers. The destruction of bacteria in a food does not depend on the total bacterial count but involves always a constant fraction per unit time interval. This logarithmic rate of destruction is the basis of the linear relationship between the radiation dose and the logarithm of the number of surviving organisms (e.g. doses of 5 kGy permit the survival of only 1 organism out of 10^6). Radiation sensitivity of bacteria is thus similar to their heat sensitivity, both processes resulting in a comparable reduction of the microbial load. Temperature and composition of the growth medium, apart from the radiation dose, are important factors in determining radiation resistance (96). Radiation resistance of vegetative bacterial forms is greater in dry or frozen systems and in vacuum or nitrogen atmosphere. Radiation sensitivity can be expressed as the number of Grays needed to reduce the bacterial count to 10% of its initial value (D_{10} value). Radiation-resistant bacteria have high D_{10} values, but are usually non-pathogenic and do not spoil food, while radiation-sensitive organisms have D_{10} values below 1 Gy (97). Table 3 sets out some D_{10} values.

Yeasts are relatively more radiation resistant than fungi and molds, but none of the eukaryotes is known to be truly radiation resistant. Viruses, however, are radiation resistant and require doses of 50–100 kGy for inactivation. Irradiation causes a loss of genes involved in survival thus making the treated organism more sensitive to heat and drying. No increase in pathogenicity after irradiation has ever been demonstrated. Irradiation is not the only means of producing mutants in fungi. Mycelial anastomoses enable heterokaryotic fungi to accept the nuclei of other strains and by this naturally occurring process to alter their characteristics.

The production of mycotoxins can be affected by irradiation under special conditions. Irradiated wheat, maize, sorghum, pearl millet, potatoes, and onions, when used in growth media, can cause an increased aflatoxin production, if the media are heat-sterilized before inoculation with the aflatoxin-producing

Table 3 Radiation dose required to reduce bacterial count by 90%

Bacterial species	Dose range (kGy)
Pseudomonas spp.	0.10–0.20
Escherichia coli (anaerobic)	0.20–0.45
Salmonella spp.	0.20–0.50
Streptococcus faecalis	0.50–1.00
Fungal spores (*Aspergillus*, etc.)	0.50–0.70
Clostridium botulinum	1.50–2.50
Micrococcus radiodurans	5.00+

Source: Modified from Langerak (97).

microorganisms. The destruction by autoclaving of the natural antifungal compounds present in these foods may have contributed to the unchecked growth of the aflatoxin producers (98,99). Irradiated cultures of *Aspergillus flavus* and *A. parasiticus* are able to produce mutants with higher mycotoxin-generating ability, but these mutants could not survive in practice because of their greater dependence on special growth conditions (100). Although a single irradiation dose temporarily damages microorganisms, it does not appear to produce new mutants, which have acquired the property of either toxin production or pathogenicity. Irradiation resistance can be induced experimentally by using cycles of alternating sublethal irradiation and growth, but such cycles cannot occur in commercial practice. This induced irradiation resistance is labile, is easily lost, and is associated with slower growth, shorter life, lower virulence, and greater sensitivity to growth conditions.

Irradiation, like UV treatment, or heat and chemical preservation, can generate mutants or act as selection pressure for mutants with desirable properties, however, these mutants are less competitive compared to the indigenous microflora and require special growth conditions for survival (101,102). The natural background irradiation produces far more mutations than food processing by irradiation. Sub-sterilizing doses, just as pasteurization, selectively destroy some organisms, allowing some strains to flourish in the absence of natural competition (65).

The natural radiation resistance of some microorganisms prevents the use of low-dose irradiation as the sole safety treatment of food. Hence combination treatments have been proposed. Irradiation has little effect on bacterial toxins present in spoiled food, but interferes essentially with the transmission of pathogens in the food chain and sensitizes the bacterial survivors to the action of heat or dryness. The suppression of spoilage organisms thus causes problems similar to those found with pasteurization, salting, or any other method of partial preservation. Easily spoiled foods, particularly fish and shellfish, still require additional preservative treatment, such as curing, smoking, or cold storage, and proper hygienic handling after irradiation pasteurization. Irradiation does not eliminate the sensory changes in smell, taste, and appearance caused by spoilage.

Irradiation of food with doses above 10 kGy involves radiation sterilisation of high-moisture foods, mostly of animal origin, of complete meals and components of meals. A second use is the decontamination of low moisture products such as spices, herbs and dried vegetables. The technology includes mild heat treatment to inactivate proteolytic enzymes (internal temperature 73–77 °C), vacuum packaging and deep-freezing prior to or during radiation processing (111). For low-acid products the radiation dose must suffice for a 12D reduction of the population of *Clostridium botulinum* spores. For some dry products doses of 30 kGy may suffice.

Elevated temperatures act synergistically to enhance the bactericidal effects of irradiation because of the damage to the bacterial DNA repair system normally operating at ambient temperatures. Bacterial spores show decreased radiation resistance as temperatures increase from 80 °C to 95 °C. Vegetative forms survive better at subfreezing temperatures probably because of restricted diffusion of radicals and decreased water activity. Anaerobic wet conditions tend to increase the resistance of vegetative bacteria. High moisture environment increases the sensitivity of microorganisms to irradiation but irradiation in the frozen state increases radiation resistance. The heat sensitivity of extremely radiation-resistant non-sporeforming vegetative bacteria suffices for destruction or injury of most cells by the inactivation heat treatment prior to high-dose irradiation. Thus *Moraxella acinetobacter*, *Deniococcus radiodurans* and *Pseudomonas radiora* which are able to survive high doses of irradiation are all very sensitive to low solute concentrations and low water activities (111). In practice no viable resistant bacteria have been detected in adequately processed high-dose irradiated products. Non-sporeforming pathogens do not survive high-dose irradiation. Similarly the combination of sequential heating and freezing plus high radiation doses will inactivate even resistant parasites. Since yeasts are heat sensitive even the most radiation-tolerant yeast is not of significance for high-dose irradiated foods. Survival of fungal contaminants in high-dose irradiated foods is not likely even with high initial burdens in dry foods provided the treated food is stored under conditions excluding an increase in moisture content (111).

Clostridium botulinum type A and B spores appear to be the most resistant organisms in radiation-sterilised food but will not grow below 1 °C. On growing these proteolytic spore types produce a conspicuous off-odor. Indirect irradiation effects seem to sensitize spores to heat inactivation and this sensitization increases with irradiation dose. To ensure that irradiation provides the same safety as thermal canning a minimum required dose must cause a 12D destruction of the most resistant spores of *Clostridium botulinum*. This is determined in inoculated pack studies, the cans being incubated for 6 months after irradiation. The dose, at which no swollen cans and no toxic samples occur, is then used for treatment. Examples are 45 kGy for uncured meats and about 30 kGy for cured meats. Irradiation above 10 kGy, if combined with sporostatic additives, reduction in pH, and reduction in water activity can ensure microbiological safety and shelf stability (111).

Temperatures above 50 °C also increase the radiation sensitivity of viruses. High-dose irradiation of moist foods would offer protection by the heat pre-treatment to inactivate proteolytic enzymes. Radiation treatment of dry commodities would still leave some viable viruses in the treated food (111). Preformed microbial toxins show a high radiation resistance and are not eliminated by high-dose irradiation (111).

OTHER BIOLOGIC ASPECTS

Irradiation with doses of the order of 0.25 kGy will kill the infectious cysts of *Entamoeba hystolytica*. Doses of 0.25–0.3 kGy will kill *Toxoplasma gondii*. The cysts of beef and pork tapeworm can be inactivated by doses up to 3 kGy, as well as through deep-freezing of the infected meat. *Trichinella spiralis* can be inactivated by doses of 0.3 kGy, while most parasitic tropical worms are inactivated by doses of 0.5 kGy (103).

ENVIRONMENTAL ASPECTS

[60]Co can only be produced deliberately through treatment in a nuclear reactor, but once created it simply adds to the total load of radioactive isotopes in the environment. [137]Cs, being a waste byproduct of the nuclear industry, requires strict control over its disposal. In either case safe use can be established through appropriate controls already existing in most countries using nuclear power or through inspection by national health or international authorities. Isotope sources thus have only a small impact on the environment, while for X-rays and electron accelerators the environmental impacts are minimal. Adequate controls, as described by the Codex Alimentarius Commission Code of Practice (12), would ensure that food irradiation is only used where it is regarded as good manufacturing practice and as an appropriate means for ensuring proper food hygiene. It could, in practice, never be employed to disguise bad and unhygienic manufacturing practices.

ECONOMIC ASPECTS

The costs of food irradiation are difficult to calculate because estimates for some of the factors are imprecise. If the irradiation facility is centralized, this involves additional transport and loading expenditures. Installing an irradiator at the end of a production line is cheaper but technically difficult. Small quantities are more economically treated in a centralized facility.

A 200 kW facility handling 50,000 tons annually involves a capital outlay, at 1980 prices in US dollars, of $10 million for a γ-ray installation. A 5 or 10 MeV linear accelerator would cost about $2 million. Converting this to cents/kg food, including depreciation, would give 7.12 cents for ^{60}Co, 5.25 cents for ^{137}Cs, and 1.65–2.03 cents for 10 MeV facilities. These costs do not include additional procedures. Low- or medium-dose irradiation would cost 1.4 cents/kg for sprout inhibition, 5.4 cents/kg for poultry decontamination, and 10.8 cents/kg for spice sterilization (21).

A facility for citrus fruit disinfestation, treating 900 tons/day with 0.25 kGy, would involve a capital outlay of $3.2 million with total annual running costs of $670,000. This would add 0.21 cents/kg to the price of the fruit. For an accelerator installation (3–6 MeV) the corresponding figures are $2.4 million, running costs for 6,000 operational hours $0.5–0.9 million. A 5 kGy treatment would cost 0.7 cents/kg and for X-ray treatment 1.5 cents/kg. Potato treatment would cost $4–12/ton, which compares with $7–10/ton at a Japanese facility (21). Other estimates for a γ-ray facility come to £400,000 for ^{60}Co, £1.25 million for the plant, and £0.75 million for the machinery. For an electron accelerator the capital cost would be £2 million (104).

REGULATORY ASPECTS

A recent review of the regulatory status of the process of food irradiation and the authorizations for irradiating foods commercially for marketing lists a total of 38 countries, of which 24 control food irradiation for commercial purposes by specific regulations and 14 by general provisions under their national food law. Thirty-one countries have authorized the irradiation of one or more specific foods, while seven countries have regulations controlling the process but have not yet granted any known authorization for commercial irradiation of any food item (105). About 12 different foods are listed as authorized for marketing as a result of these clearances, but another 28 foods have received clearance for irradiation of experimental batches for test marketing (106). About 70% of these regulatory approvals have occurred after the 1980 JECFI meeting and the subsequent adoption of the Codex General Standard of Irradiated Foods by the Codex Alimentarius Commission in 1983.

Some 50 pilot or industrial irradiators have been installed in 23 countries for the commercial irradiation of foods and food ingredients. At present, an estimated 500,000 tons of foodstuffs are likely to be irradiated each year. The role of food irradiation in developed countries is likely to be the maintenance of the quality and the microbiologic safety of food. In developing countries the role should be the reduction of postharvest losses of perishable commodities to safeguard consumption by the local population. Furthermore, it should be considered as an important public health and environmentally acceptable measure in relation to the staple foodstuffs, such as dried fish, dried fruit, nuts, and grains (107).

Standardization of national practices in irradiating food and of the regulations for commercialization and trading in irradiated foods is an essential prerequisite for establishing these products as items of international trade. As early as 1979 the General Agreement on Tariffs and Trade (GATT) on Technical Barriers to Trade provided that no technical regulations and standards be prepared, adopted, or applied by parties to the GATT that would create obstacles to international trade. Consequently the acceptance of Codex Alimentarius Standards, prepared by the Codex Alimentarius Commission, an organization representing about 136 member countries, was recommended to member countries of GATT, unless there were good reasons for not doing so. The same theme was referred to in the Tariff and Trade Negotiations in Geneva in 1987 by requesting the worldwide institution of uniform food health regulations to prevent nontariff barriers to trade in agricultural produce (108). The theme was expanded in 1989 by US proposals, submitted to GATT, for encouraging the contracting parties to bring health and sanitary measures into line with appropriate international standards (109). The

Codex Standard for Irradiated Foods prescribed a maximum overall average dose of 10 kGy absorbed radiation energy, appropriate labeling to inform the consumer and prevent unnecessary irradiation, and compliance with the Codex Code of Practice for operating radiation facilities for food irradiation (12).

Some five countries require the use of a logo as well as a text in the labeling provisions, while the three countries authorizing food irradiation require that irradiation treatment be specified on the label. The United States has specific authorizations for wheat disinfestation, for controlling microbial pathogens in fresh or frozen uncooked poultry, for controlling *T. spiralis* in pork products, and for decontaminating dry enzyme preparations, spices, vegetable seasonings, and dry aromatic vegetable substances.

Within the European Community a number of national clearances are operating in five member states. Commercial scale operations are carried out in The Netherlands and in France. irradiated food in The Netherlands covers some 11 food items, but the treated food is mainly used in the food manufacturing industry. In France some seven food items have been cleared for irradiation and marketing, particularly mechanically deboned poultry meat and spices. Although Belgium permits the commercial irradiation of seven food items, only irradiated spices, deep-frozen foods, and dehydrated vegetables are being marketed. The now-united German Federal Republic applies a strict and total ban on food irradiation.

The European Economic Commission had submitted a proposal for a Directive (COM(88)654 final) to the Council of Ministers in December 1985, but had to withdraw it because of failure to reach agreement among member states and in the European parliament. The original list of irradiated foodstuffs to be authorized contained 11 items, but the amended version of November 1989 has been shortened to eight items, and includes imports from outside the Community. Strict labeling of any irradiated foods and even ingredients is demanded, as well as special licensing of irradiation facilities. Standardization of the relevant food laws is envisioned for 1993. Meanwhile the national regulations apply.

On 22/2/1999 the Directive 1999/2/EC of the European Parliament and of the Council was published as a framework directive controlling the manufacture, marketing and importation into the Community of foods and food ingredients treated with ionizing radiation. It does not apply to foodstuffs for patients requiring sterilized diets under medical supervision. Annex I sets out the conditions for authorizing food irradiation, while Annex II stipulates the irradiation sources, e.g. ^{60}Co, ^{137}Cs, X-rays from machines operating at energies up to 5 MeV or electron sources operating at energies up to 10 MeV. Annex III sets out the dosimetry procedures.

Directive 1999/3/EC is the implementing Directive establishing an initial Community positive list of foodstuffs that may be treated with ionising radiation. Its Annex contains only one item, namely dried aromatic herbs, spices and vegetable seasonings to be irradiated with 10 kGy as the maximum overall average absorbed radiation dose. Strict labeling of any irradiated foods and even ingredients is demanded, as well as special licensing of irradiation facilities.

One of the major hurdles blocking the general commercialization of food irradiation is the failure of consumer acceptance of marketed irradiated foods. This is easy to fathom considering the negative public attitudes toward irradiation all over the world and the genuine fear of anything appearing to increase the risk of exposure to radiation. Even within the food industry there is resistance to the use of irradiated foods, as part of a firm's total palette of marketed food commodities or even as an ingredient of a food product on display. Future developments will largely depend on proper education of the public by using appropriate information and labeling.

Irradiation improves the hygienic status of the treated food, thus permitting longer storage. It does not, however, improve the nature and quality of the treated food, and cannot be portrayed as doing so (11). Before food irradiation can find universal acceptance, it must be demonstrated to bring some real benefits with no risks to the public or workers in the food industry, and the consumer must be assured that an effective system for monitoring, control, and abuse prevention has been established.

REFERENCES

1. Minck F. Zur Frage über die Einwirkung der Röntgenschen Strahlen auf Bakterien und ihre eventuelle therapeutische Verwendbarkeit. *Münch Med Wochenschr* 1896; 5:101, and 9:202.
2. British patent no. 1609 issued to J. Appleby and A.J. Banks, 1905, quoted by Diehl F-J. *Safety of irradiated foods*. New York: Marcel Dekker; 1990; 1.
3. Schwartz B. Effects of X-rays on trichinae. *J Agric Res* 1921; 20:845.
4. Hackwood S. The irradiation processing of foods. In: Thorne S, ed. *Food irradiation*. London: Elsevier Applied Science; 1991; 2.
5. Goldblith SA. Historical development of food irradiation. In: *Food irradiation. Proceedings of the international symposium on food irradiation*. STI/PUB 127. Vienna: IAEA; 1966; 3–21.
6. Scientific Committee for Food. Report on the Irradiation of Food. 18th Series. Luxembourg: Commission of the European Communities; 1989; 12–3.
7. Kampelmacher EH. Irradiation of food: a new technology for preserving and ensuring the hygiene of foods. *Fleischwirtsch* 1984; 64(3):322–7.
8. Hackwood S. The irradiation processing of foods. In: Thorne S, ed. *Food irradiation*. London: Elsevier Applied Science; 1991; 7.
9. Han YW, Catatano EA, Ciegler A. Chemical and physical properties of sugar cane bagasse irradiated with gamma rays. *J Agric Fd Chem* 1983; 31(1):34–8.
10. Paul N, Grünewald Th, Kuprianoff J. Über die Möglichkeiten einer Behandlung von Trockensuppen mit ionisierenden Strahlen. *Dtsch Lebensm Rundsch* 1969; 65(9):279–81.
11. Grünewald Th. In: l'Annunziata MF, Legg JO, eds. *Isotopes and radiation in agricultural sciences*, vol 2. New York: Academic Press; 1984; 271–301.
12. Codex Alimentarius Commission. Recommended International Code of Practice for the Operation of Radiation Facilities Used for the treatment of Foods, vol XV. Rome: FAO; 1984.

13. Herrnhut H. Anwendungsorentierte Lebensmittelbestrahlungsanlagen. deren Wirtschaftlichkeit und Energieverbrauch. *Mitt Gehiet Lebensm Hyg* 1985; 76:28–33.
14. Fraser FM. Gamma radiation processing equipment and associated energy requirements in food irradiation. In: Josephson ES, Peterson MS, eds. *Preservation of food by ionizing radiation*, vol 3. Boca Raton, FL: CRC Press; 1983; 253–7.
15. Bögl W, Heide L. Nachweis der Gewürzbestrahlung. *Fleischwirtsch* 1984; 64(9): l920–6.
16. International Atomic Energy Agency (IAEA). *Analytical detection methods for irradiated foods: a review of the current literature*. IAEA-TECDOC-587. Vienna: IAEA; 1991.
17. Stevenson MH, Crone A, Hamilton J. Irradiation detection, *Nature* 1990; 344:6263.
18. Helle N, Schreiber GA, Bögl KW. Analytical methods to identify irradiated food. In: Ehlermann DAE, Spieß WEL, Wolf N, eds. *Lebensmittellbestrahlung*. Berichte der Bundesforschungsanstalt für Ernährung. BFE-R-92-0l. Karlsruhe, Germany; 1992; 112–35.
19. Sanderson DCW, Slater C, Cams KJ. Thermoluminescence of food: origins and implications for detecting irradiation. *Rad Phys Chem* 1989; 34:915.
20. Delincée H. Experiments on the identification of radiation processed foods. In: Ehlermann DAE, Spieß WEL, Wolf N, eds. *Lebensmittellbestrahlung*. Berichte der Bundesforschungsanstalt für Ernahrung, BFE-R-92-01. Karlsruhe, Germany; 1992; 136–49.
21. Brynjolfsson A. Food irradiation in the United States. In: *Proc 26th Euro Mtg Meat Res Work*, vol I. Chicago: Amer Meat Sci Assoc; 1980; 172–7.
22. Tagekuchi CA. Regulatory and toxicological aspects of food irradiation. *J Fd Safety* l983; 5:213–17.
23. Elias PS, Cohen AJ, eds. Radiation chemistry of major food components. Amsterdam: Elsevier; 1977.
24. Urbain WM. Food irradiation. *Adv Fd Res* 1978; 24:l55–227.
25. Nawar WW. Radiation chemistry of lipids. In: Elias PS, Cohen AJ, eds. *Radiation chemistry of major food components*. Amsterdam: Elsevier; 1977; 21–61.
26. Van Straten S. *Volatile compounds in food*, 4th edn. Zeist: CIVO.TNO; 1977.
27. Dauphin JF, St Lèbe LR. Radiation chemistry of carbohydrates. In: Elias PS, Cohen AJ, eds. *Radiation chemistry of major food components*. Amsterdam: Elsevier; 1977; 131–85.
28. Dieh J-F. Radiolytic effects in foods. In: Josephson ES, Peterson MS, eds. *Preservation of food by ionizing radiation*, vol I Boca Raton, FL: CRC Press; 1983; 279–357.
29. Diehl F-J, Adam S, Delincée H, Jakerbick V. Radiolysis of carbohydrates and carbohydrate containing foodstuffs. *J Agric Fd Chem* 1978; 26:5–20.
30. Adam S. Recent developments in Radiation Chemistry of carbohydrates. In: Elias PS, Cohen AJ, eds. *Recent advances in food irradiation*. Amsterdam: Elsevier Biomedical Press; 1983; 149–70.
31. Tobback PP. Radiation Chemistry of vitamins. In: Elias PS, Cohen AJ, eds. *Radiation Chemistry of major food components*. Amsterdam: Elsevier; 1977; 187–220.
32. Diehl J. Verminderung von strahleninduzierten Vitamin E and -B$_1$ Verlusten durch Bestrahlung von Lebensmiueln bei tiefen Temperaturen und durch Ausschluß von Luftsauerstoff. *Z Lebensm Unters Forsch* 1979; 169:276–82.
33. Wilmer I, Schubert I. Mutagenicity of irradiated solutions of nucleic acid bases and nucleosides in *Salmonella typhimurium*. *Mutat Res* 1981; 88:337–42.

34. Merritt C Jr, Taub IA. Commonality and predictability of radiolytic products in irradiated meats. In: Elias PS, Cohen AJ, eds. *Recent advances in food irradiation*. Amsterdam: Elsevier Biomedical Press; 1983; 27–57.

35. Asquith J, Elias PS. Bedeutung and Berechnung des ADI Wertes. *Getreide, Mehl & Brot* 1981; 35:272.

36. Phillips BJ, Elias PS. A new approach to investigating the genetic toxicity of processed foods. *Fd Cosmer Toxicol* 1978; 16:509.

37. Elias PS. Die gesundheitliche Unbedenklichkeit bestrahlter Lebensmittel. In: Ehlermann DAE, Spieß WEL, Wolf W, eds. *Lebensmittel*. Berichte der Bundesforschungsanstalt für Emahrung. BFE-R-92-01. Karlsruhe, Germany; 1992; 16–31.

38. Groedel FM. Schneider E. Experimental studies on the question of the biological effect of X-rays. *Strahienther* 1926; 22:411

39. Elias PS. Food irradiation. In: Miller K, ed. *Toxicological aspects of food*. London: Elsevier Applied Science; 1987; 295–346.

40. USA. Statement on the Wholesomeness of Irradiated Foods by the Surgeon General of the Army. Hearings Subcommittee on Research. Development and Radiation. 89th Congress. 9–10 June. Washington, DC: US Government Printing Office; 1965; 105–6.

41. Joint FAO/IAEA/WHO Expert Committee on Wholesomeness of Irradiated Food (JECFI). *Wholesomeness of Irradiated Food*. WHO Tech Rep Ser No 659, Geneva: WHO; 1981.

42. Mafarachisi BA. Growth, fertility and tissue studies of mice fed radiolytic products arising from gamma-irradiated beef fat. PhD Thesis, Univ. Massachusetts; 1974.

43. Truhaut R, St Lèbe L. Différentes voies d'approche pour l'evaluation toxicologique de l'amidon irradié. In: *Food preservation by irradiation*, vol II. STI/PUB/470, Vienna: IAEA; 1978.

44. Rao VS. Biochemical studies on the toxicity of irradiated sugar solutions in microorganisms. PhD Thesis, Univ. Bombay, India; 1978.

45. Aiyar AS, Rao VS. Studies on mutagenicity of irradiated sugar solutions in *Salmonella typhimurium. Mutat Res* 1977; 48:17–27.

46. Varma MB, Rao KP, Nadan SD, Rao MS. Mutagenic effects of irradiated glucose in *Drosophila melanogaster. Fd Chem Toxicol* 1982; 20:947–9.

47. Varma MB, Nadan DS, Rao KP, Rao MS. Non-induction of dominant lethal mutations in mice fed irradiated glucose. *Int J Radiat Biol* 1982; 42:559–63.

48. Niemand JG, den Drijver L, Pretorius CJ, Holzapfel CW, van der Linde J. Study of the mutagenicity of irradiated sugar solutions: implication for the radiation preservation of sub-tropical fruits. *J Agric Fd Chem* 1983; 31:1016–20.

49. Wilmer J, Schubert J, Leveling H. Mutagenicity of gamma-irradiated oxygenated and deoxygenated solutions of 2-deoxy-D-ribose and D-ribose in *Salmonella typhimurium. Mutat Res* 1981; 90:385–97.

50. World Health Organization. *Wholesomeness of Irradiated Food*. WHO/FOOD ADD/77.45, Geneva: WHO; 1977.

51. World Health Organization. *Wholesomeness of Irradiated Food*. EHE/81.24, Geneva: WHO; 1981.

52. Zaitsev A. USSR submission to WHO, Geneva; 1981.

53. Van Kooij JG, Leveling HB, Schubert J. Application of the Ames mutagenicity test to food processed by physical preservation methods. In: *Food preservation by irradiation. Proc Int Symp Fd Pres Irrad*, vol 2. Vienna: IAEA; 1978; 63–71.

54. Ke-Wen Cheng, Dao-Jing Zhang. Wholesomeness studies on gamma irradiated rice. *Radiat Phys Chem* 1983; 22(3):792.

55. Barna J. *Toxicity studies of radiation- and heat-treated paprika powder in animal feeding experiments.* Central Food Res Inst. Hungary: Budapest; 1973.

56. Barna J. *Final report on preliminary studies relating to investigation of the wholesomeness of irradiated spices.* Central Food Res Inst. Hungary: Budapest; 1976.

57. Chaubey RC, Kavi BR, Chauhan PS, Sundaram K, Barna J. Cytogenetic studies with irradiated ground paprika as evaluated by the micronucleus test in mice. *Acta Alim* 1979; (2):197–201.

58. Wills ED. *Studies of Irradiated Food with Special Reference to its Lipid Peroxide Content and Carcinogenic Potential.* IFI Techn Rep Ser No. 55, Germany: Karlsruhe; 1981.

59. Anukarahanonta T, *et al.* Wholesomeness study of irradiated salted and dried mackerel in rats. Unpubl rep to IAEA, Vienna; 1980.

60. Aravindakshan M, Vakil UK, Sreenivasan A. *Nutritional and wholesomeness studies with dehydroirradiated shrimps.* Report No. 455, Bhabha Atomic Res. Centre. Bombay, India; 1973.

61. Van Logten Mi, den Tonkelaar EM, van Esch GJ, Kroes R. The wholesomeness of irradiated shrimps. *Fd Cosmet Toxicol* 1972; 10(6):781.

62. Fegley HC, Edmonds SR. *Longterm Feeding Experiment on Radiation Pasteurized Foods.* Final Report. NYO-3573-l-Unclass; 1968.

63. USA. Chronic Toxicity, Oncogenicity and Multigeneration Reproduction Study using CD-1 Mice to Evaluate Frozen, Thermally sterilized, Cobalt-60 Irradiated and 10 MeV Electron Irradiated Chicken Meat. Nat Tech Inf Serv, Order No. PB-84-187012; 1985.

64. Thayer DW, Christopher JP, Campbell LA, Ronning DC, Dahlgren RR, Thomson GM, Wierbicki E. Toxicology studies of irradiation-sterilized chicken. *J Fd Protec* 1987; 50:278.

65. Brynjolfsson A. Wholesomeness of irradiated foods: a review. *J Fd Safety* 1985; 7: 107–26.

66. Read MS, Trabosh HM, Worth WS, Kraybill HF, Witt NF. Shortterm rat feeding studies with gamma-irradiated food products. II. Beef and pork stored at elevated temperature. *Toxicol Appl Pharmacol* 1959; 1:417–23.

67. Shillinger Yu I, Kachkova VO, Maganova NB. Influence produced on the canine organism by meat food products gamma-irradiated in radio-pasteurization doses. *Voprosy Pitaniva* 1965; 24(1):19.

68. Van Logten MJ, de Vries T, van der Heyden CA, van Leeuwen EYR, Garbis-Berkvens MJM, Strijk JJTWA. Report 61740001, N.I. Public Health. The Netherlands; 1983.

69. Adamiker D. Irradiation of laboratory animal diet: A review. *Versuchstierkunde* 1978; 18:191.

70. Ley FJ. Radiation processing of laboratory animal diet. *Radiat Phys Chem* 1979; 14:677.

71. Tsuji K. Low-dose cobalt-60 irradiation for reduction of microbial contamination in raw materials for animal health products. *Fd Technol (Chicago)* 1983; 37(2):48.

72. Aker SN. On the cutting edge of dietetic science. *Nutrition Today* 1984; 19(4):24.

73. Jin W, Yuan J. *Safety evaluation of 35 kinds of irradiated food for human consumption.* IAEATECDOC-452, IAEA, Vienna; 1987; 163.

74. Thompson SW, Hunt RD, Ferrell J, Jenkins ED, Monsen H. Histopathology of mice fed irradiated foods. *J Nutr* 1965; 87:274.

75. Matschiner JT, Doisy EA Jr. Vitamin K content of ground beef. *J Nutr* 1966; 90:331.

76. Reusse U, Messow C. Geister R. Pasteurization of fish meal by irradiation. 3. The question of increased rates of mitosis after feeding radiation-pasteurized fish meal to pigs. *Zbl Vet Med* 1979; B26:500.

77. Kamaldinova ZM. Effect of culinary pretreated gamma-irradiated beef on the organism of rats. *Voprosy Pitaniya* l970; 29(7):73–7.

78. Shillinger Yu I, Osipova IN. The effect of gamma-irradiated fresh fish on the organism of white rats. *Voprosy Pitaniya* 1970; 32(6):45–50.

79. Zaitsev AN, Osipova IN. Study on mutagenic properties of irradiated fresh fish in chronic experiments. *Voprosy Piraniya* 1981; 40:53–6.

80. Zaitsev AN, Maganova NB. Effect of the diet including gamma-irradiated fish on embryogenesis and chromosomes of rats. *Voprosy Pitaniya* 1981; 40(6):61–3.

81. Bhaskaram C, Sadasivan G. Effects of feeding irradiated wheat to malnourished children. *Int J Radiat Biol* 1976; 27:93, and *Am J Clin Nutr* 1975; 28:130–5.

82. Vijayalaxmi C. Cytogenetic studies in rats fed irradiated wheat. *Int J Radiat Biol* 1975; 27:283.

83. Vijayalaxmi C. Cytogenetic studies in monkeys fed irradiated wheat. *Toxicol* 1978; 9:181–4.

84. LST. *Irradiation of Food*. Report by a Danish Working Group. Nat Fd Agency, Publ No. l, 120, Soborg, Denmark; 1986.

85. Reddi OS, Reddy PP, Ebenezer DN, Naidu NV. Lack of genetic and cytogenetic effects in mice fed on irradiated wheat. *Int J Radial Biol* 1977; 31:589–601.

86. Murphy PBK. SCE in monkeys fed irradiated wheat. *Fd Cosmet: Toxicol* 1981; 12:523.

87. Murphy PBK. Sister-chromatid exchanges in mice given irradiated wheat. *Toxicol* 1981; 20:247–9.

88. Vijayalaxmi C. Immune response in rats given irradiated wheat. *Br J Nutr* 1978; 40:535–41.

89. Kopylov VA, Osipova IN, Kuzin AM. Mutagenic effects of extracts from gamma-irradiated potato tubers on the sex cells of male mice. *Radiobiologiya* 1958; 12:58.

90. Scientific Committee for Food. Report on the Irradiation of Food. 18th Series. Commission of the European Communities. Luxembourg; 1989; 39.

91. Gower JD, Wills ED. The oxidation of benzo(a)pyrene mediated by lipid peroxidation in irradiated synthetic diets. *Int J Radiat Biol* 1986; 49:471.

92. FASEB. Evaluation of the health aspects of certain compounds found in irradiated beef. LSRO, Bethesda, MD with Suppl. I and II; 1979.

93. Diehl F-J. *Safety of irradiated foods*. New York: Marcel Dekker; 1990; 166.

94. Eggum BO. Effect of radiation treatment on protein quality and vitamin content of animal feeds. In: *Decontaminasion of animal feeds by irradiation. Proc Mtg Sofia*, Oct 1977, IAEA, Vienna; 1979; 55.

95. Thayer DW, Fox JB Jr, Lakritz L. Effects of Ionizing Radiation on Vitamins. In: Thorne S, ed. *Food irradiation*. London: Elsevier Applied Science; l991; 285–325.

96. Grecz N, Bruszew G, Amini I. In: *Combination processes in food irradiation*. Vienna: IAEA; 1981; 3–20.

97. Langerak Dis. Irradiation of foodstuffs – technological aspects and possibilities. In: *Food irradiation now*. Proc. Gammaster Symp, Ede. 21.10. l981, The Hague: Martinus Nijhoff/Dr W Junk; 1982; 40–59.

98. Priyadarshini E, Tulpule PG. Aflatoxin production in irradiated food. *Fd Cosmet Toxicol* 1976; 14:293–5.

99. Priyadarshini E, Tulpule PG. Effect of graded doses of gamma irradiation on aflatoxin production by *Aspergillus parasiticus* in wheat. *Fd Cosmet Toxicol* 1979; 19:505–7.

100. Schindler AF, Abadie AN, Simpson RE. Enhanced aflatoxin production by *Aspergillus flavus* and *Aspergillus parasiticus* after gamma irradiation of the spore inoculum. *J Fd Protec* 1980; 43:7–9.

101. Clivie DO. Unlikelihood of mutagenic effects of irradiation on viruses. *Wholesomeness of Irradiated Food*. WHO Tech Rep Ser No. 604. Annex 2, 43–44, Geneva: WHO; 1977.

102. Maxcy RB. Comparative viability of unirradiated and gamma-irradiated bacterial cells. *J Fd Sci* 1977; 42:1056–9.

103. King BL. Josephson ES. Action of radiation on protozoa and helminths. In: Josephson ES, Peterson MS, eds. *Preservation of food by ionizing radiation*, vol 2. Boca Raton, FL: CRC Press; 1983; 245.

104. Guise B. Processing practicalities. *Fed Proc* 1989; 53–4.

105. Intenational Atomic Energy Agency (IAEA). *Regulations in the Field of Food Irradiation*. IAEATECDOC-585, IAEA. Vienna; 1991.

106. Van Kooij JG. Updated list of clearances for irradiated foods in member states. IAEA, *Fd Irrad Newslett* 1985; 9(2):29–39.

107. Loaharanu P. Status of food irradiation worldwide. In: Ehlermann DAE, Spieß WEL, Wolf N, eds. *Lebensmittelbestrahlung*. Berichte der Bundesforschungsanstalt für Ernährung, BFE-R-92-0l, Karlsruhe, Germany; 1992; 76–81.

108. Forman MB, The United States – European Community Hormone treated Beef Conflict. *Harvard Int Law J* l989; 30(2):footnote 41.

109. Forman MB. The United States – European Community Hormone treated Beef Conflict. *Harvard Int Law J* 1989; 30(2):footnote 45.

110. Ehlermann DAE. Current Status of Food Irradiation in Europe. In: Thome S, ed. *Food irradation*. London: Elsevier Applied Science; 1991; 87–95.

111. World Health Organization. *High-Dose Irradiation: Wholesomeness of Food Irradiated with Doses above 10 kGy*. WHO Technical Report Series 890, Geneva: WHO; 1999.

112. Bourland CT. *NASA and Food Irradiation*. International conference on seafood irradiation. New Orleans, 16–18 June 1992.

113. Bruyn I. Application of high dose irradiation. *Proceedings of National Seminar on Food Irradiation*. Toluca: 1997; 32–40.

114. Hayashi T, Todoriki S, Otobe K, Sugiyama J. Impedance measuring techniques for identifying irradiated potatoes. *Biosc Biotech Biochem* 1992; 56:1920–32.

115. Felföldi J, László P, Barbássý S, Farkas J. Dielectric method for detection of irradiation treatment of potatoes. *Radiat Phys Chem* 1993; 41:471–80.

116. Raffi JJ, Kent M. Methods of identification of irradiated foodstuffs. In: Nollet LML, ed. *Handbook of Food Analysis*. New York: Marcel Dekker; 1996; 1889–1906.

117. Delincée H, Ehlermann DAE. Recent Advances in the identification of irradiated foods. *Radiat Phys Chem* 1989; 34:879–90.

118. Diehl J-F. *Safety of Irradiated Foods*. New York: Marcel Dekker; 1995.

119. Nesvadba P. Increased supercooling in irradiated food. *Int J Fd Sci Techn* 1991; 26:165–71.

120. Dodd NTF, Swallow AJ, Ley FJ. Use of ESR to identify irradiated food. *Radiat Phys Chem* 1985; 26:451–3.
121. Stachowicz W, Burlinska G, Michalik J, Dziedzic-Goclawska A, Ostrowski K. Application of ESR spectroscopy to radiation treated materials in medicine, dosimetry and agriculture. *Appl Radiat Isot* 1993; 44:423–7.
122. Desrosiers MF. Estimation of the absorbed dose in radiation-processed food. 2. Test of the ESR Response Function by an exponential fitting analysis. *Appl Radiat Isot* 1991; 42:617–19.
123. Bordi F, Fattibene P, Onori S, Pantaloni M. An alternative procedure for ESR identification of irradiated chicken drumsticks. *Appl Radiat Isot* 1993; 44:443–7.
124. Desrosiers MF. Gamma-irradiated seafoods: Identification and dosimetry by electron paramagnetic resonance spectroscopy. *J Agric Fd Chem* 1989; 37:96–100.
125. Desrosiers MF, McLaughlin WL. Examination of gamma-irradiated fruits and vegetables by ESR spectroscopy. *Radiat Phys Chem* 1989; 34:895–8.
126. Tabner BJ, Tabner VA. ESR spectra of γ-irradiated citrus fruit skin, skin components and stalks. *Int J Fd Sci Technol* 1994; 29:143–52.
127. Onori S, Pantaloni M. ESR technique identification and dosimetry of irradiated chicken eggs. *Int J Fd Sci Technol* 1995; 29:671–7.
128. Anderle H, Stefan I, Wilde E, Hille P. Radiolyo-chemiluminescence of bones and seafood shells – a new promising method for the detection of food irradiation. *Fresenius J Anal Chem* 1996; 354:925–8.
129. Glidewell SM, Deighton IV, Goodmann BA, Hillman JR. Detection of irradiated food: A review *J Sci Fd Agric* 1993; 61:281–300.
130. Schreiber GA, Helle W, Bögl WW. Detection of irradiated food. Methods and Routine Applications. *J Radiat Biol* 1993; 63:105–30.
131. Copin MP, Jehanno D, Bourgeois CM. Detection of irradiated deep frozen foodstuffs by comparison of DEFT and APC. *J Appl Bacteriol* 1993; 75:254–8.
132. Wirtanen G, Sjöberg AM, Boiser F, Alanko T. Microbiological screening method for indication of irradiation of spices and herbs. A BCR Collaborative Study. *JAOAC* 1993; 76:674–81.
133. Kume T. Immunochemical Identification of irradiated chicken eggs. *J Sc Fd Agric* 1994; 65:1–4.
134. Xie Z, Shi S, Li Z, Cao H, Zhang B. Study on the identification methods for the irradiated pollen. *Radiat Phys Chem* 1993; 42:413–16.
135. Taub IA, Angelini P, Merrit C Jr. Irradiated foods: validity of extrapolating wholesomeness data. *J Fd Sci* 1976; 41:942–4.
136. World Health Organization. *Safety and nutritional adequacy of irradiated food.* Geneva: WHO; 1994.

17

MECHANISTIC CONSIDERATIONS IN THE REGULATION AND CLASSIFICATION OF CHEMICAL CARCINOGENS

R. Michael McClain

CARCINOGENICITY TESTING

Most drugs, pesticides, and food and color additives require carcinogenicity testing in two rodent species prior to marketing approval. Carcinogenicity testing is the most time- and resource-intensive aspect of toxicity testing and is also the most difficult test with respect to the interpretation of the significance of positive findings. This is of considerable consequence because of the high proportion of tested compounds that are "positive" in at least one species or sex in the rodent carcinogenicity test.

Over 300 compounds have been tested since the 1960s when the National Cancer Institute (NCI) started the bioassay program which was subsequently moved to the National Toxicology Program (NTP). Approximately half of these chemicals were "positive" in at least one species or sex (1). Although this is not necessarily a random sample of chemicals, since many may have been selected based on mutagenicity or suspected chemical reactivity, as Ames and Gold have pointed out, regardless of whether a chemical is synthetic or natural approximately half of the chemicals tested are positive in the rodent bioassay (2,2a). Many of these substances have no apparent mutagenic or DNA-damaging activity, and include essential elements, vitamins, sugars, and other basic food substances. Indeed, caloric intake *per se* has a profound carcinogenic effect in rats (2).

The relevance to human risk of so many positive bioassays under high-dose test conditions has been challenged and seriously debated. At the beginning of the NCI bioassay program, only a small percentage of compounds were expected to exhibit carcinogenic activity; however, in practice many more compounds than predicted are positive in the bioassay. It has been known for many years that fundamentally different mechanisms are involved in chemical carcinogenicity and that these mechanisms differ with respect to human risk.

CAUSES OF HUMAN CANCER

Cancer is one of the most common and dreaded diseases. Death due to cancer occurs in approximately one of four individuals in the United States and thus the effort to find the cause, prevention, and treatment of human cancer is intense. Well over half of all human cancer is believed related to environmental factors and life style (3), thus the potential to reduce the human cancer burden through preventive measures is considerable. Other known causes of human cancer include viruses and a variety of recently identified genetic defects that may act alone or in concert with environmental factors in the pathogenesis of cancer (4).

The most important environmental factor is tobacco, which alone accounts for more than one-third of all human cancer. Other important environmental factors include radiation, ultraviolet (UV) exposure from sunlight, alcohol abuse, and diet (4). Certain types of occupational exposures (e.g. asbestos or vinyl chloride) account for a significant but relatively small percentage of human cancer (4). A small percentage is also caused by pharmaceutical agents (e.g. antineoplastic agents in the treatment of cancer, immunosuppressants, and drugs such as psoralen used in combination with UV light for the treatment of severe psoriasis). In the case of the pharmaceutical agents, the known risk of cancer is offset by the benefits of their therapeutic effects.

Although the importance of environmental factors in human carcinogenesis is well recognized, the actual nature of these factors is widely misperceived by the general public and sometimes misrepresented. Factors such as environmental pollution, pesticide residues, food and color additives, and, in general, synthetic organic chemicals are perceived by the public to be major causes of human cancer. This perception, however, is not supported by the epidemiologic data; in fact, as Western society has become industrialized and the production and use of synthetic chemicals has increased the incidence of cancer has declined after adjusting for the increase in tobacco-related cancers (5,6). Synthetic chemicals account for no more than a small percentage of human cancer and most of these occur under conditions of occupational exposure or other isolated unusual conditions of exposure. The misperception of risks to chemical exposure can have a negative effect by detracting attention from those factors that could have a major impact on prevention of human disease (2).

Despite the fact that synthetic chemicals do not account for much of human cancer, considerable emphasis is placed on the testing of chemicals in the rodent bioassay. Testing of chemicals is justified for two reasons. First, chemicals that pose a human risk of cancer should be identified. Second, the carcinogenic potential for new chemicals and drugs should be assessed prior to human exposure. An expansion of the list of rodent carcinogens, however, without an expansion of our understanding of the mechanism of tumor induction or an understanding of the rodent models that we use to perform these tests will not be useful for human risk assessment and will not result in significant health benefits.

MODE OF ACTION FOR CARCINOGENS

Williams and Weisburger pioneered the classification of carcinogens based on mode of action (7). They have used the term *genotoxic carcinogen* for carcinogens that interact directly with and damage DNA, with or without metabolic activation. The term *epigenetic carcinogen* refers to nongenotoxic carcinogens that produce a tumor response without a direct interaction with, or damage to, DNA. The fundamental difference between these two modes of action is that it may be possible, in the case of an epigenetic carcinogen, to establish that the effects are not cumulative and are reversible to a point. Furthermore, it may be possible to define a threshold below which there would not be a risk to exposure.

There have been a number of terms used to describe the action of carcinogens, including *genotoxic*, *direct-acting*, *initiator*, and *primary* for DNA damaging carcinogens, and *epigenetic*, *nongenotoxic*, *indirect-acting*, *promoter*, and *secondary* for carcinogens that act through a mode of action that does not involve direct DNA damage. The term *secondary mechanism of carcinogenesis* has been used frequently in both a scientific and legal context. It is a term that implies a chemical-specific mechanism that can be evaluated in the risk assessment process. The term can be defined as a tumor response caused by a factor that can be shown to exist under the conditions of the bioassay that can alter the tumor incidence independent of the test chemical *per se*. The secondary factor can involve an exaggerated pharmacologic response of a drug, as in gastrin-mediated stomach neoplasia produced with the long-acting histamine H_2 antagonists; target organ toxicity, as in bladder tumors secondary to urinary tract calculi; and physiologic disturbances, as in a variety of endocrine neoplasias demonstrated to be secondary to chemically induced hormone imbalance. In each of these examples, it is the secondary factor that is responsible for the tumor response independent of the chemical *per se*. The consideration of mode and/or mechanism of action is important since it is possible in many cases to determine if such a change occurs under conditions of human exposure and if this change would influence human tumor response when it occurs.

Thus it is critical for the interpretation of the results of a rodent carcinogenicity study to consider the mode or mechanism of action for the risk assessment and classification of carcinogens. Most current guidelines recommend the evaluation of all scientific information with respect to mechanism of action on the basis of a "weight of evidence" approach to risk assessment. These guidelines include documents such as "Risk Assessment in the Federal Government: The Process," from the National Academy of Sciences (NAS) (8); "Chemical Carcinogens: A Review of the Science and Its Associated Principles" from the Office of Science and Technology Assessment (OSTP) (9); the Environmental Protection Agency's (EPA) Guidelines for Carcinogen/Risk Assessment (10); the US Food and Drug Administration's (FDA) Criteria and Procedures for Evaluating the Safety of Carcinogenic Residues (11); and guidelines issued by the Japanese Ministry of Health (12) and the Canadian Health Protection Branch (13).

CLASSIFICATION OF CARCINOGENS

Carcinogens are classified by several agencies and international bodies, the most important of which are the EPA and the International Agency for Research on Cancer (IARC). These agencies evaluate the strength of evidence for animal and human carcinogenicity data and classify chemicals according to the probability that the chemical presents a human cancer risk. The EPA scheme (10) includes category A (*human carcinogen*), category B (*probable human carcinogen*), category C (*possible human carcinogen*), category D (*not classifiable as to human carcinogenicity*), and category E (*evidence of non-carcinogenicity*). The IARC classification (14) includes group 1 (*is carcinogenic to humans*), group 2 (2A – *is probably carcinogenic to humans*, 2B – *is possibly carcinogenic to humans*), group 3 (*is not classifiable*), and group 4 (*is probably not carcinogenic*). Recently the IARC has agreed to consider mechanism of action data for the classification of carcinogens where chemicals may be increased or decreased in the scheme in accord with the relevance of the mechanistic information (14a).

The NTP does not classify carcinogens; however, it does evaluate the results of individual animal carcinogenicity studies and determines the strength of evidence for the study conclusions. Five categories of evidence for "carcinogenic activity" are used: *clear evidence, some evidence, equivocal evidence, no evidence*, and *inadequate study*. In addition, the US Congress has mandated that an Annual Report on Carcinogens be issued by the Department of Health and Human Services for classifying chemical substances as known or reasonably anticipated to be carcinogens to which a significant number of persons are exposed. This report is prepared by the Director of the NTP. This report does not include a formal risk assessment but is significant in that a listing may trigger action under a number of State and Federal statutes.

THE REGULATION OF CARCINOGENS

Drugs, pesticides, and food and color additives are regulated, by either the FDA under the Food, Drug and Cosmetic Act (FD&C Act) or the EPA under the Federal Insecticide, Fungicide and Rodenticide Act (FIFRA). In addition, the Delaney Cancer Clauses to the FD&C Act prohibit the use of any substance found to "induce" cancer in animals or humans as a food or color additive or as an animal drug with residues in food-producing animals. The Delaney Cancer Clauses represent zero risk legislation, are essentially redundant to the general safety provisions of the FD&C Act, and have rarely been invoked.

Food and color additives

The regulation of food and color additives is especially complex. The FD&C Act contains both the general safety provisions and the Delaney Cancer Clause restriction on carcinogens passed in 1958 for food additives (Section 409) and in 1960 (Section 706) for color additives. A large number of substances are exempt from the Delaney provisions, including many common substances such as salt and sugar that are "generally recognized as safe" (GRAS). In addition, substances [e.g. nitrites, butylated hydroxyanisol (BHA)] that were granted a sanction prior to the enactment of the food additive amendments in 1958 are exempt (*prior sanctioned*). Furthermore, the FDA has resolved situations that have arisen under the Delaney provisions, including the problem of carcinogenic constituents of additives if the additive itself is not carcinogenic (*constituents policy*), which would apply to migrants from food packaging and trace impurities of additives that are themselves carcinogenic. Other exemptions from the Delaney Cancer Clauses, discussed in more detail below, are animal drugs with "no residue" in food-producing animals (*DES proviso*) and pesticide residues on raw agricultural commodities that are regulated by the EPA under the FIFRA and the Pesticide Additive Amendments (section 408).

Other approaches to the regulation of additives under the Delaney Cancer Clauses have included an unsuccessful attempt for a *de minimus* exemption. The concept of *de minimus* in a legal context means that some matters (in this case risks) are so trivial that the law is not applicable. The Commissioner of the FDA concluded that although two color additives (red 19 and orange 17) "induced" cancer the calculated risks using quantitative risk assessment methodology were trivial (less than one additional cancer in the lifetime of 1 million individuals) and, therefore, the additives were exempt from the Delaney provisions. The Court, however, ruled that this exemption was contrary to law, stating that Congress had intended to be extraordinarily strict and that the plain language of the statute prohibited a *de minimus* exception for a substance found to "induce" cancer (15).

Another approach that has been used successfully has been the secondary mechanism approach, which, as mentioned above, applies to chemicals that alter the tumor incidence in the rodent bioassay secondary to a physiologic disturbance or target organ toxicity as opposed to a direct carcinogenic effect of the chemical (16). In this approach the chain of causation is the legal and scientific issue where the chemical produces a change in the test animal and this change, or secondary factor, in turn alters the tumor incidence. From this perspective the chemical *per se* does not "induce" cancer. This approach, with some qualifications, has been successfully applied to melamine, a metabolite of an insecticide that produces bladder tumors secondary to bladder calculi, and for selenium and beverage alcohol, both of which produce liver tumors secondary to liver damage. Although the FDA stated that this was Agency policy during Congressional hearings in 1974, the secondary mechanism exclusion has not been used since even though there was general scientific and legal support for this view at the time. This issue is currently under consideration for several food and color additives. The secondary mechanism approach involves the use of scientific judgment in the interpretation of the results of the animal carcinogenicity study, and in this context the Delaney Cancer Clauses are probably as misunderstood as they are controversial. It is important to note that the Clauses are not self-executing provisions of the amendments and that they specifically apply to a scientific *decision* made by the Secretary or Commissioner of the FDA that a substance "induces" cancer. The provisions do not apply, as some would maintain, to the result of a rodent carcinogenicity study or to every substance that produces a treatment-related tumor response under any condition. The legislative history provides that full scientific judgment and discretion may be used in the decision of whether or not a substance "induces" cancer or whether a test is "appropriate." The law is silent on how this decision is made and what types of scientific data may be considered (17,18). Thus the Delaney provisions do not preclude the use of a secondary mechanism in a decision that a substance does not induce cancer and that such an approach is within the scientific discretion of the Agency (18).

Animal drugs

The last of the Delaney amendments, passed in 1968, prohibits the use of carcinogenic animal drugs in food-producing animals; however, this restriction is qualified by what is called the *DES proviso* to exempt animal drugs with "no residue." This Congressional exemption came about because of the use of anabolic steroids, including diethylstilbestrol (DES), as growth promoters in food-producing animals. Although DES and other anabolic steroids were found to be carcinogenic in animals it seemed reasonable to exempt an animal drug as long as no residue was present.

The "no residue" exemption of the DES proviso was acceptable until analytic methodology increased in sensitivity such that extremely small traces of drugs could be measured, which necessitated a redefinition of the "no residue" provision. The FDA reasoned that the "no residue" provision was fulfilled if an analytical procedure was available that could measure a residue level that was considered to represent an "insignificant risk." This policy, referred to as the *sensitivity of the method* (SOM) policy, was established in 1973 and was the advent of the widely used statistical procedures for quantitative risk assessment (QRA) (18a). The "insignificant risk" was then defined as a residue level that would represent a risk of less than one additional case of cancer in 1 million lifetimes ($<1 \times 10^{-6}$ lifetime risk). These procedures assume a linear dose response from high to low doses and the absence of a threshold. The upper 95% confidence limits of the estimate are used to establish risk (19,20).

The use of QRA has expanded since the mid-1970s and in addition to its application to animal drugs is now used to set standards for food contaminants (e.g. aflatoxin in corn and peanuts) and by the EPA to establish tolerances for pesticide residues on raw agricultural products as well as to set standards for contaminants in water and air under the Clean Water and Clean Air Acts. Although QRA is an important regulatory tool, it is important to recognize that it represents a statistical estimate of risk and that there is no basis in biology for the validity of these estimates. The statistical procedures incorporate certain biologic assumptions on the nature of the dose–response curve and sensitive species, however, these assumptions are used predominately to make the risk estimates "conservative" as opposed to enhancing the biologic validity of the estimates. The estimates are mathematically precise but not necessarily biologically relevant. The estimates are often misused or misperceived as representing the actual number of excess cancer cases that would occur in the exposed population. QRA has dominated the area of cancer risk assessment over the past two decades, often at the expense of more relevant biologic considerations, however, in recent years both the EPA and the FDA have made significant efforts to incorporate more scientific information on mechanism of action and pharmacokinetics into the cancer risk assessment process.

In the case of animal drugs, the FDA issued rules in 1987 (11) for implementing the SOM policy under the DES proviso in which they have removed the specific QRA procedure from regulations and have placed these into guidelines. The linear response from the high-to-low-dose multistage model with the 1×10^{-6} lifetime risk benchmark will continue to be used as the default method (20); however, in accord with the recommendation of the 1983 NAS risk assessment document (8) and the 1985 OSTP guidelines (9), the FDA will consider scientific data on mechanism of action. The guidelines contain a waiver provision under which a sponsor may submit scientific data and propose an alternate procedure to estimate risk [e.g. no observed effect level (NOEL) and safety factors] and therefore establish safe residue levels for an animal drug.

Pesticides

Pesticide residues come under the jurisdiction of both the FDA and the EPA. Pesticides are regulated under FIFRA and the Pesticide Chemical Amendments of 1954; the EPA is responsible for the registration of pesticides and the establishment of tolerances for pesticide residues in raw agricultural commodities, and the FDA is responsible for monitoring. The Pesticide Chemical Amendment is in Section 408 of the FD&C Act, which, unlike Section 409, does not contain a cancer clause; thus, pesticides can be regulated on a benefit-risk basis. Although the Delaney Cancer Clause does not explicitly include pesticides, it has been regulatory practice to consider pesticide residues as food additives if they concentrate in processed foods *and* exceed the tolerance established for the raw commodity. Under coordination policy, a 409 tolerance would be required after which the Delaney Cancer Clause of Section 409 would apply to pesticides. The dual jurisdiction and situation where pesticide tolerances are legal under FIFRA and Section 408 of the FD&C Act but potentially illegal under Section 409 is referred to as the "Delaney paradox" for the regulation of pesticides. Currently a number of pesticide registrations are in jeopardy under the Delaney Cancer Clause and the EPA has so far been unsuccessful in a *de minimus* approach in the courts. The *de minimus* issue with respect to pesticides differs from that with respect to the food and color additives in that Congress did not explicitly include pesticides in the Delaney Cancer Clause and they are only indirectly covered under certain circumstances as a food additive. The situation with pesticide residues is similar to that of residues of animal drugs in food-producing animals and it is likely that the "no residue" exemption for animal drugs under the DES proviso would have been extended to pesticide residues if this had been an issue at the time and an approach similar to that used for animal drug residues would be appropriate. In any event, the regulatory conundrum for pesticide residues will require either legal clarification of the applicability of the Delaney Cancer Clause to pesticide residues or legislative remediation.

Quantitative risk assessment procedures have been widely used by the EPA to establish tolerances for pesticide residues on raw agricultural commodities. As discussed above, for the regulation of animal drugs, the procedures assume a linear response from high to low doses and the absence of a threshold. The upper 95% bound confidence limits are used and calculated on the basis of 1×10^{-6} lifetime risk for cancer. The 1986 EPA guidelines for the evaluation of carcinogens allow the evaluation of additional scientific information on mechanisms, and recently an EPA Science Policy document for the evaluation of compounds that produce thyroid gland neoplasia secondary to hormone imbalance has been discussed (21,22,60). In addition, a Science Policy document for the evaluation of compounds that produce renal neoplasia in male rats secondary to α-2μ globulin induced nephropathy issued (23). These policy statements provide scientific criteria and conditions under which QRA procedures would not be invoked and pesticide tolerances could be established on the basis of a NOEL and safety factor in a manner similar to other toxic effects.

Human drugs

Human pharmaceutical agents are approved on the basis of an evaluation of the benefits of therapy vs. potential risks, including findings from rodent carcinogenicity studies. Since it has been estimated that 25–50% of all chemicals tested at the maximum tolerated dose (MTD) will be positive in one sex and species, positive tumor findings are not uncommon in drug development. Pharmaceuticals are usually screened for mutagenicity or DNA damage and with a few exceptions, obviously genotoxic chemicals would not be developed. Thus for compounds positive in rodent carcinogenicity tests, the mode of action will be predominately non-genotoxic and involve effects secondary to an exaggerated pharmacologic effect, target organ toxicity, or physiologic disturbances. Drugs are not approved on the basis of mechanism, however, mechanism information impacts the risk side of the benefit-risk evaluation and many pharmaceutical agents have been approved with rodent tumor findings. Regulatory agencies in several countries have formally incorporated the consideration of mechanism into the drug evaluation process. For example, guidelines for the evaluation of carcinogenic findings were issued by the Japanese Ministry of Health and Welfare (12) and by the Canadian Health Protection Branch (13). The Japanese guidelines state that "unless appropriate scientific evidence can be provided, the premise that a substance that has proven carcinogenic to animals has the potential of exerting the same effect in man must be taken.... However, one should not hastily regard a substance as having the risk of inducing malignant tumors in humans solely on the ground that the substance has caused the development of malignant tumors in animals." Both the Canadian and Japanese guidelines recommend supplemental studies to investigate the mechanism of the tumor response, which would include short-term studies to evaluate the genotoxic potential of the compound, additional studies to determine the mechanism for tumor formation in rodents, and studies to determine whether the proposed mechanism is operative under conditions of human exposure. All the scientific information would be evaluated to assess the potential risk vs. benefit in the approval of the pharmaceutical agent.

MECHANISTIC CONSIDERATIONS

The consideration of mode or mechanism of action information for chemicals that produce a tumor response in the rodent bioassay, as discussed above, is playing an increasingly important role in the regulation and classification of chemical carcinogens. Although this is not an easy task, there are many examples in which mechanisms have been elucidated to an extent sufficient for regulatory action.

In a survey conducted by the Pharmaceutical Manufacturers Association based on experience with over 400 bioassays, it was determined that the most frequent target sites for treatment-related tumor responses were the endocrine organs in rats and the liver in strains of mice exhibiting a high spontaneous hepatic tumor incidence. The various endocrine organs involved in rat studies were the thyroid, pituitary, adrenal, and mammary glands, testes, uterus, and ovary. Other target organs in rats were less frequently involved but included the gastrointestinal tract, urinary bladder, and pancreas. In mice, the liver was by far the most common target organ; however, the ovary and lung were involved to a lesser extent.

Since pharmaceutical agents are usually screened for mutagenic potential prior to development, the majority of the positive responses in the carcinogenicity studies would appear to be by a nongenotoxic mode of action. The following discussion of mechanism is organized according to target organs, with an emphasis on endocrine, gastrointestinal tract, urinary tract, and hepatic neoplasia. A few examples of chemical specific mechanisms are discussed for each of the organ systems.

Endocrine neoplasia

Thyroid gland

Thyroid follicular cell neoplasia is perhaps the most common drug-induced endocrine tumor observed in rat carcinogenicity studies. There are two basic mechanisms whereby chemicals produce thyroid gland neoplasia in rodents, the first of which involves chemicals that exert a direct carcinogenic effect on the thyroid gland and the other involves chemicals that, through a variety of mechanisms, disrupt thyroid gland function and produce thyroid gland neoplasia secondary to hormone imbalance. A consistent mechanism, widely accepted by many investigators, to explain the pathogenesis of thyroid tumors induced in rats treated with antithyroid drugs has been described (24,25). Antithyroid compounds initially produce a hormonal imbalance by interfering with thyroid hormone production. A sustained increase in the synthesis and secretion of thyroid stimulating hormone (TSH) occurs via the negative feedback system of the pituitary gland to stimulate thyroid function. Increased TSH stimulation produces a variety of morphologic and functional changes in the follicular cell, including follicular cell hypertrophy, hyperplasia, and ultimately neoplasia. That excessive secretion of endogenous TSH alone (in the absence of any chemical treatment) will produce a high incidence of thyroid tumors has been clearly established by experiments in which rats were fed diets deficient in iodine (26–29) or in which TSH-secreting pituitary tumors were transplanted into mice with normal thyroids (30).

Species differences

There are marked species differences in thyroid gland physiology that must be taken into account in an evaluation of the induction of thyroid gland neoplasia secondary to hormonal imbalance. The most obvious species difference between rodents and primates is the lack in the rodent and in some other species of thyroid binding globulin (TBG), which is the predominant plasma protein that binds and transports thyroid hormone in the blood (31). The absence of TBG may be one of the more important factors responsible for the much shorter half-life of T_4 (thyroxine) in the rat (12 hours vs. 5–9 days in humans) and the much higher level of TSH (31). This indicates a much higher activity in the rodent thyroid gland as compared to the primate. The relative susceptibility of rodents and humans to thyroid neoplasia secondary to hormonal imbalance or simple hypothyroidism can be assessed by comparing humans in iodine-deficient areas of endemic goiter to rats in these same areas or to rats treated with iodine-deficient diets. Relatively extensive epidemiologic studies in areas of endemic goiter have not shown a clear etiologic role in thyroid gland neoplasia (32–34). In contrast, rodents in areas of endemic goiter (35) or those fed iodine-deficient diets (26–28,36) exhibit a high incidence of thyroid gland neoplasia.

Mechanisms for altered thyroid function

Thyroid hormone synthesis, release, transport, cellular uptake, conversion of T_4 to T_3 (triiodothyronine), hormone metabolism, and the regulation of these processes by the hypothalamic-pituitary-thyroid axis and autoregulatory processes in the thyroid gland itself are complex processes that provide many ways in which chemicals can interfere with thyroid gland function via either intra-or extrathyroidal mechanisms. Intrathyroidal mechanisms can involve iodine uptake or hormone synthesis and extrathyroidal mechanisms can involve effects on hormone metabolism or disposition. A few important examples are discussed below.

HORMONE SYNTHESIS

Two classes of chemicals known to inhibit thyroid hormone synthesis are the thiourelyenes (thiourea, propylthiouracil, methimazole) and the sulfonamides (sulfadiazine, sulfamethazine). The thiourelynes are considerably more potent in inhibiting thyroid hormone synthesis than the sulfonamides, however, sufficiently high doses of many sulfonamides are goitrogenic in the rodent (37–40). Sulfamethoxazole and sulfamethazine produced goitrogenic effects and thyroid neoplasia in rats at high dosages (37,41–44); however, sulfisoxazole, which is only weakly goitrogenic in rodents, did not produce thyroid gland neoplasia (45).

The goitrogenic effect of the sulfonamides is known to be highly species specific. Some species, including rats, mice, hamsters, dogs, and swine, are sensitive, whereas no goitrogenic effect is observed in chickens, guinea pigs, or primates (37,39,40). In humans, no clinically significant effects on thyroid function have been observed with sulfonamides at therapeutic doses (46–49). Takayama and coworkers (40) studied the species difference between rats and monkeys using sulfamonomethoxine (SMM). SMM was markedly goitrogenic in rats but produced no effects on thyroid gland function or morphology in monkeys. In vitro studies showed that SMM was a potent inhibitor of microsomal thyroid peroxidase in rats but not in monkeys, which explains why sulfonamides are goitrogenic in rodents at relatively low dosages but do not produce effects in monkeys at very high dosages or in humans at therapeutic dosages.

MONODEIODINASES

The conversion of T_4 into the more active hormone T_3 occurs in many peripheral tissues and is mediated by a microsomal $5'$-monodeiodinase that removes one iodine at the $5'$ position in T_4 (50). Various iodinated organic compounds such as tetraiodofluoresceine (51), amiodarone (52) and various iodinated radiocontrast media (53) will inhibit the $5'$-monodeiodinase and disrupt the conversion of T_4 to T_3. The decrease in serum T_3 values results in a compensatory increase in pituitary TSH. Prolonged treatment with tetraiodofluoresceine at very high dosages will result in a moderate increase in thyroid follicular neoplasia in rodents (54).

THYROID HORMONE METABOLISM

The effect of chemicals on various aspects of thyroid hormone metabolism have an important impact on thyroid hormone economy in the rodent (55). The monodeiodinases are quantitatively the most important path in the disposition of T_4. In addition, T_4 is glucuronidated and T_3 is sulfated and subsequently excreted in bile. Deamination, decarboxylation, and cleavage of the ether link occur but are quantitatively of lesser importance (50). Many chemicals are hepatic microsomal enzyme inducers at high dosages and alter thyroid function in rodents by increasing the hepatic disposition of thyroid hormone (56–60). Decreased serum thyroid hormone production due to increased hepatic disposition results in a compensatory increase in pituitary TSH, which can exert a tumor-promoting effect in initiation-promotion models (61) or an increase in thyroid gland neoplasia in 2-year carcinogenicity studies (62). Studies have shown that small amounts of T_4 block the tumor-promoting effect of a microsomal enzyme inducer such as phenobarbital, thus, this effect and presumably those observed in 2-year studies are secondary to hormonal imbalance as opposed to a direct tumor-promoting or direct carcinogenic effect in the thyroid gland (59,62,63).

Clinical significance of extrathyroidal mechanisms

With respect to the chemical mentioned, tetraiodofluoresceine produced only mild effects on thyroid function in humans at very large multiples of the allowable daily intake (64). Chronic exposure to anticonvulsant drugs, many of which are enzyme inducing at therapeutic dosages, results in only mild changes in thyroid function [moderately decreased T_4 with normal T_3 and TSH values and a normal TSH response to administered thyrotropin releasing hormone (TRH)] (65). These changes are not clinically significant. There would appear to be little if any risk for apparently nongenotoxic chemicals that act secondary to hormonal imbalance at exposure levels that do not disrupt thyroid function. Further, the degree of thyroid dysfunction produced by a chemical would present a major toxicologic problem before such exposure would increase the risk neoplasia for humans.

Adrenal gland

The adrenal gland is composed of the cortex, which accounts for approximately 90% by weight, and the medulla, which makes up the remaining 10%. The adrenal cortex produces a number of steroid hormones, including mineralocorticoids and corticosteroids, the latter of which are under the control of pituitary adrenocorticotropic hormone (ACTH). Treatment-related tumors of the cortex are rarely observed; however, spontaneous proliferative lesions of the medulla are common in many strains of aging rats and the medulla is a relatively frequent site for tumor induction in rat carcinogenicity studies (66,67). The adrenal medulla is a neuroendocrine organ composed of chromaffin cells, which synthesize, store, and secrete catecholamines in response to cholinergic and peptide neurotransmitters. Tumors of the chromaffin cells are referred to as *pheochromocytomas*. The incidence of spontaneous proliferative lesions of the medulla is dependent on a variety of endogenous and exogenous factors and can vary considerably with respect to strain age, and sex (68). Exogenous factors include chemicals that affect the hypothalamic-endocrine axis, the autonomic nervous system, or calcium absorption and homeostasis (66,67). Substances that affect the hypothalamic-endocrine axis include growth hormone and estrogen (69,70), antithyroid drugs (71,72), and neuroleptic agents. Nicotine (73) and reserpine (74) have effects on the autonomic nervous system. Calcium absorption and homeostasis are affected by sugars and sugar alcohols (75), nonsteroidal anti-inflammatory agents (76), retinoids (77), and vitamin D.

Reserpine is a neuroleptic drug that produced adrenal medullary hyperplasia and neoplasia (pheochromocytomas) in a 2-year rat study (74). Reserpine is know to neurogenically mediated reflexive stimulation of catecholamine synthesis (78). Concomitantly, this stimulation produces an increase in chromaffin cell prolifer-ation at doses used in the rat carcinogenicity study (79,80). In reserpine-treated rats with a unilateral denervation of the adrenal gland, an increase in cell proliferation is observed in the intact gland; however, this effect is completely abrogated in the denervated gland (80). These results indicate that the effects of reserpine on cell proliferation are neurogenically mediated, secondary to catecholamine depletion, and support and conclusion that the tumors result from a sustained increase in chromaffin cell proliferation.

Nutritional factors have an important modulating effect on the spontaneous incidence of adrenal medullary proliferative lesions. Several sugars and sugar alcohols have produced adrenal medullary tumors at high dosages, including xylitol, sorbitol, lacitol, and lactose at concentrations of 10–20% in the diet (75,81). Although the mechanism involved is not completely understood, a role for calcium has been postulated. High dosages of slowly absorbed sugars and starches increase the absorption and urinary excretion of calcium. Hypercalcemia is known to increase catecholamine synthesis in response to stress (82) and low-calcium diets will reduce the incidence of adrenal medullary tumors in xylitol-treated rats (75). Other compounds that might act via altered calcium homeostasis include the retinoids, which will produce hypercalcemia and conditions such as progressive nephrocalcinosis in aging male rats treated with nonsteroidal anti-inflammatory agents (76).

In contrast to the high and variable spontaneous incidence of adrenal medullary tumors in various strains of rat, pheochromocytomas are very rare tumors in humans (83). In addition, the tumors observed in humans are functionally different from those observed in rodents in that human tumors contain large amounts of catecholamines, whereas the tumors in rats are mostly negative for stored catecholamines (68). Many strains of rat appear to have a genetic predisposition for adrenal medullary proliferative lesions, (67,68) and the type of response observed in rats is not observed in humans. The mouse, which has a much lower incidence of spontaneous lesions, does not usually show an increase in adrenal tumors under the same conditions under which positive findings are observed in the rat. There is no evidence that compounds that act secondary to exaggerated pharmacologic effects, altered nutritional status, or physiologic disturbances would present a risk for humans, and all of the mechanisms discussed above would be expected to exhibit a threshold.

Mammary gland

The female rat has a relatively high incidence of spontaneous mammary tumors (40% to 60%), which over the last 20 years has increased substantially along with an increase in mean body weight and food consumption over this period.

Prolactin

Several compounds have caused a treatment-related increase in mammary tumors secondary to hyperprolactinemia. Elevated prolactin stimulates tubuloalveolar growth in the mammary gland, producing mammary tumors in male and female rats. The pituitary secretion of prolactin is under the inhibitory control of dopamine released from the hypothalamus, the removal of which will result in persistent hyperprolactinemia (84). Neuroleptic drugs such as haloperidol and reserpine inhibit the synthesis and release of dopamine from the hypothalamus, resulting in hyperprolactinemia (85). In 2-year rat carcinogenicity studies, resperpine increased the incidence of mammary tumors at dosages that increased serum prolactin (74). Prolactin-mediated effects on the mammary gland can be prevented by hypophysec-tomy and interestingly bromocriptine, which is a dopamine agonist, will reduce serum prolactin in rats, resulting in a marked decrease in both spontaneous pituitary and mammary tumors (86). Estrogens are also known to decrease dopamine secre-tion in the hypothalamus, resulting in an increase in serum prolactin and an increase in mammary tumors and pituitary tumors of the prolactin-secreting cells.

Neuroleptic agents will increase serum prolactin levels in humans (87,88). Pro-lactin is not considered to be an etiologic factor in human breast cancer and human epidemiologic studies with reserpine are considered to be negative with respect to increased risk for breast cancer (88,89).

Progestins

Progestins produce mammary tumors in beagle dogs, an observation that caused major problems in the development of oral contraceptive agents. In the dog this effect is mediated by an increase in the pituitary secretion of growth hormone, which produces a marked increase in tubuloalveolar growth and eventually mam-mary neoplasia (90). These events can be prevented by hypophysectomy and appear to be species specific since in most other species tubuloalveolar growth is controlled by prolactin (90a). No increase in serum growth hormone is observed in humans treated with oral contraceptives (91).

Ovary

Ovarian neoplasia secondary to hormonal imbalance has been observed primarily in mice. The cyclical control of ovarian estrogen production is under negative feedback control, whereby the hypothalamus secretes gonadotropin releasing hormone (GRH), which stimulates the pituitary secretion of two gonadotropins, follicle stimulating hormone (FSH) and luteinizing hormone (LH). One of the classic experiments in hormonally mediated neoplasia involved the transplantation of the ovary to the spleen of mice. Under these conditions ovarian hormones are secreted directly to the liver, where they are completely metabolized, and in the absence of circulating hormone, large amounts of gonadotropin are released from the pituitary, resulting in ovarian neoplasia (92). Ovarian adenomas are also noted by 6 months of age in genetically sterile mice, which would also exhibit increased gonadotropin (93).

The administration of nitrofurantoin at 0.25% in the diet of B6C3F1 mice produce an increase in benign tubular adenomas. At this dose treated mice were sterile, having nonfunctioning atrophic ovaries, and under these conditions pituitary FSH and LH would be increased, resulting in ovarian neoplasia. Nitrofurantoin does not produce ovarian neoplasia in rats or other strains of mice in which ovarian atrophy was not produced, thus these tumors are considered to be secondary to hormonal imbalance due to ovarian atrophy as opposed to a direct carcinogenic effect of the compound (94).

Uterus

The endometrium is not a common site for tumor induction; however, bromocriptine, a dopamine agonist, will produce an increase in endometrial tumors in rats (95). As mentioned above, bromocriptine, as a dopamine agonist, will decrease the pituitary secretion of prolactin, resulting in a marked decrease in spontaneous pituitary and mammary tumors. In addition to the effect on the mammary gland, prolactin has a luteotrophic effect in rats, and leutolysis occurs as a result of decreased prolactin. This produces a state of permanent estrus, with an increase in new follicles and estrogen levels and stimulation of the endometrium. Since prolactin does not have a luteotrophic effect in humans, this mechanism would be of no relevance (90a).

Uterine leiomyomas have been produced by β-adrenergic agonists such as salbutamol, terbutaline, and ritodrine (96). This effect is receptor mediated and is an extension of their pharmacologic effect since propranolol, a β-adrenergic blocker, completely inhibits tumor induction with salbutamol (96). Furthermore, the tumor potency of salbutamol, terbutaline, and ritodrine correlates with their pharmacologic potency (97).

Testes

The spontaneous incidence of testicular tumors varies widely among strains of rat. Sprague Dawley rats have a relatively low incidence of testicular tumors, whereas the Fischer 344 rat will have a 100% incidence of testicular tumors by 2 years of age. The testis consists primarily of germ cells involved in spermatogenesis: the Sertoli cells, which serve a nutritional and support role, and the interstitial or Leydig cells, which produce testosterone. The hypothalamus produces LH releasing hormone (LHRH), which stimulates the release of LH, which in turn stimulates the Leydig cell to produce testosterone. The system is under negative feedback control, so that a decrease in testosterone will cause an increase in the plasma level of LH and stimulation of the Leydig cell. Leydig cell hyperplasia and neoplasia are observed in a variety of experimental conditions that produce a sustained increase in pituitary LH, including the administration of gonadotropins (98). Leydig cell tumors are also produced in intact rats parabiosed to castrated rats via increased LH secretion from the castrated rat (99).

Leydig cell tumors are observed with several drugs or chemicals secondary to increased serum levels of gonadotropins, such as various antiandrogen drugs used in the treatment of prostate cancer. In rats, the testosterone antagonist will produce an increase in LHRH and LH, resulting in Leydig cell hyperplasia and eventually neoplasia. Linuron, used as a selective herbicide for weed control on various crops, produced a dose–response-related increase in Leydig cell adenomas in carcinogenicity studies in rats. The compound is structurally related to flutamide, an antiandrogen, and produces an increase in serum estradiol and LH in male rats (100). Leydig cell tumors in the presence of a sustained elevation of LH have also been observed with ammonium perflurooctanoate (101) and a calcium channel blocker (102). In each of these cases the tumor response is considered secondary to hormonal imbalance and not a direct effect of the chemical.

Pancreas

The exocrine pancreas is not a very common target organ for drugs and chemicals, however, there are some interesting examples of nongenotoxic mechanisms involving corn oil (unsaturated fat) and natural trypsin inhibitors.

Unsaturated fat

Corn oil is frequently used as a vehicle for the administration of insoluble substances in rodent carcinogenicity studies. In several NTP studies using corn oil gavage, it was observed that there was a relatively high incidence of pancreatic acinar carcinomas in F344 rats as compared to untreated control rats (103). These observations led to a 2-year bioassay with corn oil in which a dose–response-related increase in acinar carcinoma was observed. The mechanism of this effect is not known; however, it is considered to be nongenotoxic since unsaturated fat acts as a tumor promoter after the administration of a direct-acting pancreatic carcinogen such as azaserine (104).

Trypsin inhibitors

Natural and synthetic trypsin inhibitors interfere with the normal process of digestion and result in an enlarged pancreas and neoplasia in long-term studies.

Raw soya flour from soybeans contains trypsin inhibitors and produces enlargement of the pancreas, with acinar cell hypertrophy and hyperplasia and adenomatous nodules and carcinomas in 2-year rat studies (105). These effects are considered to be due to cholecystokinin (CCK). Trypsin inhibition leads to a decrease in the negative feedback of CCK secretion by the pancreas, resulting in an elevation of serum CCK (106–108). CCK will increase DNA synthesis and produce acinar cell hypertrophy and hyperplasia. Aging rats are known to have a very high incidence (85%) of atypical acinar cell foci (109) and trypsin inhibitors increase acinar cell nodules, adenomas, and carcinomas via increased CCK (105). These effects are reversible up to the point of foci and nodules, but are not reversible after adenomas and carcinomas are formed. Rats appear to be more sensitive to these effects than hamsters and mice (110). In humans trypsin inhibitors will increase pancreatic secretions but there is no observed increase in serum CCK (111).

Corticosteroids

Several corticosteroids, including prednisolone, will increase pancreatic neoplasia in the rat. The mechanism for this effect is not known.

Gastrointestinal tract neoplasia

The gastrointestinal tract is not a very common site for the production of treatment-related tumors, however, there are two rather well-studied nongenotoxic mechanisms for the induction of gastric tumors in rats. The first involves tumors of the glandular stomach secondary to increased gastrin secretion in response to prolonged achlorhydria, and the second involves neoplasia of the rat forestomach secondary to chemically induced necrosis and regenerative hyperplasia at this site.

Glandular stomach

Gastric acid secretion by the parietal cells is under the control of gastrin, vagal stimulation, and histamine. Histamine stimulates gastric acid secretion through a receptor-mediated mechanism that is not blocked by classic antihistamine drugs. This receptor, designated the H_2 receptor, can be blocked by analogs developed and used clinically to reduce acid secretion for the treatment of gastric ulcers (112). Another compound, omeprazole, reduces acid secretion by a different mechanism involving an inhibition of H^+K^+-ATPase, an enzyme responsible for the production of hydrogen ions (proton pump inhibition) (113,121). In 2-year carcinogenicity studies, the long-acting H_2 antagonists [(e.g. loxtidine (114), SKF09479 (115), ICI 162846 (116), cimetidine and ranitidine (117), and the proton pump inhibitor (omeprazole) (118)] produced carcinoids referred to as "enterochromafin-like" (ECL) neoplasia in the glandular mucosa of the stomach. The ECL cells are endocrine cells under the control of gastrin. Gastrin is regulated by pH, where low pH will suppress the secretion of gastrin (119). High dosages of the long-acting H_2 antagonists or omeprazole inhibit acid secretion to the point of complete achlorhydria. Achlorhydria activates the antral gastrin cells to produce a state of hypergastrinemia, which has a trophic effect on the ECL cells (120). ECL hyperplasia is observed within a few weeks and ultimately ECL neoplasia is observed near the end of a 2-year carcinogenicity study.

A similar ECL tumor response can be produced by surgical procedures in the absence of chemical treatment (122,123). Rats have a higher ECL density and sensitivity to gastrin as compared to mice and are more susceptible to tumor induction. Although ECL neoplasia can occur in humans in conditions such as Zollinger–Ellison syndrome, characterized by sustained high levels of gastrin, omeprazole treatment did not result in a further increase in gastrin (124). Therapeutic dosages of omeprazole produce only moderate increases in serum gastrin in ulcer patients with normal gastrin levels with no observed change in endocrine cell densities (124). Thus, there would be little indication that an increased cancer risk would exist under clinical conditions.

Forestomach

There are several compounds, most notably BHA, that produce in the forestomach of rats and mice tumors associated with a sustained high level of cell proliferation at the site at which tumors are produced. BHA, an antioxidant and widely used food additive, produced papillomas and carcinomas of the forestomach at 2% in the diet, while a concentration of 0.5% was not effective (125). The oral administration of BHA produces necrosis in the forestomach, and the tumors at this site are considered to be secondary to necrosis and regenerative hyperplasia (126). In short-term studies, no histologic lesions are observed in species that do not have a forestomach, including dogs, pigs, and monkeys (127–129). The forestomach of the rodent is a modification of the esophagus for which there is no anatomic counterpart in the primate, which raises further doubts about the relevance to humans of tumors that are limited to this site.

Urinary tract neoplasia

Kidney

Spontaneous renal and bladder neoplasias are relatively rare in rats, however, a number of drugs and chemicals will produce renal tumors in male rats secondary to renal damage. One of the most extensively studied mechanisms involves renal tumors in male rats secondary to α-2μ globulin (α-2μG) nephropathy. α-2μG is a low-molecular-weight protein that is synthesized in the hepatocyte under androgen control (130). The protein is filtered by the glomerulus, partially reabsorbed by the tubules, and excreted in the urine in large amounts (131). A number of chemicals (e.g., trimethylpentane, decalin, fuels, and d-limone) can bind to α-2μG, which, when partially reabsorbed by the renal tubule, is resistant to digestion by lysosomal proteases and thus accumulates in the tubule (132,133). The chemical–protein complex is cytotoxic, causing single-cell necrosis and regenerative hyperplasia, which chronically produces renal tubular calcification and neoplasia. The α-2μG nephropathy and renal tumors are observed only in male rats. Female rats, mice of either sex, guinea pigs, and monkeys do not exhibit the nephropathy (134,135). Furthermore, the NBR rat strain, which does not synthesize the androgen-dependent form of α-2μG, does not exhibit the nephropathy, the sustained cell proliferation, or renal tumors (136,137). Although humans excrete proteins with some homology to α-2μG, a nephropathy analogous to that observed in male rats does not occur (138), thus the renal tumors secondary to α-2μG appear to be species specific and not relevant to human risk.

The chelating agent, sodium nitrilotriacetic acid (NTA), also produces renal tubular tumors in rats at high dosages (139,140). NTA, by virtue of its chelating activity, causes the excretion and systemic redistribution of large amounts of zinc. The increased tubular reabsorption of zinc is toxic to renal tubular cells, resulting in single-cell necrosis and regenerative hyperplasia leading to renal tubular neoplasia (141). This effect appears to be secondary to high-dose renal tubular toxicity of a type that would be expected to exhibit a threshold and would not present a risk at low levels of exposure.

Bladder

Nongenotoxic chemicals that produce bladder neoplasia in rats and mice do so via a sustained increase in cell proliferation (142). The most common mechanism involves conditions that lead to calculi formation in the urinary bladder, resulting in erosion and ulceration with regenerative hyperplasia. This effect can be produced with any solid pellet, including those consisting of wax, cholesterol, glass, and stainless steel (143). The fact that tumors can be induced with chemically inert pellets demonstrates that tumors can arise from mechanical damage to the urothelium independent of a direct chemical interaction. Calculi formed under conditions of chemical treatment and will elicit the same response. Substances that have produced tumors under these conditions include melamine, oxalates, uracil, xylitol, and ethylene glycol.

Melamine is a metabolite of an insecticide used in animal feed. Melamine produced bladder tumors in rats in the presence of calculi and since it is used in feed for food-producing animals was potentially subject to the Delaney Cancer Clause. It was concluded, however, that the tumors were caused by the calculi and not melamine *per se*, and since melamine did not "induce" cancer the Delaney Cancer Clause was not applicable (FDA, 1974) (16).

Xylitol is a natural five-carbon sugar alcohol that is produced in mammalian intermediary metabolism (144). Xylitol is used as a nutritive sweetener and since it does not require insulin for metabolism is useful for diabetics. Xylitol at 20% in the diet produced bladder neoplasia in mice in the presence of calculi in a 2-year study. Xylitol is metabolized to oxalic acid by a minor pathway that is more prominent in the mouse (145–147), resulting in an increase in urinary oxalate, and eventually to urinary tract calculi. There is no significant increase in urinary oxalate in the rat and no urinary tract calculi or tumors were produced under conditions similar to those of the 2-year mouse study. In addition, consumption of xylitol in humans did not cause a clinically significant increase in urinary oxylate excretion (148).

Chemicals that produce bladder neoplasia secondary to bladder stones usually do so at relatively high dosages. This mechanism would not be applicable to lower dosages that do not produce calculi, which would constitute a true threshold. In addition, there is little evidence indicating that the presence of calculi leads to an increased incidence of bladder tumors in humans (149); however, even if this were not true, the presence of bladder calculi with a drug or food additive under human exposure conditions represents a self-limiting toxicity that would preclude its use.

Bladder neoplasia has also been observed in male rats treated with high dosages of sodium salts of weak acids such as sodium saccharin. Although grossly observed calculi are not found, sodium saccharin will form silicate microcrystals in the bladder that will precipitate above pH 6.5 in the presence of large amounts of protein. Sodium saccharin produced bladder neoplasia in male rats fed at 5% in the diet. The high concentration of the α-2μG protein present in the urine of the male rats appears to be involved in a situation similar to the α-2μG nephropathy in male rats discussed above. The presence of the microcrystals produces a sustained increase in cell proliferation under these conditions. Interestingly, this appears to be a nonspecific effect of the excretion of large amounts of sodium and is observed with sodium ascorbate, erythorbate, citrate, glutamate, aspartate, bicarbonate, and chloride (142,150). Bladder neoplasia will be observed with sodium saccharin and sodium ascorbate, for example, but not with the free acid or calcium salts of these substances. Since α-2μG is involved in this nonspecific effect of sodium, the response would be expected to exhibit the same male rat only species/sex specificity as the renal neoplasia secondary to α-2μG nephropathy. Humans do not appear to be responsive to this effect, and in any event this is a process that would be expected to exhibit a threshold.

Hepatic neoplasia

The liver is, by far, the most common target tissue affected in the rodent bioassay, which may be for two reasons. First, the liver is the major site of metabolic activation of carcinogens tested and would thus be exposed to the highest concentration of carcinogens absorbed from the gastrointestinal tract. Second, a large number of apparently nongenotoxic compounds act as hepatic tumor promoters and will elicit a hepatic tumor response especially in sensitive strains of mice. The mouse liver tumor response is the most common and controversial endpoint in the rodent bioassay. More than 50% of the compounds tested since the inception of the NCI/NTP bioassay program have produced a tumor response in at least one species, and the mouse liver tumor accounts for approximately half of these response (25% of all chemicals tested) (1). Less than 50% are mutagenic in the Ames *Salmonella* test for mutagenicity (151). The controversy surrounding the relevance of this response as an indicator of human cancer risk is due to the high proportion of involved chemicals that are apparently nongenotoxic combined with the unique susceptibility of many mouse strains (including the CD-1 and B6C3F1 strains) to the development of both spontaneous and chemically induced hepatic tumors, which will be discussed in more detail below. There are four important classes of apparently nongenotoxic chemicals that produce hepatic neoplasia in rodents: the peroxisome proliferators, the necrogenic chlorinated solvents, the microsomal enzyme inducers, and hormones.

Peroxisome proliferators

The peroxisome proliferators represent a diverse group of chemicals, including the hypolipidemic drugs such as clofibrate and ciprofibrate, solvents including trichloroethylene, and various plasticizers such as di-(2-ethylhexyl phthalate). These compounds induce cytoplasmic organelles, referred to as *peroxisomes*, that contain a variety of oxidative enzymes involved in the metabolism of fatty acids. These compounds initially stimulate liver growth, inducing hyperplasia and then hypertrophy (152). Later, there is an accumulation of lipofuscin, which may be indicative of peroxidative damage, and eventually hepatocellular adenomas and carcinomas are produced in both rats and mice (153,154). The effects of the peroxisome proliferators are mediated via a receptor referred to as *peroxisome proliferator activated receptor*, which is a member of the superfamily of steroid receptors (155).

Most of the peroxisome proliferators are generally nongenotoxic in short-term tests (156,157); they do not react with DNA and do not induce unscheduled DNA synthesis, and although they are considered to be hepatic tumor promoters (158,159) the mechanism of carcinogenicity has been debated and is not completely known. Reddy and coworkers have maintained that the mechanism involves oxidative damage to DNA and have demonstrated small increases in the DNA adduct, 8-hydroxydeoxyguanosine to support this hypothesis (160). Others have maintained that the tumor response does not correlate with the amount of induced peroxisomes but does correlate with the level of sustained cell proliferation that the individual agents produce (161). Furthermore, the oxidative DNA adducts may be restricted to mitochondrial DNA since when care is taken to isolate the nucleus before DNA is extracted, no adducts of nuclear DNA are detected (Cattley RC, personal communication, 1993) (162).

The induction of peroxisomes observed in rodents exhibits species specificity in that rats and mice are most responsive, whereas, monkeys or humans are only weakly responsive (163). Human epidemiologic evidence has not indicated an increased human cancer risk (154).

407

Cytotoxic agents

Several halogenated solvents have been shown to produce hepatic tumors in mice, including chloroform, carbon tetrachloride, tetrachloroethylene, and others. Chloroform produces hepatic tumors in male and female mice (164), but only at dosages that produce hepatic necrosis. Chloroform is not obviously genotoxic (165), thus the carcinogenicity of solvents such as chloroform is considered to be secondary to hepatotoxicity and subsequent increase in regenerative hyperplasia (166–168).

Primary liver cancer in humans is very rare in Western countries where the known risk factor is cirrhosis due to hepatitis B or alcohol abuse (169). Human liver cancer under these conditions is considered secondary to regenerative hyperplasia in the cirrhotic liver. Although this situation may be similar to the regenerative hyperplasia in rodents treated with hepatotoxic doses of solvents, there is no evidence that the cytotoxic solvents are involved in human cancer. Cytotoxicity is an effect that would exhibit a threshold and would not be considered to represent a risk at low levels of exposure. In any event, the degree of hepatoxicity required to place either the animal or the human at risk would be a major problem in itself.

Microsomal enzyme inducers

The largest subclass of nongenotoxic hepatocarcinogens is the microsomal enzyme inducers. These substances would include a variety of drugs (e.g. barbiturates), the organochlorine pesticides (e.g. DDT and chlordane), the polyhalogenated biphenyls, and many other compounds (170). Phenobarbital is an extensively studied prototype for a class of hepatic microsomal enzyme inducers that induce a similar spectrum of isoenzymes involved in the metabolism of a variety of exogenous and endogenous compounds (171). The microsomal enzyme inducers produce liver enlargement at high dosages as a result of hepatocellular hyperplasia, followed by hepatocellular hypertrophy in the centrilobular area (152). Microsomal enzyme inducers will produce hepatocellular adenomas and carcinomas in male and female mice that will vary considerably depending on the strain of mouse used (172). The microsomal enzyme inducers are considered to be hepatic tumor promoters based on a large body of evidence accumulated over the last 20 years. There is no evidence to support the conclusion that compounds such as phenobarbital possess any initiating activity (173,174).

The relevance of the mouse liver tumor response has been the subject of widespread scientific debate, more so than the relevance of any other response observed in the bioassay. The controversy surrounding the mouse liver tumor response is thus due to the frequency of this response, the high proportion of compounds that are nongenotoxic, and the unique susceptibility of mouse strains to the development of both spontaneous and chemically induced hepatic tumors. With respect to the latter point, it is important to first consider the known, genetically determined problems in the liver of some strains of mouse.

Many mouse strains exhibit a high spontaneous incidence of liver tumors, which can range from 0% to 100%; male mice generally have a higher spontaneous incidence of liver tumors than do female mice (175). The B6C3F1 hybrid mouse has an extremely high (40%) and variable incidence of spontaneous liver tumors and is intermediate between the two parent strains, the C57BL/6 (0%) and C3H/He (100%) (176). The incidence can vary several-fold from study to study in the same laboratory and between different stocks of the same strain of mouse. The strain and species variation in susceptibility to phenobarbital-induced liver tumors roughly parallels the spontaneous incidence of hepatic tumors in the test animals. Those species or strains with high incidences of spontaneous liver tumors are more sensitive than species with low tumor incidences. Strain differences in response to phenobarbital were clearly demonstrated in studies by Becker (177) in which phenobarbital-treated C3H mice had a marked increase in liver tumors after 12 months of treatment, whereas no tumors were observed in the C57BL/6 mouse treated with PB for 18 months.

Studies have shown that most of the difference between mouse strains in the incidence of spontaneous tumors and susceptibility to hepatic tumor promotion is heritable and can be accounted for by a single genetic locus designated hepatocarcinogen sensitive (Hcs). The Hcs locus is known to affect the promotion and progression stage of carcinogenesis (178). The B6C3F1 strain has a high spontaneous incidence of mutated H-*ras* oncogenes (approximately 70% in spontaneous tumors) (179) and is defective in its ability to maintain normal methylation of DNA, which is a critical control of gene expression (180). The implications of this are that the hypomethylated, H-*ras* and *raf* proto-oncogenes appear to be primed for aberrant expression, which may provide an epigenetic, threshold mechanism for the induction of liver tumors (180). A rather elegant study in chimeric mice containing hepatocytes from susceptible (C3H) and resistant (C57BL) mouse strains have demonstrated that the genetic defect resides within the hepatocyte and is not due to host or environmental factors (181).

Although the mode of action of phenobarbital is clearly hepatic tumor promotion, the mechanism is not well understood. Recent research, however, suggests that there is a problem in the control of cell division and programmed cell death (apoptosis) in sensitive mouse strains. In rats, phenobarbital produces a burst of DNA synthesis followed by mitotic activity and hyperplasia. In sensitive mouse strains, increased DNA synthesis is followed by an increase in ploidy rather than cell division (182–184). After the initial burst of DNA synthesis and hyperplasia in phenobarbital-treated rats there is a decrease in the rate of cell proliferation and an inhibition of programmed cell death in normal hepatocytes (182). This corresponds to an increase in TGF-β in periportal hepatocytes, which is a potent inhibitor of cell proliferation and also plays a role in apoptosis (185). After long-term phenobarbital treatment some hepatocytes lose the ability to take up TGF-β. The loss of inhibition of cell proliferation by TGF-β would provide a growth advantage over normal hepatocytes (185). In addition, the inhibition of apoptosis in preneoplastic foci will cause a rapid expansion of foci without a persistent increase in cell proliferation (182).

Since the mouse will exhibit a hepatic tumor response for both nongenotoxic hepatic tumor promoters and genotoxic carcinogens the model lacks specificity and a promoter response will depend on the strain used. In summary, the unique genetic problems known to be present in the liver of susceptible mouse strains, including the high and variable spontaneous incidence of liver tumors, the abnormalities in critical controls for gene expression, and the high incidence of oncogene mutations, make the mouse model uninterpretable as an indicator of human cancer risk.

There is no evidence that a similar response is observed in humans exposed to microsomal enzyme inducers. A cohort of more than 8,000 patients treated with anticonvulsants admitted to a Danish epilepsy center from 1933 to 1962 has been extensively followed using registers of population, death, and cancer cases (186–188). Although it is seldom possible to obtain conclusive negative epidemiologic data, the incidence of hepatic cancer is rare, thus any increase should be readily apparent. There is no evidence for an increase in hepatic tumors in patient populations treated for prolonged periods of time with relatively high dosages of phenobarbital or other anticonvulsants. There was no indication of an increase in total cases of cancer or in cancers at other sites except for the central nervous system, which was clearly associated with the disease process rather than anticonvulsant therapy (188,189). Several expert committees have addressed the issue of the relevance of mouse liver tumors and have concluded that although there should be great concern for genotoxic carcinogens, less concern is warranted for nongenotoxic compounds that produce mouse liver tumors (190,191).

CONCLUSIONS

Based on over 25 years of experience with the rodent bioassay, the fact that 25–50% of all chemicals are rodent carcinogens when tested at the maximum tolerated dose illustrates the importance of understanding the biology underlying these responses. As we understand the mechanisms of chemical carcinogenesis we can more appropriately define the risks for human exposure. This is most important for the nongenotoxic agents since in many cases the response may be unique to the test species or involve a process with an apparent threshold below which there would be no risk to exposure.

The currently available short-term tests for mutagenicity and DNA damage can establish to a reasonable degree of certainty, whether a genotoxic or nongenotoxic mode of action is involved. In addition, tests are available in several organ systems that can distinguish between chemicals that have direct initiating activity from those that have tumor-promoting activity, and there are now a number of examples for which chemical specific mechanisms have been reasonably well defined. Now that a number of regulatory agencies have formally incorporated the use of mechanism and mode of action in the evaluation process for the classi-fication and regulation of carcinogens this should encourage further research.

REFERENCES

1. Haseman JK, Huff JE, Zeiger E, McConnell EE. Comparative results of 327 chemical carcinogenicity studies. *Environ Health Perspect* 1987; 74:229–35.
2. Ames BN, Magaw R, Gold LS. Ranking possible carcinogenic hazards. *Science* 1987; 236:271–80; (a) Gold LS, Slone TH, Bernstein L. Summary of carcinogenic potency and positivity for 492 rodent carcinogens in the carcinogenic potency database. *Environ Health Perspect* 1989; 79:259–72.
3. Doll R, Peto R. The causes of cancer: quantitative estimates of avoidable risks of cancer in the United States today. *J Natl Cancer Inst* 1981; 66:1191–308.
4. Fraumeni JF. Epidemiology of cancer. In: Brugge J, Curran T, Harlow E, McCormick F, eds. *Origins of human cancer: A comprehensive review*. Cold Spring Harbor, NY: Cold Spring Harbor Laboratory Press; 1991; 171–81.
5. National Cancer Institute. *1987 Annual cancer statistics review including cancer trends: 1950–1985*. NIH Publication no. 88-2789. Bethesda, MD: National Institutes of Health; 1988.
6. Doll R. Progress against cancer: an epidemiologic assessment. *Am J Epidemiol* 1991; 134:675–88.
7. Williams GM, Weisburger JH. Chemical carcinogenesis. In: Amdur MO, Doull J, Klaassen CD, eds. *Casarett and Doull's Toxicology. The Basic Science of Poisons*. New York: Macmillan; 1991.
8. National Academy of Science. *Risk assessment in the Federal government: Managing the process*. Washington, DC: National Research Council-National Academy Press; 1983.
9. Office of Science and Technology. Chemical carcinogens: a review of the science and its associated principles. Office of Science and Technology Policy. *Fed Regist* 1985; 50:10372–442; (a) Les v Rilley. CA 9, No. 91-70234; 1992.
10. Environmental Protection Agency. Guidelines for carcinogen risk assessment. Part II. *Fed Regist* 1986; 51:33992–4003.
11. Food and Drug Administration. Sponsored compounds in food producing animals: Criteria and procedures for evaluating the safety of carcinogenic residues; final rule. *Fed Regist* 1987; 52CFR:49572–88.
12. Japanese Ministry of Health and Welfare. *Guidelines for toxicity studies of drugs man-ual*. Chapter 6; 1990; 60–74.
13. Canada Health Protection Branch, Health and Welfare. *Carcinogen assessment*; 1991.
14. International Agency for Research on Cancer. Overall evaluation of carcinogenicity: An updating of IARC monographs, vol 1–42. (*IARC Monogr Eval Carcinog Risk Chem Hum* 1987; 7(Suppl):17–34 and 313–16); (a) International Agency for Research on Cancer. Mechanisms of carcinogenesis in risk identification. IARC Internal Tech-nical Report No. 91/002. 1991.
15. Public Citizen, *et al.*, Petitioners v. Dr. Frank Young, Commissioner FDA (DC Cir 1987). 831. *Fed Report* 2d; 1108–23.
16. Food and Drug Administration. Study of the Delaney Clause and other anticancer clauses. In: Agriculture – Environmental, and Consumer-Protection Appropriations

for 1975. Hearings before a Subcommittee of the Committee on Appropriation, House of Representatives, 93rd Cong., 2nd Sess., Part 8, at 210.

17. Merrill RA. FDA's implementation of the Delaney Clause: Repudiation of Congressional choice or reasoned adaptation to scientific progress? *Yale J Regulation* 1988; 5:1–88.

18. Merrill RA, Bohan RZ, Degnan FH, Pape SM. The FDA's authority under the Delaney Clause to consider mechanisms of action in determining whether additives "induce cancer." *Food and Drug Law J* 1992; 47:77–106; (a) FDA. Compounds used in food-producing animals. Criteria and procedures for evaluating the safety of carcinogenic residues. *Fed Regist* 1973; 38:19226–7.

19. Gaylor DW, Kodell RL. Linear interpolation algorithm for low dose risk assessment of toxic substances. *J Environ Pathol Toxicol* 1980; 4:305–12.

20. Farmer JH, Kodell RL, Gaylor DW. Estimation and extrapolation of tumor probabilities from a mouse bioassay with survival/sacrifice components, *Soc Risk Anal* 1982; 2:27–34; (a) Les v Rilley. CA9, No. 9-70234. 1992.

21. Paynter OE, Burin GJ, Jaeger RB, Gregario CA. Neoplasia induced by inhibition of thyroid gland function. (Guidance for analysis and evaluation.) Washington, DC: Hazard Evaluation Division, US Environmental Protection Agency; 1986.

22. Paymter OE, Burin GJ, Jaeger RB, Gregario CA. Goitrogens and thyroid follicular cell neoplasia. Evidence for a threshold process. *Regul Toxicol Pharmacol* 1988; 8:109–19.

23. Environmental Protection Agency. Alpha-2μ-globulin: Association with chemically-induced renal toxicity and neoplasia in the male rat. Washington, DC: US Environmental Agency; 1991.

24. Furth J. A meeting of ways in cancer research: Thoughts on the evolution and nature of neoplasms. *Cancer Res* 1959; 19:241–58.

25. Furth J. Pituitary cybernetics and neoplasia. Harvey Lectures. New York/London: Academic Press; 1969:47–71.

26. Axelrad AA, Leblond CP. Induction of thyroid tumors in rats by a low iodine diet. *Cancer* 1955; 8:339–67.

27. Bielschowsky F. Chronic iodine deficiency as cause of neoplasia in thyroid and pituitary of aged rats. *Br J Cancer* 1953; 7:203–13.

28. Isler H, Leblond CP, Axelrad AA. Influence of age and of iodine intake on the production of thyroid tumors in the rat. *J Natl Cancer Inst* 1958; 21:1065–81.

29. Leblond CP, Isler H, Axelrad A. Induction of thyroid tumors by a low iodine diet. *Can Cancer Conf* 1957; 2:248–66.

30. Furth J. Morphologic changes associated with thyrotropin-secreting pituitary tumors. *Am J Pathol* 1954; 30:421–63.

31. Dohler KD, Wong CC, Von Zur Muhlen A. The rat as a model for the study of drug effects on thyroid function: Consideration of methodological problems. *Pharmacol Ther* 1979; 5:305–18.

32. Doniach I. Aetiological consideration of thyroid carcinoma. In: Smithers D, ed. *Neoplastic diseases at various sites. Tumours of the thyroid gland*. Edinburgh/London: E&S Livingstone; 1970.

33. Pendergrast WJ, Milmore BK, Marcus SC. Thyroid cancer and thyrotoxicosis in the United States. Their relationship to endemic goiter. *J Chronic Dis* 1961; 13:22–38.

34. Saxen EA. Saxen LO. Mortality from thyroid diseases in an endemic goitre area. Studies in Finland. *Docum Med Geograph Tropica* 1954; 6:335–41.

35. Wegelin C. Malignant disease of the thyroid gland and its relationship to goitre in man and animals. *Cancer Res* 1928; 3:297–313.

36. Hellwig CA. Thyroid adenoma in experimental animals. *Am J Cancer* 1935; 23:550–5.

37. Swarm RL, Roberts GKS, Levy AC, Hines LR. Observations on the thyroid gland in rats following the administration of sulfamethoxazole and trimethoprim. *Toxicol Appl Pharmacol* 1973; 24:351–63.

38. Mackenzie JB, Mackenzie CG, McCollom EV. Effects of sulfanilguanidine on thyroid of the rat. *Science* 1941; 94:518.

39. Mackenzie CG, Mackenzie JB. Effect of sulfonamides and thioureas on the thyroid gland and basal metabolism. *Endocrinology* 1943; 32:185–209.

40. Takayama S, Aihara K, Onodera T, Akimoto T. Antithyroid effects of propylthiouracil and sulfamonomethoxine in rats and monkeys. *Toxicol Appl Pharmacol* 1986; 82:191–9.

41. Heath JE, Littlefield NA. Effect of subchronic oral sulfamethazine administration on Fisher 344 rats and B6C3F1 mice. *J Environ Pathol Toxicol Oncol* 1984; 5:201–14.

42. Fullerton FR, Kushmaul RF. Influence of oral administration of sulfamethazine on thyroid hormone levels in Fisher 344 rats. *J Toxicol Environ Health* 1987; 22: 175–85.

43. Littlefield NA, Gaylor DW, Blackwell BN, Allen RR. Chronic toxicity/carcinogenicity studies of sulphamethazine in B6C3F1 mice. *Food Chem Toxicol* 1989; 27:455–63.

44. Littlefield NA, Sheldon WG, Allen R, Gaylor DW. Chronic toxicity/carcinogenicity studies of sulphamethazine in Fischer 344/N rats: Two-generation exposure. *Food Chem Toxicol* 1990; 28:157–67.

45. National Cancer Institute. Bioassay of sulfisoxazole for possible carcinogenicity. Technical Report NCI-CG-TR-138, DHEW Publication no (NIH) 79-1393, Bethesda, MD: National Institutes of Health; 1979.

46. Koch-Weser J, Sidel VW, Dexter M, Parish C, Finer DC, Kanarek P. Adverse reactions of sulfisoxazole, sulfamethoxazole, and nitrofurantoin. Manifestations and specific reaction rats during 2,118 courses of study. *Arch Intern Med* 1971; 128: 399–404.

47. Cohen HN, Beastall GH, Ratcliffe WA, Gray C, Watson ID, Thompson JA. Effects on human thyroid function of sulphonamide and trimethoprim combination drugs. *Br Med J* 1980; 281:646–7.

48. Smellie JM, Bantok HM, Thompson BD. Co-trimoxazole and the thyroid. *Lancet* 1982; 2:96.

49. Smellie JM, Preece MA, Paton AM. Normal somatic growth in children receiving low-dose prophylactic co-trimoxazole. *Eur J Pediatr* 1983; 140:301–4.

50. Robbins J. Factors altering thyroid hormone metabolism. *Environ Health Perspect* 1981; 38:65–70.

51. Ruiz M, Ingbar SH. Effect of erythrosine (2′,4′,5′,7′-tetraiodofluoresceine) on the metabolism of thyroxine in rat liver. *Endocrinology* 1982; 110:1613.

52. Burger A, Dinichert D, Nicod P, Jenny M, Lemarchand-Beraud T, Vallotton MB. Effect of amiodarone on serum triiodothyronine, reverse triiodothyronine, thyroxine. A drug influencing peripheral metabolism of thyroid hormones. *J Clin Invest* 1976; 58:255–9.

53. Burgi H, Wimpfheimer C, Burger A, Zaunbauer W, Rosler H, Lemarchand-Beraud T. Changes of circulating thyroxine, triiodothyronine and reverse triiodothyronine after radiographic contrast agents. *J Clin Endocrinol Metab* 1976; 43:1203–10.

54. Borzelleca JF, Capen CC, Hallagan JB. Lifetime toxicity/carcinogenicity study of FD&C Red No. 3 in mice. *Food Chem Toxicol* 1987; 25:723–33.

55. Cavalieri RR, Pitt-Rivers R. The effects of drugs on the distribution and metabolism of thyroid hormones. *Pharmacol Rev* 1981; 33:55–80.

56. Oppenheimer JH, Bernstein G, Surks MI. Increased thyroxine turnover and thyroidal function after stimulation of hepatocellular binding of thyroxine by phenobarbital. *J Clin Invest* 1968; 47:1399–1406.

57. Comer CP, Chengelis CP, Levin S, Kotsonis FN. Changes in thyroidal function and liver UDP glucuronosyltransferase activity in rats following administration of a novel imidazole. *Toxicol Appl Pharmacol* 1985; 80:427.

58. Sanders JE, Eigenberg DA, Bracht LJ, Wang WR, van Zwieten MJ. Thyroid and liver trophic changes in rats secondary to liver microsomal enzyme induction by an experimental leukotriene antagonist (L-649,923). *Toxicol Appl Pharmacol* 1988; 95:378.

59. McClain RM, Levin A, Posch RC, Downing JC. The effect of phenobarbital on the metabolism and excretion of thyroxine in rats. *Toxicol Appl Pharmacol* 1989; 99:216–28.

60. Hill RN, Erdreich LS, Paynter OE, Roberts PA, Rosenthal SL, Wilkinson CF. Review: Thyroid follicular cell carcinogenesis. *Fund Appl Toxicol* 1989; 12:629–97.

61. Hiasa Y, Kitahori Y, Ohshima M, Fujita T, Yuasa T, Konishi N, Miyashiro A. Promoting effects of phenobarbital and barbital on development of thyroid tumors in rats treated with N-bis(2-hydroxypropyl)nitrosamine. *Carcinogenesis* 1982; 3:1187–90.

62. McClain RM. The significance of hepatic microsomal induction and altered thyroid function in rats: Implications for thyroid gland neoplasia. *Toxicol Pathol* 1989; 17:294–306.

63. McClain RM, Posch RC, Bosakowski T, Armstrong JM. Studies on the mode of action for thyroid gland tumor promotion in rats by phenobarbital. *Toxicol Appl Pharmacol* 1988; 94:254–65.

64. Gardner DF, Utiger RD, Schwartz SL, Witorsch B, Meyers B, Braverman LE, Witorsch RJ. Effects of oral erythrosine (2′,4′,5′,7′-tetraiodofluorescein) on thyroid function in normal men. *Toxicol Appl Pharmacol* 1987; 91:299–304.

65. Ohnhaus EE, Studer H. A link between liver microsomal enzyme activity and thyroid hormone metabolism in man. *Br J Clin Pharmacol* 1983; 15:71–6.

66. Tischler AS, DeLellis RA. The rat adrenal medulla, I. The normal adrenal. *J Am Coll Toxicol* 1988; 7:1–22.

67. Tischler AS, DeLellis RA. The rat adrenal medulla. II. Proliferative lesions. *J Am Coll Toxicol* 1988; 7:23–44.

68. Cheng L. Pheochromocytoma in rats: incidence, etiology, morphology, and functional activity. *J Environ Pathol Toxicol* 1980; 4:219–28.

69. Moon HD, Simpson ME, Li CH, Evans HM. Neoplasm in rats treated with pituitary growth hormone. II. Adrenal glands. *Cancer Res* 1950; 10:364–70.

70. Lupulescou A. Les pheochromocytomes experimentaux. *Ann Endocrinol* 1961; 22:459–68.

71. Marine D, Baumann EJ. Hypertrophy of adrenal medulla of white rats in chronic thiouracil poisoning. *Am J Physiol* 1945; 144:69–73.

72. Stoll R, Favconnau N, Maraud R. Evolution dela medullo-surrenale du rat soumis a un traitement generateur du syndrome de Sipple. *C R Soc Biol* 1982; 176:166–70.

73. Eränkö O. Nodular hyperplasia and increase of noradrenaline content in the adrenal medulla of nicotine treated rats. *Acta Pathol Microbiol Scand* 1955; 36:210–18.

74. Dept of Health, Education and Welfare. Bioassay of reserpine for possible carcino-genicity. DHEW publication no. NIH 79-1749; 1979.

75. Baer A. Sugars and adrenomedullary proliferative lesions: The effects of lactose and various polyalcohols. *J Am Coll Toxicol* 1988; 7:71–82.

76. Mosher AH, Kircher CH. Proliferative lesions of the adrenal medulla in rats treated with zomepirac sodium. *J Am Coll Toxicol* 1988; 7:83–94.

77. Kurokawa Y, Hayashi Y, Maekawa A, Takahashi M, Kukubo T. High incidences of pheochromocytomas after long-term administration of retinol acetate to F344/DuCrj rats. *J Natl Cancer Inst* 1985; 74:715–23.

78. Sietzen M, Schober M, Fischer-Colbrie R, Scherman D, Sperk G, Winkler H. Rat adrenal medulla. Levels of chromogranins, enkephalins, dopamine beta-hydroxylase and the amine transporter are changed by nervous activity and hypophysectomy *Neuroscience* 1987; 22:131.

79. Tischler AS, DeLellis RA, Nunnemacher G, Wolfe HJ. Acute stimulation of chromaf-fin cell proliferation in the adult rat adrenal medulla. *Lab Invest* 1988; 58:733.

80. Tischler AS, McClain RM, Childers H, Downing J. Neurogenic signals regulate chro-maffin cell proliferation and mediate the mitogenic effect of reserpine in the adult rat adrenal medulla. *Lab Invest* 1991; 65:374–6.

81. Roe FJC, Baer A. Enzootic and epizootic adrenal medullary proliferative disease of rats; Influence of dietary factors which affect calcium absorption. *Hum Toxicol* 1985; 4:27–52.

82. Sowers JR, Barrett JD. Hormonal changes associated with hypertension in neoplasia-induced hypercalcemia. *Am J Physiol* 1982; 242:E330–4.

83. Symington T. Functional pathology of the human adrenal glands. Edinburgh and Lon-don: E&S Livingstone; 1969; 244.

84. MacLeod RM. Regulation of prolactin secretion. *Front Neuroendocrinol* 1976; 4:169.

85. Horowski R, Graf KJ. Influence of dopaminergic agonists and antagonists on serum prolactin concentration in the rat. *Neuroendocrinology* 1976; 22:273–86.

86. Horowski R, Graff KJ. Neuroendocrine effects of neuropsychotropic drugs and their possible influence on toxic reactions in animals and man – the role of the dopamine-prolactin system. *Arch Toxicol* 1979; 2(Suppl):93–104.

87. Lee PA, Kelly MR, Wallin JD. Increased prolactin levels during reserpine treatment of hypertensive patients. *JAMA* 1976; 235:2316–17.

88. Ross RR, Paganini-Hill A, Krailo MD, Gerkins VR, Henderson BE, Pike MC. Effects of reserpine on prolactin levels and incidence of breast cancer in postmenopausal women. *Cancer Res* 1984; 44:3106–8.

89. Feinstein AR. Scientific standards in epidemiologic studies of the menace of daily life. *Science* 1988; 242:1257–63.

90. El Etreby MF, Gräf KJ, Neumann F. Evaluation of effects of sexual steroids on the hypothalamic pituitary system of animals and man. *Arch Toxicol* 1979; 2(Suppl): 11–39; (a) Neumann G. Early indicators for carcinogenesis in sex-hormone-sensitive organs. *Mutation Research* 1991; 248:341–56.

91. Hausmann L, Goebel KM, Zehner J. Influence of an oral contraceptive (ethinyl estra-diol/d-norgestrel) on growth hormone and insulin secretion. *Schweiz Med Wschr* 1976; 106:1470–4.

92. Biskind MS, Biskind GS. Development of tumors in the rat ovary after transplantation into the spleen. *Proc Soc Exp Biol Med* 1944; 55:176–9.

93. Murphy ED. Hyperplastic and early neoplastic changes in the ovaries of mice after genetic deletion in germ cells. *J Natl Cancer Inst* 1972; 48:1283–95.

94. Stitzel KA, McConnell RF, Dierckman TA. Effects of nitrofurantoin on the primary and secondary reproductive organs of female B6C3F1 mice. *Toxicol Pathol* 1989; 17:774–81.

95. Griffith RW. Bromocriptine and neoplasia. *Br Med J* 1977; 2:1605.

96. Jack D, Poynter D, Spurling NW. Beta-adrenoreceptor stimulants and mesovarian leiomyomas in the rat. *Toxicology* 1983; 2:315–20.

97. Colbert WE, Wilson BF, Williams PD, Williams GD. Relationship between in vivo relaxation of the costo-uterine smooth muscle and mesovarial leiomyoma formation in vivo by β-receptor agonists. *Arch Toxicol* 1991; 65:575–9.

98. Christensen AK, Peacock KC. Increase in Leydig cell number in testes of adult rats treated chronically with an excess of human chorionic ganadotrophin. *Biol Reprod* 1980; 22:383–91.

99. Brown CE, Warren S, Chute RN, Ryan KJ, Todd RB. Hormonally induced tumors of the reproductive system of parabiosed male rats. *Cancer Res* 1979; 39:3971–6.

100. Cook JC, Mullin LS, Frame ST, Biegel LB. Investigation of a mechanism for Leydig Cell Tumorigenesis by linuron in rats. *Toxicol Appl Pharmacol* 1993; 119: 195–204.

101. Cook JC, Murray SR, Frame SR, Hurtt ME. Induction of Leydig cell adenomas by ammonium perfluorooctanoate: A possible endocrine-related mechanism. *Toxicol Appl Pharmacol* 1992; 113:209–17.

102. Roberts SA, Nett TM, Hartman HA, Adams TE, Stoll RE. SDZ 200-110 induces Leydig cell tumors by increasing gonadotrophins in rats. *J Am Coll Toxicol* 1989; 8:487–505.

103. Landers RE, Norvell MJ, Bieber MA. Oil gavage test-compound administration effects in NTP carcinogenesis-toxicity testing. In: *Dietary fat and cancer*. New York: Alan R. Liss; 1986; 357–74.

104. Roebuck BD. Effects of high levels of dietary fats on the growth of azaserine-induced foci in the rat pancreas. *Lipids* 1986; 21:281–4.

105. McGuinness EE, Morgan RGH, Levison DA, Frape DL, Hopwood D, Wormsley KG. The effects of long-term feeding of soya flour on the rat pancreas. *Scand J Gastroenterol* 1980; 15:497–502.

106. Liddle RA, Goldfine ID, Williams JA. Bioassay of plasma cholecystokinin in rats: effects of food, trypsin inhibitor, and alcohol. *Gastroenterology* 1984; 87:542–9.

107. Göke B, Printz H, Koop I. Endogenous CCK release and pancreatic growth after feeding a proteinase inhibitor (camostate). *Pancreas* 1986; 1:509–15.

108. Lovie DS, May D, Miller P, Owyang C. Cholecystokinin mediates feedback regulation of pancreatic enzyme secretion in rats. *Am J Physiol* 1986; 250:G252–9.

109. Longnecker DS, Morgan RGH. Diet and cancer of the pancreas: epidemiological and experimental evidence. In: Reddy BS, Cohen LA, eds. *Nutrition and Cancer: A Critical Evaluation*. New York: CRC Press; 1986; 12–25.

110. Liener IE, Hasdai A. The effect of long-term feeding of raw soy flour on the pancreas of the mouse and hamster. *Adv Exp Med Biol* 1986; 199:189–97.

111. Adler G, Mullenhoff A, Bozkurt T, Göke B, Koop I, Arnold R. Comparison of the effect of single and repeated administrations of a protease inhibitor (camostate) on pancreatic secretion in man. *Scand J Gastroenterol* 1988; 23:158–62.

112. Black JW, Duncan WAM, Durant CJ, Ganellin CR, Parsons EM. Definition and antagonism of histamine H_2-receptors. *Nature* 1972; 236:385–90.

113. Fellenius E, Berglindh T, Sachs G, Olbe L, Elander B, Sjostrand SE, Wallmark B. Substituted benzimidazoles inhibit acid secretion by blocking $(H^+ + K^+)$ATPase. *Nature* 1981; 290:159–61.

114. Poynter D, Pick CR, Harcourt SAM, Selway SA, Ainge G, Harman IW, Spurling NW, Fluck PA, Cook JL. Association of long lasting unsurmountable histamine H_2 blockade and gastric carcinoid tumours in the rat. *Gut* 1985; 26:1284–95.

115. Betton GR, Salmon GK. Neuroendocrine (carcinoid) tumours of the glandular stomach of the rat following treatment with an H_2-receptor antagonist. *Arch Toxicol* 1986; 9:471.

116. Streett CS, Robertson JL, Crissman JW. Morphologic stomach findings in rats and mice treated with the H_2 receptor antagonists ICI 125,211 and ICI 162,846. *Toxicol Pathol* 1988; 16:299–304.

117. Havu N, Mattsson H, Ekman L, Carlsson E. Enterochromaffin-like cell carcinoids in the rat gastric mucosa following long term administration of ranitidine. *Digestion* 1990; 45:189–95.

118. Ekman L, Hansson E, Havu N, Carlsson E, Lundberg G. Toxicological studies on omeprazole. *Scand J Gastroenterol* (Suppl) 1985; 108:20, 53–69.

119. Becker HD, Reeder DD, Thompson JC. The effect of changes in antral pH on the basal release of gastrin. *Proc Soc Exp Biol Med* 1973; 143:238–40.

120. Konturek SJ, Cieszkowski M, Kwiecien N, Konturek J, Tasler J, Bilski J. Effects of omeprazole, a substituted benzimidazole on gastrointestinal secretions, serum gastrin, and gastric mucosal blood flow in dogs. *Gastroenterology* 1984; 86:71–7.

121. Larsson H, Carlsson E, Junggren U, Olbe L, Sjostrand SE, Skanberg I, Sundell G. Inhibition of gastric acid secretion by omeprazole in the dog and rat. *Gastroenterology* 1983; 86:900–7.

122. Alumets J, El Munshid HA, Håkanson R, Liedberg G, Oscarson J, Rehfeld JF, Sundler F. Effect of antrum exclusion on endocrine cells of rat stomach. *J Physiol* 1979; 286:145–55.

123. Alumets J, El Munshid HA, Håkanson R, Hedenbro J, Liedberg G, Oscarson J. Gastrin cell proliferation after chronic stimulation: effect of vagal denervation or gastric surgery in the rat. *J Physiol* 1980; 298:557–69.

124. Helander HF. Oxyntic mucosa histology in omeprazole-treated patients suffering from duodenal ulcer or Zollinger–Ellison syndrome. *Digestion* 1986; 35:123–9.

125. Ito N, Fukushima S, Hagiwara A, Shibata M, Ogiso T. Carcinogenicity of butylated hydroxyanisole in Fischer 344 rats. *J Natl Cancer Inst* 1983; 70:343–52.

126. Clayson DB, Iverson F, Nera EA, Lok E. The significance of induced forestomach tumours. *Annu Rev Pharmacol Toxicol* 1990; 30:441–63.

127. Ikeda GJ, Stewart JE, Sapienza PP, Peggins JO, Michel TC, Olivito V, Alam HZ, O'Donnell MW. Effect of subchronic dietary administration of butylated hydroxyanisole on canine stomach and hepatic tissue. *Food Chem Toxicol* 1986; 24:1201–21.

128. Tobe M, Furuya T, Kawasaki I, Naito K, Sekita K, Matsumoto K, Ochiai T, Usui A, Kukobo T, Kanno J, Hayashi Y. Six month toxicity study of butylated hydroxyanisole in beagle dogs. *Food Chem Toxicol* 1986; 24:1223–8.

129. Olsen P. The carcinogenic effect of butylated hydroxyanisole in the stratified epithelium of the stomach in rat vs. pig. *Cancer Lett* 1983; 21:115–16.

130. Roy AK, Neuhaus OW, Harmison CR. Preparation and characterization of sex-dependent rat urinary protein. *Biochim Biophys Acta* 1966; 127:72–81.

131. Neuhaus OW, Leserth DS. Dietary control of the renal reabsorption and excretion of alpha2-microglobulin. *Kidney Int* 1979; 16:409–15.

132. Charbonneau M, Strasser J, Borghoff SJ, Swenberg JA. In vitro hydrolysis of [^{14}C]-alpha-2-microglobulin isolated from male rat kidney. *Toxicologist* 1988; 8:537.

133. Lehman-McKeeman LD, Rivera-Torres MI, Caudill D. Lysosomal degradation of alpha2-microglobulin and alpha2-microglobulin-xenobiotic conjugates. *Toxicol Appl Pharmacol* 1990; 103:539–48.

134. Alden CL. A review of unique male rat hydrocarbon nephropathy. *Toxicol Pathol* 1986; 14:109–11.

135. Swenberg JA, Short BG, Borghoff SJ, Strasser J, Charbonneau M. The comparative pathobiology of alpha2-microglobulin nephropathy. *Toxicol Appl Pharmacol* 1989; 97:35–46.

136. Chatterjee B, Demyan WF, Song CS, Garg BD, Roy AK. Loss of androgenic induction of α_{2u}-globulin gene family in the liver of NIH Black rats. *Endocrinology* 1989; 125:1385–8.

137. Dietrich DR, Swenberg JA. NBR males fail to develop renal disease following exposure to agents that induce alpha2-microglobulin (A2μ) nephropathy. *Toxicologist* 1990; 10:1064.

138. Akerstroem B, Loedgberg L, Babiker-Mohamed H, Lohmander S, Rask L. Structural relationship between α_1-microglobulin from man, guinea-pig, rat, and rabbit. *Eur J Biochem* 1987; 170:143–8.

139. Li JL, Okada S, Hamazaki S, Ebina Y, Midorikawa O. Subacute nephrotoxicity and induction of renal cell carcinoma in mice treated with ferric nitrilotriacetate. *Cancer Res* 1987; 47:1867–9.

140. Anderson RL, Bishop WE, Campbell RL. A review of the environmental and mammalian toxicology of nitrilotriacetic acid. *CRC Rev Toxicol* 1988; 15:1.

141. Anderson RL, Alden CL, Merski JA. The effects of nitrilotriacetate on cation disposition and urinary tract toxicity. *Food Chem Toxicol* 1982; 20:105–22.

142. Cohen SM, Ellwein LB, *et al.* Cell proliferation in carcinogenesis. *Science* 1990; 249:1007–11.

143. Cohen SM, Ellwein LB. In: Banbury Report 25: Nongenotoxic Mechanisms in Carcinogenesis. Cold Spring Harbor Laboratory; 1987; 55–67.

144. Touster O. The metabolism of polyols. In: Sipple HL, McNutt KW, eds. *Sugars in nutrition.* New York: Academic Press; 1974; 229–39.

145. Hannett B, Thomas DW, Chalmers AH, Rofe AM, Edwards JB, Edwards RG. Formation of oxalate in pyridoxine or thiamine deficient rats during intravenous xylitol infusions. *J Nutr* 1977; 107:458–65.

146. Hauschildt S, Brand K. [^{14}C]-oxalate formation from [U^{14}C]-glucose and [U-$^{14-}$C]-xylitol in rat liver homogenate. *Biochem Med* 1979; 21:55–61.

147. Rofe AM, Thomas DW, Edwards RG, Edwards JB. [^{14}C]-oxalate synthesis from [U-^{14}C]-xylitol: in vivo and in vitro studies. *Biochem Med* 1977; 18:440–51.

148. Federation of American Societies for Experimental Biology. Health aspects of sugar alcohols and lactose. Bethesda, MD: Life Sciences Research Office, FASEBJ; 1986.

149. Matanoski GM, Elliott EA. Bladder cancer epidemiology. *Epidemiol Rev* 1981; 3:203–29.

150. Ellwein LB, Cohen SM. The health risks of saccharin revisited. *Crit Rev Toxicol* 1990; 20:311–26.

151. Ashby J, Tennant RW. Chemical structure, *Salmonella* mutagenicity and extent of carcinogenicity as indicators of genotoxic carcinogenesis among 222 chemicals tested in rodents by the US NCI/NTP. *Mutation Res* 1988; 204:17–115.

152. Schulte-Hermann R. Induction of liver growth by xenobiotic compounds and other stimuli. *CRC Crit Rev Toxicol* 1974; 3:97–158.

153. Reddy JK, Rao MS, Moody DE. Hepatocellular carcinomas in acatalasemic mice treated with nafenopin, a hypolipidemic peroxisome proliferator. *Cancer Res* 1976; 36:1211–17.

154. Conway JG, Cattley RC, Popp JA, Butterworth BE. Possible mechanisms in hepato-carcinogenesis by the peroxisome proliferator di(2-ethylhexyl)phthalate. *Drug Metab Rev* 1989; 21:65–102.

155. Issemann I, Green S. Activation of a member of the steroid hormone receptor super-family by peroxisome proliferators. *Nature* 1990; 347:645–50.

156. Butterworth BE, Loury DJ, Smith-Oliver T, Cattley RC. The potential role of chem-ically induced hyperplasia in the carcinogenic activity of the hypolipidemic carcino-gens. *Toxicol Ind Hlth* 1987; 63:129–49.

157. Cattley RC, Smith-Oliver T, Butterworth BE, Popp JA. Failure of the peroxisome proliferator WY-14, 643 to induce unscheduled DNA synthesis in rat hepatocytes following in vivo treatment. *Carcinogenesis* 1988; 9:1179–83.

158. Schulte-Hermann R, Timmermann-Trosiener I, Schuppler J. Promotion of spontan-eous preneoplastic cells in rat liver as a possible explanation of tumor production by nonmutagenic compounds. *Cancer Res* 1983; 43:839–44.

159. Cattley RC, Popp JA. Differences between the promoting activities of the peroxi-some proliferator Wy-14,643 and phenobarbital in rat liver. *Cancer Res* 1989; 49: 3246–51.

160. Kasai H, Okada Y, Nishimura S, Rao MS, Reddy JK. Formation of 8-hydroxydeoxy-guanosine in liver DNA of rats following long-term exposure to a peroxisome prolif-erator. *Cancer Res* 1989; 49:2603–5.

161. Marsman DS, Cattley RC, Conway JG, Popp JA. Relationship of hepatic peroxi-some proliferation and replicative DNA synthesis to the hepatocarcinogenicity of the peroxisome proliferators di(2-ethylhexyl) phthalate and [4-chloro-6-(2,3-xylidino)-2-pyrimidinylthio]acetic acid (Wy-14,643) in rats. *Cancer Res* 1988; 48: 6739–44.

162. Hegi ME, Ulrich D, Sagelsdorff P, Richter C, Lutz WK. No measurable increase in thymidine glycol or 8-hydroxydeoxyguanosine in liver DNA of rats treated with nafenopin or choline-devoid low-methionine diet. *Mut Res* 1990; 238:325–9.

163. Eacho PI, Foxworthy PS, Johnson WD, Hoover DM, White SL. Hepatic peroxisomal changes induced by a tetrazole-substituted alkoxyacetophenone in rats and compari-son with other species. *Toxicol Appl Pharmacol* 1986; 83:430–7.

164. National Cancer Institute. Carcinogenesis Bioassay of Chloroform, National Tech. Inform. Service No. PB264018/AS. Bethesda MD: National Cancer Institute; 1976.

165. Rosenthal SL. A review of the mutagenicity of chloroform. *Environ Mol Mutagen* 1987; 10:211–26.

166. Reitz RH, Fox TR, Quast JF. Mechanistic considerations for carcinogenic risk estimation: Chloroform. *Environ Health Perspect* 1982; 46:163–8.

167. Butterworth BE, Goldsworthy TL. The role of cell proliferation in multistage car-cinogenesis. *Proc Soc Exp Biol Med* 1991; 198:683–7.

168. Larson JL, Wolf DC, Butterworth BE. Acute hepatotoxic and nephrotoxic effects of chloroform in male F-344 rats and female B6C3F1 mice. *Fund Appl Toxicol* 1993; 20:302–15.

169. Higginson J. The epidemiology of Human Liver tumors. *Toxicol Pathol* 1982; 10:121–6.

170. Moore MA, Kitagawa T. Hepatocarcinogenesis in the rat: The effect of promoters and carcinogens in vivo and in vitro. *Int Rev Cytol* 1986; 101:125–73.

171. Conney AH. Pharmacological implications of microsomal enzyme induction. *Pharmacol Rev* 1967; 19:317–66.

172. McClain RM. Mouse liver tumors and microsomal enzyme-inducing drugs: Experimental and clinical perspectives with phenobarbital. *Prog Clin Biol Res* 1990; 331: 345–65.

173. Pitot HC, Sirica AE. The stages of initiation and promotion in hepatocarcinogenesis. *Biochim Biophys Acta* 1980; 605:191–215.

174. Peraino C, Staffeldt E, Haugen DA, Lombard LS, Stevens FJ, Fry RJ. Effects of varying the dietary concentration of phenobarbital on its enhancement of 2-acetyl-aminofluorene-induced hepatic tumorigenesis. *Cancer Res* 1980; 40:3268–73.

175. Grasso P, Hardy J. Strain differences in natural incidence and response to carcinogens. In: Butler WH, Newberne PM, eds. *Mouse hepatic neoplasia*. Amsterdam: Elsevier; 1974; 111–42.

176. Tarone RE, Chu KC, Ward JM. Variability in the rates of some common naturally-occurring tumors in Fischer 344 rats and (C57BL/6N × C3H/HeN)F1 (B6C3F1) mice. *J Natl Cancer Inst* 1981; 66:1175–81.

177. Becker FF. Characterization of spontaneous and chemically induced mouse liver tumors. In: Popp JA, ed. *Mouse liver neoplasia: Current perspectives*. Washington, DC: Hemisphere; 1984; 145–59.

178. Drinkwater NR, Ginsler JJ. Genetic control of hepatocarcinogenesis in C57BL/6J and C3H/HeJ inbred mice. *Carcinogenesis* 1986; 7:1701–7.

179. Fox TR, Watanabe PG. Detection of a cellular oncogene in spontaneous liver tumors of B6C3F1 mice. *Science* 1985; 228:596–7.

180. Vorce RL, Goodman JI. Hypomethylation of ras oncogenes in chemically-induced and spontaneous B6C3F1 mouse liver tumors. *Mol Toxicol* 1989; 2:99–116.

181. Lee G-H, Nomura K, Kanda H, Kusakabe M, Yoshiki A, Sakakura T, Kitagawa T. Strain specific sensitivity to diethylnitrosamine-induced carcinogenesis is maintained in hepatocytes of C3H/HeN-C57BL/6N chimeric mice. *Cancer Res* 1991; 51:3257–60.

182. Bursch W, Schulte-Hermann R. Synchronization of hepatic DNA synthesis by scheduled feeding and lighting in mice treated with the chemical induced of liver growth α-hexachlorocychlohexane. *Cell Tissue Kinet* 1983; 16:125–34.

183. Böhm N, Moser B. Reversible hyperplasia and hypertrophy of the mouse liver induced by a functional charge of phenobarbital. *Beitr Pathol* 1976; 157:283–300.

184. Böhm N, Noltemeyer N. Excessive reversible phenobarbital induced nuclear DNA-polyploidization in the growing mouse liver. *Histochemistry* 1981; 72:63–71.

185. Jirtle RL, Meyer SA. Liver tumor promotion: Effect of phenobarbital on EGF and protein kinase C signal transduction and transforming growth factor-β-1 expression. *Dig Dis Sci* 1991; 36:659–68.

186. Clemmesen J, Hjalgrim-Jensen S. On the absence of carcinogenicity to man of phenobarbital. *Acta Pathol Microbiol Scand* 1977; 261 (Suppl):38–50.

187. Clemmesen J, Hjalgrim-Jensen S. Is phenobarbital carcinogenic? A follow-up of 8078 epileptics. *Ecotoxicol Environ Safety* 1978; 1:457–70.

188. Olsen JH, Boice JD, Jensen JP, Fraumeni JF. Cancer among epileptic patients exposed to anticonvulsant drugs. *J Natl Cancer Inst* 1989; 81:803–8.

189. White SJ, McLean AE, Howland C. Anticonvulsant drugs and cancer: A cohort study in patients with severe epilepsy. *Lancet* 1979; 2:458–61.

190. European Chemical Industry Ecology and Toxicology Center. Hepatocarcinogenesis in laboratory rodents: Relevance for man. Brussels; 1982.

191. Nutrition Foundation. The Relevance of Mouse Liver Hepatoma to Human Carcinogenic Risk: A Report of the International Expert Advisory Committee to the Nutrition Foundation, Washington, DC; 1983.

INDEX

Authors' articles appear where page numbers are **bold**.

422